Remote Sensing in Applied Geophysics

Remote Sensing in Applied Geophysics

Editors

Chiara Colombero
Cesare Comina
Alberto Godio

MDPI • Basel • Beijing • Wuhan • Barcelona • Belgrade • Manchester • Tokyo • Cluj • Tianjin

Editors
Chiara Colombero
Politecnico di Torino,
Department of Environment,
Land and Infrastructure
Engineering (DIATI)
Italy

Cesare Comina
Università degli Studi di Torino,
Department of Earth Sciences (DST)
Italy

Alberto Godio
Politecnico di Torino,
Department of Environment,
Land and Infrastructure
Engineering (DIATI)
Italy

Editorial Office
MDPI
St. Alban-Anlage 66
4052 Basel, Switzerland

This is a reprint of articles from the Special Issue published online in the open access journal *Remote Sensing* (ISSN 2072-4292) (available at: https://www.mdpi.com/journal/remotesensing/special_issues/Applied_Geophysics).

For citation purposes, cite each article independently as indicated on the article page online and as indicated below:

LastName, A.A.; LastName, B.B.; LastName, C.C. Article Title. *Journal Name* **Year**, *Volume Number*, Page Range.

ISBN 978-3-03943-733-7 (Hbk)
ISBN 978-3-03943-734-4 (PDF)

© 2020 by the authors. Articles in this book are Open Access and distributed under the Creative Commons Attribution (CC BY) license, which allows users to download, copy and build upon published articles, as long as the author and publisher are properly credited, which ensures maximum dissemination and a wider impact of our publications.

The book as a whole is distributed by MDPI under the terms and conditions of the Creative Commons license CC BY-NC-ND.

Contents

About the Editors vii

Chiara Colombero, Cesare Comina and Alberto Godio
Special Issue "Remote Sensing in Applied Geophysics"
Reprinted from: *Remote Sens.* **2020**, *12*, 3413, doi:10.3390/rs12203413 1

Edward R. Henry, Alice P. Wright, Sarah C. Sherwood, Stephen B. Carmody, Casey R. Barrier and Christopher Van de Ven
Beyond Never-Never Land: Integrating LiDAR and Geophysical Surveys at the Johnston Site, Pinson Mounds State Archaeological Park, Tennessee, USA
Reprinted from: *Remote Sens.* **2020**, *12*, 2364, doi:10.3390/rs12152364 5

Luca Piroddi, Sergio Vincenzo Calcina, Antonio Trogu and Gaetano Ranieri
Automated Resistivity Profiling (ARP) to Explore Wide Archaeological Areas: The Prehistoric Site of Mont'e Prama, Sardinia, Italy
Reprinted from: *Remote Sens.* **2020**, *12*, 461, doi:10.3390/rs12030461 35

Rita Deiana, David Vicenzutto, Gian Piero Deidda, Jacopo Boaga and Michele Cupitò
Remote Sensing, Archaeological, and Geophysical Data to Study the Terramare Settlements: The Case Study of Fondo Paviani (Northern Italy)
Reprinted from: *Remote Sens.* **2020**, *12*, 2617, doi:10.3390/rs12162617 57

Giuseppe Stanghellini, Fabrizio Del Bianco and Luca Gasperini
OpenSWAP, an Open Architecture, Low Cost Class of Autonomous Surface Vehicles for Geophysical Surveys in the Shallow Water Environment
Reprinted from: *Remote Sens.* **2020**, *12*, 2575, doi:10.3390/rs12162575 79

Giorgio Cassiani, Elena Bellizia, Alessandro Fontana, Jacopo Boaga, Andrea D'Alpaos and Massimiliano Ghinassi
Geophysical and Sedimentological Investigations Integrate Remote-Sensing Data to Depict Geometry of Fluvial Sedimentary Bodies: An Example from Holocene Point-Bar Deposits of the Venetian Plain (Italy)
Reprinted from: *Remote Sens.* **2020**, *12*, 2568, doi:10.3390/rs12162568 99

Ahmed Gaber, Adel Kamel Mohamed, Ahmed ElGalladi, Mohamed Abdelkareem, Ahmed M. Beshr and Magaly Koch
Mapping the Groundwater Potentiality of West Qena Area, Egypt, Using Integrated Remote Sensing and Hydro-Geophysical Techniques
Reprinted from: *Remote Sens.* **2020**, *12*, 1559, doi:10.3390/rs12101559 123

Alice Vacilotto, Rita Deiana and Paolo Mozzi
Understanding Ancient Landscapes in the Venetian Plain through an Integrated Geoarchaeological and Geophysical Approach
Reprinted from: *Remote Sens.* **2020**, *12*, 2973, doi:10.3390/rs12182973 147

Chiara Colombero, Cesare Comina, Emanuele De Toma, Diego Franco and Alberto Godio
Ice Thickness Estimation from Geophysical Investigations on the Terminal Lobes of Belvedere Glacier (NW Italian Alps)
Reprinted from: *Remote Sens.* **2019**, *11*, 805, doi:10.3390/rs11070805 171

Xiangbin Cui, Jamin S. Greenbaum, Shinan Lang, Xi Zhao, Lin Li, Jingxue Guo and Bo Sun
The Scientific Operations of Snow Eagle 601 in Antarctica in the Past Five Austral Seasons
Reprinted from: *Remote Sens.* **2020**, *12*, 2994, doi:10.3390/rs12182994 . . . . **191**

Zejun Dong, Xuan Feng, Haoqiu Zhou, Cai Liu, Zhaofa Zeng, Jing Li and Wenjing Liang
Properties Analysis of Lunar Regolith at Chang'E-4 Landing Site Based on 3D Velocity Spectrum of Lunar Penetrating Radar
Reprinted from: *Remote Sens.* **2020**, *12*, 629, doi:10.3390/rs12040629 . . . . **207**

Maya Ilieva, Łukasz Rudziński, Kamila Pawłuszek-Filipiak, Grzegorz Lizurek, Iwona Kudłacik, Damian Tondaś and Dorota Olszewska
Combined Study of a Significant Mine Collapse Based on Seismological and Geodetic Data—29 January 2019, Rudna Mine, Poland
Reprinted from: *Remote Sens.* **2020**, *12*, 1570, doi:10.3390/rs12101570 . . . . **221**

Xiaoran Lv, Falk Amelung, Yun Shao, Shu Ye, Ming Liu and Chou Xie
Rheology of the Zagros Lithosphere from Post-Seismic Deformation of the 2017 Mw7.3 Kermanshah, Iraq, Earthquake
Reprinted from: *Remote Sens.* **2020**, *12*, 2032, doi:10.3390/rs12122032 . . . . **235**

Guoli Wu, Hefeng Dong, Ganpan Ke and Junqiang Song
Shear-Wave Tomography Using Ocean Ambient Noise with Interference
Reprinted from: *Remote Sens.* **2020**, *12*, 2969, doi:10.3390/rs12182969 . . . . **249**

Minao Sun and Shuanggen Jin
Multiparameter Elastic Full Waveform Inversion of Ocean Bottom Seismic Four-Component Data Based on A Modified Acoustic-Elastic Coupled Equation
Reprinted from: *Remote Sens.* **2020**, *12*, 2816, doi:10.3390/rs12172816 . . . . **269**

Sixin Liu, Qi Lu, Hongqing Li and Yuanxin Wang
Estimation of Moisture Content in Railway Subgrade by Ground Penetrating Radar
Reprinted from: *Remote Sens.* **2020**, *12*, 2912, doi:10.3390/rs12182912 . . . . **293**

About the Editors

Chiara Colombero has served as Researcher at the Department of Environment, Land and Infrastructure Engineering of Politecnico di Torino since March 2018. She graduated in Geology Applied to Engineering and Environment at the Earth Science Department of Università degli Studi di Torino in 2012. From this same department, she received her PhD in Earth Sciences in March 2017, with her thesis in Applied Geophysics on microseismic and ambient noise monitoring of potentially unstable rock masses. She won the AGLC 2016 prize for a paper related to her PhD thesis. She currently works in the Applied Geophysics research group and is lecturer/teaching assistant for the courses Innovation Lab for Climate Change, Applied Geophysics, Exploration Geophysics, and Geophysics. Her research interests encompass seismic and non-seismic methods applied to a wide variety of fields, from rock mass stability to glacier characterization and monitoring, from surface-wave analyses for the detection of subsurface anomalies to geological and hydrogeological investigations in water-covered environments, and from engineering geology purposes to archaeological research and cultural heritage preservation and restoration. She is author of more than 40 publications in international journals and national/international conference proceedings and serves as guest editor and reviewer.

Cesare Comina is Associate Professor in Applied Geophysics at Università degli Studi di Torino. He has also served as Research Associate at Politecnico di Torino where he previously graduated in Civil Engineering and was awarded his PhD in Geotechnical Engineering. In the past years, he has carried out his research at the Department of Structural and Geotechnical Engineering of Politecnico di Torino, where he collaborated on numerous funded research projects and participated in the scientific activities of the Geotechnical section. He is currently affiliated with the Department of Earth Sciences of Università degli Studi di Torino and collaborates with the Applied Geology group. He has teaching experience in both Geophysics and Geotechnics. His research involves the use of different geophysical methods for geotechnical and geological characterization. He has been particularly involved in developing innovative testing apparatuses for geophysical monitoring of laboratory tests and has addressed the use of surface wave tests for dynamic site characterization. He is currently involved in research related to geothermal energy usage and exploration and has also worked on waterborne geophysical surveys for the characterization of riverbed or lake sediments. On these topics, he is author and co-author of more than 70 scientific and technical papers both in international journals and conferences.

Alberto Godio holds a degree in Mining Engineering (1988) and a PhD in Georesource Engineering from Politecnico di Torino (1993). He is currently Professor at Politecnico di Torino in Geophysics, Applied Geophysics and Design for Environmental Engineering. His main research interests are focused on the application of electrical and electromagnetic methods for prospection and monitoring in environmental, engineering, hydrogeology and geology, and glaciology. The main topics deal with numerical methods applied to the processing of geophysical data. He has been responsible for research projects within the framework of EU-funded programs; he was responsible for Life Project, a bilateral Italy–India project (2018–2020), and several national programs funded by the Italian government (PRIN, FIRB) and regional agencies. He has been responsible for industrial projects with partners such as Italcementi, Teksid, and San Pellegrino. As the Delegate of the Rector for student mobility (2015–2018), he has promoted international cooperation among European universities in the higher educational framework; nowadays, he is the Rector's Advisor for student mobility and Erasmus+ program. He is author and co-author of 70 papers in peer-reviewed journals and more than 120 contributions at national and international conferences.

Editorial

Special Issue "Remote Sensing in Applied Geophysics"

Chiara Colombero [1,*], Cesare Comina [2] and Alberto Godio [1]

1. Department of Environment, Land and Infrastructure Engineering (DIATI), Politecnico di Torino, 10129 Torino, Italy; alberto.godio@polito.it
2. Department of Earth Sciences (DST), Università degli Studi di Torino, 10125 Torino, Italy; cesare.comina@unito.it

* Correspondence: chiara.colombero@polito.it

Received: 9 October 2020; Accepted: 12 October 2020; Published: 18 October 2020

The Special Issue "Remote Sensing in Applied Geophysics" is focused on recent and upcoming advances in the combined application of remote sensing and applied geophysics techniques, sharing the advantages of being non-invasive research methods, suitable for surface and near-surface high-resolution investigations of even wide and remote areas.

Applied geophysics analyzes the distribution of physical properties in the subsurface for a wide range of geological, engineering and environmental applications at different scales. Geophysical surveys are usually carried out deploying or moving the appropriate instrumentation directly on the ground surface. However, recent technological advances have brought to the development of innovative acquisition systems more typical of the remote sensing community (e.g., airborne surveys and unmanned aerial vehicle systems). At the same time, while applied geophysics mainly focuses on the subsurface, typical remote sensing techniques have the ability to accurately image the Earth's surface with high-resolution investigations carried out by means of terrestrial, airborne, or satellite-based platforms. The integration of surface and subsurface information is often crucial for several purposes, including the georeferencing and processing of geophysical data, the characterization and time-lapse monitoring of surface and near-surface targets, and the reconstruction of highly detailed and comprehensive 3D models of the investigated areas.

Contributions to the issue showing the added value of surface reconstruction and/or monitoring in the processing and interpretation of geophysical data, integration and cross-comparison of geophysical and remote sensing techniques were required to the research community. Contributions discussing the results of pioneering geophysical acquisitions by means of innovative remote systems were also addressed as interesting topics.

The Special Issue received great attention in the combined community of applied geophysicists and remote sensing researchers. A total of 15 papers are included in the Special Issue, covering a wide range of applications. This is one of the highest numbers of papers among the *Remote Sensing* Special Issues, showing great interest in the proposed topic. The relevant number of contributions also highlights the relevance and increasing need for integration between remote sensing and ground-based geophysical exploration or monitoring methods.

In particular, one of the main fields of research showing the potential integration of the geophysical and remote sensing techniques is archaeological exploration. Indeed, archaeologists often use near-surface geophysics or remote sensing in their research. However, these surface and near-surface data are rarely integrated to offer a robust understanding of the complex historical setting characterizing archaeological sites. Several research efforts need therefore to be addressed to reach a more efficient integration of the techniques in the archaeological field. In this Special Issue, research efforts are particularly related to: integrating aerial and terrestrial remote sensing programs to identify different uses of archaeological sites through time and across space [1]; building high-resolution maps of electrical resistivity using computer-assisted and remote sensed acquisition tools [2]; multidisciplinary integration of remote sensing and archaeological data with electric resistivity tomography and frequency domain electromagnetic measurements to provide new useful and interesting information for archaeologists [3].

Several research efforts were also devoted to the development of innovative acquisition systems integrating remote sensing and geophysical techniques with a precise georeferencing of the acquired data. These contributions are related to: the development of innovative low-cost autonomous vehicles for geological/geophysical studies of shallow water environments [4]; efficient acquisition of frequency domain electromagnetic data with a specifically designed georeferenced wooden carriage to allow for integration with other remote sensing data [5]; automated resistivity profiling with constant GPS referencing and combined Digital Terrain Model to obtain correct data positioning and topographic reconstruction [2].

The integration of remote sensing and applied geophysics techniques has also been applied in environmental and geomorphological conditions of test sites requiring special attention. This can be the case of wide research areas, where the integration of localized geophysical data and wide remote sensing imagery is essential for a more accurate reconstruction of the environment, particularly in the hydrogeological framework [6,7], or in remote areas where acquisition and interpretation of geophysical measurements is challenging and feasible only with the help of remote sensing approaches. This is particularly true for the cryosphere, e.g., inland glaciers [8] and Antarctic ice sheet [9], or for the exploration of other planets and satellites [10]. In this last context, remote sensing techniques clearly play a major role in the effective acquisition of geophysical data.

The abovementioned applications appear to be the research fields where an effort in the integration of remote sensing and geophysical methodologies can be more successful, resulting in an increased comprehension of the investigated sites and in the improved efficiency of the combined surveys. Other papers in the Special Issue could nevertheless suggest additional interesting development topics; this is the case for papers focusing on the combination of geodetic and seismological observations for an increased understanding of local subsidence events [11] or wider scale post-seismic deformations [12].

As a conclusion, we can foresee a growing interest in the collaboration between the research communities of remote sensing and applied geophysics, with many potential overlapping research topics and undeniable benefits in the integration of different research approaches. Particularly, remote sensing data and data acquisition approaches are still of great interest and support for applied geophysics researchers since they can further help in the correct georeferencing of the acquired geophysical data and in a better comprehension of the test site settings, which could significantly improve the interpretation of geophysical evidence.

Author Contributions: The authors contributed equally to all aspects of this Editorial. All authors have read and agreed to the published version of the manuscript.

Funding: This research received no external funding.

Acknowledgments: We would like to sincerely thank all the authors who contributed to this volume and all the reviewers for their valuable and constructive work. We are grateful to *Remote Sensing* Editorial Staff for providing us this opportunity and constant support. We especially thank Traey Wu for the substantial work and essential help in all the processes of this Special Issue.

Conflicts of Interest: The authors declare no conflict of interest.

References

1. Henry, E.R.; Wright, A.P.; Sherwood, S.C.; Carmody, S.B.; Barrier, C.R.; Van de Ven, C. Beyond Never-Never Land: Integrating LiDAR and Geophysical Surveys at the Johnston Site, Pinson Mounds State Archaeological Park, Tennessee, USA. *Remote Sens.* **2020**, *12*, 2364. [CrossRef]
2. Piroddi, L.; Calcina, S.V.; Trogu, A.; Ranieri, G. Automated Resistivity Profiling (ARP) to Explore Wide Archaeological Areas: The Prehistoric Site of Mont'e Prama, Sardinia, Italy. *Remote Sens.* **2020**, *12*, 461. [CrossRef]

3. Deiana, R.; Vicenzutto, D.; Deidda, G.P.; Boaga, J.; Cupitò, M. Remote Sensing, Archaeological, and Geophysical Data to Study the Terramare Settlements: The Case Study of Fondo Paviani (Northern Italy). *Remote Sens.* **2020**, *12*, 2617. [CrossRef]
4. Stanghellini, G.; Del Bianco, F.; Gasperini, L. OpenSWAP, an Open Architecture, Low Cost Class of Autonomous Surface Vehicles for Geophysical Surveys in the Shallow Water Environment. *Remote Sens.* **2020**, *12*, 2575. [CrossRef]
5. Cassiani, G.; Bellizia, E.; Fontana, A.; Boaga, J.; D'Alpaos, A.; Ghinassi, M. Geophysical and Sedimentological Investigations Integrate Remote-Sensing Data to Depict Geometry of Fluvial Sedimentary Bodies: An Example from Holocene Point-Bar Deposits of the Venetian Plain (Italy). *Remote Sens.* **2020**, *12*, 2568. [CrossRef]
6. Gaber, A.; Mohamed, A.K.; ElGalladi, A.; Abdelkareem, M.; Beshr, A.M.; Koch, M. Mapping the Groundwater Potentiality of West Qena Area, Egypt, Using Integrated Remote Sensing and Hydro-Geophysical Techniques. *Remote Sens.* **2020**, *12*, 1559. [CrossRef]
7. Vacilotto, A.; Deiana, R.; Mozzi, P. Understanding Ancient Landscapes in the Venetian Plain through an Integrated Geoarchaeological and Geophysical Approach. *Remote Sens.* **2020**, *12*, 2973. [CrossRef]
8. Colombero, C.; Comina, C.; De Toma, E.; Franco, D.; Godio, A. Ice Thickness Estimation from Geophysical Investigations on the Terminal Lobes of Belvedere Glacier (NW Italian Alps). *Remote Sens.* **2019**, *11*, 805. [CrossRef]
9. Cui, X.; Greenbaum, J.S.; Lang, S.; Zhao, X.; Li, L.; Guo, J.; Sun, B. The Scientific Operations of Snow Eagle 601 in Antarctica in the Past Five Austral Seasons. *Remote Sens.* **2020**, *12*, 2994. [CrossRef]
10. Dong, Z.; Feng, X.; Zhou, H.; Liu, C.; Zeng, Z.; Li, J.; Liang, W. Properties Analysis of Lunar Regolith at Chang'E-4 Landing Site Based on 3D Velocity Spectrum of Lunar Penetrating Radar. *Remote Sens.* **2020**, *12*, 629. [CrossRef]
11. Ilieva, M.; Rudziński, Ł.; Pawłuszek-Filipiak, K.; Lizurek, G.; Kudłacik, I.; Tondaś, D.; Olszewska, D. Combined Study of a Significant Mine Collapse Based on Seismological and Geodetic Data— 29 January 2019, Rudna Mine, Poland. *Remote Sens.* **2020**, *12*, 1570. [CrossRef]
12. Lv, X.; Amelung, F.; Shao, Y.; Ye, S.; Liu, M.; Xie, C. Rheology of the Zagros Lithosphere from Post-Seismic Deformation of the 2017 Mw7.3 Kermanshah, Iraq, Earthquake. *Remote Sens.* **2020**, *12*, 2032. [CrossRef]

Publisher's Note: MDPI stays neutral with regard to jurisdictional claims in published maps and institutional affiliations.

© 2020 by the authors. Licensee MDPI, Basel, Switzerland. This article is an open access article distributed under the terms and conditions of the Creative Commons Attribution (CC BY) license (http://creativecommons.org/licenses/by/4.0/).

Article

Beyond Never-Never Land: Integrating LiDAR and Geophysical Surveys at the Johnston Site, Pinson Mounds State Archaeological Park, Tennessee, USA

Edward R. Henry [1,2,*], Alice P. Wright [3], Sarah C. Sherwood [4], Stephen B. Carmody [5], Casey R. Barrier [6] and Christopher Van de Ven [4]

1. Department of Anthropology and Geography, Colorado State University, Fort Collins, CO 80523-1787, USA
2. Center for Research in Archaeogeophysics and Geoarchaeology (CRAG), Colorado State University, Fort Collins, CO 80523-1787, USA
3. Department of Anthropology, Appalachian State University, Boone, NC 28608-2016, USA; wrightap2@appstate.edu
4. Department of Earth and Environmental Systems, The University of the South, Sewanee, TN 37383, USA; sherwood@sewanee.edu (S.C.S.); chvandev@sewanee.edu (C.V.d.V.)
5. Department of Social Sciences, Troy University, Troy, AL 36082, USA; scarmody@troy.edu
6. Department of Anthropology, Bryn Mawr College, Bryn Mawr, PA 19010-2899, USA; cbarrier@brynmawr.edu
* Correspondence: edward.henry@colostate.edu

Received: 18 June 2020; Accepted: 21 July 2020; Published: 23 July 2020

Abstract: Archaeologists often use near-surface geophysics or LiDAR-derived topographic imagery in their research. However, rarely are the two integrated in a way that offers a robust understanding of the complex historical palimpsests embedded within a social landscape. In this paper we present an integrated aerial and terrestrial remote sensing program at the Johnston Site, part of the larger Pinson Mounds landscape in the American MidSouth. Our work at Johnston was focused on better understanding the history of human landscape use and change so that we can begin to compare the Johnston Site with other large Middle Woodland (200 BC–AD 500) ceremonial centers in the region. Our research allowed us to examine the accuracy of an early map of the Johnston Site made in the early 20th century. However, our integrated remote sensing approach allows us to go well beyond testing the usefulness of the map; it helps identify different uses of the site through time and across space. Our research emphasizes the importance of an integrated remote sensing methodology when examining complex social landscapes of the past and present.

Keywords: archaeological prospection; near-surface geophysics; LiDAR; magnetic gradiometry; surface magnetic susceptibility; electromagnetic induction; Middle Woodland period; Hopewell archaeology

1. Introduction

Independently, applying topographic imagery derived from LiDAR (light detection and ranging) or geophysical remote sensing methods in archaeological research is well-established in archaeology [1–21]. However, they are increasingly being applied together to create more robust understandings of social landscapes—including the emergence and long-term modification of built environments in the archaeological past (cf. [22–24]). This integration of aerial and terrestrial remote sensing methods has the potential to help tease apart the complexity of ever-evolving landscape palimpsests [25–27]. Archaeologists interrogate these large units of archaeological analysis (i.e., landscapes) at a given point in time, but they form over many millennia as a result of diverse human and natural processes that can build up, cut away, and rearrange the earth in ways that no singular remote sensing method can adequately elucidate. Moreover, from an anthropological perspective of remote sensing [28], the integration of LiDAR-derived imagery and near-surface geophysical applications only enhances the ability of archaeologists to propose and explore new research questions and hypotheses apart from, or in conjunction with, excavations [29].

Our recent work at the Johnston Site in western Tennessee, USA (Figure 1) illustrates the efficacy of integrating these multi-scalar remote sensing tools to explore anthropological questions pertaining to Middle Woodland era (200 BC–AD 500) hunter-gatherer-gardener societies of the North American Midsouth, and to formulate new questions based on the results of such multi-scalar work. The Johnston Site, a satellite property of the better-known Pinson Mounds State Archaeological Park (PMSAP), is a large multi-mound center that has received very little attention by professional archaeologists since the site and the terrace on which it is located was first mapped in 1917 by E. G. Buck, a local civil engineer, hired by William Myer, a research associate of the Smithsonian Institution [30]. The integrated remote sensing approach and limited test excavations we used at Johnston resulted in a thorough evaluation of this 1917 map [31]. Our results afford us the ability to identify proxies for human-landscape interactions and environmental change in this area. This includes identifying areas where erosion has impacted the site and its monuments, as well as discovering shifts in monumentality at Johnston. Beginning to trace these changes allows us to lay the foundations for a landscape biography [32] of the Pinson Mounds vicinity that can be further developed with future research in this region.

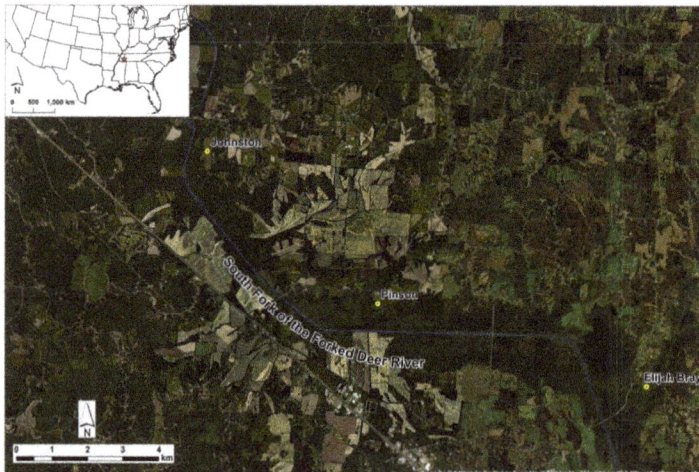

Figure 1. Location of the Pinson Mounds Landscape in western Tennessee, USA (inset) and locations of Middle Woodland mound centers situated along the South Fork of the Forked Deer River discussed herein (primary map).

2. The Johnston Site within the Middle Woodland Era Pinson Mounds Landscape

The Middle Woodland period in eastern North America is characterized by the florescence of a near continent-wide social movement evidenced by novel religious rituals, elaborate craft production and exchange, and the rise of monumental ceremonial centers [33–42]. Alongside these changes came an increase in the importance of domesticated plant crops, some of which were associated with mortuary and other rituals, while a reliance on foraging wild plant foods and hunting was maintained; archaeologists refer to this subsistence pattern as the Eastern Agricultural Complex [43–48]. The Johnston site is part of the larger PMSAP, the largest Middle Woodland period ceremonial center in the southeastern United States (Figure 1). Spanning roughly 160 ha., Pinson Mounds exhibit a wide range of earthen monuments including Sauls Mound, the second tallest earthen monument ever constructed in North America at 22 m tall [49]. Other monuments include a large rounded geometric enclosure with a diameter of almost 340 m at its widest point, and at least 13 mounds comprised of low (ca. 1 m) and tall (ca. 10 m) rectilinear platforms, as well as small (ca. <1 m) and large (ca. 6.5 m) conical burial mounds. Aside from the impressive organization of labor and engineering required to construct the earthen monuments at Pinson, evidence for complex mortuary practices and the recovery of elaborate artifacts resembling those found in the Ohio Hopewell core area has positioned Pinson Mounds among the most important Middle Woodland centers for religious ceremonies, exchange, and pilgrimage in the eastern United States [49–51]. Even Hopewellian scholars working in Ohio have commented that Pinson is the, "premier Hopewellian center in the Southeast" because it was such an important destination for Middle Woodland societies [52].

However, Pinson represents only one, albeit the largest, collection of earthen mounds in this section of the South Fork of the Forked Deer River (SFFDR) in western Tennessee. It is centrally positioned amongst a landscape of three Middle Woodland ceremonial centers in the region that encompasses nearly 100 km^2, an unusual collection of sites for this region of the U.S. Using Sauls Mound as the center of Pinson, the Elijah Bray mound site is situated roughly 8 km upstream from Pinson on a terrace overlooking the confluence of Clarks Creek and the SFFDR. This comparatively small site is comprised of at least two conical burial mounds measuring 5.5 and 3 m respectively, with an associated artifact scatter spanning roughly 4 ha [44] (p. 15). The focus of this research, the Johnston site, is much larger than Elijah Bray. Johnston covers roughly 48 ha of a terrace overlooking the SFFDR 6 km northwest of Pinson (Figures 2 and 3). The site is characterized by a collection of 10 rectilinear platform and conical mounds. This concentration of mound centers in a condensed stretch of a major tributary river to the Lower Mississippi River Valley is exceptional relative to the absence of mound centers in neighboring drainages across west Tennessee. This calls into question the historical development of this landscape during the Middle Woodland period, as well as the possible situational nature of pre-Contact American Indian use of these separate mound centers. Questions revolving around the unknown social, historical, and environmental contingencies that certainly influenced the indigenous use of this landscape motivated our archaeological research in this area of West Tennessee.

Previous Research and Cartography at the Johnston Site

Eastern North America has a long history of naturalists, antiquarians, and early professional archaeologists mapping indigenous earthen monuments [53–57]. Sometimes these people had professional backgrounds in surveying, sometimes they did not. It was not uncommon for some well-funded researchers to hire local surveyors to conduct mapping projects. Today, archaeologists using modern technologies like GIS software and aerial and terrestrial remote sensing methods are documenting the mixed successes of these early site surveys [2,22,58,59]. LiDAR, and geophysical surveys across eastern North America have shown that sites mapped more than 150 years ago were sometimes quite accurate. However, sometimes features were drawn differently than we might be able to discern today. These differences may relate to simple mistakes, generalizations, or overactive imaginations.

Like Pinson, the Johnston site was initially investigated by William Myer, an associate researcher of the Smithsonian Institution, who hired local civil engineer E. G. Buck to produce maps of both sites in 1917 [30] (p. 32), [49] (p. 52). However, unlike Pinson, the early map of Johnston arranged by Myer [60] has not been sufficiently reexamined using modern methods (e.g., [49,59]) to determine what this landform looked like at the time of early European expansion into West Tennessee. This is important to understand the broader Middle Woodland landscape along the SFFDR because the work of Mainfort and colleagues [59] identified numerous discrepancies in Myer's 1922 map of Pinson. For instance, they argue the elaborate enclosure walls that span the exterior boundaries of Pinson, as well as the "Inner Citadel" and their associated intersecting mounds, might not have existed, and thus may have been embellished or severely impacted by plowing. Identifying such discrepancies led to the title of their article *"Mapping Never-Never Land"*, from which we derive the title of our article.

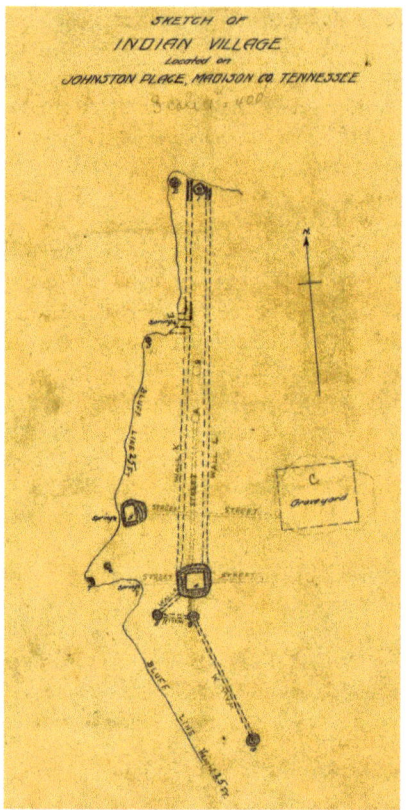

Figure 2. The 1917 map of the Johnston Site by E.G. Buck presented in Myer [31] and first published by Kwas and Mainfort [30]. Shown here courtesy of the National Anthropological Archives, National Museum of Natural History.

The work of Mainfort et al. [59] provides a cautionary lesson on cartographic 'artifacts' and the potential for embellished earthen architecture at Johnston. Nevertheless, to begin understanding earthen monuments at Johnston, and identifying how they might have changed since their initial mapping, we had to begin by assessing the original map of the site as first published by Kwas and Mainfort [30]. We use the 1917 map as a comparative documentation of the site prior to more than 100 years of agricultural impact. The full sketch map of Johnston depicts 10 mounds situated on a north-south oriented terrace (Figure 2). The details of this map are described in an unpublished manuscript by Myer [31] and discussed in detail in an article by Kwas and Mainfort [30].

The two platform mounds (Mounds 4 and 5) at the center of the site are the largest and most visible today (see also Figure 3). A pair of conical burial mounds (Mounds 1 and 2) are drawn at the northern edge of the terrace. Only Mound 1 is currently still visible. Three additional small conical mounds are situated along the bluff line that rises above the SFFDR floodplain (Mounds 3, 6, and 7), while three more small conical mounds are drawn south of Mound 4 (Mounds 8, 9, and 10). The dimensions of these monuments as they appeared in 1917 were recorded (Table 1) and provide baseline measurements that we can compare to the results of our research presented here. Low-lying parallel embankments (Walls K and L) are discussed as once being 3 m wide and 0.75 m tall and depicted on the Johnston map to have extended from Mound 4 north to Mound 1, leaving an open turn west near Mound 3 that led to a spring at the base of the bluff. Myer discussed the embankments being most visible in 1917 at either side of Mound 1, where they are drawn as solid constructions. Elsewhere on the map the Quadrature-phase (QP), which represents apparent soil conductivity, and the In-phase (IP), which represents apparent volume magnetic susceptibility. Some EMI instruments can do this simultaneously. For instance, the Geonics EM38-MK2 (Geonics Limited, Mississauga, ON, Canada) we used in this study measures QP and IP at two different depths simultaneously (Figure 4c). This instrument has electromagnetic coil separations of 0.5 and 1 m, resulting in an approximate maximum depth penetration of 0.75 m and 1.5 m for conductivity data, measured in millisiemens per meter (mS/m), and 0.3 m and 0.6 m for magnetic susceptibility data, measured in parts per thousand (ppt) when operated, as we did, in the vertical dipole mode. Data were collected every 0.5 m along transects spaced 0.5 m apart. Data were downloaded and processed using the TerraSurveyor software package, with typical application of despike and either high-pass or low-pass filter operations applied before being interpolated to 0.25 m pixels and exported to ArcGIS.

3.5. Test Excavations of Geophysical Anomalies

After our LiDAR-derived imagery and geophysical datasets were processed and analyzed, we examined a non-random sample of the identified topographical and geophysical anomalies through test excavations. Excavations were conducted in a range of trench sizes, by hand, in arbitrary 10 cm levels until intact features were identified. We screened a 10–25 percent random sample of all excavated non-feature soil matrix (e.g., plowzone) using 6.35 mm (0.25") mesh. The entirety of all features we identified and excavated were processed using water flotation. This excavation methodology ensured a sufficient sample of displaced non-feature artifacts, and the total recovery of artifacts and ecofacts from feature contexts. The test excavations are presented in detail elsewhere [83]. The excavations we discuss here are intended to provide the reader examples of how our integrated remote sensing approach led to success in identifying surface and subsurface archaeological features that help us enhance our understanding of this important site, and better situate it within the Pinson ritual landscape [83].

4. Results

In this section, we present the results of our analyses of LiDAR-derived imagery and geophysical surveys at the Johnston Site. We begin with a comparison of the LiDAR-derived imagery and the 1917 map of the Johnston Site. We then move to discuss the results of the large-scale magnetic gradiometer and magnetic susceptibility surveys before discussing the EMI results. Test excavations of topographic and geophysical anomalies are presented when applicable.

4.1. Analysis of LiDAR-Derived Imagery Compared with the 1917 Map of the Johnston Site

Overlaying the 1917 map based on E. G. Buck's survey work at Johnston over our LiDAR imagery allowed us to compare Buck's cartography and Myer's descriptions with the current state of topographic relief at the site (Figure 5a). It also allowed us to examine the Johnston landscape for topographic features not included on the 1917 map. Using the three clearly visible mounds at the site as anchor points for georeferencing the 1917 map to real world coordinates, we were able to confidently determine correlations and differences in topographic features between the maps. After redrawing Buck's features as polygons, we could determine where a given mound or embankment should be in relation to topographic features identifiable in LiDAR-derived imagery (Figure 5b).

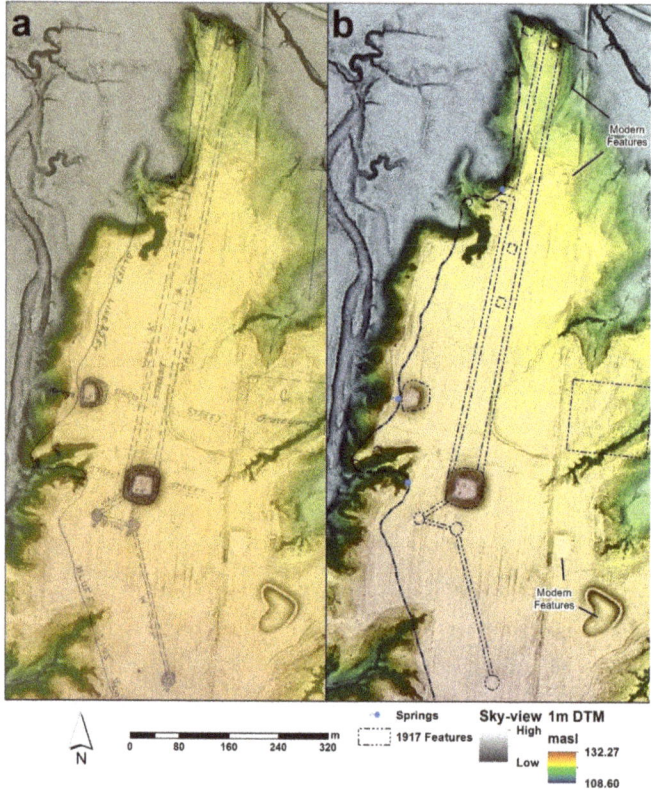

Figure 5. Results of the 1917 Myer/Buck map of Johnston overlay with LiDAR imagery. (**a**) The 1917 Johnston map overlaid on LiDAR imagery with 60% opacity; (**b**) Features from the 1917 Johnston map as polygons on blended LiDAR imagery (Sky-view and color stretched DTM), showing fit of 1917 map overlay.

The first discrepancy we can easily identify is the 1917 bluff line that undulates along the western edge of the terrace landform. In some areas the bluff edge is depicted 60 m or more east of where the bluff is. In others, the 1917 bluff line is depicted west of the current bluff line, now dominated by incised erosional gullies. While the former can be attributed to 'sketching in' the bluff line as a feature relative to other earthen monuments, there may have been areas where the 1917 bluff line was more accurately depicted. If this were the case, we could assume that the erosional gullies present in areas that were depicted as level terrace in 1917 provide a proxy for landform evolution since the map was drawn. These two scenarios are hard to assess given the identified discrepancies between the Buck/Myer map and current topography at Pinson [59].

According to the 1917 map, the northern extent of Johnston supposedly contained Mounds 1–3 and some of the best-preserved portions of the parallel embankments. Our LiDAR-derived imagery shows that Mound 1 is still present at Johnston, conical in shape, but smaller in height and base diameter than described in 1917 (Figure 6; A to A'). Mound 2 is hard to evaluate. A topographic rise that is roughly 0.2 m in height is present in the approximate location of Mound 2 but there is no clear shape to this feature. This may be because much of Mound 2 area has eroded into the gully to the north. However, it is equally possible that this is a relic of natural topography and was interpreted as a built earthen feature in 1917. Determining whether this topographic feature is natural or cultural will require future investigations and maybe excavation.

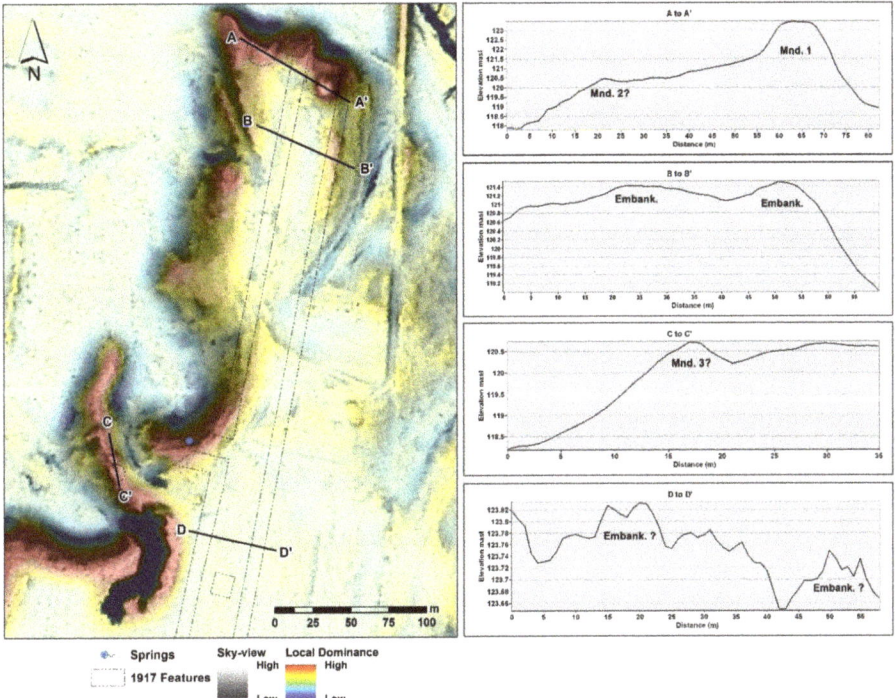

Figure 6. LiDAR-derived imagery (blended image using Sky-view factor and Local dominance) and topographic profiles from the northern portion of the Johnston Site with features from the 1917 map overlaid.

Descriptions of the parallel embankments indicate the best-preserved sections extended south of Mounds 1 and 2. Our analyses of LiDAR-derived imagery and topographic profiles shows that two parallel rises exist directly south of Mounds 1 and 2 (Figure 6; B to B'). These features are both close to 15.5 m wide and 0.3 m in height. They extend approximately 80 m to the south before no longer being clearly visible in elevation profiles or other visualization methods. However, an elevation profile across the purported location of these features in a former agricultural field 300 m south of Mound 1 (Figure 6; D to D') does suggest very subtle parallel rises are present but only 0.04 or 0.06 m tall.

The location of Mound 3 is slightly off from our georeferencing of the 1917 map. However, it is still visible as a conical mound, albeit smaller in height and diameter than the reported dimensions in 1917. Our results estimate that it is roughly 7 m in diameter and 0.3 m tall (Figure 6; C to C'). Mound 3 is currently in a precarious position, situated on a relict finger-like landform directly between two deep erosional gullies.

The central portion of Johnston has the two largest mounds (4 and 5) recorded at the site, in addition to four small conical mounds situated near them (Figure 7). Mounds 4 and 5 are still easily visible on the ground and in our data visualizations. Mound 5 was listed as a polygon in the 1917 map but appears more like a deflated low-lying rectangular mound in our topographic visualizations. The impact of agricultural plowing may have caused, or at least contributed, to this slight change in shape. Currently, Mound 5 is roughly 3.6 m tall and is 39.3 × 45.3 m in size at its base, while the surface spans 22.9 × 25 m (Figure 7; E to E'). Mound 4, the largest mound at Johnston, is a classic rectangular platform mound measuring 5.8 m tall, 57 × 59.5 m at the base, and exhibiting a surface extent of 32.8 × 34.6 m (Figure 7; F to F'). Multiple circular depressions on the surface of Mound 4 are visible, suggesting either looter attempts to recover artifacts from the mound, or potentially concavities from tree falls.

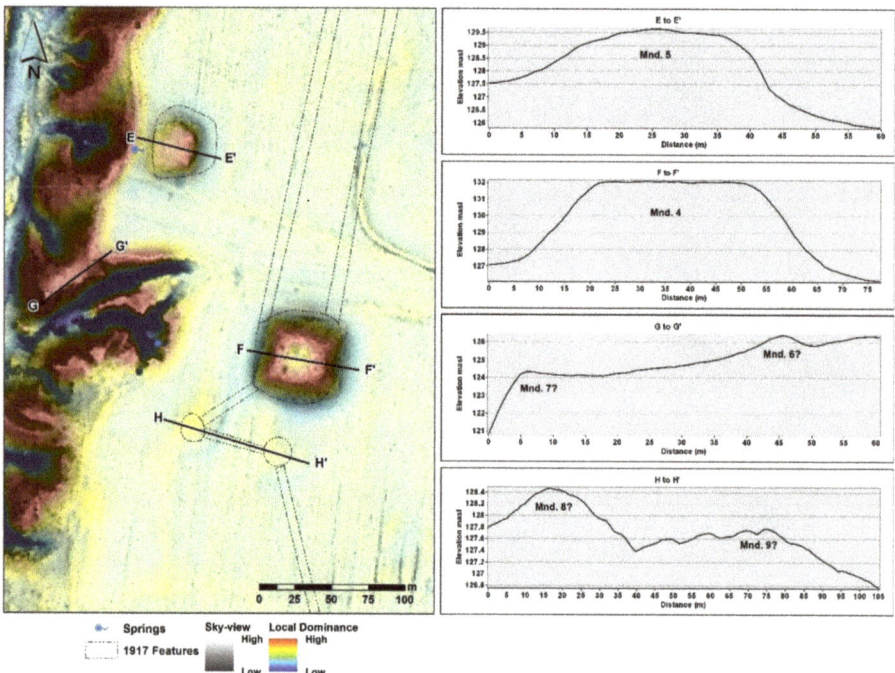

Figure 7. LiDAR-derived imagery (blended image using Sky-view factor and Local dominance) and topographic profiles from the central portion of the Johnston Site with features from the 1917 map overlaid.

On a narrow finger-like residual landform that extends west toward the SFFDR floodplain between Mounds 4 and 5 is the location of two small conical mounds. These are labeled Mounds 6 and 7 in the 1917 map. Mound 6 is the more visible of the two, potentially because it is situated at the center of this landform. Mound 6 is about 5.9 m in diameter at the base, and roughly 0.4 to 0.5 m in height. Mound 7 is harder to evaluate. It is positioned at the edge of the landform, surrounded by erosional gullies. The 1917 map describes it as a 'half oval' or crescent. We can distinguish a roughly 0.3 m rise at the edge of this landform but if this is the remnants of an earthen mound, very little remains intact. The small rise extends 4 m from the edge of the landform and measures roughly 7 m north to south (Figure 7; G to G').

Mounds 8 and 9 on the 1917 map are situated south of Mound 4 and connected via purported earthen 'walls'. We can detect no topographic evidence of earthen walls that connect these three mounds to one another. Moreover, it is not clear that the subtle topographic relief associated with the location of these two mounds confirm human construction. If they are built earthen architecture, then of the two, Mound 8 is the most visible in our imagery analyses. It measures 20 m in diameter and is roughly 0.6 m in height. Mound 9 is much harder to discern in LiDAR-derived visualizations and topographic profiles. Still, a topographic rise is present in the area where the 1917 map denotes the location of this feature. This rise is 20 m in diameter and 0.2 m in height (Figure 7; H to H').

According to the 1917 map of Johnston, the southern portion of the site contains an earthen wall that extends to a conical mound, Mound 10. Like the walls connecting Mounds 4, 8, and 9, no surficial evidence of this earthen wall exists. However, there is a clear topographic rise associated with the purported location of Mound 10 (Figure 8a; J to J'). In this case, our LiDAR-derived visualizations suggest that this rise associated with this area is 0.4 m tall and roughly 25 m in diameter. Like Mounds 8 and 9, it is not clear from topographic analyses alone that this is a human-made feature. We discuss below how we confirmed it as such. An exciting and unexpected discovery in our examination of LiDAR-derived imagery of the southern portions of the Johnston Site comes in the form of a non-mound feature that was not in the historic survey of the site. West of Mound 8 and 9 we discovered a depression encircled with an embankment. This feature is morphologically like the feature designated the "Duck's Nest" at PMSAP (Figure 8a; I to I'; Figure 8b; K to K') [49] (pp. 155–158).

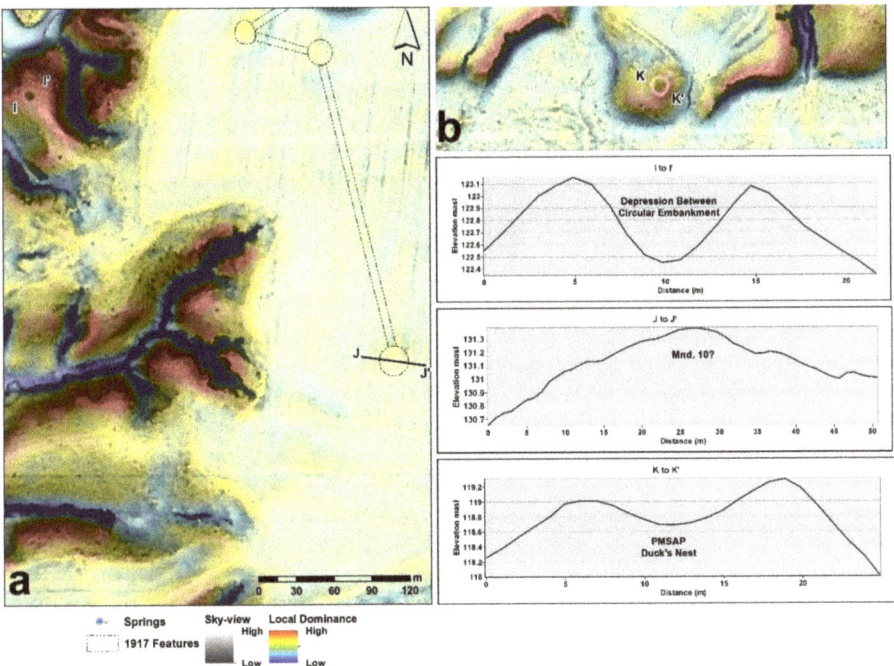

Figure 8. LiDAR-derived imagery (blended image using Sky-view factor and Local dominance) and topographic profiles from the southern portion of the Johnston Site with features from the 1917 map overlaid, with comparison of the Duck's Nest at PMSAP. (**a**) LiDAR-derived imagery from the Johnston Site. The profile line from I to I' is absent to better show the circular embankment feature; (**b**) LiDAR-derived imagery from the Duck's Nest at PMSAP set to the same scale, with profile line K to K' removed to better show the feature.

4.2. Gradiometer Results from the Johnston Site

Our gradiometer survey of Johnston was limited to the central and southern portions of the Johnston Site, where large open spaces could easily accommodate our cart-based instrument. There are also areas with large amounts of metal in the northern portions of the survey area that would make a gradiometer survey unsuccessful. Viewing the full extent of our magnetic coverage brings into focus the magnitude of agricultural plowing and drainage modifications in this portion of the site in the historic era (Figure 9). Intensive plow scars are seen in the data trending north-south, with the exception of the extreme southern portion of the survey area, where scars trend east-west—an indication of a former field boundary. Some plow scars are intense, potentially reflecting early deep-chisel plowing. The long history of agricultural practices seen in the gradiometer data undoubtedly impacted the topography of small features like the purported 1917 parallel embankments and smaller mounds. We can also identify the remnants of several lightning strikes across our survey area, something more archaeologists are discussing in recent research (see [84]).

Figure 9. Gradiometer coverage at the Johnston Site with features from the 1917 map overlaid (in yellow). Numbers denote extant mounds. A clear example of a lightning strike is present in the southern extent of the image.

We focus our discussion on results from selected magnetic features examined in test excavations that offer insight for our research questions outlined here. To begin, in the northern half of our gradiometer survey we were able to delineate numerous clusters of pit features denoted by rounded and spatially distinct magnetic highs (Figure 10a). Limited test excavations of a sample of these features have shown that they are either hickory nut (*Carya* sp.) roasting pits or storage pits with little to no cultural material but are filled with dark organically enriched material. Several pits are aligned in a row to the north of Mound 4, suggesting not only the spatial contemporaneity, but potentially association with Mound 4 as well. Test excavations of one pit in this group revealed a dark fill but no diagnostic cultural artifacts to associate the feature with a specific time of use (Figure 11a). We have several charred botanical samples that have been submitted for AMS 14C dating, but those dates are the focus of another article we are preparing.

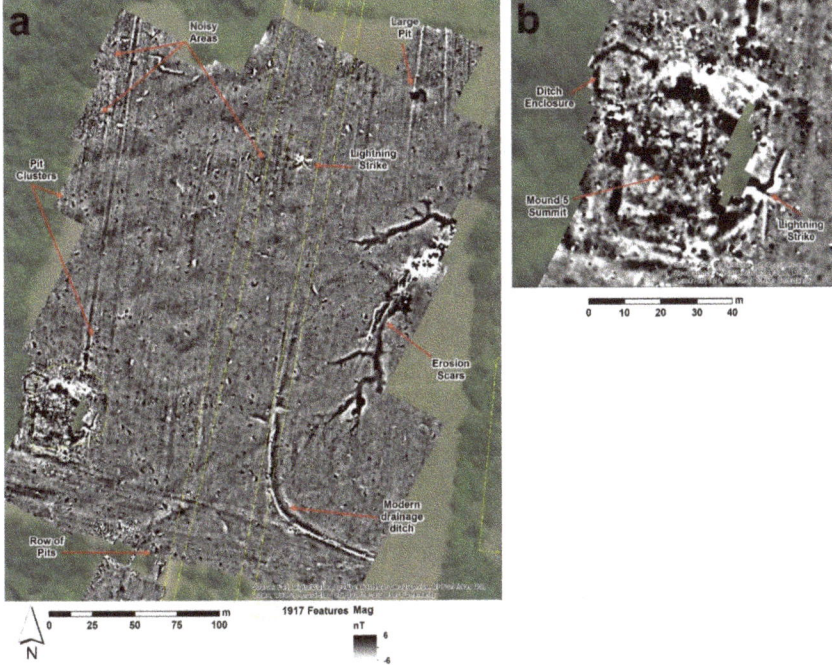

Figure 10. Gradiometer coverage north of Mound 4 at the Johnston Site with features from the 1917 map overlaid (in yellow). (**a**) Birds-eye view of gradiometer data north of Mound 4 exhibiting magnetic features discussed in the text; (**b**) Close-up of gradiometer data from areas around Mound 5 depicting the summit of Mound 5 and the buried ditch enclosure associated with the mound.

A large irregular feature measuring nearly 11 m in diameter is visible in the northeastern portion of our gradiometer data. Test excavations revealed a deep (>1 m) pit refilled with comparatively dark organic sediment (Figure 11b). Unfortunately, this feature contained no diagnostic artifacts to denote when the feature was used or refilled. In between the area where parallel embankments were mapped in 1917, and where the potential 'structure A' was situated (evidenced by burned daub present in plowed soil), we discerned the magnetic remnants of an isolated lightning strike roughly 12 m in diameter. Soils below plow zone were examined through a test excavation in this area and were documented as a reddish clayey sand. Plow scars cutting into this reddish horizon were visible below the plow zone. Therefore, it is possible that the structural feature 'A' mapped in this locale in 1917 was the result of a lightning strike. However, further research must be conducted to confirm this hypothesis. It is interesting to note that a noisy area of magnetic gradient borders the lightning strike to the west and southwest. This dense area of subsurface magnetic features likely represents a concentrated area of past human activity. Additional areas of seemingly 'noisy' magnetic phenomena are represented by non-patterned high and low magnetic features visible to the northwest of the locale inside the parallel embankments mapped in 1917. These can be characterized by high magnetic susceptibility due to the intensity of the response to plowing in this area. No magnetic response correlates with the location of the parallel embankments, leaving us wondering whether subsurface remnants of these features still exist.

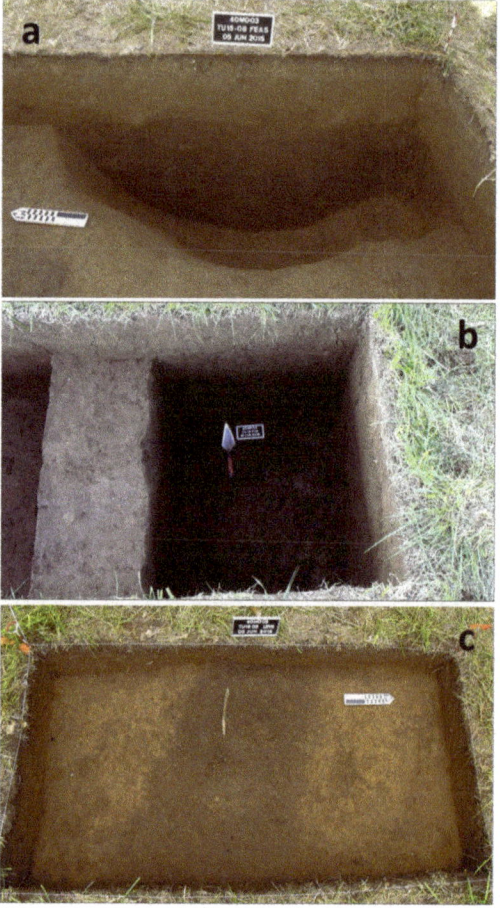

Figure 11. Test excavation of gradiometer features identified in the northern portions of the survey area. (**a**) Profile of a pit in the row situated north of Mound 4. Profile shows dark feature fill absent of cultural material beneath plowzone; (**b**) Photograph of nearly black feature fill from the large deep pit in the northeastern portion of the survey area; (**c**) Planview of the enclosure ditch beneath Mound 5 prior to full excavation.

Gradiometer results associated with Mound 5 are particularly intriguing. Our results identify a rectangular platform shape to this mound that contradicts the 1917 descriptions of the mound as a polygon. This discrepancy is likely the result of the intensive plowing at the site. The summit of Mound 5 exhibits numerous magnetic features potentially associated with platform activities. A lightning strike is visible on the eastern flank of the mound. However, most interesting to this area is a small ditched enclosure that we identified under the northern flank of Mound 5. This feature is morphologically like those identified in the Middle Ohio Valley and associated with Hopewellian ritual activity. The ditch encloses approximately 185 m^2 and exhibits a magnetically enhanced feature 3 m in diameter at its center. A test unit cross-cutting a small section of the ditch confirmed this interpretation and revealed the ditch contained burned earth and dense amounts of charcoal, suggesting the termination of the enclosure was intentional and likely occurred prior to the construction of Mound 5 (Figure 11c).

Gradiometer results from the fields south of Mound 4 raise additional questions about the indigenous landscape. Gradiometer coverage over Mounds 8 and 9 are noisy and reflect isolated higher magnetic susceptibility in these areas, visible in the non-uniform magnetic characteristics of the plow scars between the mounds (Figure 12a). However, no clear association is visible between the mound locations and our data, leaving questions to whether the mounds mapped in 1917 are now less visible due to plowing, or if they are relicts of natural topography. There are no clear clusters of pits south of Mound 4 like we see north of the mound. However, some individual pits can be discerned southeast of Mound 4. An area of low magnetic gradient is present in the southern field with four potential pits situated at the center of the area (Figure 12a; Low nT Area). We mapped numerous lightning strikes across this open area south of Mound 4.

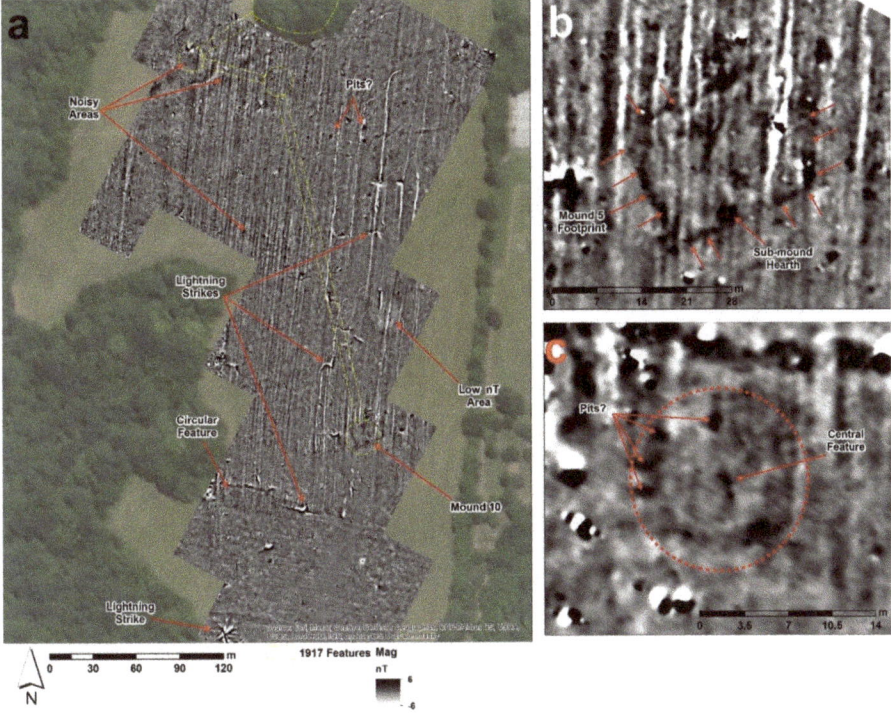

Figure 12. Gradiometer coverage south of Mound 4 at the Johnston Site with features from the 1917 map overlaid (in yellow). (**a**) Birds-eye view of gradiometer data south of Mound 4 exhibiting magnetic features discussed in the text; (**b**) Close-up of gradiometer data from areas around Mound 10 depicting the shape of the monument and internal subsurface features; (**c**) Close-up of gradiometer data from the circular feature southwest of Mound 10 depicting associated high magnetic features.

The area mapped historically as Mound 10, which has no topographic relief, was among the significant outcomes of our assessment of the magnetic data. Here we identified a rectangular feature and a round pit, both tested in a 1 × 10 m long trench (Figure 12a,b). These results suggest there was a mound present, but it was likely rectangular and not round or conical. The magnetic response to Mound 10's base dimensions may reflect magnetically enhanced topsoil that eroded over time to the bottom of the mound, revealing a 27 × 20.5 m base dimension.

Test excavations confirmed the presence of at least two different mound fills related to Mound 10's construction and identified a small basin shaped hearth under the mound fill. Analyses continue to assess how these features relate to one another. (Figure 13). A small clay cooking ball was recovered from the mound fill. Artifacts like these imply a Late Archaic (ca. 4000–1000 BC) presence on the site [85]. Additional excavations and a robust radiocarbon dating program is needed to confirm the construction age of Mound 10.

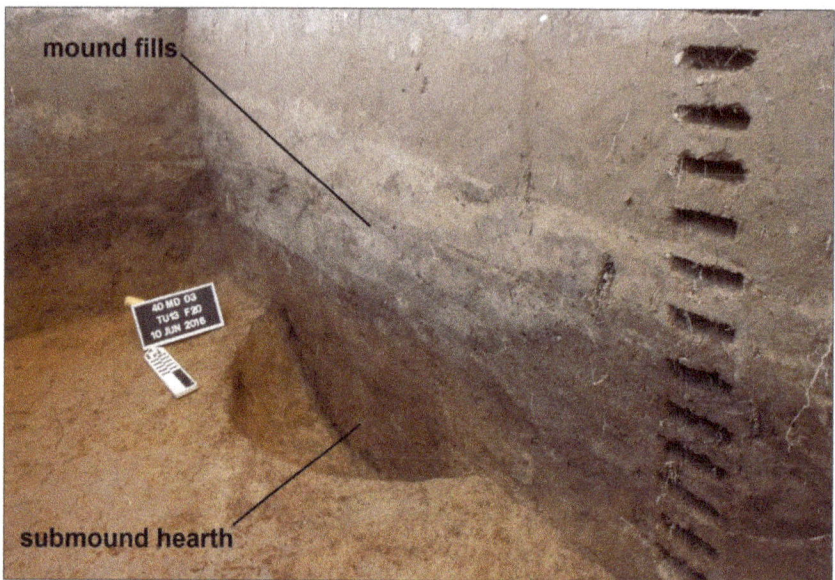

Figure 13. Profile of test excavations at Mound 10 indicating intact construction fills and sub-mound hearth.

Southwest of Mound 10 we identified a potential circular post structure 10 m in diameter (Figure 12c). This feature appears similar to other post-enclosures or paired-post structures in the Middle Ohio Valley [2,22,36,86–91]. We have yet to test this feature with excavations; however, the exterior of the circular feature appears to have numerous possible pits associated with it. A central magnetic feature is also present within this circular structure. This may be a central hearth, a depositional or refuse feature, or a central post. Excavations should clarify the nature of this structure.

To summarize, the results of our gradiometer survey at Johnston revealed numerous subsurface magnetic features that exhibit no topographic relief. The identification of such features allows us to discern activity areas across space in ways LiDAR-derived imagery do not. The gradiometer data have little correlation with some of the built features outlined on the 1917 map, such as the parallel embankments north of Mound 4 and the 'streets' or 'walls' that connect Mounds 8 and 9 to Mound 4. Moreover, there is little evidence that Mounds 8 and 9 exhibit clear magnetic signatures, leading us to question the nature of their existence. The gradiometer data over Mound 5 reveals the original shape of the mound to be rectilinear and shows that a uniquely different monument, a small ditched enclosure, preceded its construction.

4.3. Results from Large-Scale Surface Magnetic Susceptibility Surveys at the Johnston Site

Our large-scale surface magnetic susceptibility survey at Johnston was intended to complement the results of our gradiometer and LiDAR imagery survey and analyses. By examining the magnetic nature of the near-surface across the site, we hoped to better understand the accuracy of the 1917 map, interrogate the noisier portions of the gradiometer survey, and examine areas of the Johnston Site that we were not able to survey with our cart-based gradiometer or that contained large amounts of metal.

The results of our magnetic susceptibility survey exposed multiple areas of high susceptibility likely related to past human activities. We consider these 'activity areas' in the broadest sense of the term. North of Mound 4, the highest readings and largest activity area spatially coincides with the western opening in the parallel embankments from the 1917 map (Figure 14b). Probably not coincidentally, this entrance area leads to one of the springs denoted on the 1917 map. This suggests either a long history of pre-Contact indigenous people accessing this natural feature, or a short and intensive period of activity around the spring. Other large activity areas north of Mound 4 identified by spatially-distinct areas of high magnetic susceptibility correlated with the 'A' and 'B' purported structures inside the parallel embankments on the 1917 map, as well as an area situated along the tree line to the west of these structures. While we do not have gradiometer data over the activity area associated with the northern-most structure 'B' on the 1917 map, we do have comparative gradiometer data to assess correlations between the two magnetic datasets and the activity area associated with structure 'A' on the 1917 map and the area to the west at the tree line. The activity area associated with structure 'A' exhibits a large lightning strike right over the mapped structure, in addition to an area of elevated background gradiometer readings and isolated magnetic highs likely related to archaeological features (e.g., posts and pits) (see Figure 10a). The activity area at the tree line exhibited what we characterize as high levels of 'noise' in the gradiometer data, stemming from intense plow scars and numerous isolated high and low magnetic features (see Figure 10a). Therefore, we consider both areas as important locales exhibiting strong evidence for past human use of the Johnston Site. Moreover, the correlation with highly magnetic plow scars in this area indicates that highly magnetic plow scars elsewhere at the Johnston Site, as well as at other sites in the eastern U.S., may be an indicator for areas that have enhanced magnetic susceptibility from pre-Contact human occupation [29].

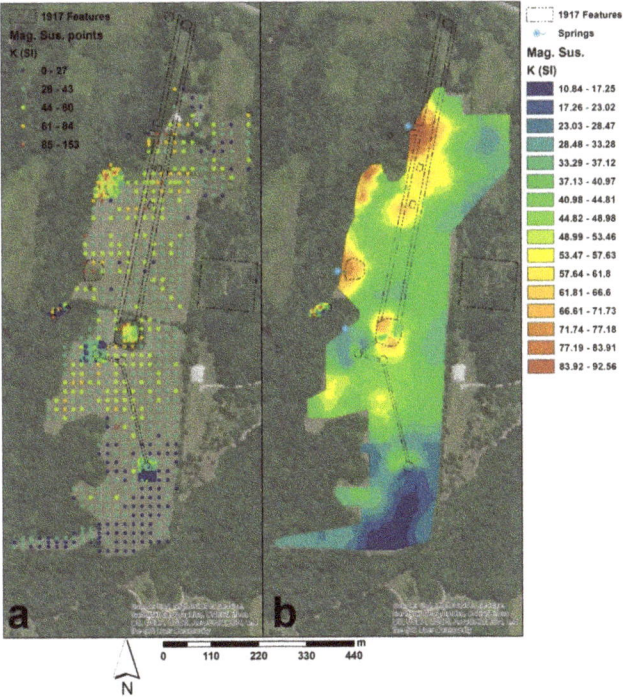

Figure 14. Surface magnetic susceptibility data from the Johnston Site. (**a**) Location and intensity of magnetic susceptibility readings at Johnston; (**b**) Gridded raster (kriging, 2.2 m cell size) of magnetic susceptibility data classified into 16 natural classes.

Magnetic susceptibility data over Mounds 4 and 5 exhibit high readings associated with the mounds themselves and with areas directly adjacent to them. For Mound 5, high levels of magnetic susceptibility extend to the west of the current mound boundaries. This may be the result of historic plowing that has displaced mound fills westward, or these data may be indicating off-mound activity areas. Magnetic susceptibility data around Mound 4 show high values extending to the northwest and southeast of the mound. These values may represent erosion of mound fill in these directions. We do not have the gradiometer coverage over the northwestern areas of Mound 4 for comparison, but data to the southeast exhibits highly magnetic plow scars similar to the activity area at the tree line in the northwest portion of our magnetic survey. However, in this locale there are not as many isolated magnetic features in our gradiometer data, lending support to the hypothesis that high magnetic susceptibility around Mound 4 may be related to eroded mound fills, and the spreading of those sediments during historic plowing.

A series of magnetic susceptibility readings in tree cover over Mounds 6 and 7 show good correlations with the topographic remnants of these features and high magnetic susceptibility values. However, we also identified an oval area of high magnetic susceptibility in between the mounds that suggest a subsurface activity area or cluster of archaeological features.

South of Mound 4 our magnetic susceptibility survey shows a large area of moderately high readings. This includes the area over Mounds 8 and 9; however, a more isolated area of moderately high magnetic susceptibility is associated with Mound 8. Two isolated areas of high magnetic susceptibility lie south of where Mounds 8 and 9 were mapped in 1917. Like our survey results north of Mound 4, these two areas correlate with plow scars in the gradiometer data that appear more magnetic than those elsewhere. Therefore, we consider this as good evidence for activity areas.

Magnetic susceptibility data associated with Mound 10 was moderately high only north of the mound remnants identified through the gradiometer data, despite an increased sampling density in this area. Highly magnetic plow scars are present in this area as well, suggestive of an off-mound activity area associated with Mound 10. Low to moderately low magnetic susceptibility readings are correlated with the circular feature west of Mound 10 in the gradiometer data. This may indicate that this feature was not used for a long period of time or peripheral to more intensive activity areas in the core of the site.

The results of our large-scale surface magnetic susceptibility survey identify areas at Johnston that were likely used by humans but were absent in other remote sensing imagery (surface or geophysical). As such, it provides an important supplement to understand the spatial relationships between human activities and monumental architecture at the site.

4.4. Results from an Electromagnetic Induction Survey of Mound 8

After assessing the LiDAR-derived imagery, gradiometer data, and surface magnetic susceptibility data for the Johnston Site, we were still unsure of whether a few of the small conical mounds at the site were indeed pre-Contact indigenous mounds rather than relict topography. However, we only had the opportunity to survey one with a slingram EMI meter, so we decided to focus on Mound 8.

Mound 8 has four datasets. The magnetic susceptibility data from the 0.5 and 1 m coil separations (ca. 0.3 and 0.6 m depth) show parallels with the gradiometer and surface magnetic susceptibility data. In data from both depths, this includes higher background magnetic susceptibility around the locale where Mound 8 was mapped in 1917 and the identification of isolated magnetic features that likely represent intact subsurface archaeological features associated with the topographic rise we can identify in the LiDAR-derived visualizations (Figure 15a,b). The data from the 0.5 m coil shows a transitional arc of high and low magnetic susceptibility in the southeastern portion of our survey block that mirrors the bend of Mound 8 as it was mapped in 1917.

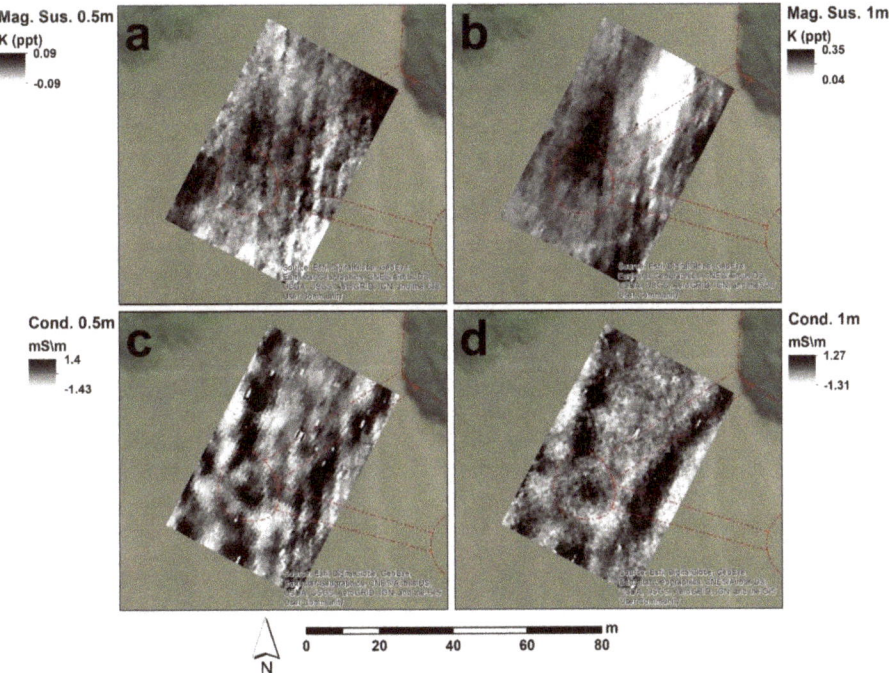

Figure 15. EMI data from the Mound 8 locale with the 1917 features overlaid in red. (**a**) Magnetic susceptibility data from the 0.5 m coil separation; (**b**) Magnetic susceptibility data from the 1 m coil separation; (**c**) Conductivity data from the 0.5 m coil separation; (**d**) Conductivity data from the 1 m coil separation.

The conductivity results of our EMI survey over Mound 8 depict no clear high values correlating with the complete coverage of the supposed Mound 8 location (Figure 15c,d). A high conductivity pattern is common among plowed down earthen mounds in the eastern U.S. because the remnants of clay-rich sediments used to build mounds often retain more moisture than surrounding soils (see [78,79,87]). However, in both the 0.5 and 1 m coil datasets (ca. 0.75 and 1.5 m depth), a high conductivity feature is present and spatially centralized over the topographic rise denoted as Mound 8. This may suggest that intact mound remnants are located here and are retaining more moisture, or there is a centralized non-mound feature under this 'rise'. In either case, the sum of data in this locale supports an interpretation that Mound 8 is anthropogenic, although this needs to be confirmed through excavation.

5. Discussion

Our integrated application of LiDAR-derived imagery and multiple geophysical survey techniques at the Johnston Site is formulated to help untangle the complex palimpsest of pre-Contact American Indian use and construction of the landscape. Simultaneously, we implemented these techniques to interrogate the validity and preservation of built features as mapped in 1917, and in doing so, also tried to better understand the impact of modern American agricultural practices on this important site. Our results pushed us toward a more detailed understanding of the Johnston Site that no single method we employed could have offered on its own. This multi-staged approach allows us to revise the reported dimensions and forms of mounds at the Johnston Site and work toward new research questions for future work. From this perspective, in the face of increasing erosion around the edges of the Johnston site, we can use our results to provide new information toward effective site conservation.

5.1. Toward a New Map of the Johnston Site

One of the primary goals of our research was to better situate the Johnston Site within understandings of the broader Middle Woodland landscape along the SFFDR. This required a formal interrogation of the 1917 map based on the work of Buck and Myer. Because of heavy vegetation, our use of LiDAR-derived imagery offered the only way to reassess the built environment at the northern-most portion of the site. In this area, we have strong topographic evidence that Mound 1 is both authentic and similar in size to the dimensions reported in 1917 (see Tables 1 and 3). We identified Mound 1 as nearly 2 m smaller in the base diameter but nearly the same height. One explanation for these differences in size could be attributable to the erosion of the mound surface, but this should lead to a larger base diameter. Therefore, our measurements may just be more accurate than those made in 1917, or the surveyor Buck may have directed less attention to mapping Mound 1 because it was not as large as other mounds at the site (e.g., Mounds 4 and 5).

Table 3. Revised dimensions and shapes of mounds at the Johnston Site based on work presented here.

Mound No.	Shape	Height (m)	Surface Dimensions (m)	Base Dimensions (m)
1	conical	2.03	n/a	18.41 diameter
2	conical	0.2	n/a	<5 diameter
3	conical	0.76	n/a	7 diameter
4	rectangular	5.8	32.8 × 34.6	57 × 59.5
5	rectangular	3.6	22.9 × 25	39.3 × 45.3
6	conical	0.5	n/a	5.9 diameter
7	half oval	0.3	n/a	4 × 7
8	conical	0.6	n/a	20 diameter
9	conical	0.46	n/a	20 diameter
10	rectangular	0.4	n/a	20.5 × 27

Differences we can identify with Mound 2 are more complex to describe. A small rise is present in the location where this mound was mapped in 1917; however, the rise and the base dimensions of this rise are much smaller than what was reported previously. The placement on the edge of the Johnston terrace may indicate that much of this mound was lost as the terrace retreated. Such active erosion is visible today [83]. Alternatively, this topographic feature may be a natural relict of the landscape and its proximity with Mound 1 may have led Buck and Myer to interpret it as a. Further research is needed to confirm the nature of this rise.

The parallel embankment walls that are described from Mounds 1 and 2 south to Mound 4 are also hard to explain. In the northern reaches of the site there is strong topographic evidence for the eastern wall, and good evidence for the western wall. However, as the walls extend south into areas where we have gradiometer and magnetic susceptibility coverage, the topographic evidence is minimal and there is no correlation to geophysical signatures where the walls should be. This leaves us to question whether the history of modern agricultural plowing at the site has destroyed topographic evidence for the embankments in most unforested portions of the site. If the embankments are present near Mounds 1 and 2, but are not clearly visible further south, the early map may have continued the walls for consistency. The area northeast of Mound 3 and associated with the natural spring and western opening of the embankments shows extremely high magnetic susceptibility values for the entire site. This lends some validity to there being an important activity area in this portion of the site, but offers little support for walls or an opening to any walls here.

The location of Mound 3 as described in 1917 exhibits a clear conical rise with a nearly identical height, but like Mound 1 the base dimensions are off by nearly 3 m. This implies an error in the 1917 base measurements for this mound since any erosion should have led to a potentially larger, rather than smaller, mound base area. The two structures on the 1917 map labeled 'A' and 'B' corelate with high magnetic susceptibility. In the case of 'A', a lightning strike is visible in the gradiometer data. This suggests that the surveyors may have seen reddish soil discoloration in these areas and mapped those patterns as archaeological features. This said, we cannot definitively assign such an interpretation to 'B' because we only know that it is associated with high magnetic susceptibility. Extending our gradiometer coverage north will offer more insights into whether the 'B' structure is also related to a lightning strike.

There is no topographic or geophysical evidence for any of the streets or walls that connect Mounds 4 and 5, or those associated with mounds further south. However, our work does offer new insights into Mound 5, specifically its history and shape. Our topographic imagery shows that the base of Mound 5 is 2–3 m smaller than previously recorded. In contrast, the surface area is larger by about 3–4 m. This likely relates to the impact of plowing around and over this small platform. Our gradiometer data offers clear evidence for a ditch enclosure present before the construction of Mound 5. Our test excavations confirmed this feature and show preliminary evidence that it was refilled with anthropogenic materials. If the deconstruction of the enclosure and the construction of Mound 5, a platform mound, marks an important shift in the use of this space, then we can begin to build new research questions for the Mound 5 locale. For instance, Middle Woodland enclosures have been interpreted as collective monuments that imply an internal exclusivity for those who enter and use their interior spaces. These monuments are built to enclose an area, interpreted as creating a perception of 'us and inside' versus 'them and outside', and have been considered a place where people from diverse geographic scales participated in specialized ritual events [35,92–96]. Alternatively, Middle Woodland and Late Woodland platform mounds, while also considered monuments emphasizing collective notions of society, are commonly interpreted as socially inclusive because of their association with the remains of feasting [41,97–99]. We reference Late Woodland (ca. AD 600–1000) platform mounds here also because we have no relative or absolute chronological information for Mound 5; we only know that it post-dates a Middle Woodland enclosure.

The current topographic data for Mound 4 shows that the base dimensions are less by 3 and 4 m, while the surface dimensions are greater by 3 to almost 5 m. This may be a combined result of erosion and deflation of the mound surface, leading to a larger surface platform, while plowing around the mound would move mound sediments around the field, causing the base to decrease in size rather than increase from the accumulation of eroded surface sediments. Magnetic susceptibility from the surface of Mound 4 shows a combination of high and low values. These differences may relate to variation in mound fill used to construct Mound 4 or it may represent features associated with the use of the mound summit.

Our examination of Mounds 6 and 7 suggest that remnants of these features are present and can be represented geophysically by high magnetic susceptibility. Topographic signatures from these features show they have eroded measurably, with Mound 7 in danger of vanishing completely. The combined topographic and geophysical data for Mounds 8 and 9 are hard to interpret. The LiDAR-derived imagery shows small rises in these locations that are slightly lower than the 1917 heights and roughly 20 cm larger in base diameters, suggestive of deflation from plowing. The gradiometer data show several isolated magnetic highs associated with these small rises. The EMI data over Mound 8 identified a high conductivity feature at the center of the rise but none of the datasets clarify whether these are mounds.

Work related to Mound 10 allowed us to revise the footprint of the monument, from round to rectangular; however, its height is reduced by approximately 30 cm, probably related to plowing indicated by the several plow scars traversing the mound in the gradiometer data. The gradiometer survey also revealed internal features that our preliminary excavations have confirmed were associated with use of the Johnston landscape before Mound 10 was constructed.

Other aspects of our work that allow us to create a new map of Johnston include the circular embankment feature along the western edge of the Johnston landform and several activity areas denoted by both sets of high surface magnetic susceptibility values. Clusters of isolated high magnetic features that likely represent subsurface pit remnants or large posts also adds to our new map of the Johnston Site and the broader understanding of land use. The potential circular structure west of Mound 10 is another important feature we have identified at the site. Our new awareness of the Johnston landscape, illuminated by our integrated remote sensing approach, allows us to formulate new research questions for the site.

5.2. Beyond Never-Never Land: Developing Future Questions for the Johnston Site

Our research has identified several differences between the 1917 map of Johnston and its present condition. However, for the most part, we can correlate topographic rises or measurable geophysical trends with features mapped at the site in 1917. Only the 'streets' and 'walls' linking Mounds 4, 5, 8, 9, and 10 are indistinguishable using the methods we employed. Therefore, we think it is important to move beyond questions that undermine the validity of Buck and Myers' work and move forward with new research questions for the site. To this end, we propose that additional geophysical survey methods have the potential to help elucidate issues related to the parallel embankments running north-south between Mounds 1 and 2 and Mound 4, as well as the streets and walls we cannot identify at all. Recent large-area GPS-guided EMI surveys have shown to be effective in teasing apart buried archaeological features that are not detectible using magnetometry [100]. The ability to map patterns of earthen conductivity across the entire Johnston Site might provide some evidence of the parallel embankments if they were built using soils high in clay. Even residual clays from an embankment construction would retain more moisture than surrounding soils today, potentially making them detectable via conductivity surveys. Additional geophysical surveys, employing methods like ground-penetrating radar (GPR), over the summits of mounds at Johnston has the potential to identify features associated with the uses of the summits. This would be particularly useful at Mounds 4 and 5 to determine if any mound-top structures are evident, or to identify features related to feasting activities.

Beyond additional geophysical surveys, we find the rate of erosion evident along the northern and western edges of the site troubling. If Mounds 2 and 7 are indeed indigenous constructions, some amount of these earthen mounds have probably eroded away already. To both determine whether these mounds are indigenous and examine how much is left, we propose that future 'cut-bank' examinations would be important. Cutbank geophysics that employ magnetic susceptibility have already proved to be successful in identifying cultural layers and archaeological features eroding into river courses [101]. Adding cutbank geophysics to well-known geoarchaeological methods like sequential loss-on-ignition, particle size analyses, and soil micromorphology is likely to confirm the nature of Mounds 2 and 7.

An additional avenue of future research we propose is important for better understanding the Johnston Site, as well as the larger SFFDR landscape, which relates to the temporality. The mounds, activity areas, pit clusters, and subsurface structural remnants we have identified here appear to represent a long history of landscape use in this area of western Tennessee. There are good reasons to believe that a lot of this evidence for human occupation at Johnston is not all Middle Woodland in age. For instance, the hickory nut roasting pits we have identified north of Mound 5 may relate to a pre-Middle Woodland hunter-gatherer use of Johnston. The clay cooking ball recovered from construction fills at Mound 10 is indicative of Late Archaic occupations at the site. Moreover, we have already identified a relative chronology for shifting forms of space relating to the change from enclosure to platform mound at Mound 5. However, we do not yet know when that shift took place, or how long each monument was in use. Therefore, untangling the uses of Johnston through time should reveal significant changes related to social complexity and the palimpsestic history of human occupation on this landform.

6. Conclusions

The goal for our research at the Johnston Site was to situate the site within the broader context of Middle Woodland activity along this stretch of the SFFDR. This included contextualizing the Johnston Site in relation to other mounds centers nearby, like Pinson Mounds and the Elijah Bray site. In doing so, we wanted to lay the foundations for creating a landscape biography of the area that focused on exploring the indigenous history of this region. The application of a robust remote sensing approach that integrated LiDAR-derived visualizations and geophysical methods allowed us to non-invasively examine the Johnston landscape from various scales of analysis. In doing so, we were able to build upon the earlier archaeological cartography of Johnston and the work of E.G. Buck and William Myer, as well as the interpretations by Kwas and Mainfort [30]. The results of our work at Johnston suggest that there were most likely occupations that were precursors to the Middle Woodland occupation of Pinson Mound. In this sense, Kwas and Mainfort's assessment is probably correct. However, we would note that the majority of research at Pinson Mounds has focused on mounds and that the similarities between Johnston and Pinson in terms of their location on terrace landforms overlooking the SFFDR, suggests that Pinson too probably has a vast pre- and post-Middle Woodland occupation.

Our research shows the importance of integrating both aerial and terrestrial remote sensing methods. In our case study presented here, this amalgamation of remote sensing methods provided insights that no singular method could offer. Our methodology allowed us to critically assess the 1917 map of the Johnston site and identify a variety of surface and subsurface features beyond the original map. This integrative approach also permitted us to identify portions of the 1917 map that may have been embellished, although further geophysical surveys, archaeological excavations, and soil analyses should be performed to confirm this notion.

Author Contributions: Conceptualization, E.R.H., A.P.W., S.C.S., S.B.C., C.R.B. and C.V.d.V.; methodology, E.R.H., A.P.W., S.C.S., S.B.C., C.R.B. and C.V.d.V.; formal analysis, E.R.H., A.P.W. and C.V.d.V.; investigation, E.R.H., A.P.W., S.C.S., S.B.C., C.R.B. and C.V.d.V.; resources, E.R.H., A.P.W., S.C.S., S.B.C., C.R.B. and C.V.d.V.; data curation, E.R.H.; writing—Original draft preparation, E.R.H. and A.P.W.; writing—Review and editing, E.R.H., A.P.W., S.C.S., S.B.C., C.R.B. and C.V.d.V.; visualization, E.R.H., A.P.W. and C.V.d.V.; supervision, E.R.H., A.P.W., S.C.S., S.B.C., C.R.B. and C.V.d.V.; project administration, S.C.S.; funding acquisition, S.C.S. and A.P.W. All authors have read and agreed to the published version of the manuscript.

Funding: This research was funded by: A University of the South Faculty Development Grant to S.C.S. funded logistical needs for the field work. The magnetic susceptibility equipment and travel costs were funded by the American Philosophical Society, Franklin Research Grant awarded to A.P.W.

Acknowledgments: We thank our respective institutions for support throughout the duration of this project. The Tennessee Division of Archaeology (TDOA) provided access to the Johnston Site. We thank TDOA archaeologist Bill Lawrence, who helped facilitate our project logistics and aided in our fieldwork. Thanks are due to Tim Poole, site manager for PMSAP, for arranging access to the Group Camp at PMSAP. This project would not have been possible without Tristram R. Kidder, Director of the Geoarchaeology Laboratory at Washington University in St. Louis, who provided access to geophysical instruments and processing software.

Conflicts of Interest: The authors declare no conflict of interest.

References

1. Bewley, R.H.; Crutchley, S.P.; Shell, C.A. New Light on an Ancient Landscape: LiDAR Survey in the Stonehenge World Heritage Site. *Antiquity* **2005**, *79*, 636–647. [CrossRef]
2. Burks, J.; Cook, R.A. Beyond Squier and Davis: Rediscovering Ohio's Earthworks Using Geophysical Remote Sensing. *Am. Antiq.* **2011**, *76*, 667–689. [CrossRef]
3. Chase, A.F.; Chase, D.Z.; Fisher, C.T.; Leisz, S.J.; Weishampel, J.F. Geospatial Revolution and Remote Sensing LiDAR in Mesoamerican Archaeology. *Proc. Natl. Acad. Sci. USA* **2012**, *109*, 12916–12921. [CrossRef] [PubMed]
4. Conyers, L.B. *Ground-Penetrating Radar for Archaeology*; AltaMira Press: Walnut Creek, CA, USA, 2004; ISBN 0-7591-0772-6.

5. Cowley, D.; Standring, R.A.; Abicht, M.J. (Eds.) *Landscapes through the Lens: Aerial Photographs and Historic Environment*; Oxbow Books: Oxford, UK, [Distributed in the US by]; David Brown Book Co.: Oakville, CT, USA, 2010; ISBN 978-1-84217-981-9.
6. Eppelbaum, L.V.; Khesin, B.E.; Itkis, S.E. Prompt Magnetic Investigations of Archaeological Remains in Areas of Infrastructure Development: Israeli Experience. *Archaeol. Prospect.* **2001**, *8*, 163–185. [CrossRef]
7. Evans, D.H.; Fletcher, R.J.; Pottier, C.; Chevance, J.-B.; Soutif, D.; Tan, B.S.; Im, S.; Ea, D.; Tin, T.; Kim, S.; et al. Uncovering Archaeological Landscapes at Angkor Using LiDAR. *Proc. Natl. Acad. Sci. USA* **2013**, *110*, 12595–12600. [CrossRef]
8. Gaffney, C.F.; Gater, J. *Revealing the Buried Past: Geophysics for Archaeologists*; Tempus: Stroud, UK, 2003; ISBN 0-7524-2556-0.
9. Goodman, D.; Piro, S. *GPR Remote Sensing in Archaeology*; Springer: Berlin/Heidelberg, Germany, 2013; ISBN 978-3-642-31856-6.
10. McKinnon, D.P.; Haley, B.S. (Eds.) *Archaeological Remote Sensing in North America: Innovative Techniques for Anthropological Applications*; University of Alabama Press: Tuscaloosa, AL, USA, 2017; ISBN 978-0-8173-1959-5.
11. Henry, E.R.; Laracuente, N.R.; Case, J.S.; Johnson, J.K. Incorporating Multistaged Geophysical Data into Regional-Scale Models: A Case Study from an Adena Burial Mound in Central Kentucky. *Archaeol. Prospect.* **2014**, *21*, 15–26. [CrossRef]
12. Howey, M.C.L.; Sullivan, F.B.; Tallant, J.; Kopple, R.V.; Palace, M.W. Detecting Precontact Anthropogenic Microtopographic Features in a Forested Landscape with LiDAR: A Case Study from the Upper Great Lakes Region, AD 1000–1600. *PLoS ONE* **2016**, *11*, e0162062. [CrossRef]
13. Johnson, J.K. (Ed.) *Remote Sensing in Archaeology: An Explicitly North American Perspective*; University of Alabama Press: Tuscaloosa, AL, USA, 2006; ISBN 978-0-8173-5343-8.
14. Johnson, K.M.; Ouimet, W.B. Rediscovering the Lost Archaeological Landscape of Southern New England Using Airborne Light Detection and Ranging (LiDAR). *J. Archaeol. Sci.* **2014**, *43*, 9–20. [CrossRef]
15. Kvamme, K.L. Geophysical Surveys as Landscape Archaeology. *Am. Antiq.* **2003**, *68*, 435–457. [CrossRef]
16. Opitz, R.S.; Cowley, D. Interpreting Archaeological Topography: Lasers, 3D Data, Observation, Visualisation and Applications. In *Interpreting Archaeological Topography: Airborne Laser Scanning, 3D Data, and Ground Observation*; Opitz, R.S., Cowley, D.C., Eds.; Oxbow Books: Oxford, UK, 2013; pp. 1–12.
17. Pluckhahn, T.J.; Thompson, V.D. Integrating LiDAR Data and Conventional Mapping of the Fort Center Site in South-Central Florida: A Comparative Approach. *J. Field Archaeol.* **2012**, *37*, 289–301. [CrossRef]
18. Riley, M.A.; Tiffany, J.A. Using LiDAR Data to Locate a Middle Woodland Enclosure and Associated Mounds, Louisa County, Iowa. *J. Archaeol. Sci.* **2014**, *52*, 143–151. [CrossRef]
19. VanValkenburgh, P.; Walker, C.P.; Sturm, J.O. Gradiometer and Ground-penetrating Radar Survey of Two Reducción Settlements in the Zaña Valley, Peru. *Archaeol. Prospect.* **2015**, *22*, 117–129. [CrossRef]
20. VanValkenburgh, P.; Cushman, K.C.; Butters, L.J.C.; Vega, C.R.; Roberts, C.B.; Kepler, C.; Kellner, J. Lasers Without Lost Cities: Using Drone Lidar to Capture Architectural Complexity at Kuelap, Amazonas, Peru. *J. Field Archaeol.* **2020**, *45*, S75–S88. [CrossRef]
21. Venter, M.L.; Shields, C.R.; Ordóñez, M.D.C. Mapping Matacanela: The Complementary Work of LiDAR and Topographical Survey in Southern Veracruz, Mexico. *Anc. Mesoam.* **2018**, *29*, 81–92. [CrossRef]
22. Henry, E.R.; Shields, C.R.; Kidder, T.R. Mapping the Adena-Hopewell Landscape in the Middle Ohio Valley, USA: Multi-Scalar Approaches to LiDAR-Derived Imagery from Central Kentucky. *J. Archaeol. Method Theory* **2019**, *26*, 1513–1555. [CrossRef]
23. Thompson, V.D.; Marquardt, W.H.; Walker, K.J. A Remote Sensing Perspective on Shoreline Modification, Canal Construction and Household Trajectories at Pineland along Florida's Southwestern Gulf Coast: Remote Sensing at Pineland. *Archaeol. Prospect.* **2014**, *21*, 59–73. [CrossRef]
24. Thompson, V.; DePratter, C.; Lulewicz, J.; Lulewicz, I.; Roberts Thompson, A.; Cramb, J.; Ritchison, B.; Colvin, M. The Archaeology and Remote Sensing of Santa Elena's Four Millennia of Occupation. *Remote Sens.* **2018**, *10*, 248. [CrossRef]
25. Alizadeh, K.; Ur, J.A. Formation and Destruction of Pastoral and Irrigation Landscapes on the Mughan Steppe, North-Western Iran. *Antiquity* **2007**, *81*, 148–160. [CrossRef]
26. Mlekuž, D. Messy Landscapes: LiDAR and the Practices of Landscaping. In *Interpreting Archaeological Topography: Lasers, 3D Data, Observation, Visualisation and Applications*; Cowley, D.C., Opitz, R.S., Eds.; Oxbow Books: Oxford, UK, 2013; pp. 90–101.

27. Johnson, K.M.; Ouimet, W.B. An Observational and Theoretical Framework for Interpreting the Landscape Palimpsest Through Airborne LiDAR. *Appl. Geogr.* **2018**, *91*, 32–44. [CrossRef]
28. Thompson, V.D.; Arnold, P.J.; Pluckhahn, T.J.; Vanderwarker, A.M. Situating Remote Sensing in Anthropological Archaeology. *Archaeol. Prospect.* **2011**, *18*, 195–213. [CrossRef]
29. Horsley, T.; Wright, A.; Barrier, C. Prospecting for New Questions: Integrating Geophysics to Define Anthropological Research Objectives and Inform Excavation Strategies at Monumental Sites. *Archaeol. Prospect.* **2014**, *21*, 75–86. [CrossRef]
30. Kwas, M.L.; Mainfort, R.C., Jr. The Johnston Site: Precursor to Pinson Mounds? *Tenn. Anthropol.* **1986**, *11*, 30–41.
31. Myer, W.E. *Stone Age Man in the Middle South n.d.*; Manuscript available from the Tennessee Division of Archaeology; Tennessee Division of Archaeology: Nashville, TN, USA, 1967.
32. Kolen, J.; Renes, J.; Hermans, R. (Eds.) *Landscape Biographies: Geographical, Historical and Archaeological Perspectives on the Production and Transmission of Landscapes*; Amsterdam University Press: Amsterdam, The Netherlands, 2015.
33. Carr, C.; Case, D.T. (Eds.) *Gathering Hopewell: Society, Ritual, and Interaction*; Kluwer Academic/Plenum Publishers: New York, NY, USA, 2005.
34. Charles, D.K.; Buikstra, J.E. (Eds.) *Recreating Hopewell*; University Press of Florida: Gainesville, FL, USA, 2006; ISBN 0-8130-2898-1.
35. Henry, E.R. Earthen Monuments and Social Movements in Eastern North America: Adena-Hopewell Enclosures on Kentucky's Bluegrass Landscape. Ph.D. Dissertation, Washington University St. Louis, St. Louis, MO, USA, 2018.
36. Henry, E.R.; Barrier, C.R. The Organization of Dissonance in Adena-Hopewell Societies of Eastern North America. *World Archaeol.* **2016**, *48*, 87–109. [CrossRef]
37. Redmond, B.G.; Ruby, B.J.; Burks, J. (Eds.) *Encountering Hopewell in the Twenty-First Century, Ohio and Beyond: Volume One: Monuments and Ceremony*; University of Akron Press: Akron, OH, USA, 2019; ISBN 978-1-62922-102-1.
38. Redmond, B.G.; Ruby, B.J.; Burks, J. *Encountering Hopewell in the Twenty-First Century, Ohio and Beyond: Volume Two: Settlements, Foodways, and Interaction*; University of Akron Press: Akron, OH, USA, 2020; ISBN 978-1-62922-103-8.
39. Thompson, V.D.; Pluckhahn, T.J. Monumentalization and Ritual Landscapes at Fort Center in the Lake Okeechobee Basin of South Florida. *J. Anthropol. Archaeol.* **2012**, *31*, 49–65. [CrossRef]
40. Wallis, N.J. *The Swift Creek Gift: Vessel Exchange on the Atlantic Coast*; University of Alabama Press: Tuscaloosa, AL, USA, 2011; ISBN 978-0-8173-5629-3.
41. Wright, A.P. Local and "Global" Perspectives on the Middle Woodland Southeast. *J. Archaeol. Res.* **2017**, *25*, 35–83. [CrossRef]
42. Wright, A.P.; Henry, E.R. (Eds.) *Early and Middle Woodland Landscapes of the Southeast*; University Press of Florida: Gainesville, FL, USA, 2013; ISBN 0-8130-4460-X.
43. Gremillion, K.J. The Development and Dispersal of Agricultural Systems in the Woodland Period Southeast. In *The Woodland Southeast*; Anderson, D.G., Mainfort, R.C., Eds.; University of Alabama Press: Tuscaloosa, AL, USA, 2002; pp. 483–501.
44. Mueller, N.G. *Mound Centers and Seed Security: A Comparative Analysis of Botanical Assemblages from Middle Woodland Sites in the Lower Illinois Valley*; Springer: New York, NY, USA, 2013.
45. Mueller, N.G.; Fritz, G.J.; Patton, P.; Carmody, S.; Horton, E.T. Growing the lost crops of eastern North America's original agricultural system. *Nat. Plants* **2017**, *3*, 1–5. [CrossRef] [PubMed]
46. Mueller, N.G. The earliest occurrence of a newly described domesticate in Eastern North America: Adena/Hopewell communities and agricultural innovation. *J. Anthropol. Archaeol.* **2018**, *49*, 39–50. [CrossRef]
47. Smith, B.D. Low-Level Food Production. *J. Archaeol. Res.* **2001**, *9*, 1–43. [CrossRef]
48. Struever, S. Implications of vegetal remains from an Illinois Hopewell site. *Am. Antiq.* **1962**, *27*, 584–587. [CrossRef]
49. Mainfort, R.C., Jr. *Pinson Mounds: Middle Woodland Ceremonialism in the Midsouth*; University of Arkansas Press: Fayetteville, AR, USA, 2013.
50. Mainfort, R.C., Jr. Middle Woodland Ceremonialism at Pinson Mounds, Tennessee. *Am. Antiq.* **1988**, *53*, 158–173. [CrossRef]

51. Stoltman, J.B. *Ceramic Petrography and Hopewell Interaction*; University of Alabama Press: Tuscaloosa, AL, USA, 2015; ISBN 978-0-8173-1859-8.
52. Carr, C. Rethinking Interregional Hopewellian "Interaction". In *Gathering Hopewell: Society, Ritual, and Interaction*; Carr, C., Case, D.T., Eds.; Kluwer Academic/Plenum Publishers: New York, NY, USA, 2005; pp. 575–623.
53. Rafinesque, C.S. *Map of the Lower Alleghanee Monuments on North Elkhorn Creek 1820*; University of Kentucky Special Collections Library: Lexington, KY, USA, 1820.
54. Rafinesque, C.S. *A Life of Travels and Researches in North America and South Europe*; Turner: Philadelphia, PA, USA, 1836.
55. Squire, E.G.; Davis, E.H. *Ancient Monuments of the Mississippi Valley*, 150th anniversary ed.; Smithsonian Books: Washington, DC, USA, 1998; ISBN 1-56098-898-3.
56. Thomas, C. *The Circular, Square, and Octagonal Earthworks of Ohio*; Bulletin; Smithsonian Institution, Bureau of American Ethnology: Washington, DC, USA, 1889.
57. Thomas, C. Report on Mound Explorations of the Bureau of Ethnology. In *Twelfth Annual Report of the Bureau of Ethnology to the Secretary of the Smithsonian Institution, 1890–1891*; Powell, J.W., Ed.; Bureau of American Ethnology: Washington, DC, USA, 1894; pp. 3–742.
58. Henry, E.R. A Multistage Geophysical Approach to Detecting and Interpreting Archaeological Features at the LeBus Circle, Bourbon County, Kentucky. *Archaeol. Prospect.* **2011**, *18*, 231–244. [CrossRef]
59. Mainfort, R.C., Jr.; Kwas, M.L.; Mickelson, A.M. Mapping Never-Never Land: An Examination of Pinson Mounds Cartography. *Southeast. Archaeol.* **2011**, *30*, 148–165. [CrossRef]
60. Myer, W.E. Recent Archaeological Discoveries in Tennessee. *Art Archaeol.* **1922**, *14*, 141–150.
61. Kokalj, Ž.; Somrak, M. Why Not a Single Image? Combining Visualizations to Facilitate Fieldwork and On-Screen Mapping. *Remote Sens.* **2019**, *11*, 747. [CrossRef]
62. Zakšek, K.; Oštir, K.; Kokalj, Ž. Sky-View Factor as a Relief Visualization Technique. *Remote Sens.* **2011**, *3*, 398–415. [CrossRef]
63. Sampson, C.P.; Horsley, T.J. Using Multistaged Magnetic Survey and Excavation to Assess Community Settlement Organization: A Case Study from the Central Peninsular Gulf Coast of Florida. *Adv. Archaeol. Pract.* **2020**, *8*, 53–64. [CrossRef]
64. Crutchley, S.; Crow, P. *The Light Fantastic: Using Airborne Laser Scanning in Archeological Survey*; Historic England: Swindon, UK, 2009.
65. Opitz, R.S. An Overview of Airborne and Terrestrial Laser Scanning in Archaeology. In *Interpreting Archaeological Topography: Airborne Laser Scanning, 3D Data, and Ground Observation*; Opitz, R.S., Cowley, D.C., Eds.; Oxbow Books: Oxford, UK, 2013; pp. 13–31.
66. Challis, K.; Forlin, P.; Kincey, M. A Generic Toolkit for the Visualization of Archaeological Features on Airborne LiDAR Elevation Data: Visualizing Archaeological Features in Airborne LiDAR. *Archaeol. Prospect.* **2011**, *18*, 279–289. [CrossRef]
67. Devereux, B.J.; Amable, G.S.; Crow, P. Visualisation of LiDAR Terrain Models for Archaeological Feature Detection. *Antiquity* **2008**, *82*, 470–479. [CrossRef]
68. Mayoral, A.; Toumazet, J.-P.; Simon, F.-X.; Vautier, F.; Peiry, J.-L. The Highest Gradient Model: A New Method for Analytical Assessment of the Efficiency of LiDAR-Derived Visualization Techniques for Landform Detection and Mapping. *Remote Sens.* **2017**, *9*, 120. [CrossRef]
69. Kokalj, Ž.; Hesse, R. *Airborne Laser Scanning Raster Data Visualization: A Guide to Good Practice*; Založba ZRC: Ljubljana, Yugoslavia, 2017; ISBN 978-961-254-984-8.
70. Kokalj, Ž.; Zakšek, K.; Oštir, K.; Pehani, P.; Čotar, K.; Somrak, M. Relief Visualization Toolbox, ver. 2.2.1 Manual. *Remote Sens.* **2016**, *3*, 389–415.
71. Aspinall, A.; Gaffney, C.F.; Schmidt, A. *Magnetometry for Archaeologists*; AltaMira Press: Lanham, MD, USA, 2008; ISBN 0-7591-1348-3.
72. Kvamme, K.L. Magnetometry: Nature's Gift to Archaeology. In *Remote Sensing in Archaeology: An Explicitly North American Perspective*; Johnson, J.K., Ed.; University of Alabama Press: Tuscaloosa, AL, USA, 2006; pp. 205–234.
73. Dalan, R.A. Magnetic Susceptibility. In *Remote Sensing in Archaeology: An Explicitly North American Perspective*; Johnson, J.K., Ed.; University Alabama Press: Tuscaloosa, AL, USA, 2006; pp. 161–203.

Article

Automated Resistivity Profiling (ARP) to Explore Wide Archaeological Areas: The Prehistoric Site of Mont'e Prama, Sardinia, Italy

Luca Piroddi *, Sergio Vincenzo Calcina, Antonio Trogu and Gaetano Ranieri

Department of Civil Engineering, Environmental Engineering and Architecture, University of Cagliari, 09123 Cagliari, Italy; sergiocalcina@virgilio.it (S.V.C.); atrogu@unica.it (A.T.); granieri@unica.it (G.R.)
* Correspondence: lucapiroddi@yahoo.it

Received: 11 December 2019; Accepted: 30 January 2020; Published: 1 February 2020

Abstract: This paper deals with the resistivity continuous surveys on extensive area carried out at the Mont'e Prama archaeological site, in Sardinia (Italy). From 2013 to 2015, new research was performed using both non-destructive surveys and traditional archaeological excavations. The measurements were done in order to find geophysical anomalies related to unseen buried archaeological remains and to define the spatial extension of the ancient necropolis. The electrical resistivity of soils was measured by means of the Automated Resistivity Profiling (ARP©) system. This multi-pole method provided high-resolution maps of electrical resistivity in the whole investigated area using a computer-assisted acquisition tool, towed by a small vehicle. Through this acquisition layout, a surface of 22,800 m² was covered. The electrical resistivity data were derived in real time with centimetric horizontal precision through a differential GPS positioning system. Thanks to the simultaneous acquisition of ARP and GPS data, the rigorous georeferencing of the tridimensional experimental dataset was made possible, as well as the reconstruction of a detailed Digital Terrain Model. Here, the experimental results are analyzed and critically discussed by means of the integration of the results obtained by a high-resolution prospection performed with a multi-channel Ground Penetrating Radar system and taking into account other information derived from previous geological and archaeological studies. Geophysical results, jointly with topographic reconstruction, clearly permitted the identification of more interesting areas where future archaeological investigations could be focused.

Keywords: archaeological prospection; automated resistivity profiling ARP; electrical resistivity survey; multi-channel ground penetrating radar; geophysical methods integration

1. Introduction

Modern archaeological surveys require non-invasive techniques that can be employed on wide areas in order to study the shallow subsoil, highlighting the presence of physical anomalies that could be related to buried archaeological remains. These techniques should be able to identify the physical properties of the soils by means of measuring systems positioned on a mobile vehicle in order to reduce the time of acquisition compared to other more traditional methods [1].

Electrical resistivity prospection specifically proved to be an effective method to characterize the subsoil and materials in various fields of study, such as applied geology [2–4], hydrogeology [5,6], engineering [7–9], environmental sciences [10,11], and agriculture [12–14]. This method is widely used for archaeological surveys: many past research has shown that the variations of electrical resistivity are often related to walls, floorings, paving and burials [15–22]. These structures are generally characterized by different electrical properties compared to the geological layers that, over time, have buried them.

Electrical methods based on fixed arrays permit the creation of resistivity models of the medium through a set of electrodes arranged along linear profiles or along two-dimensional configurations [23].

In the case of two-dimensional Electrical Resistivity Tomographies (ERT) the surveys provide a vertical section of the medium. The distance between the electrodes is constant. Each measurement is related to a specific distance between current and potential electrodes. Loops of electrodes with various geometries are used to perform tri-dimensional electrical resistivity tomographies in order to obtain a 3D model of the investigated medium [24–34].

Different systems for measuring and mapping electrical resistivity have been developed in order to overcome the main limitations of the fixed-array systems [35]. In particular, multi-polar mobile systems allow the performance of continuous resistivity profiling without a fixed device with an effective reduction in the time of acquisition. The first solution proposed a quadrupolar mobile device, provided with wheels equipped with electrodes for measuring the electrical resistivity of the soil at a constant theoretical depth [36]. The depth of investigation increased by varying the space between the wheels using an extendable support. Different solutions based on multi-polar mobile systems were developed and described in Panissod et al. [37]. These authors proposed an instrument based on eight electrodes named MUlti-pole Continuous Electrical Profiling (MUCEP), positioned along a V-shaped array (also called "Vol-de-canards" array) and located on a mobile measurement system. This equipment was used to explore shallow layers up to 3 meters in depth. In the same context, the comparison among resistivity prospections, obtained by the use of electrodes arrays towed by a vehicle, showed both advantages and limitations based on Pole–Pole arrays with a single depth of investigation with regards to the multi-pole systems. Multi-polar configuration provides a three-dimensional model of the subsoil with a significant reduction in acquisition time [38]. Further studies [35] have improved these multi-pole systems in order to reduce the noise effects on the raw data. In particular, the authors propose the equipment known as Automated Resistivity Profiling (ARP). This acquisition instrument is composed of a transversal current dipole and of three transversal receiving dipoles, set at an increasing distance from the transmitting one. This electrode array is designed to obtain simultaneously measurements of apparent resistivity simultaneously at three different subsoil depths. Papadopoulos et al. [39,40] described a three-dimensional inversion of resistivity data acquired with the ARP system in order to obtain a reliable numerical model capable of describing the spatial distribution of the electrical resistivity in the subsoil of the investigated area. The inversion process described by the authors is based on the least squares method and establishes smoothness constraints in order to consider the instability of the model and the non-uniqueness of solutions [41].

Several studies described multi-pole acquisition systems with the application of precision agriculture and soil characterization [42–45], archaeological prospection [46], and non-destructive evaluation of pavement and concrete infrastructures [47]. The ARP system was used by some authors [48] to evaluate the reduction of soil characterization costs for viticulture: the authors showed how the ARP system represents an effective method to reduce the cost of soil analysis, collecting a minor number of soil samples for direct tests. Many results of archaeological surveys carried out with the ARP system are extensively described in [1].

Integrated prospection with complementary geophysical methods permits to better define the physical model of the investigated subsoil [49]. One of the most integrated approaches in archaeology is the combination of electrical and Ground Penetrating Radar (GPR) methods [19,50].

In this paper, we present the results of surveys carried out by means of the ARP method at the archaeological site of Mont'e Prama (Sardinia, Italy). Geophysical investigations were designed to support future excavation activities and to define the spatial extension of the site in the northern side of an area in which archaeological digs took place in the 1970s. The analysis of electrical resistivity anomalies, both due to the buried archaeological remains and the geomorphological and geo-pedological spatial variability, permitted the evaluation of the effectiveness of this method for the site of study. Resistivity data were compared with the results achieved by extensive high-resolution surveys performed with multi-channel GPR at the same site. The archaeological interpretation of the experimental data was supported by the integration of different complementary geophysical methods and GPS data.

2. Materials and Methods

2.1. The Archaeological Site of Mont'e Prama

The study area of *Mont'e Prama* is one of the most important archaeological areas of the western Mediterranean [51]. The site is located in the Sinis peninsula (western-central Sardinia, Italy) as shown in Figure 1. The first archaeological remains were accidentally discovered in 1974 during soils ploughing when a local farmer found a fragment of a Nuragic sculpture. In 1975, 1977 and 1979 various campaigns of archaeological excavation were performed and a necropolis with characteristic monumental sculptures from the early Iron Age was studied. Many baetyls, models of "nuraghe" and numerous fragments of sculptures (usually called "Giants") were collected [51,52]. The last archaeological trials of these campaigns did not find new evidence and excavations were suspended.

Figure 1. (a) Location of the Sinis Peninsula in Sardinia (Italy); (b) geographic localization of the archaeological site of Mont'e Prama (red point); (c) detailed aerial image of the site of study where the surveyed area is white-bordered and highlighted.

Recently, new archaeological and geophysical investigations were carried out from 2013 to 2015 in the framework of a joint project between the Universities of Cagliari and Sassari and the

Archaeological Superintendency of Cagliari and Oristano, in which context other significant discoveries were made [53–56]. Overall, 28 statues representing figures of boxers, archers, warriors, measuring up to 2.20 m in height, 16 models of "nuraghe" and baetyls were pieced together from over 5000 fragments of limestone. Archaeologists identified more than 50 shaft graves. Figure 1a,b indicates the geographic location of the site of study and the area with main archaeological digs carried out in 2015 (Figure 1c). After this campaign, the full extent of the archaeological site appears significantly greater than that supposed by the first researchers in 1970s.

2.2. Geological Setting

The geology of the Sinis peninsula, the wide region including the Mont'e Prama site, consists of both sedimentary and volcanic rocks [57]. In particular, the geographic sector closest to prospection area is mainly characterized by sedimentary lithologies. These formations range from messinian fenestrae-rich limestones and breccias of supratidal and intertidal environment with typical benthic fossil content to evaporitic fine-grained limestones, sub-litoral marls and bioclastic limestones (*Calcari Laminati del Sinis, Capo S. Marco* formation). Quaternary conglomerates, sands and mud deposits, organized in river terraces and alluvial cones, are recognized in the upper part of the sequence.

In the context of 2014 archaeological digs, the stratigraphic sequence in correspondence to the excavations was reconstructed [58]: it is characterized by the presence of geological units related to the sedimentary complex of the Sinis peninsula. In particular, the site presents a thin layer of silt-sandy soils (Pleistocene and Olocene) with a maximum thickness between 40 cm and 50 cm. The soil layer covers a yellowish thick carbonate crust of pedogenetic origin. The hard crust is probably derived from salt precipitation as a result of capillarity action in past climatic conditions and reaches 1 m of maximum thickness. This horizon is observed with good spatial continuity in all over Sinis peninsula and covers previous messinian sediments, mainly composed by carbonate sandstones and marlstones. The area of geophysical surveys is expected to present mean characteristics similar to these results with a certain spatial variability.

2.3. Methods and Instrumental Layout

2.3.1. ARP Survey

The electrical surveys were carried out using the multi-polar ARP03 device (ARP®, Automated Resistivity Profiling, Géocarta SA, Paris, France), a measuring system made of four pairs of rolling electrodes arranged in a characteristic V-shaped configuration. Figure 2a shows a picture of the ARP system. The electrodes located on the first pair of wheels represent the current dipole, through which an electrical current of 10 mA is flowing. The other three couples of rolling electrodes are three potential dipoles designed to measure the electrical potential with increasing depth (Figure 2b). The longitudinal distance between the potential dipole electrodes and the current electrodes corresponds to 0.50 m, 1.0 m and 1.70 m, respectively. Moreover, the three potential dipoles constitute of pairs of wheels with axial distances of 0.5 m, 1.0 m and 2.0 m, respectively. Therefore, the resistivity measurements are related to three different depths of investigation, conventionally indicated as 0.5 m, 1.0 m and 1.70 m.

The sensor allows the acquisition of multiple measurements of apparent resistivity through the three pairs of rolling electrodes with a sampling frequency of 22.7 Hz. A radar doppler system is used to measure the distance. Depending on the vehicle speed, raw data can be acquired up to a 10 cm sampling interval.

Moreover, the positioning system, equipped with a differential correction DGPS, enables georeferencing of the acquired data and showing raw resistivity data in real time.

The speed of data acquisition is mainly due to the morphological features of the investigated area (slope, irregular topography, shallow obstacles, plan shape and irregularities, etc.). Usually, the employment of a quad bike to tow the carriage allows a maximum speed of 15 km/h.

The acquisition on the ground is typically performed along parallel profiles with variable length. For high-resolution surveys, such as archaeological investigations, the acquisition pattern is composed of profiles positioned at a 1 m distance.

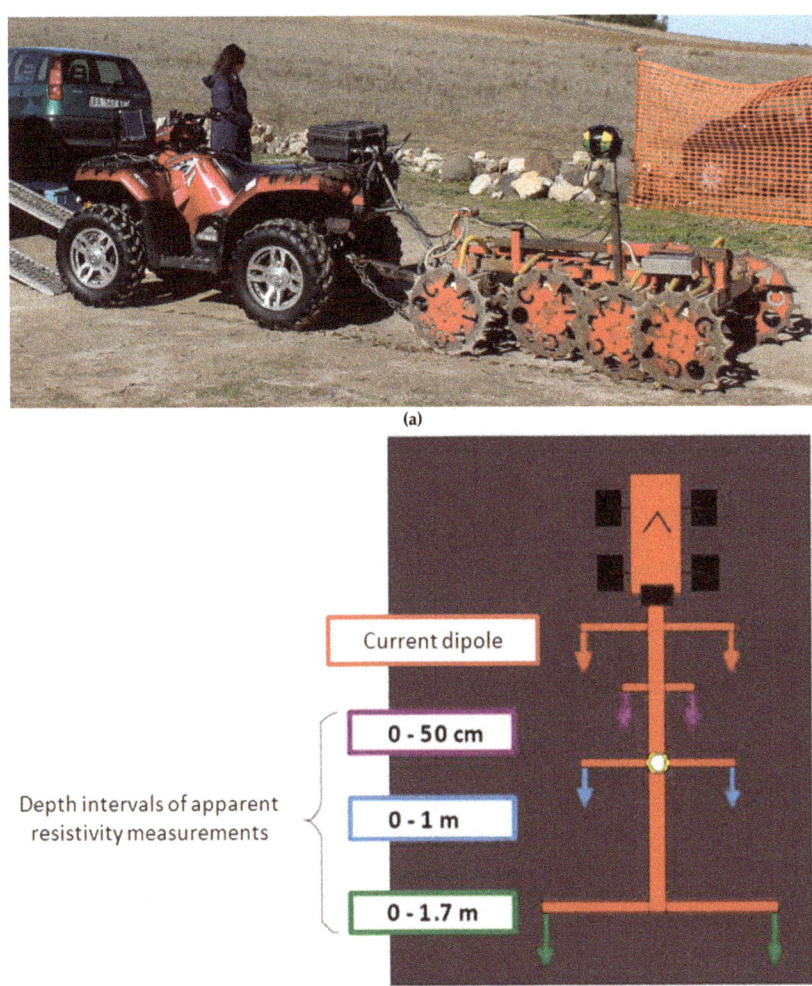

Figure 2. (a) Image of the Automated Resistivity Profiling (ARP) system towed by a quad; (b) scheme of ARP instrumental layout.

The ARP survey at the site was carried out on November 21st, 2014, on a surface of about 2.28 hectares (Figure 1c). Two series of measurements were done:

1. the first dataset was collected in the sector located in the north of the area involved in the archaeological excavations (large 20,155 m^2);
2. the second dataset was acquired in the area around the open archaeological digs (2730 m^2).

Data were collected in autumn and on a day characterized by optimal weather conditions in order to perform the measurements with humidity conditions of the outermost shallow layers able to guarantee low contact resistances and high-quality measurements.

The acquisition allowed the detection of 277,319 punctual measurements of apparent electrical resistivity along a 27.8 km long route and with a velocity of 7.53 km/h along profiles oriented in an easterly direction. Table 1 summarizes the main acquisition parameters.

Table 1. Summary of the main parameters of the ARP prospection.

	Full Duration of the Acquisition	3 h 39 m 25 s
	Mapping Coverage	2.28 ha
	Covered distance	27,536.81 m
	Medium speed	7.53 km/h – 2.09 m/s
	Frequency of acquisition of GPS signal	1 Hz
	Average distance between two next measurements along one profile	0.098 m
	Number of measuring points	277,319
Geometric factor K	Dipole 1 (0.5 m)	4.62
	Dipole 2 (1.0 m)	10.71
	Dipole 3 (1.7 m)	51.28
Intensity of injected DC		10 mA

The spatial distribution of the measurement points is shown in Figure 3a. From this representation, it is possible to observe two distinctive points densities between the two main datasets, partially due to the different acquisition velocities, which are mainly related to the availability of undisturbed wide paths for the towing vehicle but mostly due to the overlap configurations; in fact, the smaller area was surveyed along shorter and orthogonal profiles, even because of superficial archaeological features partially prohibiting vehicle motion, which produced a denser points distribution. While in the northern free area it was possible to perform the acquisition with optimal sampling parameters, for the southern most difficult area, the requirement of spatial coverage and reliability of the local dataset entailed a little higher redundancy of data (rough density of measurements about 29 point/m^2).

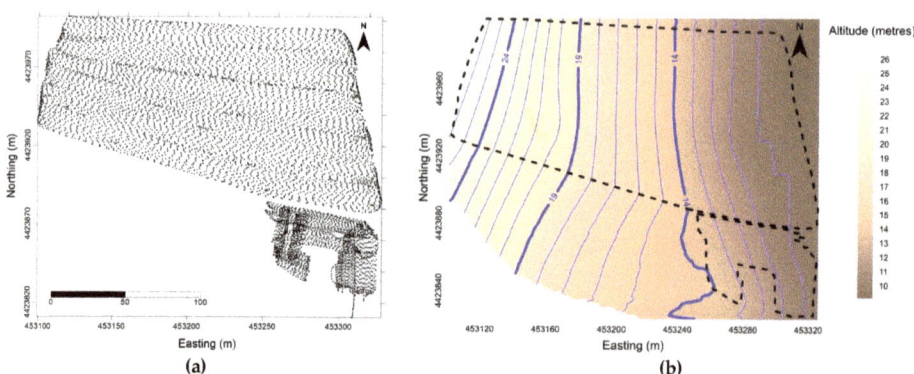

Figure 3. (a) Planimetric spatial distribution of apparent resistivity data. (b) Elevation contour map of the digital elevation model (DTM) obtained by integrated acquisition of GPS data over surveyed areas (dashed bordered areas).

The GPS data continuously and automatically acquired during the measurements at 1 Hz sample rate, integrated with isolated points external to the ARP acquisition area, also allowed the creation of a detailed digital terrain model (DTM) of the surveyed area, which provides useful information for the analysis and interpretation of ARP data. Based on DTM results, the site presents a regular topography with a weak slope angle (mean dip angle 4.3°) indicatively oriented from west (top) to east (bottom) for a difference in height of about 16.5 m (Figure 3b).

In order to delete both negative resistivity values and outliers, raw data were filtered using a bi-dimensional median filter operating on a mobile spatial window with a 2 meter radius centered on each measurement point. The filtering algorithm removes resistivity values that are higher than the reference threshold function of the median of values measured within an area centered on every measuring point. Once filtered, the experimental data are interpolated using 2D cubic spline functions over plan surface to permit to obtain a regular grid that can be used to support a successive processing phase or the interpretation of the geophysical anomalies.

2.3.2. GPR Survey

Ground Penetrating Radar methods study both electric and magnetic properties of materials by means of the interaction of microwave signals with the investigated bodies [59]. Electromagnetic waves, transmitted using different superficial probes, can be refracted, reflected and diffracted at the electromagnetic discontinuities and are absorbed passing through the medium. These effects are a function of material properties (e.g., electrical permittivity, magnetic permeability) and signal features (e.g., wavelength, waveform, etc.). These techniques have been widely used in recent decades in engineering, geology and archaeology [60].

In order to validate and interpret ARP data, a comparison with some GPR profiles is proposed. The Ground Penetrating Radar survey was performed in the context of the same research project on a wide area of study, even including the sector investigated by means of the ARP method [31–34]. The GPR prospection was done using a 15-channels (16 antennas) GPR system (STREAM-X, by IDS), with a 200 MHz central frequency that allows the acquisition of synchronous recordings of 15 parallel radargrams with a fast acquisition speed (about 10–15 km/h). Each radar profile is spaced 0.12 m from the others. Therefore, by means of this instrumental setup, it was possible to collect a big dataset characterized by a high spatial density (horizontal: 12 cm perpendicularly to the moving direction and 9 cm along it: vertical: less than 1 cm). A differential GPS antenna was used for positioning the measurements with a horizontal shift less than 5 cm. This equipment with respect to standard single/dual channels GPR systems includes different advantages such as time efficiency, fixed distance between the fifteen profiles (12 cm), absolute parallelism between all synchronous profiles and 3D highlighting of buried targets.

GPR data were processed using Reflexw software (by Sandmeier®). Processing flow includes background removal, de-wow filtering, gain, bandpass filtering (100–300 MHz) and time–depth conversion.

3. Results

3.1. ARP Results

The geophysical surveys carried out by means of the ARP system permitted to simultaneously obtain three apparent resistivity maps, referring to the three depth intervals. Their measured global resistivity values ranged between 11 and 154 $\Omega \cdot m$, with different responses between each depth map (Figure 4). First, the shallowest map (Figure 4a), referred to a 0–0.50 m layer, reports the apparent resistivity values ranging from 50 to 117 $\Omega \cdot m$ with a data distribution showing maximum frequency at about 80 $\Omega \cdot m$ but generally more shifted towards low than high resistivity values for less numerous classes (Figure 4d). The intermediate map, the 0–1.00 m layer (Figure 4b), presents apparent resistivity values ranging from 11 $\Omega \cdot m$ to 56 $\Omega \cdot m$ with maximum concentration around 25.5 $\Omega \cdot m$ and a general symmetrical distribution (Figure 4). The 1.00 m map appears spatially more noisy than 0.50 m one but it is possible to note a good coherence between their resistivity patterns. The deepest map, 0–1.70 m layer (Figure 4c) shows apparent resistivity data ranging from 30 to 154 $\Omega \cdot m$ with maximum concentration of around 60 $\Omega \cdot m$, significantly lower than the mean value (92 $\Omega \cdot m$) distributed with a lower spreading (Figure 4f).

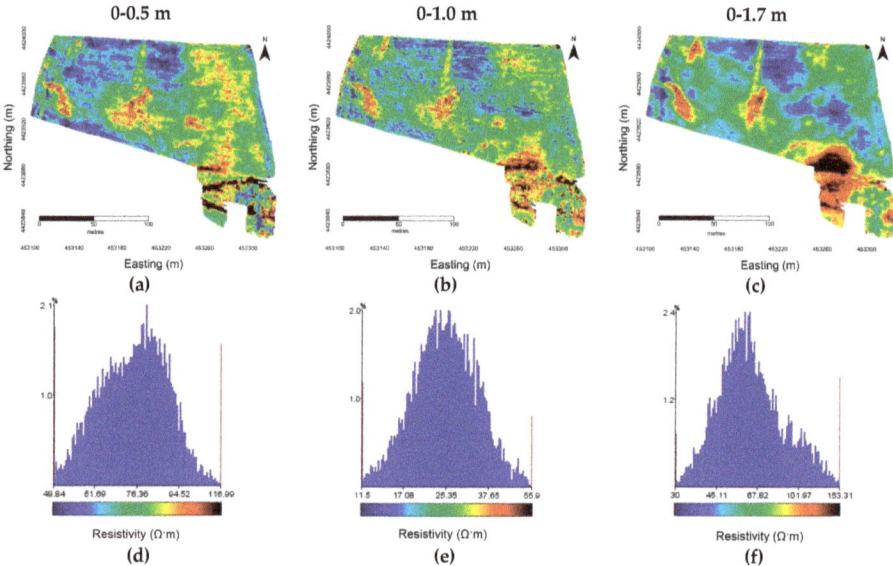

Figure 4. Apparent Resistivity maps obtained with ARP at 0.5 m (**a**), 1.0 m (**b**) and 1.7 m (**c**); data frequency histograms referred to three maps above: 0.5 m (**d**), 1.0 m (**e**) and 1.7 m (**f**).

As is predictable, the results of the 1.70 m map are the smoothest obtained, with clear readability of resistivity patterns and anomalies at a bigger spatial wavelength and a lower presence of both spatially high-frequency signals and noise. Vertical gradients of resistivity ranges are probably related to soil layering properties, in particular, a possible compaction of deepest volumes due to the proximity to bedrock and a moisture gradient for the two shallowest layers due to superficial drying phenomena or different clay concentration with depth. In all the cases, the apparent resistivity patterns show no relation with the topography gradient that characterizes the investigated sector of the site. The result of tillage operations, carried out in the recent past, appear in the form of stripes east–west oriented and in some route traces, are recognizable as double parallel curves on the shallowest map (0.5 m), Figure 5a,b.

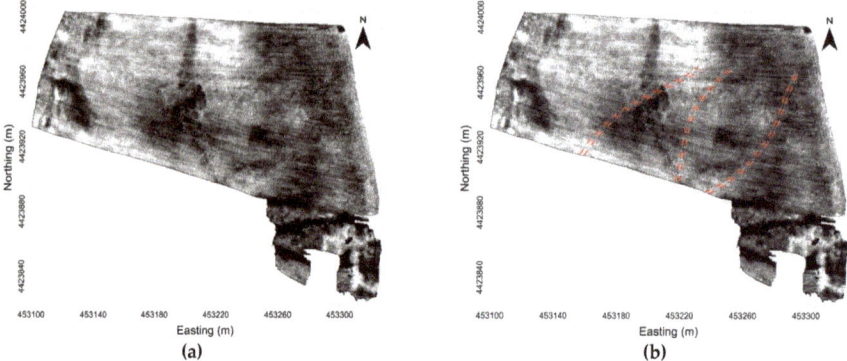

Figure 5. (**a**) Normalized apparent resistivity maps in gray color scale (0.5 m nominal depth); (**b**) traces of farm vehicles clearly readable in the ARP resistivity map referred to 0.5 m depth (dashed red lines).

Looking for archaeological purposes at apparent resistivity patterns, the most significant ones seem to be resistive areas (red tones), which are highlighted in the following maps. Nevertheless, some conductive patterns (blue tones) are also evidenced and reported as signals of potential interest, especially for the top and bottom layers.

In Figure 6, the resistivity map at 0.5 m nominal depth is reported with the indication, bordered in red, of some interesting resistive areas which have vertical continuity even in other two maps. The same areas are reported also at a 1.0 m depth (Figure 7) and 1.7 m (Figure 8).

Figure 9 reports the most interesting conductive areas: in particular, the most superficial map (0.5 m, Figure 9a) shows, in its southern part, some conductive regular patterns (indicated as **I1**), just inside the protected area where digs were concentrated, which could have archaeological relevance such as tombs or similar geometries filled with moist silty or clay soil. The 0.5 m map presents other conductive larger regions which have dimensions not compatible with archeological features and are most probably interpretable as superficial geological variations: this is the case of the **G1** region that is confirmed by deeper layers of the dataset and interpreted as a continuous geological body. Indeed, the map at 1.7 m shows wide conductive areas which could have importance to reconstruct paleoenvironment and paleo-landscape of the area (Figure 9b, **F3**, **G3** and **H3**); area **I3** of the same map shows that at higher depths, the conductive anomalies indicated before as **I1** inside the protected are less evident and probably their low effect could be due to the fact that the deeper map still includes all the layers between 0.0 m and 1.7 m.

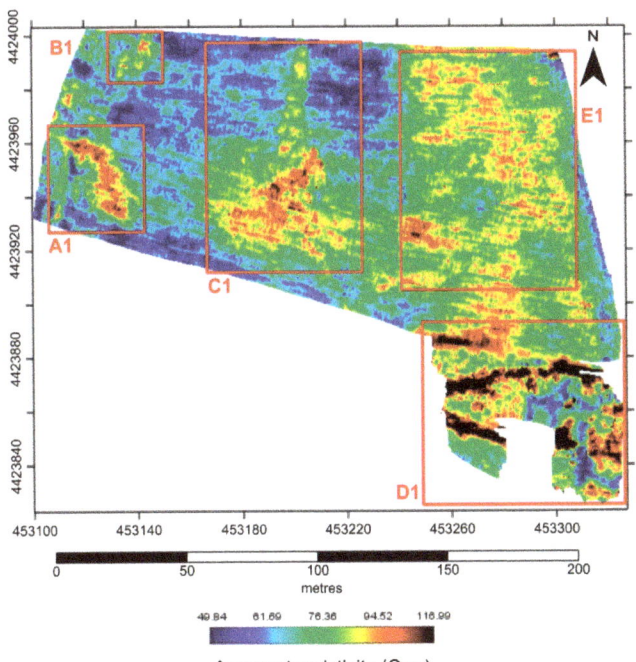

Figure 6. Identification of the main resistive patterns in the ARP map at 0–0.5 m depth interval.

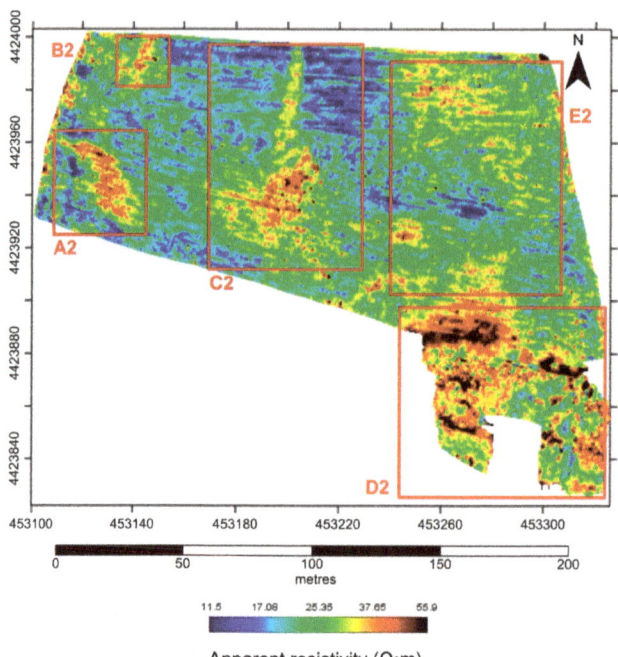

Figure 7. Identification of the main resistive patterns in the ARP map at 0–1 m depth interval.

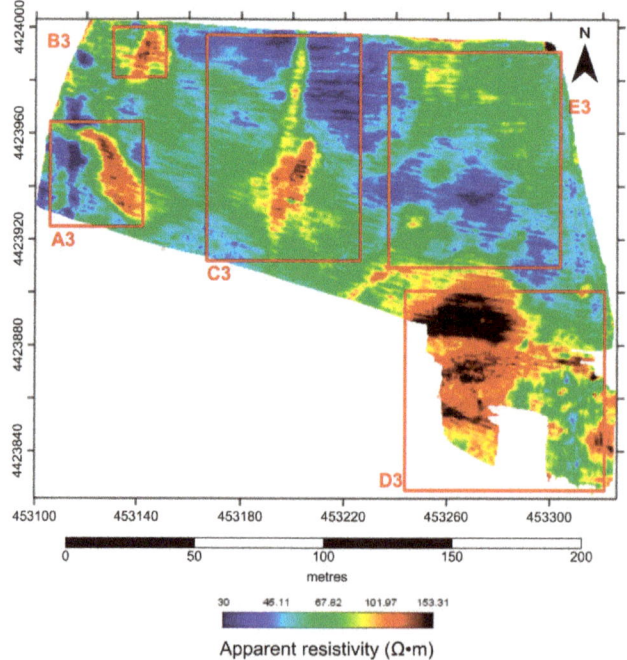

Figure 8. Identification of the main resistive patterns in the ARP map at 0–1.7 m depth interval.

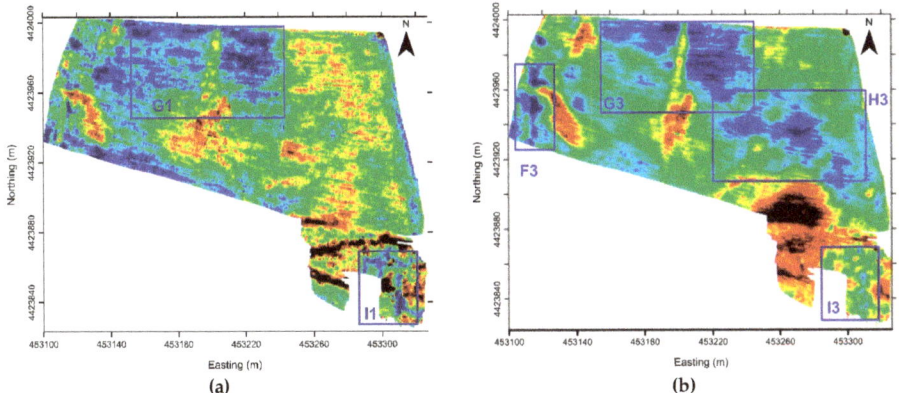

Figure 9. (a) Shallow conductive anomalies located in the southern part of the site of study (0–0.5 m interval depth); (b) wide conductive sectors at the northern part of the surveyed area (0–1.7 m interval depth).

3.2. Vertical Analysis of ARP Anomalies and Comparison with GPR Data

To explore the three-dimensional features of the ARP results, in the next paragraph, the data are analyzed by means of their vertical pseudo-sections in correspondence to the most important planimetric patterns. ARP pseudo-sections are also compared with GPR sections acquired over the same paths (Figure 10). Five vertical profiles have been selected over areas where coverage of both datasets was available: three are located at the north-western part of the ARP-surveyed area, intersecting three resistive anomalies A-B-C in Figures 6–8; the other two profiles are located at the eastern part of the ARP-surveyed area, just outside the protected area partially interested by archaeological excavations, in correspondence of the northern part of the D resistive anomalous pattern of Figures 6–8.

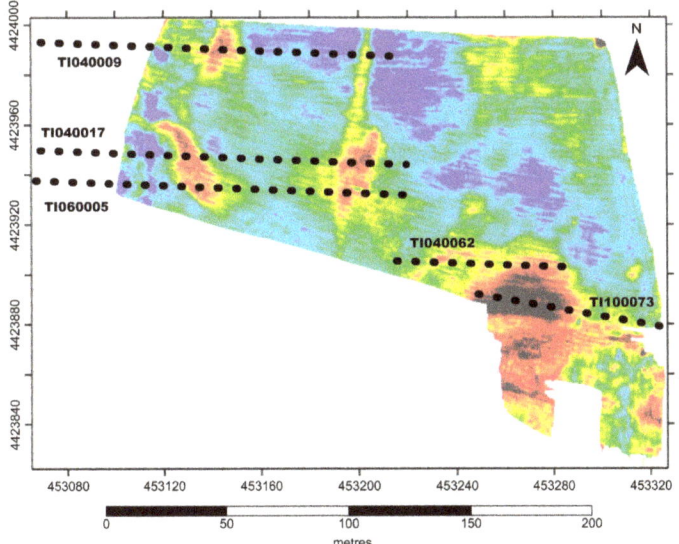

Figure 10. Location of Ground Penetrating Radar profiles.

In Figures 11a–c and 12a,b the comparison of radargrams and resistivity pseudo-sections is reported. As also observed in Figure 11a, looking at pseudo-sections, it is possible to notice the difference of apparent resistivity ranges at different depths: mean-resistive data on the top layer; a more conductive layer at middle-depth in which relatively resistive points vertically correspond with the top layer ones; the most resistive layer is the deeper one in which vertical correspondence of relatively high resistivity data with the two top layers is confirmed. This behavior leads to the consideration that the resistive bodies should have a certain vertical continuity from bottom to top, even in the intermediate layer where a probable higher humidity of soil partially influences apparent resistivity ranges shifting all values towards more conductive ones. The three nominal depths of ARP raw data are indicated with the black horizontal dotted lines on pseudo-sections.

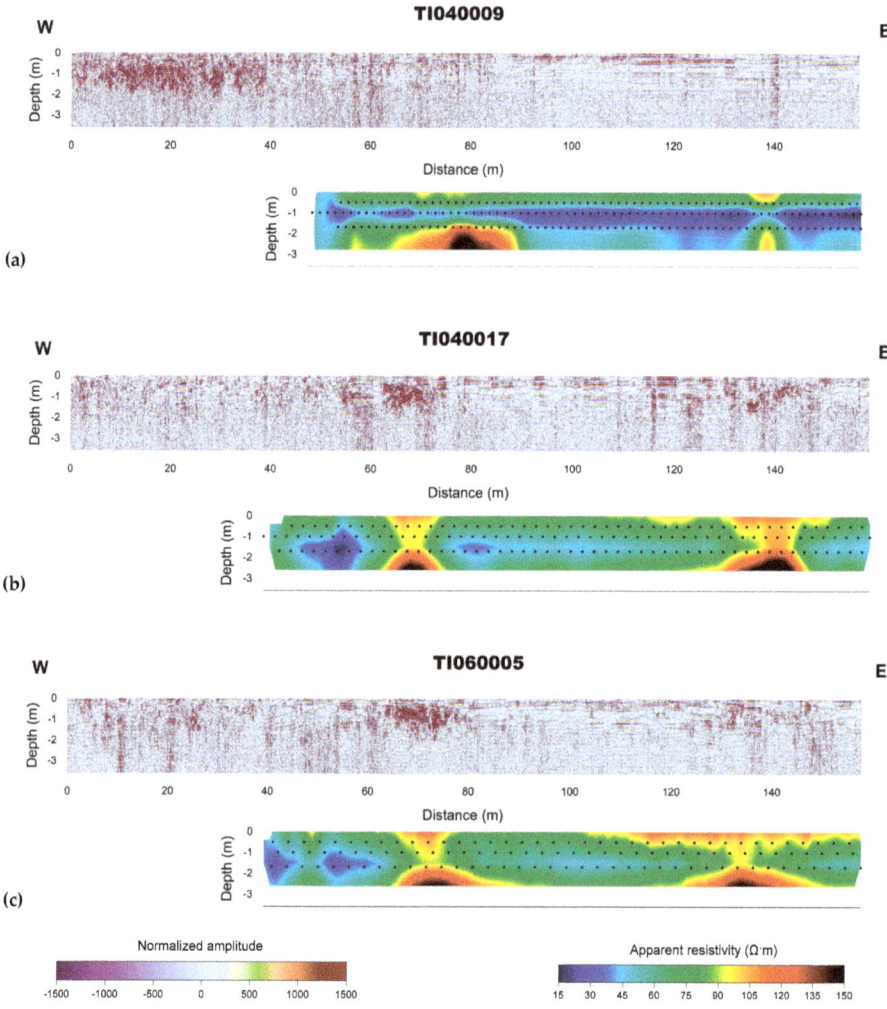

Figure 11. Comparison of ARP pseudo-sections with GPR profiles: (**a**) TI040009; (**b**) TI040017; (**c**) TI060005. Four times vertical amplification.

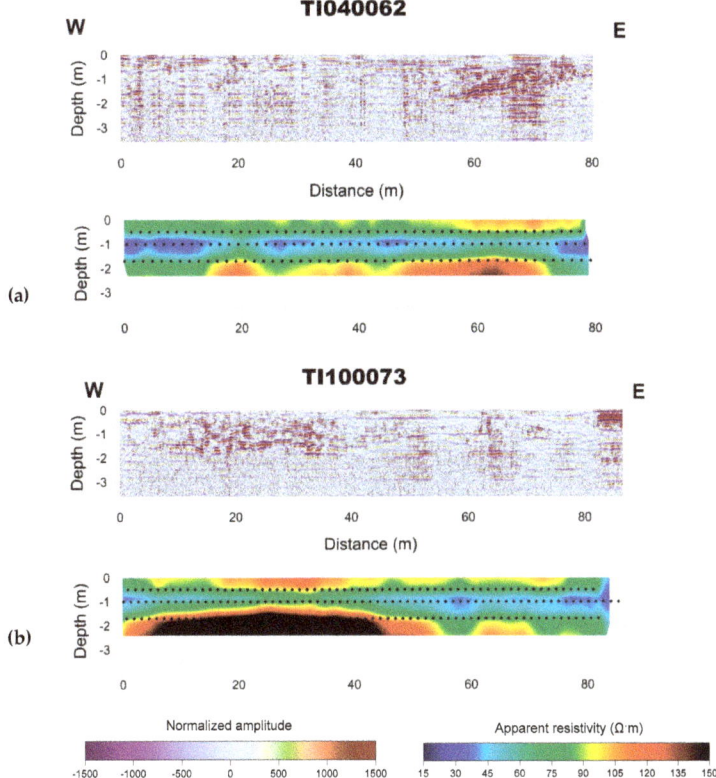

Figure 12. Comparison of ARP pseudo-sections with GPR profiles: (**a**) TI040062; (**b**) TI100073. Four times vertical amplification.

In Figure 11a, the northern profile (**TI040009**) is reported. ARP pseudo-section presents two resistive patterns belonging to regions **B** and **C** in Figures 6–8. The bigger one, belonging to region **B**, is located at progressive horizontal coordinates range of 65–90 m and corresponds to parts of the radargram with an intermediate amplitude signal. Moving to the right in the radargram, it is possible to notice the signal weakening in correspondence of conductive areas where microwaves are more absorbed. From 135 m to 140 m, the second resistive anomaly appears as a section of the top linear resistive pattern in region **C**; at the same coordinates, GPR data present the most noisy traces.

In Figure 11b, the ARP profile and the radargram acquired along the **TI040017** line in the south-western side of the study area are represented. The profile crosses two wide resistive anomalies, which are identified as **A** and **C** in Figures 6–8. In correspondence of both resistive regions, the radar profile is characterized by the strong amplitudes of the signals. Moving from west to east, a strong reflective region is found from 62 m to 75 m, with a globally complex shape due to the superimposition of multiple diffraction signals. This anomalous region, at depth from 0.5 m to 2.0 m, is spatially related to the first resistive body of the **TI040017** pseudo-section, clearly detected also in ARP maps. The second significant radar-reflective region is located from 90 m to 110 m. This signal pattern is related to targets with a decreasing depth from left to right approximately ranging from 1.5 m to 0.5 m. The same trend is observable in the resistivity pseudo-section where the top layer resistive region is right-shifted compared to that of the bottom layer.

Figure 11c reports the comparison between the **TI060005** radar section and the resistivity pseudo-section at the same position. This line is parallel to the **TI040017** radar profile, previously presented (Figure 11b) and is located about 10–12 m south of this. As Figure 10 shows, this profile intersects the same resistive areas (**A** and **C** in Figures 6–8) close to their southern border. The pseudo-section shows the same resistive volumes but with slightly lower resistivity values and wider spatial extents. The GPR profile is characterized by the presence of diffuse intermediate signals with two regions where signal amplitude increases: the first one is particularly strong and concentrated between the 60 m and 80 m progressive coordinates, in correspondence with the western resistive body. The second region, located between 125 and 150 m, presents weaker GRP signals than the first one but is also weaker than the corresponding second anomaly in the **TI040017** profile. A preferential direction of radar signal is less evident in this case with regards to the **TI040017** GPR profile. The depth ranges are the same as in the **TI040017** profile for both regions.

The last comparisons of ARP and GPR data, Figure 12a,b, mainly concern the most resistive and widest area of ARP maps which was indicated with **D** in Figures 6–8 in the southern part of the prospection. This highly apparent resistivity region is partially located in the restricted archaeological area in the south, and in the bordering part of the free northern area. The two comparative profiles are placed in the free northern area, the first one over the northern limit of the wide resistive pattern and the second one crossing the central higher resistivity part.

In Figure 12a, the first profile, **TI040062**, is plotted. The pseudo-section is characterized by a large resistive region at its bottom layer, ranging from 15 m to 75 m progressive horizontal coordinate, with the highest apparent resistivity values concentrated between 45 m and 75 m. On the top layer, resistive values are right-shifted and less extended, located between 60 m and 80 m. the GPR profile is in very good agreement with ARP pseudo-section, presenting an intense group of signals between 60 and 80 m with a preferential tilted direction climbing up towards the right. These reflected signals come from variable depths ranging between 0.5 m and 1.8 m.

Figure 12b reports the **TI100073** comparison profile, which crosses the **D** in Figures 6–8 at its highest values position. This profile is almost parallel and east-shifted of about half length from the previous one, **TI040062**. Because of this, anomalous patterns linked to the same body, which are still present, are located at different horizontal coordinates. The deeper ARP anomaly presents the highest resistivity values of all the datasets, saturating the color-scale at 150 $\Omega \cdot m$ between 5 m and 45 m, but resistive area can be also considered between 0 m and 55 m. The top resistive area is a bit wider than the range from 20 m to 40 m. Even for this profile, GPR data are in good agreement with the ARP pseudo-section: in fact, a large region with clear sub-horizontal GPR signals is present between 5 m and 40 m, at an approximative depth of 1-2 m.

4. Discussion

Interpretation of the geophysical results is suggested in Figure 13, where surveyed areas are classified in different color classes by evidence of their archaeological potential:

- dark gray color: areas with highly evident ARP anomalies without GPR comparison;
- red color: areas with high ARP resistivity contrast to background and consistent high amplitude GPR signals;
- orange color: areas with ARP anomalies with good accordance to strong GPR reflections;
- yellow color: areas with significant features of ARP anomalies with only partial accordance with GPR signals;
- light gray color: areas with relatively weak ARP spatial patterns without clear agreement or comparison with GPR data.

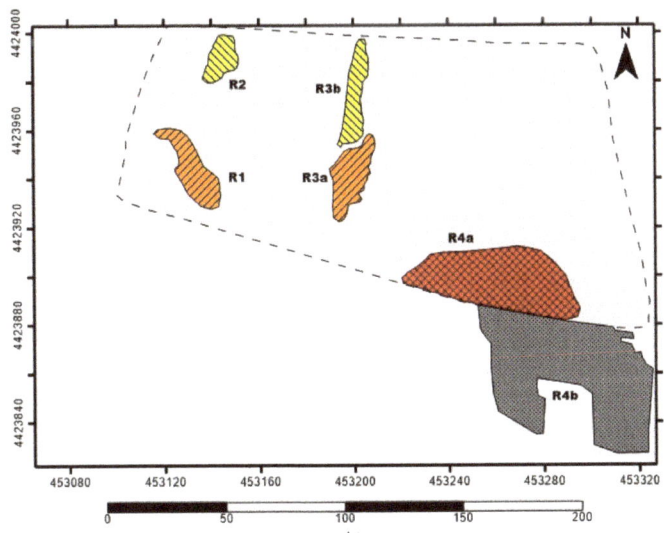

Figure 13. Archaeological interpretation of joint geophysical results in five classes: areas with simultaneous presence of strong ARP and GPR anomalies (**R1** and **R3a**, orange); significant ARP anomalies with locally poor, weak or too noisy, GPR signals (**R2** and **R3b**, yellow); simultaneous presence of strongest ARP and GPR signals (**R4a**, red); areas with important ARP anomalies without GPR comparison (**R4b**, dark gray); areas with relatively weak, sometimes coherent, ARP spatial patterns without clear agreement or comparison with GPR data (remaining surveyed area, light gray).

After the joint analysis of ARP and GPR data, it was possible to classify some areas in order of potential archaeological importance. In particular, the most significant areas of the ARP survey could be divided into five classes on the basis of the agreement of apparent resistivity anomalies and the GPR signal behavior or the absence of GPR data. One of the most promising areas is the southern sector internal to the restricted archaeological area, indicated with **R4b** in Figure 13 (dark gray color, 2730 m^2). This complex pattern of conductive and resistive anomalies is characterized by linear and sub-rectangular shapes of dimensions which can be attributed to macro-archaeological features. Nevertheless, the absence of higher resolution data, such as GPR, does not allow to fully analyze tridimensional relationships of the different ARP signals and the sector is classified as very promising but not completely explored with complementary geophysical tools.

The other most important group of anomalies, indicated with **R4a** in Figure 13 (red color), is contained in a semi-elliptical shape with the main dimensions of 75 m and 25 m (1460 m^2), in spatial continuity with the previous anomaly. The area is characterized by a cluster of very high resistivity values with a main dimension of about 40–50 m at the deeper level of ARP measures (150 Ω·m at 1.70 m nominal depth). In this region GPR profiles were able to highlight parts of bigger importance inside the wider anomaly which correspond to the most resistive elements (Figure 12a, b). The area is likely to contain archaeological remains, in particular, because of the complex patterns of GPR signals which are difficult to interpret as geological features. The wide extension of this anomaly and its topographic position at the lower altitude of the agricultural field suggest the interpretation as a possible accumulation of stone pieces, even of possible archaeological importance, due to ancient ploughing.

The third most important anomaly is indicated with **R1** in Figure 13 (orange color). It is an area 390 m^2 in extent, with 40 m by 13 m main axes, with apparent resistivity values up to 120 Ω·m at 1.7 m nominal depth. After the integrated analysis of ARP and GPR data (Figure 11b,c), the anomaly could be interpreted, even for this case, as a localized concentration of stone bodies and fragments. In fact,

GPR reflections and diffractions are characterized by high amplitude and spatial concentration without lateral continuity, at least at depths with an acceptable Signal to Noise ratio. Similar properties are recognizable for area **R3a**, which has an extension of 380 m^2, 16 m wide and 38 m long. The comparison of ARP and GPR data (right anomaly in Figure 11b,c) emphasized a correspondence, even if weak, between the two datasets. Resistivity values are on the same order as anomaly **R1** (up to 120 Ω·m) and GPR signals present high amplitude in the northern section of the anomaly but are still readable in the southern one. Especially at the northern section, GPR anomalies do not show lateral continuation outside the resistive volumes. Because of these considerations, anomalies in areas **R1** and **R3a** can be potentially considered for fruitful archaeological investigations.

To interpret the other two regions bordered in Figure 13, **R2** (187 m^2, 22 m × 12 m) and **R3b** (284 m^2, 47 m × 8 m), yellow color, the joint analysis of ARP and GPR data was less effective, because in correspondence of the area **R2**, radar signal was present but not intense as in the other cases, while in correspondence of norther section of area **R3b**, it was affected by the high noise level. Nevertheless, lateral apparent resistivity contrasts and spatial features suggest the opportunity for further investigations with other complementary geophysical methods and archaeological trials.

The planimetric analysis of apparent resistivity data in the remaining areas (light gray color in Figure 13, 17,830 m^2) evidences a complex scattered distribution of spatial patterns with a low resistivity contrast which could be indicative of the presence of in situ archaeological features, less sensitive than possible accumulation of stone fragments to volume analysis, such as ARP survey.

Aiming at the estimation of the effective investigation depth of the ARP survey, an inversion procedure, based on a tetrahedral Finite Elements structure, was performed on two apparent resistivity profiles by means of the commercial software ERTLab. Pseudo-sections TI040017 and TI040062 were selected by considering the good concordance between apparent resistivity anomalies and GPR signals. The resistivity models were set up with 3D cells of 20 cm edge dimensions and inverted with smoothness constraints to achieve the final resistivity models. The tomographic results confirm the location of resistive bodies and the presence of an intermediate conductive layer (Figure 14). Some artifacts are present due to the scarcity (only three levels) of data distributed across the pseudo-sections: this is most evident for a shallow layer over the top of the first apparent resistivity level where most resistive values are distributed and characterized by a very high discontinuity along the horizontal profile. The same effect is quite common, even for traditional acquisition ERTs, typically with minor thicknesses [61]. The deepest bodies under the intermediate conductive layer still have a resistive behavior, but values are significantly lower than in the case of pseudo-sections.

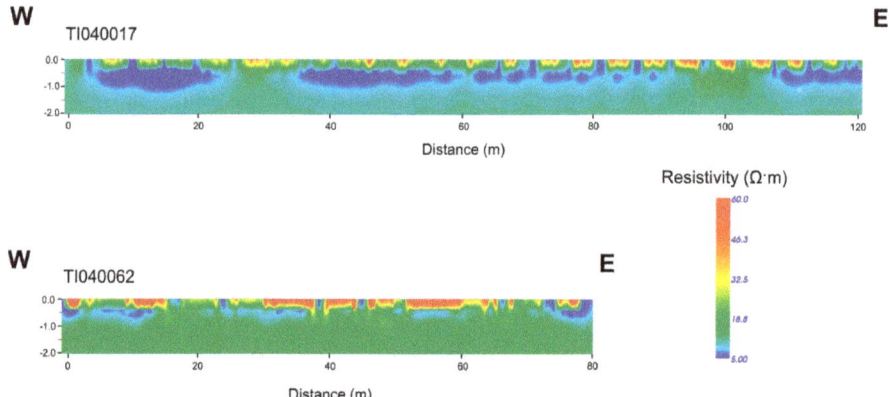

Figure 14. Electrical resistivity tomographies obtained from ARP data: TI040017 and TI040062 pseudo-sections.

The spatial distribution of inverted data substantially illustrates the same patterns of pseudo-sections but most resistive values were obtained on the top soils instead of the bottom ones where the most intense resistive anomalies were located in the pseudo-sections: this opposing behavior was probably caused by data density and distribution which bring ERTs to very smooth variations at middle-bottom depths and to apparently noisy variations on the top depths. Comparing the ERT results and pseudo-sections, it is possible to observe a good agreement of depths for apparent interfaces and spatial resistivity transients, with a slight overestimation of depths under conductive volumes. A similar behavior is also described by Papadopoulos et al. [39,40]. Combining ARP, 0.50 m depth, with DTM data (Figure 15), it is possible to evaluate the independence and spatial relations of the widest ARP anomalies to each other. No clear spatial link is evidenced between the most promising areas identified in Figure 13. The only weak relationship which could be notice is on the eastern bottom areas where the strong linear anomalies are in connection with a northern wide area characterized by less intense resistive anomalies (D1 and E1 in Figure 6). The same connection is no longer present at deeper depths where the northern area (E2 in Figure 7 and E3 in Figure 8) gradually becomes less resistive and disconnects from the strongest anomalies to the south. This feature, although the low contrast of apparent resistivities, could be a marker of a geological transition in the investigated volumes, rather than archaeological remains. From the same three-dimensional comparison, no connection is recognizable along the eastward direction between resistive bodies under areas located at different altitudes.

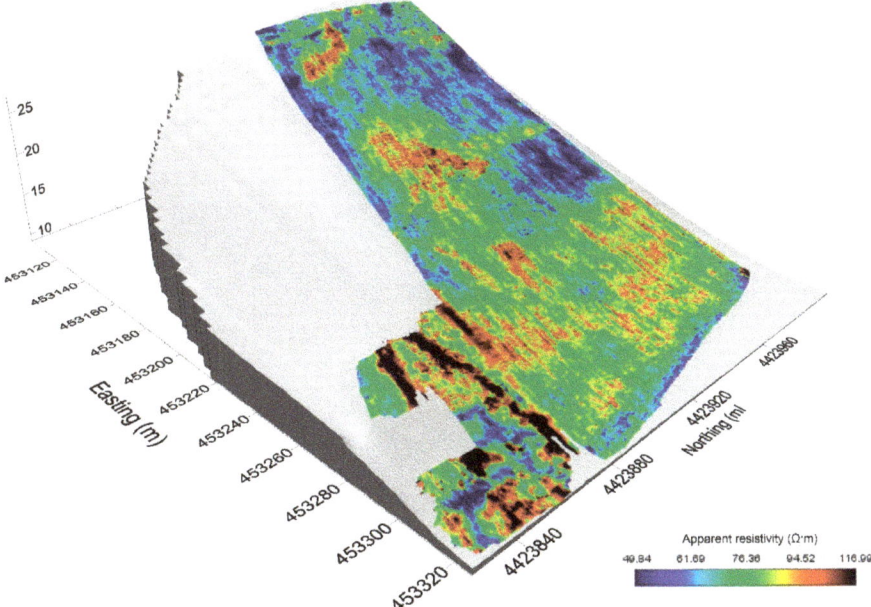

Figure 15. Tridimensional visualization of 0.5 m depth ARP data integrated with digital terrain model information (three times vertical exaggeration).

5. Conclusions

A geophysical prospection of a prehistoric archaeological site in Sardinia, Italy, is proposed. In particular, results from an Automated Resistivity Profiling (ARP) survey over a 2.28 hectares area are analyzed and discussed in comparison with Ground Penetrating Radar (GPR) sections over the same area. The purpose of the prospection was to reveal the archaeological potential of a wide area surrounding a site accidentally discovered and partially excavated in the 1970s when a necropolis and related archaeological findings (e.g., fragments of large stone statues and funeral monuments) were found. The site was abandoned until the joint archaeological-geophysical scientific project (2013–2015) in which the ARP and GPR prospections were carried out.

The ARP survey covered a wide area localized at the northern side of the historic archaeological site. Apparent resistivity data were simultaneously acquired at three different depths of investigation (0.5 m, 1 m and 1.7 m) and exhibit a wide range of variability (from 15 $\Omega \cdot$m to 250 $\Omega \cdot$m). The most intense planimetric patterns of the apparent resistivity maps were characterized by wide resistive regions with irregular shapes and linear spatial extensions of the order of tens of meters, but even some conductive anomalies were found which were characterized by sub-regular shape and dimensions of possible archaeological interest. Some small apparent resistivity patterns presented spatial coherence but weak local resistivity contrast; therefore, it is difficult to assign them to archaeological features without more detailed measurements of the possible artifacts.

In order to verify preliminary interpretations, the tridimensional distribution of apparent resistivity measures was compared with high resolution GPR data acquired along linear profiles. In most cases, the results show a good correspondence between apparent resistivity anomalies and GPR signals. Based on the agreement between ARP and GPR datasets, it was possible to define the surveyed area in five classes which varies in function of the two kind of signal intensities and their complementary spatial features. Two very promising and neighboring wide areas were delimited for an approximate total extension of 4190 m^2 in the southern part of the surveyed region, corresponding to the limit of the archaeological restricted area (inside and outside). GPR comparison was possible only for one of them (and was positive), but their continuity and ARP data parameters (intensity and spatial features) are evidence of their global archaeological potentiality. Of the other four large resistive regions (in the range of 190–390 m^2), two were characterized by the spatial correspondence of strong ARP and GPR anomalies, while the others, presenting high apparent resistivity contrast and shapes potentially attributable to non-geological origin, presented GPR signal of medium amplitude or locally covered by strong noise. Consequently, they were classified as susceptible in the first case to include buried archaeological evidence, and in the second case, suggested further investigations with other complementary geophysical methods and archaeological trials. The remaining wide area (17,830 m^2) showed a complex scattered distribution of planimetric patterns with low resistivity contrast which could be indicative of the presence of in situ archaeological features, less sensitive than possible accumulation of stone fragments to volume analyses, such as ARP.

In order to validate ARP apparent resistivity patterns and nominal map depth, two sample pseudo-sections were inverted, substantially confirming their validity but obtaining few differences in terms of hierarchy of most resistive values between top and bottom subsoil and partially in terms of depths in correspondence with the most conductive layers. The depth values of the pseudo-sections and the ERTs bottom interfaces are still compatible with the geological investigations summarized in Section 2.2 considering the ordinary spatial variability of geological features.

In the present case study, the ARP method was confirmed as a reliable and expeditious geophysical technique able to investigate wide regular surfaces and to identify areas likely to contain archaeological features. Thanks to the simultaneous acquisition of positioning data by means of a differential GPS system mounted in the middle of the electrical array, ARP data were instantly geopositioned and an accurate georeferencing was automatically performed. Very good performances were evidenced in indicating large subsoil volumes with a high apparent resistivity contrast. Some spatially coherent patterns were also recognizable with weak apparent resistivity contrast for which further investigations could be useful. Overall, ARP proved to be an effective geophysical technique for the preliminary assessment of the archaeological potential of investigated areas and to direct more concentrated, expensive and time-consuming higher resolution prospections. Furthermore, joint interpretation of ARP and GPR measurements could be used to differentiate potential archaeological zones, demonstrating the effectiveness of the integration of both extensive non-destructive methods to address preliminary researches concerning archaeological studies.

Author Contributions: Conceptualization, L.P., S.V.C., A.T. and G.R.; Data processing, L.P., S.V.C. and A.T.; Funding acquisition, G.R.; Investigation, L.P., S.V.C., A.T. and G.R.; Methodology, L.P., S.V.C., A.T. and G.R.; Validation, L.P., S.V.C., A.T. and G.R.; Writing—original draft, L.P. and S.V.C.; Writing—review and editing, L.P. and S.V.C. All authors have read and agreed to the published version of the manuscript.

Funding: This research was funded by *Regione Autonoma della Sardegna*, LR 7/2007. The APC was funded by *Fondazione di Sardegna*.

Acknowledgments: The authors are grateful to Luigi Noli and Mario Sitzia for their essential technical support during the surveys and GEOCARTA © for the acquisition of ARP data. The authors wish to thank the Archaeological Superintendency of Cagliari and Oristano (*Soprintendenza Archeologia, belle arti e paesaggio per la città metropolitana di Cagliari e le province di Oristano e Sud Sardegna*) for the courtesy and for the permission to carry out the surveys in the site of study and to publish the results of geophysical researches.

Conflicts of Interest: The authors declare no conflict of interest.

References

1. Campana, S. Archaeological site detection and mapping: Some thoughts on differing scales of detail and archaeological 'non-visibility'. In *Seeing the Unseen. Geophysics and Landscape Archaeology*; Campana, S., Piro, S., Eds.; CRC Press, Taylor & Francis Group: London, UK, 2009; pp. 5–26.
2. Drahor, M.G.; Göktürkler, G.; Berge, M.A.; Kurtulmuş, T. Özgür Application of electrical resistivity tomography technique for investigation of landslides: A case from Turkey. *Environ. Earth Sci.* **2006**, *50*, 147–155.
3. Boyle, A.; Wilkinson, P.B.; Chambers, J.E.; Meldrum, P.I.; Uhlemann, S.; Adler, A. Jointly reconstructing ground motion and resistivity for ERT-based slope stability monitoring. *Geophys. J. Int.* **2017**, *212*, 1167–1182. [CrossRef]
4. Perrone, A.; Lapenna, V.; Piscitelli, S. Electrical resistivity tomography technique for landslide investigation: A review. *Earth-Sci. Rev.* **2014**, *135*, 65–82. [CrossRef]
5. Mastrocicco, M.; Vignoli, G.; Colombani, N.; Abu Zeid, N. Surface electrical resistivity tomography and hydrogeological characterization to constrain groundwater flow modeling in an agricultural field site near Ferrara (Italy). *Environ. Earth Sci.* **2009**, *61*, 311–322. [CrossRef]
6. Watlet, A.; Kaufmann, O.; Triantafyllou, A.; Poulain, A.; Chambers, J.E.; Meldrum, P.I.; Wilkinson, P.B.; Hallet, V.; Quinif, Y.; Van Ruymbeke, M.; et al. Imaging groundwater infiltration dynamics in the karst vadose zone with long-term ERT monitoring. *Hydrol. Earth Syst. Sci.* **2018**, *22*, 1563–1592. [CrossRef]
7. Braga, A.C.O.; Malagutti, F.W.; Dourado, J.C.; Chang, H.K. Correlation of Electrical Resistivity and Induced Polarization Data with Geotechnical Survey Standard Penetration Test Measurements. *J. Environ. Eng. Geophys.* **1999**, *4*, 123–130. [CrossRef]
8. Cosenza, P.; Marmet, E.; Rejiba, F.; Cui, Y.J.; Tabbagh, A.; Charlery, Y. Correlations between geotechnical and electrical data: A case study at Garchy in France. *J. Appl. Geophys.* **2006**, *60*, 165–178. [CrossRef]
9. Soupios, P.M.; Georgakopoulos, P.; Papadopoulos, N.; Saltas, V.; Andreadakis, A.; Vallianatos, F.; Sarris, A.; Makris, J.P. Use of engineering geophysics to investigate a site for a building foundation. *J. Geophys. Eng.* **2007**, *4*, 94–103. [CrossRef]
10. Chambers, J.; Ogilvy, R.; Meldrum, P.; Nissen, J. 3D resistivity imaging of buried oil- and tar-contaminated waste deposits. *Eur. J. Environ. Eng. Geophys.* **1999**, *4*, 3–15.
11. Chambers, J.E.; Kuras, O.; Meldrum, P.I.; Ogilvy, R.D.; Hollands, J. Electrical resistivity tomography applied to geologic, hydrogeologic, and engineering investigations at a former waste-disposal site. *Geophysics* **2006**, *71*, B231–B239. [CrossRef]
12. Corwin, D.L.; Lesch, S.M. Application of soil electrical conductivity to precision agriculture: theory, principles and guidelines. *Agron. J.* **2003**, *95*, 455–471. [CrossRef]
13. Garre, S.; Javaux, M.; VanderBorght, J.; Pagès, L.; Vereecken, H. Three-Dimensional Electrical Resistivity Tomography to Monitor Root Zone Water Dynamics. *Vadose Zone J.* **2011**, *10*, 412–424. [CrossRef]
14. Beff, L.; Günther, T.; Vandoorne, B.; Couvreur, V.; Javaux, M. Three-dimensional monitoring of soil water content in a maize field using Electrical Resistivity Tomography. *Hydrol. Earth Syst. Sci.* **2013**, *17*, 595–609. [CrossRef]
15. Drahor, M.G. Integrated geophysical studies in the upper part of Sardis archaeological site, Turkey. *J. Appl. Geophys.* **2006**, *59*, 205–223. [CrossRef]

16. Cardarelli, E.; Fischanger, F.; Piro, S. Integrated geophysical survey to detect buried structures for archaeological prospecting. A case-history at Sabine Necropolis (Rome, Italy). *Near Surf Geophys.* **2008**, *6*, 15–20. [CrossRef]
17. Gaffney, C. Detecting trends in the prediction of the buried past: a review of geophysical techniques in archaeology. *Archaeometry* **2008**, *50*, 313–336. [CrossRef]
18. Trogu, A.; Ranieri, G.; Calcina, S.V.; Piroddi, L. The Ancient Roman Aqueduct of Karales (Cagliari, Sardinia, Italy): Applicability of Geophysics Methods to Finding the Underground Remains. *Archaeol. Prospect.* **2014**, *21*, 157–168. [CrossRef]
19. Ranieri, G.; Godio, A.; Loddo, F.; Stocco, S.; Casas, A.; Capizzi, P.; Messina, P.; Orfila, M.; Cau, M.; Chávez, M.; et al. Geophysical prospection of the Roman city of Pollentia, Alcúdia (Mallorca, Balearic Islands, Spain). *J. Appl. Geophys.* **2016**, *134*, 125–135. [CrossRef]
20. Ranieri, G.; Trogu, A.; Loddo, F.; Piroddi, L.; Cogoni, M. Digital Museum from Integrated 3D Aerial Photogrammetry, Laser Scanner and Geophysics Data. In Proceedings of the 24th European Meeting of Environmental and Engineering Geophysics, Porto, Portugal, 9–13 September 2018; pp. 1–5.
21. Piroddi, L.; Vignoli, G.; Trogu, A.; Deidda, G.P. Non-destructive Diagnostics of Architectonic Elements in San Giuseppe Calasanzio's Church in Cagliari: A Test-case for Micro-geophysical Methods within the Framework of Holistic/integrated Protocols for Artefact Knowledge. In Proceedings of the 2018 IEEE International Conference on Metrology for Archaeology and Cultural Heritage, Cassino, Italy, 22–24 October 2018; pp. 17–21.
22. Yilmaz, S.; Balkaya, Ç.; Çakmak, O.; Oksum, E. GPR and ERT explorations at the archaeological site of Kılıç village (Isparta, SW Turkey). *J. Appl. Geophys.* **2019**, *170*, 103859. [CrossRef]
23. Samouëlian, A.; Cousin, I.; Tabbagh, A.; Bruand, A.; Richard, G. Electrical resistivity survey in soil science: A review. *Soil Tillage Res.* **2005**, *83*, 173–193. [CrossRef]
24. Chávez, G.; Tejero, A.; Alcantara, M.A.; Chavez, R.E. The 'L-Array', a tool to characterize a fracture pattern in an urban zone. In Proceedings of the 17th Near Surface Geophysics meeting, European Section Meeting, Leicester, UK, 12–14 September 2011.
25. Berge, M.A.; Drahor, M.G. Electrical Resistivity Tomography Investigations of MultiLayered Archaeological Settlements: Part I - Modelling. *Archaeol. Prospect.* **2011**, *18*, 159–171. [CrossRef]
26. Santarato, G.; Ranieri, G.; Occhi, M.; Morelli, G.; Fischanger, F.; Gualerzi, D. Three-dimensional Electrical Resistivity Tomography to control the injection of expanding resins for the treatment and stabilization of foundation soils. *Eng. Geol.* **2011**, *119*, 18–30. [CrossRef]
27. Trogu, A.; Ranieri, G.; Fischanger, F. 3D Electrical Resistivity Tomography to Improve the Knowledge of the Subsoil below Existing Buildings. *Environ. Semeiot.* **2011**, *4*, 63–70. [CrossRef]
28. Argote-Espino, D.; Tejero-Andrade, A.; Cifuentes-Nava, G.; Iriarte, L.; Farías, S.; Chávez, R.E.; López, F. 3D electrical prospection in the archaeological site of El Pahñú, Hidalgo State, Central Mexico. *J. Archaeol. Sci.* **2013**, *40*, 1213–1223. [CrossRef]
29. Chavez, R.E.; Vargas, D.; Cifuentes-Nava, G.; Hernandez-Quintero, J.E.; Tejero, A. Tri-Dimensional Electric Resistivity Tomography (ERT-3D) Technique, an Efficient Tool to Unveil the Subsoil of Archaeological Structures. Available online: https://ui.adsabs.harvard.edu/abs/2014AGUFMNS33A3945C/abstract (accessed on 30 January 2020).
30. Loddo, F.; Ranieri, G.; Piroddi, L.; Trogu, A.; Cogoni, M. On the Use of Electrical Resistivity Tomography in Shallow Water Marine Environment for Archaeological Research. In Proceedings of the Near Surface Geoscience 2016—22nd European Meeting of Environmental and Engineering Geophysics, Barcelona, Spain, 4–8 September 2016. [CrossRef]
31. Al-Saadi, O.S.; Schmidt, V.; Becken, M.; Fritsch, T. Very-high-resolution electrical resistivity imaging of buried foundations of a Roman villa near Nonnweiler, Germany. *Archaeol. Prospect.* **2018**, *25*, 209–218. [CrossRef]
32. Piroddi, L.; Loddo, F.; Calcina, S.V.; Trogu, A.; Cogoni, M.; Ranieri, G. Integrated Geophysical Survey to Reconstruct Historical Landscape in Undug Areas of the Roman Ancient Town of Nora, Cagliari, Italy. In Proceedings of the 2018 IEEE International Conference on Metrology for Archaeology and Cultural Heritage, Cassino, Italy, 22–24 October 2018; pp. 244–248.
33. Tejero-Andrade, A.; Argote-Espino, D.L.; Cifuentes-Nava, G.; Hernández-Quintero, E.; Chávez, R.E.; García-Serrano, A. 'Illuminating' the interior of Kukulkan's Pyramid, Chichén Itzá, Mexico, by means of a non-conventional ERT geophysical survey. *J. Archaeol. Sci.* **2018**, *90*, 1–11. [CrossRef]

34. Fischanger, F.; Catanzariti, G.; Comina, C.; Sambuelli, L.; Morelli, G.; Barsuglia, F.; Ellaithy, A.; Porcelli, F. Geophysical anomalies detected by electrical resistivity tomography in the area surrounding Tutankhamun's tomb. *J. Cult. Heritage* **2019**, *36*, 63–71. [CrossRef]
35. Dabas, M. Theory and practice of the new fast electrical imaging system ARP©. In *Seeing the Unseen. Geophysics and Landscape Archaeology*; Campana, S., Piro, S., Eds.; CRC Press, Taylor & Francis Group: London, UK, 2009; pp. 105–126.
36. Pittau, L. La prospezione geofisica nella ricerca archeologica. Studio delle capacità risolutive e valutazione di opportunità di impiego. Ph.D. Thesis, "Ingegneria Geologico Ambientale - X ciclo". Politecnico di Torino – Università degli Studi di Cagliari, Cagliari, Italy, 1994.
37. Panissod, C.; Dabas, M.; Jolivet, A.; Tabbagh, A. A novel mobile multipole system (MUCEP) for shallow (0-3 m) geoelectrical investigation: the 'Vol-de-canards' array. *Geophys. Prospect.* **1997**, *45*, 983–1002. [CrossRef]
38. Panissod, C.; Dabas, M.; Hesse, A.; Jolivet, A.; Tabbagh, J.; Tabbagh, A. Recent developments in shallow-depth electrical and electrostatic prospecting using mobile arrays. *Geophysics* **1998**, *63*, 1542–1550. [CrossRef]
39. Papadopoulos, N.G.; Tsokas, G.N.; Dabas, M.; Yi, M.-J.; Tsourlos, P. 3D Inversion of Automated Resistivity Profiling (ARP) Data. *ArchéoSciences* **2009**, 329–332. [CrossRef]
40. Papadopoulos, N.G.; Tsokas, G.N.; Dabas, M.; Yi, M.-J.; Kim, J.-H.; Tsourlos, P. Three-dimensional inversion of automatic resistivity profiling data. *Archaeol. Prospect.* **2009**, *16*, 267–278. [CrossRef]
41. Sasaki, Y. 3-D resistivity inversion using the finite-element method. *Geophysics* **1994**, *59*, 1839–1848. [CrossRef]
42. Dabas, M.; Aubry, L.; Rouiller, D.; Larcher, J.M. Caractérisation de la variabilité spatiale intraparcellaire des sols agricoles Méthode MUCEP: une gestion prédictive des rendements dans le cadre de l'agriculture de précision. In Proceedings of the 2ème Colloque de Géophysique des Sols et des Formations Superficielles, Orléans, France, 21–22 September 1999; pp. 125–129.
43. Dabas, M.; Cassassolles, X. Characterization of Soil Variability and Its Application to the Management of Vineyard (Arp System). Available online: http://www.liendelavigne.org/ANG/RapportsANG/11--2002ANG/LDV_021122_Dabas_en.pdf (accessed on 14 January 2012).
44. Rossi, R.; Pollice, A.; Diago, M.-P.; Oliveira, M.; Millan, B.; Bitella, G.; Amato, M.; Tardáguila, J. Using an Automatic Resistivity Profiler Soil Sensor On-The-Go in Precision Viticulture. *Sensors* **2013**, *13*, 1121–1136. [CrossRef] [PubMed]
45. Buvat, S.; Thiesson, J.; Michelin, J.; Nicoullaud, B.; Bourennane, H.; Coquet, Y.; Tabbagh, A. Multi-depth electrical resistivity survey for mapping soil units within two 3ha plots. *Geoderma* **2014**, *232*, 317–327. [CrossRef]
46. Dabas, M.; Hesse, A.; Tabbagh, J. Experimental resistivity survey at Wroxeter archaeological site with a fast and light recording device. *Archaeol. Prospect.* **2000**, *7*, 107–118. [CrossRef]
47. Chouteau, M.; Vallières, S.; Toe, E. A multi-dipole mobile array for the non-destructive evaluation of pavement and concrete infrastructures: A feasability study. In Proceedings of the International Symposium Non-Destructive Testing in Civil Engineering (NDT-CE 2003), Berlin, Germany, 16–19 September 2003.
48. Andrenelli, M.; Magini, S.; Pellegrini, S.; Perria, R.; Vignozzi, N.; Costantini, E. The use of the ARP© system to reduce the costs of soil survey for precision viticulture. *J. Appl. Geophys.* **2013**, *99*, 24–34. [CrossRef]
49. Reynolds, J.M. An Introduction to Applied and Environmental Geophysics. Available online: https://bit.ly/3b2dchN (accessed on 30 January 2020).
50. Masini, N.; Capozzoli, L.; Chen, P.; Chen, F.; Romano, G.; Lu, P.; Tang, P.; Sileo, M.; Ge, Q.; Lasaponara, R. Towards an Operational Use of Geophysics for Archaeology in Henan (China): Methodological Approach and Results in Kaifeng. *Remote. Sens.* **2017**, *9*, 809. [CrossRef]
51. Bedini, A.; Tronchetti, C.; Ugas, G.; Zucca, R. *Giganti di Pietra. Monteprama. L'heroon che cambia la storia del Mediterraneo*; Fabula Editore: Cagliari, Italy, 2012.
52. Lilliu, G. *La grande statuaria nella Sardegna nuragica. Atti Accademia dei Lincei*; Memorie Scienze Morali Storiche Filologiche: Roma, Italy, 1997.
53. Ranieri, G.; Zucca, R.; Trogu, A.; Calcina, S.V.; Piroddi, L.; Usai, A. Multi-channel GPR Prospection in the Archaeological Site of Monte Prama (Cabras, Italy). In Proceedings of the 20th Annual Meeting of European Association of Archaeologists, Istanbul, Turkey, 10–14 September 2014; pp. 187–188.
54. Ranieri, G.; Zucca, R. *Monte Prama I. Ricerche 2014. Collana Sardegna Archeologica, Scavi e Ricerche Vol. 12*; Carlo Delfino Editore: Sassari, Italy, 2015.

55. Ranieri, G.; Trogu, A.; Loddo, F.; Piroddi, L.; Zucca, R. Geophysics-An Essential Tool for Modern Archaeology. A Case from Monte Prama (Sardinia, Italy). In Proceedings of the Near Surface Geoscience 2015—21st European Meeting of Environmental and Engineering Geophysics, Torino, Italy, 6–10 September 2015; pp. 1–5.
56. Trogu, A.; Ranieri, G.; Piroddi, L.; Loddo, F.; Cogoni, M. New GPR Data from the Archeological Site of Mont'e Prama (Sardinia, Italy). In Proceedings of the Near Surface Geoscience 2016—22nd European Meeting of Environmental and Engineering Geophysics, Barcelona, Spain, 4–8 September 2016; pp. 1–5.
57. Lecca, L.; Carboni, S. The Tyrrhenian of San Giovanni di Sinis (central-western Sardinia): a stratigraphyc signal of a single high stand. *Riv. Ital. Paleontol. S.* **2007**, *113*, 509–523.
58. Carboni, S. Il contesto geologico. In *Monte Prama 1. Ricerche 2014, Colluna Sardegna Archeologica, Scavi e Ricerche*; Ranieri, G., Zucca, R., Eds.; Carlo Delfino Editore: Sassari, Italy, 2015; Volume 12.
59. Persico, R. *Introduction to Ground Penetrating Radar: Inverse Scattering and Data Processing*; IEEE Press Series on Electromagnetic Wave Theory; Wiley-IEEE Press: Hoboken, NJ, USA, 2014; p. 392.
60. Jol, H. Ground Penetrating Radar Theory and Applications Elsevier Science. Available online: https://bit.ly/2tcRgPI (accessed on 30 January 2020).
61. Loke, M.H. Tutorial: 2-D and 3-D electrical imaging surveys. Geotomo Software, accessed January 17, 2020. Available online: https://www.geotomosoft.com/downloads.php (accessed on 30 January 2020).

© 2020 by the authors. Licensee MDPI, Basel, Switzerland. This article is an open access article distributed under the terms and conditions of the Creative Commons Attribution (CC BY) license (http://creativecommons.org/licenses/by/4.0/).

Article

Remote Sensing, Archaeological, and Geophysical Data to Study the Terramare Settlements: The Case Study of Fondo Paviani (Northern Italy)

Rita Deiana [1],*, David Vicenzutto [1], Gian Piero Deidda [2], Jacopo Boaga [3] and Michele Cupitò [1]

1. Department of Cultural Heritage, University of Padova, Piazza Capitaniato 7, 35139 Padova, Italy; david.vicenzutto@unipd.it (D.V.); michele.cupito@unipd.it (M.C.)
2. Department of Civil and Environmental Engineering and Architecture, University of Cagliari, Via Marengo 2, 09123 Cagliari, Italy; gpdeidda@unica.it
3. Department of Geosciences, University of Padova, Via Gradenigo 6, 35129 Padova, Italy; jacopo.boaga@unipd.it
* Correspondence: rita.deiana@unipd.it

Received: 22 June 2020; Accepted: 6 August 2020; Published: 13 August 2020

Abstract: During the Middle and Recent Bronze Age, the Po Plain and, more broadly Northern Italy were populated by the so-called "Terramare", embanked settlements, surrounded by a moat. The buried remains of these archaeological settlements are characterized by the presence of a system of palaeo-environments and a consequent natural gradient in soil moisture content. These differences in the soil are often firstly detectable on the surface during the seasonal variations, with aerial, satellite, and Laser Imaging Detection and Ranging (LIDAR) images, without any information on the lateral and in-depth extension of the related buried structures. The variation in the moisture content of soils is directly related to their differences in electrical conductivity. Electrical resistivity tomography (ERT) and frequency domain electromagnetic (FDEM), also known as electromagnetic induction (EMI) measurements, provide non-direct measurements of electrical conductivity in the soils, helping in the reconstruction of the geometry of different buried structures. This study presents the results of the multidisciplinary approach adopted to the study of the Terramare settlement of Fondo Paviani in Northern Italy. Remote sensing and archaeological data, collected over about 10 years, combined with more recent ERT and FDEM measurements, contributed to the analysis of this particular, not yet wholly investigated, archaeological site. The results obtained by the integrated multidisciplinary study here adopted, provide new useful, interesting information for the archaeologists also suggesting future strategies for new studies still to be conducted around this important settlement.

Keywords: electrical resistivity tomography (ERT); frequency domain electromagnetic (FDEM); archaeology; terramare; bronze age

1. Introduction

In the broad range of geophysical measurements applied to archaeology [1–6], electrical resistivity tomography (ERT) is today one of the most popular methods. Multichannel systems and new inversion techniques, developed between the 1980s and the 1990s, rapidly increased, in fact, the application and popularity of this technique in different archaeological contexts [7–15]. During the last 10 years, thanks to the development of 3D ERT surveys and new tools for 3D data inversion, this method enhances the possibility to reconstruct the spatial distribution and the shape of the archaeological targets, both in small or large areas, as well as in rural or urban context [16–22]. In this wide range of applications and possible uses of the ERT, undoubtedly, the main advantage offered by this method is highlighted in the contexts in which the distribution of buried large natural or artificial systems must be defined, extended not only laterally but also in-depth. Theoretically, the penetration depth of the electrical signal depends on the total length of the ERT line, the power source, but also the electrical conductivity of the investigated systems. The use of ERT, often combined with frequency-domain electromagnetic (FDEM), also known as electromagnetic inductance (EMI) measurements, is recently mostly documented in agriculture, geomorphological, and sedimentary research applications [23–37]. Both methods are, in fact, susceptible to the differences in electrical conductivity of the soils, strictly related to their different soil moisture content. This capability is fundamental, for example, in the identification of buried channels and corresponding palaeo-environments, or related structures, often also interesting from the archaeological point of view. The differences in the soil moisture content, existing between these buried structures and the hosting system, produce visible crop marks or surface evidence, identifiable by aerial and satellite images. In recent years, several scientific studies, both in geological and archaeological contexts, moving from preliminary information collected by remote sensing data, used geophysical prospection to define these buried structures [38–44]. This paper, considering these promising recent applications, shows a multidisciplinary approach carried out to characterize the fortified system of Late Bronze Age Terramara settlement of Fondo Paviani. The term "Terramara" identifies the archaeological remains of the fortified settlements diffused in Northern Italy, more specifically in the Po Plain, between the central phase of the Middle Bronze Age and the end of the Recent Bronze Age (1600/1550–1175/1150 BC) [45,46]. A palisade or a rampart, and a wide ditch connected with a watercourse surrounded the Terramare (Figure 1).

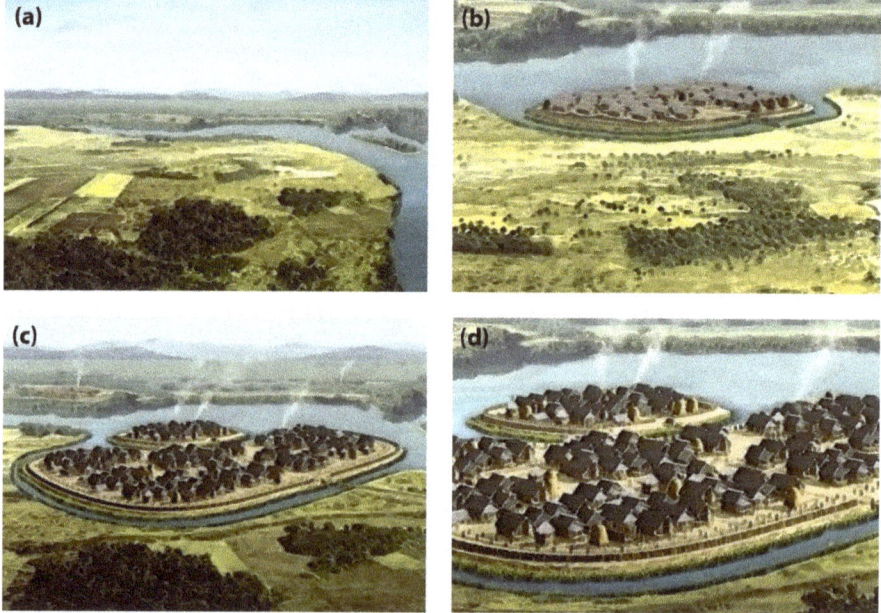

Figure 1. Reconstruction of a Terramara development (Santa Rosa di Poviglio, [47], modified by authors). (**a**) Landscape view before the settlement; (**b**)" Villaggio Piccolo": a little settlement surrounded by ditch and palisade, later destroyed and covered by a rampart. (**c**) "Villaggio Grande": expansion of the small village with new fortified part with ditch and rampart. (**d**) Particular of the settlement at its maximum development, when small and big villages coexisting.

An intensive agro-pastoral economy characterized the Terramare culture. Almost every site was surrounded by a vast structured agrarian hinterland with channeling systems. In the Middle Bronze Age, the social order of the communities was tribal, while in the Recent Bronze Age, it seems that the social model was comparable to an evolved chiefdom [48–50]. At the end of the Recent Bronze Age, a general crisis, due to demographic and environmental factors, involved Terramare Culture. In the Southern Po Plain, the crisis led to a full abandonment of the settlements, while in the Northern Po Plain has registered only a contraction of the population [51,52].

The buried remains of the Terramara, thanks to the presence of palaeo-environments and related natural differences in soil moisture content with the hosting system, represents an ideal target for ERT and FDEM measurements [53,54]. The first step in this recent multidisciplinary study of the Fondo Paviani settlement (Figure 2) moved from the pieces of evidence provided by the aerial photographs (Figures 2c and 3a–c).

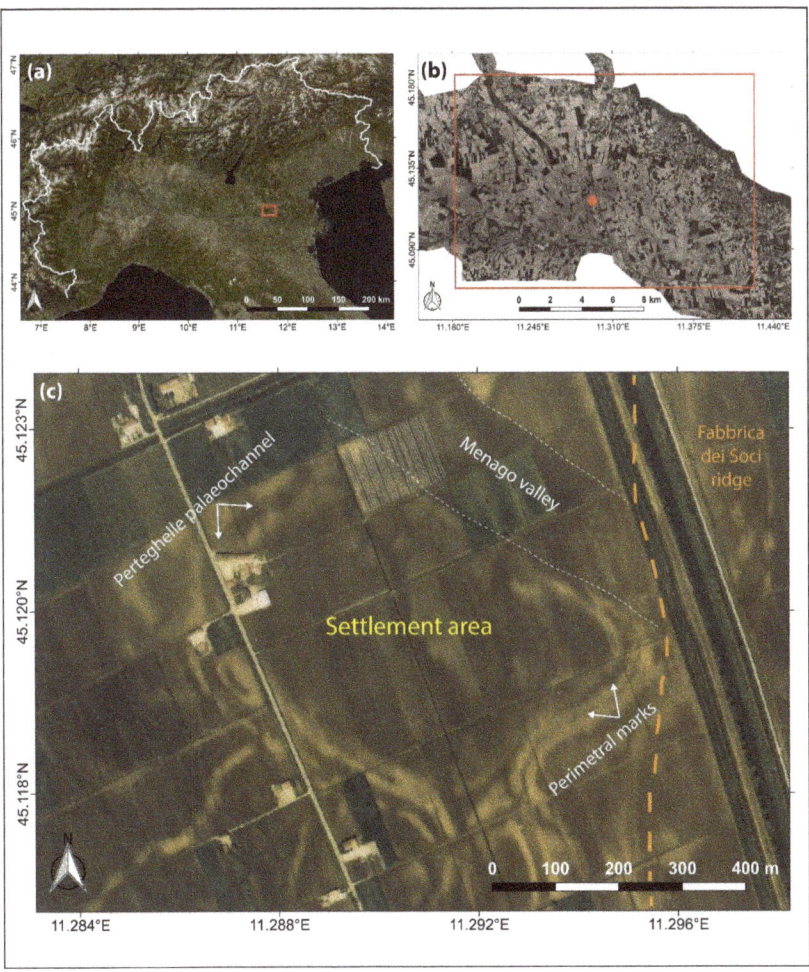

Figure 2. (**a**) Location of the embanked site of Fondo Paviani in Northern Italy (2020 aerial view from Maps-Bing modified by authors). (**b**) Fondo Paviani within the Valli Grandi Veronesi landscape (2012 photomosaic of C.G.R. aerial frames modified by authors). (**c**) Detail of the settlement landscape (1990 C.G.R. aerial frame modified by authors).

Table 1. List of aerial photo frames used for the photo interpretation.

Flight	Date	Elevation (m a.s.l.)	Focal Length (mm)	Spatial Resolution (Pixels)	Band Spectra Image	Single Frame Codes
IGMI. GAI.	24 May 1955	5000	154.17	6051 × 5960 (23 × 23 cm photo)	Visible Light (B/W)	VV GAI M 26 AMS; 5276
SCAME.	16 March 1983	2500	153.26	6000 × 5598 (23 × 23 cm photo)	Visible Light (B/W)	Re.Ven.-VR/2 1983; 12/5070
CGR.	13 April 1990	3000	151.77	5833 × 5993 (23 × 23 cm photo)	Visible Light (Colors)	Re.Ven. 90; 26A/21
ROSSI	27 July 1997	2500	153.04	6078 × 5718 (23 × 23 cm photo)	Visible Light (Colors)	Re.Ven.-VR-SUD; 16/89
CGR.	17 April 1999	2500	153.26	6097 × 5631 (23 × 23 cm photo)	Visible Light (Colors)	Re.Ven. Veneto Prov.Ro 99; 63/1648
CGR.	26 June 2004	2500	153.28	6218 × 5626 (23 × 23 cm photo)	Visible Light (Colors)	Re.Ven. VR-SUD-1-VOLO-2004; 550

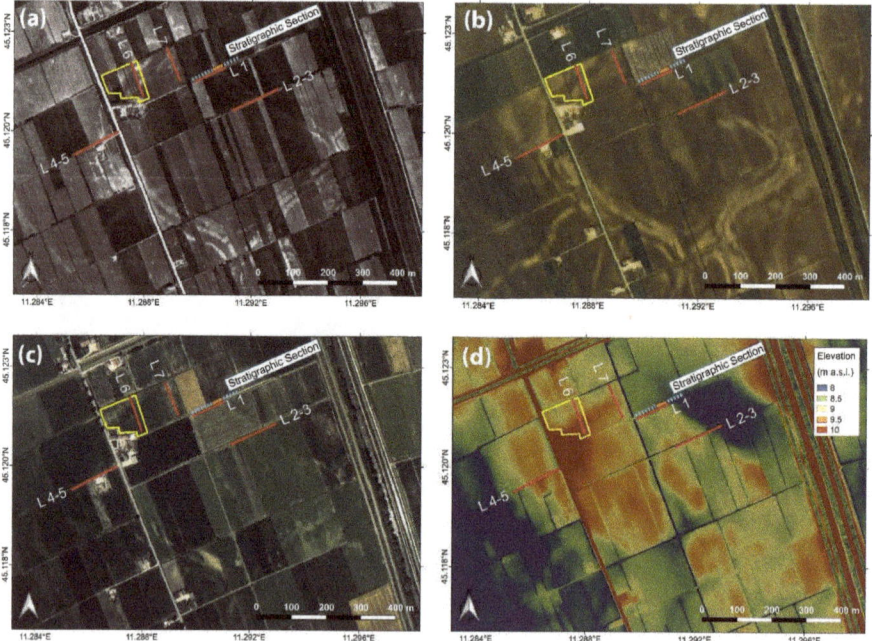

Figure 3. Location of geophysical measurements (red lines: ERT; yellow area: FDEM) and stratigraphic section (dotted cyan line: stratigraphic section; orange line: segment intercepting the fortification structures). (a) 1955 I.G.M.I. G.A.I aerial frame (Table 1) modified by authors. (b) 1990 C.G.R aerial frame (Table 1) modified by authors. (c) 2004 C.G.R aerial frame (Table 1) modified by authors. (d) DTM from LIDAR data modified by authors.

The comparison between an old aerial image (Figure 3a) taken in 1955 and more recent images (Figure 3b,c) of the same area, respectively taken in 1990 and 2004, highlights the progressive attenuation of the northern boundaries of the settlement.

The relative recent intensive agricultural practice, consisting of the regularization of the soil surfaces, in some cases, determines the removal of the upper part of the soil. This practice, also adopted at the Fondo Paviani in the last 40 years, could definitively compromise the archaeological deposit of the Terramare in some parts of the farm. The elevation data obtained by LIDAR (Figure 3d) confirms local topographical modifications in the northern part of the site. Moving from these preliminary indications, a series of ERT lines collected in different parts of Fondo Paviani site help in the analysis of the actual condition and extension of these palaeo buried structures. The combination of ERT and FDEM measurements, also supported, in a test area, in the identification of the multitemporal, complex palaeo-systems spatial distribution. The integration between ERT and FDEM measurements, remote sensing data, and some pieces of evidence made by archaeological excavations, provided new interesting information about the studied area. The results of this multidisciplinary approach also support new hypotheses about the original structure of the Fondo Paviani embankment system, also suggesting the best way to continue the study around this important and not wholly investigated archaeological site.

Study Site Background

The broad fortified Terramara of Fondo Paviani, located in the Verona low plain, since 2007, is the focus of a multidisciplinary project [55] directed by the Prehistoric-Protohistoric research team at the Department of Cultural Heritage of the University of Padova. The settlement, developed between the last decades of the 14th and the 11th centuries BC, represents one of the most critical contexts for understanding the Late Bronze Age historical events of the Po Plain and, more broadly, in Northern Italy. Fondo Paviani was the central place of the complex political-territorial system known as Valli Grandi Veronesi polity. Between the 13th and the beginning of the 12th century BC, the settlement became the central keystone of the relations and trades connecting the Terramare Culture on the one hand with the Alpine area and with peninsular Italy, and on the other hand with Mycenean Greece, Cyprus, and the Levant [55].

Fondo Paviani also was one of the few centers that survived in the 12th century BC at the crisis producing the collapse of the Terramare Culture. Then it became the model for the development of the later population pattern and the new international trade network focused on the central place of Frattesina, on the Po river [56–61]. The development of the Fondo Paviani settlement reflects the typical model of the evolution of the Terramare-type settlements [62,63]. In the first phase, Fondo Paviani was surrounded only by a palisade and by a small ditch without remarkable water flows. During the 13th and the beginning of the 12th century BC a wide rampart was built and a ditch connected to the wetlands of the Menago river valley surrounded the site.

At that time, the settlement, which finally took the Terramara-type shape, reached the surface of 20 ha (16 ha of inner area and 4 ha of perimetral structures), an exceptional size for a Bronze Age settlement in Italy. The new fortified system, one of the biggest of the entire Terramare world, involved the active exploitation of pre-existing river channels and alluvial ridges, implying modifications and adaptations of the related landscape. Between the second half of the 12th and the 11th century BC, Fondo Paviani suffered a general crisis [60]. Broad areas of the settlement, for a long time occupied by houses and related infrastructures, were converted into cultivated spaces. In the last phase of the life of the settlement and many centuries after its abandonment, the wide rampart, although collapsed in several points and damaged by diggings, retained its primary function. The rampart represented an essential shield between the development of the stratification of the inner part of the settlement and the growing clogging of the ditch linking to the Menago valley [47].

2. Materials and Methods

2.1. Geomorphological Setting

The settlement of Fondo Paviani is part of the Valli Grandi Veronesi system, a lowland of Adige alluvial fan, on an alluvial ridge inside the valley of the Menago river belonging to the Bassa Pianura Veronese. The valley flows from the Adige alluvial fan through the Valli Grandi Veronesi, where it gets the shape of an N–W/S–E large depression with slight escarpments. The Menago valley is blocked by a large Late Pleistocene ridge ("Fabbrica dei Soci" ridge) immediately S–E of the site of Fondo Paviani (Figure 2). This ridge, since the Holocene, has prevented the drainage of the valley, gradually changing the whole depression in a wetland, without covering the high bump of the settlement [64].

The Valli Grandi Veronesi area is well known [64–68] for the high preservation of the Bronze Age landscape. The depth of archaeological remains in this area is very shallow (between 0.2 and 1.4 m) [69,70]. Sometimes, the topography of the ancient structures corresponds to that of the current landscape [70], where this high preservation is the result of a lucky sequence of hydrogeological and anthropic events in the area from the end of Bronze Age to nowadays [65–71].

The main preservation factor of the Valli Grandi Veronesi ancient landscape is due to the presence of an alluvial/marshy clay layer [65], which gradually had covered the whole area from the Early Middle Age to the Modern Age. The presence of this layer and the general low drainage had made the Valli Grandi Veronesi a swamp almost uninhabited until the end of the 19th century when the reclamations of the area began.

For these reasons, the Valli Grandi Veronesi represents an optimal context both for archaeological research and for non-invasive analysis as aerial photo interpretation, DTM data from LIDAR, and geophysical measurements.

The hydrographic network contemporary to Fondo Paviani Terramara mainly consisted of spring-fed rivers flowing through the Menago Valley [65,66]. The sedimentary contribution of these rivers started between the end of the Early Bronze Age and the central phases of the Middle Bronze Age (between the 17th and the 16th century BC), is also at the origin of the formation of the ridge where, about two centuries later, the Fondo Paviani settlement developed [65]. A spring-fed river and Perteghelle channel, a watercourse certainly active between the Late Pleistocene and the beginning of the Holocene surround the N–W and S–W sides of the site. Probably, the Perteghelle palaeo-channel was not wholly inactive contemporarily to the life cycle of Fondo Paviani, although carrying slight flows of water [67].

2.2. Aerial Photograph Interpretation

The analysis and the consequent interpretation of aerial photographs represented the first step of the whole research. The aerial photograph interpretation helps in the identification of some hypothetical anthropic structures and natural features of the buried landscape, allowing in the plan of the field investigation the direct or non-direct verification of these features.

Firstly, a broad photo set of the area of Fondo Paviani settlement and its nearest hinterland allow the identification of the aerial frames of interest. Six different years and four companies were selected: I.G.M.I. (Istituto Geografico Militare Italiano) and G.A.I. (Gruppo Aeronautico Italiano) 1955 flights, S.C.A.M.E. (Società Cartografica Aero Fotogrammetrica Meridionale di Scambia) 1983 flight, C.G.R. (Compagnia Generale Riprese aeree) 1990, 1999, and 2004 flights, and ROSSI (ROSSI s.r.l.) 1997 flight (Table 1). The details of each aerial shot are reported in Table 1. Different years and months were considered to analyze the same area in different seasonal and cultivation conditions (Table 1).

The high quality of the frames and the high visibility of the buried landscape reduce the processing to the enhancement and object classification steps [72,73]. The piece of evidence related to the archaeological and natural landscape features have been selected and compared, frame by frame, with different soil marks and crop marks [74]. A previous study of the Valli Grandi Veronesi landscape, based on aerial photo-interpretation [75], attributes light colors at features consisting of draining sediments (sand and coarse silt), often related to elevation structures, that can correspond to alluvial ridges or artificial ramparts, and dark colors at limited drainage areas and, more broadly, wetlands or fillings of abandoned water channels.

The aerial frames and the vector data of the photo interpretation were then loaded on Quantum GIS (release 2.18) and georeferenced. The contribution of GIS is essential to measure the features, to define their geometry better, and to determine their spatial relationships. Through GIS it was also possible to integrate the aerial photo interpretation with DTM data derived from LIDAR defining the geographical position of the features to plan further field investigations.

2.3. LIDAR

Thanks to the initiative of Consorzio di Bonifica Veronese, a LIDAR survey was carried out by C.G.R. (Compagnia Generale Riprese aeree) in Spring 2012. The survey was made from an elevation of 900 m using an airborne laser scanner equipped with Optech "Pegasus" sensors (elevation accuracy: 0.1 m; planimetric accuracy: 0.2 m; point density: 3 p/m^2). A DTM with a resolution of 0.5 m per cell was extracted from this dataset by C.G.R. These DTM results, therefore, already processed from LIDAR data, were used in this research. DTM data related to Fondo Paviani and its hinterland were then processed through GIS in order to better isolate the features recognized by aerial frame analysis. These data were also used to identify other elements of the ancient landscape that are not clearly evident through the photo interpretation (e.g., slight morphological discontinuities and slope breaks). At a later time, the analysis focused on the parts of the settlement fortified system that DTM can display, that is to say moat and rampart (remains of palisades cannot be readable through DTM). As for the moats, considering their negative shape, it is possible to have information from the DTM in terms of visibility and morphology, but not in terms of state of preservation. Instead, considering that ramparts are elevated structures that in larger Terramara settlements—as Fondo Paviani—could reach a width of 20 m and a height of 6 m, through DTM analysis it is possible to register not only data about presence and morphology, but also information about state of preservation. Moreover, DTM analysis has allowed to better plan the field investigations.

2.4. Stratigraphic Analysis

The stratigraphic analysis of the archaeological layers, carried out both in open area and in section, is one of the main topics of the Fondo Paviani project [55]. The open area investigations involved two excavation sectors in the North-Eastern part of the settlement: sector 1 (48 m^2), located between the inner part of the village and the rampart, and sector 2/2.1, positioned in the inner part of the village 38 m south-west from sector 1. Anthropogenic deposits, between 0.40 and 0.60 m deep from the ground level, were covered by the agrarian topsoil and by a sandy alluvial deposit dating to the first phases of the Iron Age. While sector 1 is still under investigation, the studies in sector 2/2.1, that ended in 2018, have allowed to reconstruct the living sequence, the housing structures, and other facilities, and more generally the activities, including handcrafting, carried out in this part of the settlement [55,61]. Current data from open area excavations do not allow to elaborate a complete planimetry of the dwellings and to add information on the fortified system, excepts in terms of chronological excursus; for this reason, open area investigations were not considered in this work.

The analysis of the E–W stratigraphic section (Figure 3), here discussed, and studied before in the frame of other projects [64], is fundamental in the comprehension of the fortified system of the Fondo Paviani settlement [66]. This archaeological section, 90 m long from a current drainage ditch crossing from E to W in a large part of the settlement (Figure 3), intercepts from W to E the inner stratification of the settlement, the perimetral fortification system, and a part of the wetlands of the Menago valley. The stratigraphic analysis of this section, carried out between 2007 and 2012 using a "genetic-process approach" [76], allowed to identify a complex sequence of natural and anthropogenic layers [55,66,77,78] covered by 0.4–0.6 m of agrarian topsoil. After this analysis, the stratigraphic sequence was measured and geographically positioned (Figure 3). Pottery sherds and samples of organic material were collected from the section in order to date, both typologically and with 14C dating, the sequence, where many other samples supported micromorphological, archaeobotanical, and malacological analysis [55,66,78].

2.5. Electrical Resistivity Tomography (ERT)

In 2013 and 2015, an extensive field campaign of ERT acquisitions was planned at the archaeological site of Fondo Paviani to verify the perimeter of the settlement, its preservation, and its real extension. In total, five ERT sections were collected, the location of which, based on the pieces of evidence offered by remote data, is shown in the images in Figure 3. Three positions of interest were selected for the tomographies acquired in the SW–NE direction and two in the N–S direction (Figure 3). The resistivity measurements were performed using an IRIS Syscal Pro 72 resistivity meter (Figure 4a). The acquisition dataset was optimized to take full advantage of the 10 physical channels available in this instrument. For the acquisition, a complete skip 4 (i.e., dipole spacing of five electrodes) dipole-dipole scheme was adopted, setting as a measurement criterion a pulse duration of 250 ms for each cycle and a target of 50 mV for potential readings for each current injection. In particular, L1, L6, L7 lines (Figure 3) were acquired using 48 electrodes with 2 m spacing, for a total length of 94 m and an investigation depth of approximately 18 m. Lines L2-3 and L4-5 (Figure 3) are instead the results of two acquisitions with 72 electrodes, spacing 2 m, for a total length of 142 m, and an investigation depth of about 28 m. For each ERT line, direct and reciprocal measurements were registered by swapping current with potential electrodes, to estimate the errors in the dataset [79].

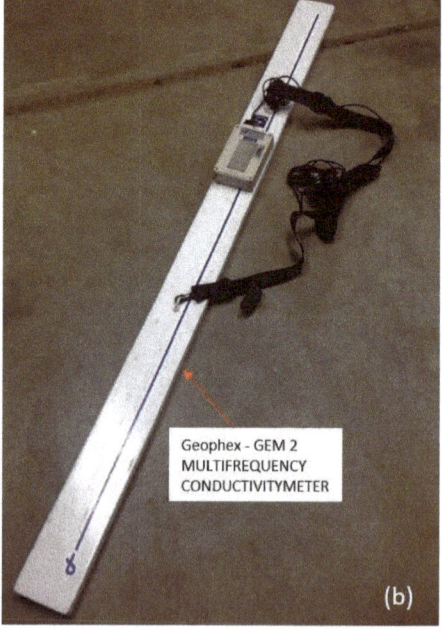

Figure 4. (a) ERT instrumentation during an acquisition at Fondo Paviani site; (b) Multifrequency conductivitymeter used for the measurements at Fondo Paviani site.

A quality factor "Q" (the difference between cycle results) equal to 5% was selected. For the ERT inversion, a regularized weighted least squares approach was adopted, according to the 'Occam's' rule [80] described in detail by LaBrecque et al. [81]. The smoothness of the resistivity distribution here calculated depends on the errors in the data set.

Binley et al. [82] demonstrated that a good evaluation of errors might be obtained by the analysis of these in direct and reciprocal measurements. Theoretically, these measurements shall be equal, providing the same resistance value. Any deviation may be interpreted as an error estimate, quantifying the error parameters useful for the inversion. In the present study, the inverted datasets present errors smaller than 5%. The visualization of the inverted data was done using Surfer 10 software (Golden Software).

2.6. Frequency Domain Electromagnetic (FDEM)

In the frame of the geophysical measurements carried out over a few years at Fondo Paviani, in 2015 a small area was selected with clear traces of potential structures belonging to the perimeter system of the embanked site, visible from aerial and satellite images (Figure 3).

In this area, an extensive FDEM multifrequency measurement was made and integrated by the L6 ERT line acquisition, as shown in Figure 3.

For FDEM data collection, a Geophex GEM-2 multifrequency conductivity meter was used (Figure 4b). The instrument operates with a fixed source-receiver separation of 1.66 m at multiple frequencies ranging from 330 Hz to 48 kHz. Both quadrature and in-phase responses were recorded carrying the instrument in the vertical-dipole configuration at a height of 1 m above the ground surface and using seven frequencies: $f_1 = 775$ Hz, $f_2 = 3775$ Hz, $f_3 = 6275$ Hz, $f_4 = 10,475$ Hz, $f_5 = 17,275$ Hz, $f_6 = 29,025$ Hz, and $f_7 = 47,025$ Hz. The FDEM data were collected with the GEM-2 ski oriented in-line to the walking path, at about 0.5-m sampling interval along with parallel profiles, approximately 2 m apart, in a northwest–southeast direction. Synchronizing the GEM-2 device with a differential GPS receiver (Trimble 5800), UTM coordinates of each FDEM measurement point were recorded with sub-meter accuracy. Raw data were preliminarily analyzed for detecting and removing possible DC (static) shifts, outliers, and short-wavelength noises, which usually adversely affect the quality of the inversions, causing spiking and very different solutions with adjacent soundings. Since the in-phase component of the data revealed too noisy, the only quadrature component of the filtered data was inverted to produce a 1D electrical conductivity profile below each measurement point. Then, the 1D models obtained were stitched together to build a pseudo-3D volume of the investigated area. Inversion was performed using the FDEMtools [83], a free MATLAB software package implementing the numerical algorithms mainly discussed by Deidda et al. [84–86]. A layered starting model consisting of 30 layers, to a depth of 3.5 m, was used to invert the electromagnetic data. During the inversion, the magnetic permeability of the layers, as well as the depth of their interfaces, remain fixed.

The Nonlinear Forward Problem and the Inversion Procedure

The forward modeling used to calculate the nonlinear EM response of a layered half-space (Figure 5) for dipole source excitation is well known [87–89]. It is based on Maxwell's equations, suitably simplified thanks to the cylindrical symmetry of the problem, since the magnetic field sensed by the receiver coil is independent of the rotation of the instrument around the vertical axis. When the axes of the coils are aligned vertically to the ground, the model prediction $M(\sigma, \mu)$, defined as the ratio of the secondary (H_S) to the primary (H_P) EM field, can be expressed in terms of Hankel transforms of order zero:

$$\frac{H_S}{H_P} = \int_0^\infty \lambda^2 e^{-2h\lambda} R_0(\lambda) J_0(r\lambda) d\lambda, \qquad (1)$$

where λ is a variable of integration with no particular physical meaning, h is the height of the instrument above the ground, r the coil separation, J_0 the Bessel function of order 0, and $R_0(\lambda)$ the response kernel, which is a complex value function of the parameters that describe the layered subsurface (i.e., for the k-th layer: the electrical conductivity σ_k, the magnetic permeability μ_k, and the layer thickness d_k) besides the frequency and λ.

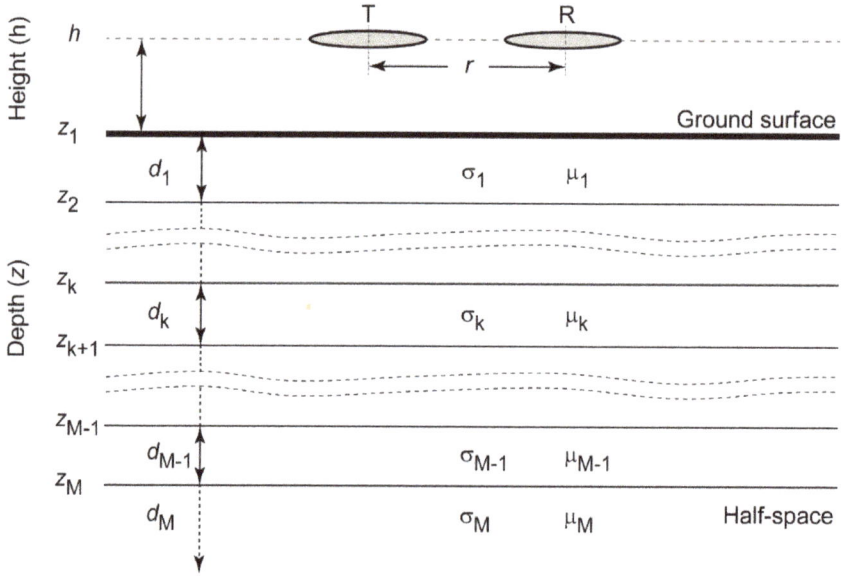

Figure 5. Schematic representation of the subsoil discretization and parameterization along with the coils of the measuring device above the ground.

The nonlinear inversion procedure proposed by Deidda et al. [84–86] is a general procedure that allows the estimation of the electrical properties (electrical conductivity and magnetic permeability) of the subsurface by inverting the complex multi-depth response of different electromagnetic devices designed to record data at multiple coil spacings, using a single frequency, or at multiple frequencies with a fixed coil spacing. Let us suppose that the aim of the survey is estimating the vertical distribution of the electrical conductivity, using a multifrequency electromagnetic dataset (e.g., data recorded with the GEM-2 device). Fixing the magnetic permeability μ_k ($k = 1, \ldots, M$), the best approximation of conductivity values σ_k ($k = 1, \ldots, M$) can be found minimizing the Euclidean norm of the complex residual $\mathbf{r}(\sigma)$ between the data $\mathbf{b}(\omega_i)$ (where ω_i with $i = 1, \ldots, N$, are the operating angular frequencies) and the forward model prediction based on Equation (1), that is,

$$\widehat{\sigma} = \arg\min_{\sigma \in \mathbb{R}^M} \frac{1}{2}\|\mathbf{r}(\sigma)\|^2 = \arg\min_{\sigma \in \mathbb{R}^M} \frac{1}{2}\|\mathbf{b}(\omega_i) - \mathcal{M}(\sigma)\|^2. \tag{2}$$

Alternatively, either the in-phase (real) or the quadrature (imaginary) component of the residual $\mathbf{r}(\sigma)$ could also be minimized. The nonlinear minimization problem is solved with an algorithm based on a damped regularized Gauss–Newton method. Assuming the Fréchet differentiability of $\mathbf{r}(\sigma)$, the problem is linearized at each iteration by means of a first order Taylor expansion with the analytical Jacobian (sensitivity matrix), which makes the computation faster and more accurate than using a finite difference approximation [84,85]. The regularized solution to each linear subproblem is then computed by the truncated generalized singular value decomposition (TGSVD) [90], employing different regularization operators (first and second derivatives).

3. Results

Aerial photo interpretation allowed us to preliminary define both anthropic and natural elements that probably delimited the settlement area of Fondo Paviani. Stratigraphic analysis of the archaeological section confirmed that the light and straight mark, visible from aerial photos in the N-E part of the settlement, correspond to the structure of an impressive artificial rampart. The dark areas E of the rampart correspond to the Menago valley wetlands (Figure 6).

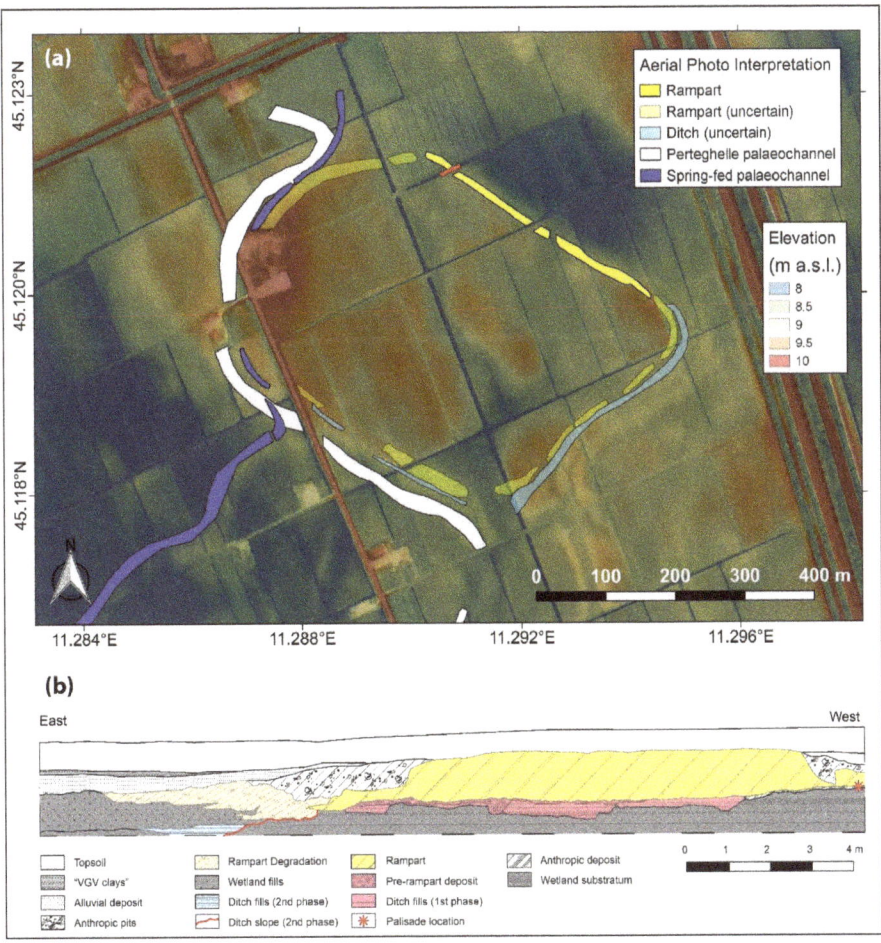

Figure 6. (a) Reconstruction of the fortified Fondo Paviani settlements with probable recent Bronze Age hydrographic network identified from aerial photo-interpretation (orange line: stratigraphic section segment intercepting the perimeter structures (overlapping of DTM obtained from LIDAR with 1990 C.G.R aerial frame). (b) The segment of the stratigraphic section intercepting the perimeter structures.

Even the elevation model (DEM) obtained from LIDAR confirms the coherence of these data with the presence of both the rampart and the Menago valley depression. However, only the stratigraphic analysis of the archaeological section allowed us to fully understand the characteristics of the fortified system and its evolution during the centuries in the first phase, which started in the last decades of the 14th century BC [47], the settlement was not enclosed by a rampart, but probably only by a palisade and a small ditch (Figure 6b) [55]. The wide rampart was built between the 13th and the beginning of the 12th century BC [47]. The structure of this one consisted of a silt-sandy basal core and a top body of fine-textured sediments, built using an imposing internal wooden frame (Figure 6b) [55]. Thanks to the aerial photo interpretation, an alternating sequence of light and dark polylinear marks have been identified along the S–E and S–W sides of the settlement. The outline and shape of these marks are compatible both with anthropic structures, such as ramparts and ditches, and with natural morphologies, such as alluvial ridges and channels. The absence of recent ground checks in these sectors of the settlement prevented from establishing the effective origin (natural or artificial) of these marks. However, the internal light marks, having dimensions comparable to the traces of the N–E rampart, can correspond to the remains of an artificial embankment. The N–W side of the settlement seems to be protected by the natural alluvial ridge embankment offered by the so-called Perteghelle palaeo-channel, visible in the aerial frames, and highlighted by the DTM extracted from LIDAR data.

The twisting linear mark, visible both by aerial photo interpretation and DTM analysis and that cross the settlement from North to South, can be interpreted as a palaeo-channel. Focused stratigraphic analysis has allowed to date this channel to Middle Ages.

The results of the ERT measurements are shown in the images in Figure 7.

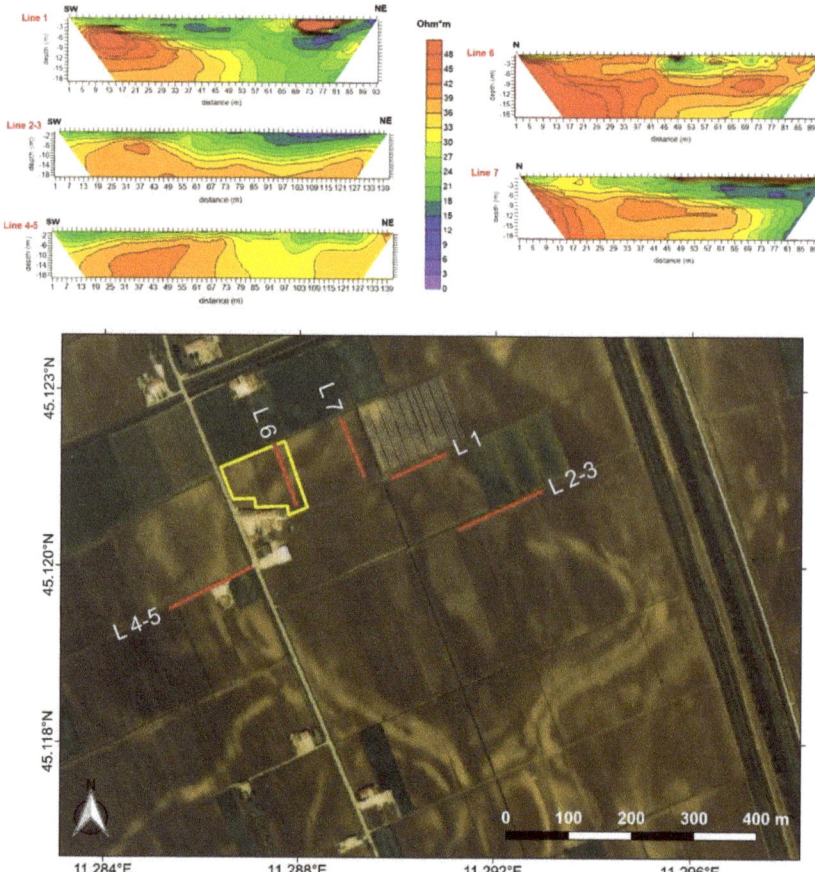

Figure 7. ERT lines results normalized in the range 0–50 Ohm*m with their field localization.

To allow smooth and fast comparison between the different tomographies, the resistivity range of the ERT sections, acquired in different seasons and years, was standardized between 0 and 50 Ohm*m. The ERT sections in Figure 7 show the resistivity distribution in five areas along the perimeter of the Fondo Paviani settlement. The ERT results highlight the current state of the studied embanked system and its high heterogeneity, both lateral and in-depth, from area to area. L1 was collected corresponding to the position of the documented stratigraphic section. In Figure 8, the higher resistivity anomaly (red), localized in the NE part of the L1 section, identifies the preserved rampart. The L1 section was then used as a marker or term of comparison to evaluate, using other ERT sections, the degree of preservation of the boundary system in different areas. In general, ERT L1 (Figures 7 and 8) identifies a transition from more resistive to low resistive values moving from SW to NE in the last part of the section. In particular, a clear conductive deep incision is here registered in the NE part, below the identified high resistive anomaly corresponding to the preserved rampart. The ERT L2-3 section (Figure 7), also collected in the E sector, south of the L1 line and almost parallel to it, shows a low lateral variation of the resistivity compared to L1 with a shallow conductive part. The two other ERT sections L6 and L7 were acquired parallel to each other (Figure 7). These describe a general coherent system, with a more resistive area in the northern part, and the distribution of shallow conductive bodies in the southern part of the section.

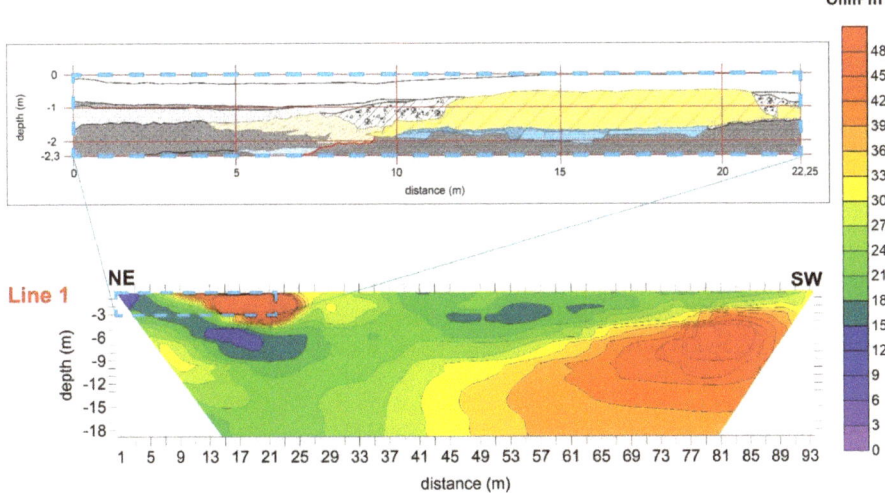

Figure 8. Comparison between mirrored L1 ERT line and stratigraphic section with preserved rampart.

Finally, ERT section L4-5 (Figure 7), collected in the supposed western boundary of the settlement, describes a low lateral variation of the resistivity, similar to this registered in the section L2-3, except to an apparent resistive deep incision. No evidence about the presence of a preserved rampart are registered in all ERT lines from L2-3 to L6. The inversion of FDEM data made it possible to extract a depth map of the distribution of electrical conductivity 2 m below the surface in the area where the ERT L6 was acquired. Figure 9 shows the location of the investigated area with the FDEM method and the relative position of the ERT L6 line. The electrical conductivity data of the FDEM map was then converted into resistivity values to allow us a direct comparison with the ERT L6 line (Figure 9), using its real data range (0–100 Ohm*m) to evaluate the degree of information of these two different methods and their consistency.

The resistivity pattern visible in the FDEM map obtained by the inversion of this dataset, highlights and confirms the presence of the palaeo-system visible from the satellite image (Figure 9a). The ERT L6 section, entirely consistent with the FDEM data, completes with more details the information provided by the EM method about lateral extension and total depth of these buried geomorphological palaeo-structures.

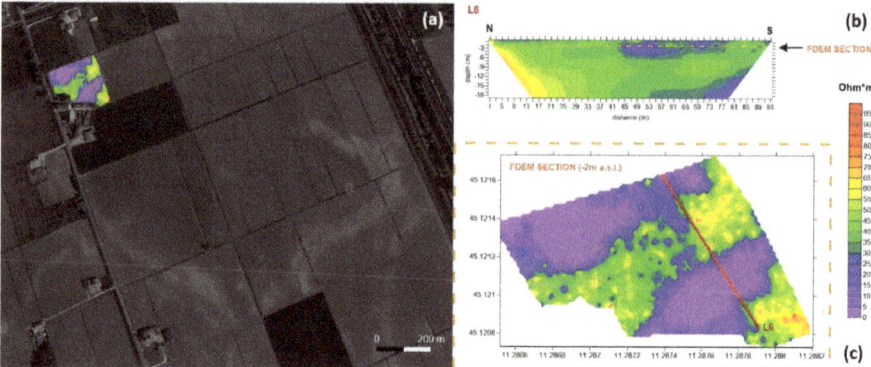

Figure 9. (a) Overlap of FDEM results with ERT L6 position in Google Earth image (2005); (b) ERT L6 and FDEM map (c) at −2 m a.s.l. (with default range of resistivity).

4. Discussion

The archaeological questions underlying the multidisciplinary study conducted in the Bronze Age embanked site of Fondo Paviani and described in this contribution can be summarized in the following points:

(a) Check the state of preservation of the perimetric system in different parts of the site;

(b) Verify the degree of information available by the combination of different geophysical methods measuring the differences in the electrical conductivity of the soils and therefore defining the geometry (size and depth) and the spatial distribution of the palaeo-channels that characterize this embanked system;

(c) Identify the stratigraphic/sedimentation relationships describing the reactivation sequence of older river systems with more recent ones;

(d) Define a multidisciplinary protocol applicable to the whole site, capable of providing clear information on the position of the structures of interest.

To answer the question of point (a), we started from the evidence made by the analysis of some aerial photographs relating to a chronological span of about 40 years, which have also guided the excavations and archaeological reconnaissance for over a decade. Over time, aerial photographs have shown a marked decrease in the contrast, which was previously visible between the traces of paleochannels in the northern perimeter of the site and the hosting system. The stratigraphic section drawn up following the excavation carried out in a part of the perimeter of the system, which is preserved today, constituted the first direct useful data to plan the field strategy for the geophysical measurements carried out with the use of the ERT and the FDEM methods.

The ERT sections acquired in different points of the embankment system of Fondo Paviani allow here to formulate some interesting considerations, made possible thanks to the comparison and calibration of the no-direct data with the direct ones provided by the stratigraphic section. In general, ERT measurements find a wide application and allows us to obtain excellent results in contexts and on systems where there is a marked gradient of the soil water content between the target and the hosting system.

The archaeological site of Fondo Paviani, characterized by the presence of the remains of a housing settlement consisting of a perimeter system, with a moat, characterized by river deposits, and a bright contrast to the hosting system, represents in this sense the ideal case-study for these types of non-invasive measurements. Thanks to the information obtained from the L1 section (Figures 7 and 8), with the identification of the response of the preserved rampart, detected by the stratigraphic section made in the same position (Figure 8), it was, in fact, possible to verify what was hypothesized by the analyses carried out on aerial images and by the LIDAR data. The ERT sections, mainly made in the northern sector of the embanked settlement, confirm the absence of the preserved rampart in this sector in the shallow subsoil, where one would expect to find the same, as in section L1.

Therefore, it can be reasonably said that in the sections corresponding to the ERT lines from L2-3 to L7 the highest and therefore most shallow system relating to the rampart has been removed due to the agricultural practices that have taken place in the last 40 years. This data would also be confirmed by the topographical data extracted by LIDAR (Figure 3), where mean lower altitudes are registered corresponding to the areas in which the ERT sections were acquired. However, from the comparison between the DTM data extracted from LIDAR and the ERT sections, it is possible to make further considerations on the possible correlation between the topographical surface data and the real buried deep geomorphological data. In particular, despite the apparent identical topographic information relating to the NE sector at the two ERT sections L1 and L2-3, parallel to each other, the geophysical data shows that the L1 section intercepts a deep conductive incision where instead the section L2-3 intersects a shallower system. From these real data, it is, therefore, possible to affirm that the topographical data and the overall view, albeit informative of the DTM data, do not always correspond to the real deep-buried structure. The ERT sections undoubtedly provided additional information of extreme interest regarding the geomorphology of the area. In particular, only the lines L1 and L7 seem to intercept some deep incisions characterized by conductive deposits for the whole investigation depth. In contrast, in the other ERT sections, the conductive structures, attributable to the presence of river deposits related to the perimeter system of the dam site, appear less in-depth.

In the areas of the settlement not investigated by geophysical measurements, the DTM analysis allowed us to monitor the state of preservation of the elevated structures (natural or anthropogenic) on a preliminary basis. In particular, in north-eastern and south-western corners of the settlement, an abrupt and regular lowering of the surfaces is visible. These anomalies, considering that they follow exactly agricultural field borders, can be interpreted as the outcome of a strong agricultural impact with possible partial destruction of the ancient landscape features.

Considering the point b), the ERT method, capable of providing data on the lateral and in-depth extension of the targets, could be integrated into vast areas by faster soil electrical conductivity measurements. These can be obtained thanks to the use of multi-frequency FDEM systems that allow the inversion of the data providing detailed information at different depths. Multi-frequency FDEM measurements carried out in a test area, in which the L6 ERT section was acquired, thanks to the inversion of the data, demonstrate absolute consistency with the ERT measurements and, therefore, the possibility of integrating this different kind of data. ERT combined with inverted FDEM data, thus confirming the possibility to quickly and precisely verify the extension of the structures only locally identified with ERT (Figure 9).

The integration between ERT and FDEM methods allows us to also answer question c positively. Undoubtedly, the two geophysical methods allow us to detect the different targets in the subsoil and, therefore, to define the real stratigraphic differences not detectable from the aerial photo images, where the surface signal does not precisely define the nature and the stratigraphic relationship between the different targets.

Finally, the integrated multidisciplinary approach adopted in this study, answers to point (d), highlighting how the information provided by every single method used, although fundamental, may be limited in the absence of a multidisciplinary approach as adopted here. The positive evidence obtained thanks to this type of integration allows us to outline future investigation strategies to be adopted to complete the study of this important embanked site.

5. Conclusions

The multidisciplinary study carried out on the Terramara of Fondo Paviani, with the contribution of remote sensing data, geophysical data, and the analysis of a stratigraphic section, provides new interesting information about this significant settlement of the Late Bronze Age. The first advantage of the combination of remote sensing data with geophysical data is the possibility of reconstructing the real geometry of the buried structures and their spatial distribution and the consequent stratigraphic relationship between them. It is clear that the evidence of the presence of the palaeo-channels, readable with remote sensing data, loses the information about the stratigraphy and, therefore, historical information about the time sequence of these structures. The singular information provided from remote sensing data, especially in the archaeological context, such as the one analyzed here, inserted in an active agricultural system, also does not allow us to define the real state of preservation of the buried structures. Additionally, the remote sensing data clarifies the development dynamics of the settlement even less, based on the exploitation of natural waterways and partly on the construction of artificial structures, that can be clarified only with the contribution of archaeological stratigraphic analysis and partially with geophysical data. The results of the multidisciplinary study carried out at the Terramara of Fondo Paviani, here presented, therefore, allows us to obtain important information on the state of preservation of this settlement, also suggesting some preliminary considerations on its fortification structure and evolution, and proposing new perspectives and new strategies for future research on this important archaeological site.

Author Contributions: Conceptualization, R.D. and M.C.; methodology, R.D., G.P.D., and D.V.; software, R.D. and G.P.D.; validation, R.D. and M.C.; investigation, R.D. and J.B.; resources, R.D., G.P.D. and D.V.; data curation, R.D., G.P.D. and D.V.; writing—original draft preparation, R.D., G.P.D., M.P. and D.V.; writing—review and editing, R.D., G.P.D., M.C., J.B., D.V.; visualization, D.V., R.D., G.P.D.; supervision, R.D. and M.C.; project administration, M.C.; funding acquisition, M.C. and R.D. All authors have read and agreed to the published version of the manuscript.

Funding: This research was funded by national PRIN 20085T5KYN project, by the Department of Cultural Heritage of the University of Padova and by the Municipality of Legnago.

Acknowledgments: The authors are grateful to Elisa Dalla Longa for her scientific support and to Eng. S. De Pietri from Consorzio di Bonifica Veronese—Veneto Region—for granting the use of LIDAR data and for the support during the archeological investigations.

Conflicts of Interest: The authors declare no conflict of interest.

References

1. Clark, A.J. *Seeing Beneath the Soil: Prospecting Methods in Archaeology*; BT Batsford Ltd.: London, UK, 1990; ISBN 0-203-16498-9.
2. Schmidt, A. *Geophysical Data in Archaeology: A Guide to Good Practice*; Oxbow Books Ltd.: Oxford, UK, 2001; ISBN 9781900188715.
3. Kvamme, K.L. Geophysical surveys as landscape archaeology. *Am. Antiq.* **2003**, *68*, 435–457. [CrossRef]
4. Witten, A. *Handbook of Geophysics in Archaeology*; Routledge, Taylor & Francis Group: New York, NY, USA, 2014; ISBN 978-1-904768-60-9.
5. Campana, S.; Piro, S. *Seeing the Unseen. Geophysics and Landscape Archaeology*; Taylor & Francis Group: London, UK, 2009; ISBN 978-0-415-44721-8.
6. Sala, R.; Garcia, E.; Tamba, R. Archaeological Geophysics—From Basics to New Perspectives. In *Archaeology. New Approaches in Theory and Techniques*; Ollich-Castanyer, I., Ed.; IntechOpen Ltd.: London, UK, 2012; pp. 133–166, ISBN 978-953-51-0590-9.
7. Noel, M. Multielectrode Resistivity Tomography for Imaging Archaeology. In *Geoprospection in the Archaeological Landscape*; Spoerry, P., Ed.; Oxbow Books Ltd.: Oxford, UK, 1992; pp. 89–99, ISBN 9780946897421.
8. Walker, A.R. Multiplexed Resistivity Survey at the Roman Town of Wroxeter. *Archaeol. Prospect.* **2000**, *7*, 119–132. [CrossRef]
9. Thacker, P.T.; Ellwood, B.B. Detecting palaeolithic activity areas through electrical resistivity survey: An assessment from Vale de Obidos, Portugal. *J. Archaeol. Sci.* **2002**, *29*, 563–570. [CrossRef]
10. Papadopoulos, N.G.; Tsourlos, P.; Tsokas, G.N.; Sarris, A. Two-dimensional and three-dimensional resistivity imaging in archaeological site investigation. *Archaeol. Prospect.* **2006**, *13*, 163–181. [CrossRef]
11. Burkart, U.; Günther, T.; Rücker, C. Electrical resistivity tomography methods for archaeological prospection. In *Layer of Perception, Proceedings of the 35th International Conference on Computer Applications and Quantitative Methods in Archaeology (CAA), Berlin, Germany, 2–6 April 2007*; Posluschny, A., Lambers, K., Herzog, I., Eds.; Dr. Rudolf Habelt GmbH: Bonn, Germany, 2008.

12. Tsokas, G.N.; Tsourlos, P.I.; Stampolidis, A.; Katsonopoulou, D.; Soter, S. Tracing a major Roman road in the area of Ancient Helike by resistivity tomography. *Archaeol. Prospect.* **2009**, *16*, 251–266. [CrossRef]
13. Mol, L.; Preston, P. The writing's in thewall: A review of new preliminary applications of electrical resistivity tomography within archaeology. *Archaeometry* **2010**, *52*, 1079–1095.
14. Callistus, N.; Akwasi, A.; Danuor, S.; Reginald, N. Delineation of graves using electrical resistivity tomography. *J. Appl. Geophys.* **2016**, *126*, 138–147.
15. Fischanger, F.; Catanzariti, G.; Comina, C.; Sambuelli, L.; Morelli, G.; Barsuglia, F.; Ellaithy, A.; Porcelli, F. Geophysical anomalies detected by electrical resistivity tomography in the area surrounding Tutankhamun' tomb. *J. Cult. Herit.* **2018**, *36*, 63–71. [CrossRef]
16. Papadopoulos, N.; Apostolos, S.; Myeong-Jong, Y.; Jung-Ho, K. Urban archaeological investigations using surface 3D Ground Penetrating Radar and Electrical Resistivity Tomography methods. *Explor. Geophys.* **2009**, *40*, 56–68. [CrossRef]
17. Argote-Espino, D.L.; Lopez-García, P.A.; Tejero-Andrade, A. 3D-ERT geophysical prospecting for the investigation of two terraces of an archaeological site northeast of Tlaxcala state, Mexico. *J. Archaeol. Sci. Rep.* **2016**, *8*, 406–415. [CrossRef]
18. Fernandez, J.P.; Rubio-Melendi, D.; Quirós Castillo, J.A.; González Quirós, A.; Cimadevilla-Fuente, D. Combined GPR and ERT exploratory geophysical survey of the Medieval Village of Pancorbo Castle (Burgos, Spain). *J. Appl. Geophys.* **2017**, *144*, 86–93. [CrossRef]
19. Casas, A.; Cosentino, P.L.; Fiandaca, G.; Himi, M.; Macias, J.M.; Martorana, R.; Muñoz, A.; Rivero, L.; Sala, R.; Teixell, I. Non-invasive Geophysical Surveys in Search of the Roman Temple of Augustus Under the Cathedral of Tarragona (Catalonia, Spain): A Case Study. *Surv. Geophys.* **2018**, *39*, 1107–1124. [CrossRef]
20. Deiana, R.; Bonetto, J.; Mazzariol, A. Integrated Electrical Resistivity Tomography and Ground Penetrating Radar Measurements Applied to Tomb Detection. *Surv. Geophys.* **2018**, *39*, 1081–1105. [CrossRef]
21. Monik, M.; Lend'áková, Z.; Ibáñez, J.; MuñizÁlvarez, J.; Borrell, F.; Iriarte, E.; Teira, L.; Kuda, F. Revealing early villages—Pseudo-3D E.R.T. geophysical survey at the pre-pottery Neolithic site of Kharaysin, Jordan. *Archaeol. Prospect.* **2018**, *25*, 339–346. [CrossRef]
22. Kenady, S.L.; Lowe, K.M.; Ridd, P.V.; Ulm, S. Creating volume estimates for buried shell deposits: A comparative experimental case study using ground-penetrating radar (GPR) and electrical resistivity under varying soil conditions. *Archaeol. Prospect.* **2018**, *25*, 121–136. [CrossRef]
23. McNeill, J.D. *Electrical Conductivity of Soils and Rocks*; Technical Note TN-5; Geonics Limited: Mississauga, ON, Canada, 1980.
24. Baines, D.; Derald, G.S.; Froese, D.G.; Bauman, P.; Grant, N. Electrical resistivity ground imaging (ERGI): A new tool for mapping the lithology and geometry of channel–Belts and valley–Fills. *Sedimentology* **2002**, *49*, 441–449. [CrossRef]
25. Dabas, M.; Tabbagh, A. A comparison of EMI and DC methods used in soil mapping—Theoretical considerations for precision agriculture. In *Precision Agriculture*; Stafford, J., Werner, A., Eds.; Wageningen Academic Publishers: Wageningen, The Netherlands, 2003; pp. 121–127, ISBN 978-90-76998-21-3.
26. Corwin, D.L.; Lesch, S.M. Apparent soil electrical conductivity measurements in agriculture. *Comput. Electron. Agric.* **2005**, *46*, 11–43. [CrossRef]
27. Maillet, G.; Rizzo, E.; Vella, C. High Resolution Electrical Resistivity Tomography (ERT) in a Transition Zone Environment: Application for Detailed Internal Architecture and Infilling Processes Study of a Rhône River Paleo-channel. *Mar. Geophys. Res.* **2005**, *26*, 317–328. [CrossRef]
28. Bersezio, R.; Giudici, M.; Mele, M. Combining sedimentological and geophysical data for high-resolution 3-D mapping of fluvial architectural elements in the Quaternary Po plain (Italy). *Sediment. Geol.* **2007**, *202*, 230–248. [CrossRef]
29. Zhu, Q.; Lin, H.; Doolittle, J. Repeated electromagnetic induction surveys for improved soil mapping in an agricultural landscape. *Soil Sci. Soc. Am. J.* **2010**, *74*, 1763–1774. [CrossRef]
30. Scapozza, C.; Laigre, L. The contribution of Electrical Resistivity Tomography (ERT) in Alpine dynamics geomorphology: Case studies from the Swiss Alps. *Géomorph. Relief Process. Environ.* **2014**, *20*, 27–42. [CrossRef]
31. Van Dam, R.L. Landform characterization using geophysics—Recent advances, applications, and emerging tools. *Geomorphology* **2012**, *137*, 57–73. [CrossRef]

32. Fidolini, F.; Ghinassi, M.; Aldinucci, M.; Billi, P.; Boaga, J.; Deiana, R.; Brivio, L. Fault-sourced alluvial fans and their interaction with axial fluvial drainage: An example from the Plio-Pleistocene Upper Valdarno Basin (Tuscany, Italy). *Sediment. Geol.* **2013**, *289*, 19–39. [CrossRef]
33. Papadopoulos, N.G.; Sarris, A.; Parkinson, W.A.; Gyucha, A.; Yerkes, R.W.; Duffy, P.R.; Tsourlos, P. Electrical Resistivity Tomography for the Modelling of Cultural Deposits and Geomophological Landscapes at Neolithic Sites: A Case Study from Southeastern Hungary. *Archaeol. Prospect.* **2014**, *21*, 169–183. [CrossRef]
34. Bianchi, V.; Ghinassi, M.; Aldinucci, M.; Boaga, J.; Brogi, A.; Deiana, R. Tectonically driven deposition and landscape evolution within upland incised valleys: Ambra Valley fill, Pliocene-Pleistocene, Tuscany, Italy. *Sedimentology* **2015**, *62*, 897–927. [CrossRef]
35. Deiana, R.; Dieni, I.; Massari, F.; Perri, M.T.; Rossi, M.; Brovelli, A. A multidisciplinary study of deformation of the basaltic cover over fine-grained valley fills: A case study from Eastern Sardinia, Italy. *Int. J. Earth Sci.* **2016**, *105*, 1245–1255. [CrossRef]
36. Van Der Kruk, J.; Von Hebel, C.; Brogi, C.; Kaufmann, M.S.; Tan, X.; Weihermüller, L.; Huisman, J.A.; Vereecken, H.; Mester, A. *Calibration, Inversion, and Applications of Multiconfiguration Electromagnetic Induction for Agricultural Top- and Subsoil Characterization*. S.E.G.; Technical Program Expanded Abstracts; Society of Exploration Geophysicists: Tulsa, OK, USA, 2018; pp. 2546–2550.
37. Khan, I.; Sinha, R. Discovering 'buried' channels of the Palaeo-Yamuna river in NW India using geophysical evidence: Implications for major drainage reorganization and linkage to the Harappan Civilization. *J. Appl. Geophys.* **2019**, *167*, 128–139. [CrossRef]
38. Lasaponara, R.; Coluzzi, R.; Gizzi, F.T.; Masini, N. On the LiDAR contribution for the archaeological and geomorphological study of a deserted medieval village in Southern Italy. *J. Geophys. Eng.* **2010**, *7*, 155–163. [CrossRef]
39. Alexakis, D.; Agapiou, A.; Hadjimitsis, D.; Sarris, A. Remote Sensing Applications in Archaeological Research. In *Remote Sensing*; Escalante, B., Ed.; IntechOpen Ltd.: London, UK, 2012; pp. 435–462, ISBN 978-953-51-0651-7.
40. Kasprzak, M.; Traczyk, A. LiDAR and 2D Electrical Resistivity Tomography as a Supplement of Geomorphological Investigations in Urban Areas: A Case Study from the City of Wrocław (SW Poland). *Pure Appl. Geophys.* **2014**, *171*, 835–855. [CrossRef]
41. Elmahdy, S.; Mohamed, M.M. Remote sensing and geophysical survey applications for delineating near surface paleochannels and shallow aquifer in the United Arab Emirates. *Geocarto Int.* **2015**, *30*, 723–736. [CrossRef]
42. Křivánek, R. Comparison Study to the Use of Geophysical Methods at Archaeological Sites Observed by Various Remote Sensing Techniques in the Czech Republic. *Geosciences* **2017**, *7*, 81. [CrossRef]
43. Cozzolino, M.; Longo, F.; Pizzano, N.; Rizzo, M.L.; Voza, O.; Amato, V. A Multidisciplinary Approach to the Study of the Temple of Athena in Poseidonia-Paestum (Southern Italy): New Geomorphological, Geophysical and Archaeological Data. *Geosciences* **2019**, *9*, 324. [CrossRef]
44. Monterroso-Checa, A.; Teixidó, T.; Gasparini, M.; Peña, J.A.; Rodero, S.; Moreno, J.C.; Morena, J.A. Use of Remote Sensing, Geophysical Techniques and Archaeological Excavations to Define the Roman Amphitheater of Torreparedones (Córdoba, Spain). *Remote Sens.* **2019**, *11*, 2937. [CrossRef]
45. Bernabò Brea, M.; Mutti, A. " … *Le Terramare di scavano per concimare i prati* … ". *La nascita dell'archeologia preistorica a Parma nella seconda metà dell'Ottocento*; Silva: Parma, Italy, 1994; ISBN 978-8877650474.
46. Bernabò Brea, M.; Cardarelli, A.; Cremaschi, M. *Le Terramare, La più Antica Civiltà Padana, 1997*; Catalogo della Mostra: Milano, Italy, 1994.
47. Available online: https://www.tes.com/lessons/u_uWq0a1bZw7Ew/i-popoli-italici-prima-dei-romani (accessed on 15 May 2020).
48. Peroni, R. *Introduzione Alla Protostoria Italiana*; Editori Laterza: Roma-Bari, Italy, 1994.
49. Peroni, R. *L'Italia Alle Soglie Della Storia*; Editori Laterza: Roma-Bari, Italy, 1996.
50. Cardarelli, A.; Vanzetti, A. L'approccio di Renato Peroni allo studio delle società protostoriche dalla fine degli anni '60 del XX secolo. In *150 anni di Preistoria e Protostoria in Italia, "Studi di Preistoria e Protostoria"* 1; Guidi, A., Ed.; Istituto Italiano di Preistoria e Protostoria: Firenze, Italy, 2014; ISBN 978-8860450555.
51. Cardarelli, A. The collapse of the Terramare Culture and growth of new economic and social systems during the Late Bronze Age in Italy. *Sci. dell'Antichità* **2010**, *15*, 449–520.

52. Cupitò, M.; Leonardi, G. Il Veneto tra Bronzo antico e Bronzo recente. In *Preistoria e Protostoria del Veneto*; Leonardi, G., Tinè, V., Eds.; Studi di Preistoria e Protostoria 2; Istituto Italiano di Preistoria e Protostoria: Firenze, Italy, 2015; pp. 201–240.
53. Mele, M.; Cremaschi, M.; Giudici, M.; Bassi, A.; Pizzi, C.; Lozej, A. Reconstructing Hidden Landscapes. DC and EM Prospections in the Terramara Santa Rosa (Bronze Age Settlement—Northern Italy). Geophysical Research Abstracts; EGU: Vienna, Austria, 2012; Volume 14, p. 5269.
54. Mele, M.; Cremaschi, M.; Giudici, M.; Lozej, A.; Pizzi, C.; Bassi, A. The Terramare and the surrounding hydraulic structures: A geophysical survey of the Santa Rosa site at Poviglio (Bronze Age, Northern Italy). *J. Archaeol. Sci.* **2013**, *40*, 4648–4662. [CrossRef]
55. Cupitò, M. Micenei in Italia settentrionale. In *Le Grandi Vie Delle Civiltà. Relazioni e Scambi Fra Mediterraneo e il Centro Europa Dalla Preistoria Alla Romanità*; Marzatico, F., Gebhard, R., Gleirscher, P., Eds.; Catalogo della Mostra, Trento, Castello del Buonconsiglio, 1 July–13 November 2011; Museo Castello Buonconsiglio: Milano, Italy, 2011; pp. 193–197, ISBN 978-8890090981.
56. Cupitò, M.; Leonardi, G.; Dalla Longa, E.; Nicosia, C.; Balista, C.; Dal Corso, M.; Kirleis, W. Fondo Paviani (Legnago, Verona): Il central place della polity delle Valli Grandi Veronesi nella tarda Età del bronzo. Cronologia, aspetti culturali, evoluzione delle strutture e trasformazioni paleoambientali. In *Preistoria e Protostoria del Veneto*; Leonardi, G., Tiné, V., Eds.; Studi di Preistoria e Protostoria: Firenze, Italy, 2015; pp. 357–375.
57. Bettelli, M.; Cupitò, M.; Levi, S.T.; Jones, R.; Leonardi, G. Tempi e modi della connessione tra mondo egeo e area padano-veneta. Una riconsiderazione della problematica alla luce delle nuove ceramiche di tipo miceneo di Fondo Paviani (Legnago, Verona). In *Preistoria e Protostoria del Veneto*; Leonardi, G., Tiné, V., Eds.; Studi di Preistoria e Protostoria: Firenze, Italy, 2015; pp. 377–387.
58. Bettelli, M.; Cupitò, M.; Jones, R.; Leonardi, G.; Levi, S.T. The Po Plain, Adriatic and Eastern Mediterranean in the Late Bronze Age: Fact, fancy and plausibility. In *Hesperos. Aegean Seen from the West, Proceedings of the 16th International Aegean Conference, Ioannina, Greece, 18–21 May 2016*; Fotiadis, M., Laffineur, R., Lolos, Y., Vlachopoulos, A., Eds.; Aegeum; Annales ligéoises et PASPiennes d'archéologie égéenne, 2017, 41; Peeters Publisher: Leuven-Liége, Belgium, 2017; pp. 165–172, ISBN 978-90-429-3562-4.
59. Cupitò, M.; Leonardi, G. Il sito arginato di Fondo Paviani e la polity delle Valli Grandi Veronesi prima e dopo il collasso delle terramare. Nuovi dati per una riconsiderazione del probleme. In *Preistoria e Protostoria dell'Emilia Romagna*; BernabòBrea, M., Ed.; Studi di Preistoria e Protostoria: Firenze, Italy, 2018; Volume II, pp. 175–186.
60. Cupitò, M.; Dalla Longa, E.; Donadel, V.; Leonardi, G. Resistances to the 12th century BC crisis in the Veneto region: The case studies of Fondo Paviani and Montebello Vicentino. In *Collapse or Continuity? Environment and Development of Bronze Age Human Landscapes, Proceedings of the International Workshop Socio-environmental Dynamics over the last 12,000 Years: The Creation of Landscapes II, Kiel, Germany, 14–18 March 2011*; Kneisel, W., Dal Corso, M., Taylor, N., Tiedtke, V., Eds.; Universitätforschungen zur Prähistorischen Archäologie; Dr. Rudolf Habelt GmbH: Bonn, Germany, 2011; pp. 55–70.
61. Cupitò, M.; Dalla Longa, E.; Balista, C. From "Valli Grandi Veronesi system" to "Frattesina system". Observations on the evolution of the exchange system models between Veneto Po Valley area and the Mediterranean world during the Late Bronze, Rivista di Scienze Preistoriche—Numero speciale, Italia tra Mediterraneo ed Europa. *Mobilità Interazioni Scambi* **2016**, 285–302, in press.
62. Bernabò Brea, M.; Cremaschi, M. La terramara di S. Rosa di Poviglio: Le strutture. In *Le Terramare: La Più Antica Civiltà Padana*; Bernabò Brea, M., Cardarelli, A., Cremaschi, M., Eds.; Mondadori Electa: Milano, Italy, 1997; pp. 196–212, ISBN 978-8843560622.
63. Cupitò, M. Dinamiche costruttive e di degrado del sistema aggere-fossatodella terramara di Castione dei Marchesi (Parma). Rilettura e reinterpretazione dei dati ottocenteschi. *Riv. Sci. Preist.* **2012**, *62*, 231–248.
64. Balista, C.; De Guio, A. Ambiente ed insediamenti dell'età del bronzo nelle Valli Grandi Veronesi. In *Le Terramare. La Più Antica Civiltà Padana*; Bernabò Brea, M., Cardarelli, A., Cremaschi, M., Eds.; Mondadori Electa: Milano, Italy, 1997; pp. 137–195, ISBN 978-8843560622.
65. Balista, C. Le risposte del sistema paleoidrografico di risorgiva delle Valli Grandi Veronesi meridionali alle fluttuazioni climatiche tardo-oloceniche e agli impianti antropici legati ai cicli insediativi dell'età del bronzo, di età romana e di età tardorinascimentale-moderna. *Padusa* **2009**, *45*, 73–131.

66. Dalla Longa, E.; Dal Corso, M.; Vicenzutto, D.; Nicosia, C.; Cupitò, M. The Bronze Age settlement of Fondo Paviani (Italy) in its territory. Hydrography, settlement distribution, environment and in-site analysis. *J. Archael. Sci. Rep.* **2019**, *28*, 102018. [CrossRef]
67. Balista, C. La fine dell'età del bronzo ed i processi di degrado dei suoli innescati dai reinsediamenti della prima età del ferro e dai deterioramenti climatici del Sub-atlantico al margine settentrionale delle Valli Grandi Veronesi (il caso studio di Perteghelle di Cerea-VR). *Padusa* **2006**, *42*, 45–127.
68. De Guio, A.; Whitehouse, R.; Wilkins, J. Il progetto Alto-Medio Polesine-Basso Veronese. In *Dalla Terra al Museo, Mostra di reperti Preistorici e Protostorici degli ultimi dieci anni di ricerca dal Territorio Veronese*; Belluzzo, G., Salzani, L., Eds.; Fondazione Fioroni: Legnago, Italy, 1996; pp. 283–285.
69. Balista, C.; De Guio, A. Il sito di Fabbrica dei Soci (Villabartolomea—VR): Oltre la superficie. *Padusa* **1990**, *26–27*, 9–85.
70. Balista, C.; Bortolami, F.; Marchesini, M.; Marvelli, S. Terrapieni a protezione dei campi dall'invasione delle torbiere nelle Valli Grandi Veronesi nell'età del Bronzo Medio-Recente. *Ipotesi Preist.* **2016**, *8*, 53–102.
71. De Guio, A.; Baldo, M.; Balista, C.; Bellintani, P.; Betto, A. Tele-Frattesina: Alla ricerca della firma spettrale della complessità. *Padusa* **2009**, *45*, 133–167.
72. Gonzales, R.C.; Woods, R.E. *Digital Image Processing*, 3rd ed.; Prentice Hall: Upper Saddle River, NJ, USA, 2008; ISBN 978-0131687288.
73. Campana, S.; Forte, M. *Remote Sensing in Archaeology. XI ciclo di lezioni sulla Ricerca applicata in Archeologia*; All'insegna del Giglio: Firenze, Italy, 2001.
74. Lock, G.R. *Using Computers in Archaeology. Towards Virtual Pasts*; Taylor & Francis Group: New York, NY, USA, 2003.
75. Balista, C.; De Guio, A.; Ferri, R.; Vanzetti, A. Geoarcheologia delle Valli Grandi Veronesi e Bonifica Padana (Rovigo): Uno scenario evolutivo. In *Tipologia di Insediamento e Distribuzione Antropica Nell'area Veneto-Istriana Dalla Protostoria all'Alto Medioevo*; Cassola Guida, P., Borgna, E., Pettarin, S., Eds.; Edizioni della Laguna: Mariano del Friuli, Italy, 1992; pp. 111–123, ISBN 8885296335.
76. Leonardi, G. Processi Formativi Della Stratificazione Archeologica. In Proceedings of the Atti del Seminario Internazionale Formation Processes and Excavation Methods in Archaeology: Perspectives, Padova, Italy, 15–27 July1991; Università degli Studi di Padova: Padova, Italy, 1992. Saltuarie del Piovego 3, Imprimitur s.n.c., Padova.
77. Balista, C.; Cupitò, M.; Dalla Longa, E.; Leonardi, G.; Nicosia, C. Il Sito Arginato Dell'età Del Bronzo di Fondo Paviani (Legnago). Campagna di scavo 2011. In *Quaderni di Archeologia del Veneto*; All'Insegna del Giglio: Sesto Fiorentino, Italy, 2012; Volume 18, pp. 91–96.
78. Dal Corso, M.; Nicosia, C.; Balista, C.; Cupitò, M.; Dalla Longa, E.; Leonardi, G.; Kirleis, W. Bronze Age crop processing evidence in the phytolith assemblages from the ditch and fen around Fondo Paviani, Northern Italy. *Veg. Hist. Archaeobotany* **2016**, *26*, 5–24. [CrossRef]
79. Daily, W.A.; Ramirez, A.; Binley, A.; LaBrecque, D. Electrical resistivity tomography. *Lead. Edge* **2004**, *23*, 438–442. [CrossRef]
80. DeGroot-Hedlin, C.; Constable, S. Occam's inversion to generate smooth, two-dimensional models from magnetotelluric data. *Geophysics* **1990**, *55*, 1613–1624. [CrossRef]
81. LaBrecque, D.J.; Morelli, G.; Daily, W.; Ramirez, A.; Lundegard, P. Occam's inversion of 3D ERT data. In *Three-Dimensional Electromagnetics*; Spies, B., Ed.; SEG: Tulsa, OK, USA, 1999; pp. 575–590, ISBN 978-1-56080-079-8.
82. Binley, A.; Ramirez, A.; Daily, W. Regularised image reconstruction of noisy electrical resistance tomography data. In *Process Tomography, Proceedings of the 4th Workshop of the European Concerted Action on Process Tomography, Bergen, Norway, 6–8 April 1995*; Beck, M.S., Hoyle, B.S., Morris, M.A., Waterfall, R.C., Williams, R.A., Eds.; University of Manchester Institute of Science and Technology: Manchester, UK, 1995; pp. 401–410.
83. Deidda, G.P.; Díaz de Alba, P.; Fenu, C.; Lovicu, G.; Rodriguez, G. FDEMtools: A MATLAB package for FDEM data inversion. *Numer. Algorithms* **2020**, 1313–1327. [CrossRef]
84. Deidda, G.P.; Fenu, C.; Rodriguez, G. Regularized solution of a nonlinear problem in electromagnetic sounding. *Inverse Probl.* **2014**, *30*, 125014. [CrossRef]
85. Deidda, G.P.; Díaz de Alba, P.; Rodriguez, G. Identifying the magnetic permeability in multi-frequency EM data inversion. *Electron Trans. Numer. Anal.* **2017**, *47*, 1–17. [CrossRef]

86. Deidda, G.P.; Díaz de Alba, P.; Rodriguez, G.; Vignoli, G. Inversion of Multiconfiguration Complex EMI Data with Minimum Gradient Support Regularization: A Case Study. *Math. Geosci.* **2020**, 1–26. [CrossRef]
87. Wait, J.R. *Geo-Electromagnetism*; Academic Press: New York, NY, USA, 1982; ISBN 978-0-12-730880-7.
88. Ward, S.H.; Hohmann, G.W. Electromagnetic theory for geophysical applications Electromagnetic Methods in Applied Geophysics. Volume 1: Theory. In *Investigation in Geophysics*; Nabighian, M.N., Ed.; Society of Exploration Geophysicists: Tulsa, OK, USA, 1987; Volume 3, pp. 131–311, ISBN 978-0-931830-51-8.
89. Hendrickx, J.M.H.; Borchers, B.; Corwin, D.L.; Lesch, S.M.; Hilgendorf, A.C.; Schlue, J. Inversion of soil conductivity profiles from electromagnetic induction measurements. *Soil Sci. Soc. Am. J.* **2002**, *66*, 673–685.
90. Díaz de Alba, P.; Rodriguez, G. Regularized inversion of multi-frequency EM data in geophysical applications. In *Trends in Differential Equations and Applications*; Ortegón, G.F., Redondo, N.M., Rodríguez, G.J., Eds.; Springer: Basel, Switzerland, 2016; Volume 8, pp. 357–369, ISBN 978-3-319-32012-0.

© 2020 by the authors. Licensee MDPI, Basel, Switzerland. This article is an open access article distributed under the terms and conditions of the Creative Commons Attribution (CC BY) license (http://creativecommons.org/licenses/by/4.0/).

Article

OpenSWAP, an Open Architecture, Low Cost Class of Autonomous Surface Vehicles for Geophysical Surveys in the Shallow Water Environment

Giuseppe Stanghellini [1], Fabrizio Del Bianco [2] and Luca Gasperini [1,*]

1. ISMAR-CNR, Istituto di Scienze Marine, U.O. Geologia Marina, CNR, Via Gobetti 101, 40129 Bologna, Italy; giuseppe.stanghellini@ismar.cnr.it
2. Proambiente Scrl, Via Gobetti 101, 40129 Bologna, Italy; f.delbianco@consorzioproambiente.it
* Correspondence: luca.gasperini@ismar.cnr.it

Received: 14 July 2020; Accepted: 8 August 2020; Published: 11 August 2020

Abstract: OpenSWAP is a class of innovative open architecture, low cost autonomous vehicles for geological/geophysical studies of shallow water environments. Although they can host different types of sensors, these vehicles were specifically designed for geophysical surveys, i.e., for the acquisition of bathymetric and stratigraphic data through single- and multibeam echosounders, side-scan sonars, and seismic-reflection systems. The main characteristic of the OpenSWAP vehicles is their ability of following pre-defined routes with high accuracy under acceptable weather and sea conditions. This would open the door to 4D (repeated) surveys, which constitute a powerful tool to analyze morphological and stratigraphic changes of the sediment/water interface and of the shallow substratum eventually caused by sediment dynamics (erosion vs. deposition), slumps and gravitative failures, earthquakes (slip along seismogenic faults and secondary effects of shaking), tsunamis, etc. The low cost and the open hardware/software architectures of these systems, which can be modified by the end users, lead for planning and execution of cooperative and adaptive surveys with different instruments not yet implemented or tested. Together with a technical description of the vehicles, we provide different case studies where they were successfully employed, carried out in environments not, or very difficultly accessed through conventional systems.

Keywords: Autonomous Surface Vehicles (ASV); marine geophysics; shallow water environments; repeated 4D surveys; NAIADI Project (New Autonomous/automatIc systems for the study AnD monitoring of aquatic envIronments)

---

1. Introduction

Natural or artificial shallow water environments, such as harbors, coastal areas, waterways, lakes and lagoons, are in general affected by anthropogenic pressures. For this reason, they would require periodic monitoring, to mitigate the effects of environmental crises caused by human activity or natural processes. However, to date, geophysical studies in shallow water areas (shallower than a few meters) are not a consolidated practice for various reasons, including the following: they present difficult access, even using small boats, in absence of accurate bathymetric maps, the shallow water represents an efficient waveguide for acoustic and ultrasonic noises that limits penetration of the signals into the substrate and the quality of echographic and seismic data, the effect of noise due to propellers, or other natural and artificial causes, is amplified, and the rapidity of environmental changes would require repeated investigations (4D), which is not economically viable with conventional methods.

The economic and social importance of shallow water environments, therefore, calls for the development of new technologies and methods, that could open their study to a wider range of researchers and environmental protection agencies: progresses and developments in the field of marine robotics could be an interesting opportunity to achieve this goal. In fact, the relatively recent availability of miniaturized although accurate sensors, as well as the development of innovative hardware architectures (Arduino®, Raspberry™, etc.) simplify design and implementation of low cost but highly performing Autonomous Surface Vehicles (ASV), which can operate in a variety of aquatic environments. This is the case of SWAP (Shallow Water Autonomous Prospector) a class of vehicles developed by ISMAR-CNR and Proambiente Scrl, characterized by limited size, high versatility, and low cost, and operating with a variety of different payloads. OpenSWAP is the follow-up of that original project, developed under NAIADI (New Autonomous/automatIc systems for the study AnD monitoring of aquatic envIronments, https://www.consorzioproambiente.it/en/projects/terminati/51-naiadi-por-fesr-2014-2020), in the frame of a POR-FESR Emilia Romagna initiative [1]. The intensive use of "open" technologies and software packages for data acquisition and processing [2–4], as well as the low cost of production, have the potential to extend the use of these techniques and methods to a growing public of scientists studying geological processes in these rapidly changing environments. Although to date several ASV are available on the market, we believe that OpenSwap is innovative in many respects, as summarized in Table 1.

Table 1. Main characteristics of the OpenSwap vehicles.

Open HW/SW architectures, widely documented and easily modified by the end-users
Innovative hulls design which optimize navigation and data acquisition
Embedded geophysical instrumentation, including a single-beam echosounder (with full echogram recording), and a "chirp" sub-bottom profiler
High accuracy (within +/−30 cm) in repeating programmed navigation lines
Modular HW/SW interfaces which enable for integrating thirty-party instruments, both for navigation and data acquisition

Together with a technical description of the vehicles, we present here some examples of data acquisition, which include single-beam echograms, side-scan sonar images, seismic reflection profiles, as well as multibeam data from different shallow water environments.

2. OpenSWAP, Philosophy and Motivations

OpenSWAP is a class of ASVs (autonomous surface vehicles) developed with the aim of providing flexible and easily operating autonomous aquatic vehicles (Figure 1), from both hardware and software point of view, allowing to perform data acquisition in the shallow water environment.

Figure 1. An OpenSWAP vehicle during a test phase.

Although these ASVs are suitable for a number of different payloads, including video cameras, current-meters, chemical and physical water sensors, water samplers, etc., we focused their design on acquisition of marine bathymetric and seismo-stratigraphic data. Our first target was implementing the ability of performing repeated surveys, i.e., following navigation paths with centimeters accuracy during subsequent runs. In fact, such performances are mandatory to analyze and monitor time-variant environmental processes and variables. Another functionality considered important was the possibility of planning in advance every practical aspect of a survey, such as: type of sensor employed and coverage to be obtained in a study area; routes to be followed by the vehicle during the run, avoiding obstacles; acquisition parameters (data sampling rate, maximum depths, etc.); time needed to complete the survey, in relation to batteries duration. This would allow for optimizing workflows in logistically difficult environments, reducing risks and deployment time. The development of the OpenSWAP vehicles was focused from the very beginning considering implementation of two embedded geophysical sensors: (1) a single-beam echosounder (SBE), to perform bathymetric (repeated) surveys, and (2) a chirped sub-bottom profiler (SBP), allowing for the acquisition of high-resolution stratigraphic data.

The vehicle was designed small enough to be easily transported and deployed, also in difficult conditions (steep shores; absence of docks; waters too shallow; etc.), but large enough to host suitable payloads and batteries. In a short summary, the design of the ASV was performed considering the following characteristics: housing as many batteries as possible, in relation to floating and navigation performances, to ensure long, self-running surveys with minimum returns to the base point; hosting in a safe waterproof container the on-board electronics; gathering implementation of different propulsion systems (i.e., waterborne vs. aerial propellers); providing additional space for hosting most common third-party sensors. Other features considered important were: low acoustic noise and low water turbulence induced by propellers and hulls close to the acoustic sensors; compatibility with AI (artificial intelligence) stay-on-route algorithms, to gather repeated surveys with centimeters errors, also in presence of water and air turbulence caused by waves, currents, and/or strong wind under acceptable weather and sea conditions. During our tests, we have found that navigation paths are followed up to Beaufort 4, but data quality under such conditions is very poor. We conclude that Beaufort 3 would be the limit for single-beam acoustic surveys.

The final version of the vehicles should have been easily customizable, allowing for installation of additional proprietary sensors, such as commercially available multibeam echosounders (MBES), SBP, water samplers etc.

To obtain such performances, the main electronic board has been developed to include several low-level serial RS232 I/O ports, as well as higher level network supports (i.e., Ethernet, Wi-Fi etc.), allowing for interfacing any proprietary instruments with the in-house positioning systems in the frame of a local-area network architecture.

The software and firmware routines were developed in form of separate modules, to allow the end-users implementing new functionalities through the use of internal scripting. This feature was improved to enable low-level software customizations, including new navigation algorithms, to enhance, for example, navigation accuracy under special condition/environment, or to adapt the current course to incoming parameters (wind speed and direction, wave height, etc.). Finally, the choice of an open hardware architecture for the electronics allowed to reduce costs, favoring distribution of those parts not available on the consumer market, also in the form of self-assembled kits (see next section).

3. Design of the Vehicles

3.1. Nautical Aspects

For several reasons, we decided to base the design of the vehicles on a multi hull scheme, and in particular a catamaran. In fact, compared to mono hull boats, catamarans are characterized by: wider beams, which increase stability; shallower draught, a major plus in shallow waters; smaller hydrodynamic resistance, implying less power delivered to the propellers. The design was carried out also considering some specific requirements of geophysical surveys, which employ acoustic and ultrasonic signals generated and received by transducers that should be perfectly coupled with the water. For these reasons, the vehicles would have been able to minimize the tradeoff between "speed of the vehicle", which influences the time required to complete a survey, and "low acoustic noise in the water", caused by turbulence generated by propellers and viscous drags along the hull. The first technical solution to limit this negative effect, particularly evident in shallow waters, was installing the acoustic sensors between the two hulls of the catamaran, in an area characterized by relatively low water turbulence and shielded by the bubble carpet produced by propellers.

The main task of the OpenSWAP vehicles is performing geophysical surveys for study morphology and stratigraphy of the seafloor and penetrate the first tens of meters in unconsolidated or poorly consolidated sediments. For such objectives, acoustic and ultrasonic transducers are generally used, with alternating emission and detection of acoustic signal towards the bottom and below. Two instruments constitute very basic sensors for such surveys: the SBE (10^1–10^2 kHz), for accurate determination of depth and bottom reflectivity; the SBP, generating lower frequency (10^0–10^1 kHz) impulsive or frequency-modulated (chirped) signals, penetrating the subbottom and being reflected by acoustic impedance contrasts.

These instruments, particularly the SBP, are very sensitive to acoustic and electric noise, as well as to the presence of air bubbles in the water originated by turbulence close to the hulls or by propellers. To minimize such effects, we developed a catamaran with asymmetric hulls. Once designed, the fluid dynamic behavior was CFD (computational fluid dynamics) simulated, to iteratively optimize their shapes and to determine positions affected by minimum noise. As shown in Figure 2, a CDF plot of turbulence simulated at 7 km/h (3.7 knots) speed, the region between the hulls is very silent, and hence the best place for deploying the acoustic transducers. At higher speeds, noise slightly increases, but the region between the hulls remains the most favorable.

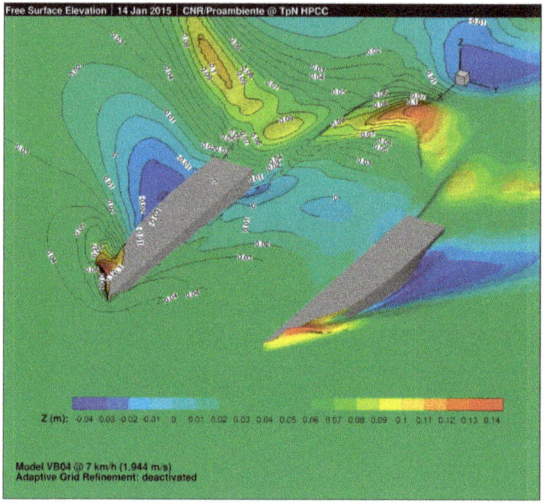

Figure 2. CFD simulation of the hulls behavior during the design phase of the vehicle at 7 km/h of speed. Note minimum turbulence between the two hulls.

The drawback of this design is that a catamaran with asymmetric hulls shows a slight stronger friction relative to the symmetric ones, but a very low noise between the hulls (Figure 2).

The design of hulls and vessel containing the electronics was carried out through the implementation of different standalone vehicles, whose tests carried out during different acquisition trials led to a final design implementing an electric powered plastic catamaran, made by linear low-density polyethylene (LLDPE) with two asymmetrical lateral hulls and a central case housing the electronics (Figure 3). The frame is made of aluminum profiles, which provide connection between different parts of the vehicle and could be used as supports to deploy other instruments and sensors within the low noise area between the hulls.

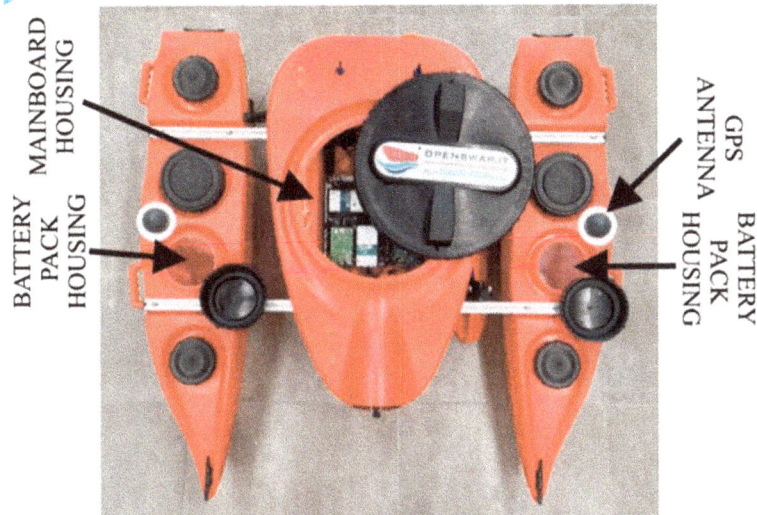

Figure 3. Photo and technical details of the latest OpenSWAP vehicle.

3.2. Electronic Boards

The "core" of the OpenSWAP hardware is constituted by a single-board multilayer PCB (printed circuit board), displayed in Figure 4, designed and developed to host embedded subsystems and boards (Arduino®, Raspberry™, Xbee, etc.). It provides all connections and power supplies to electronic boards and devices, as well as signal conditioning circuits and I/O ports. Several jumper pins are included to quickly modify settings and parameters according to special setups and/or payloads. This is the only electronic component not available in the consumer market and distributed by Proambiente (https://www.openswap.it/).

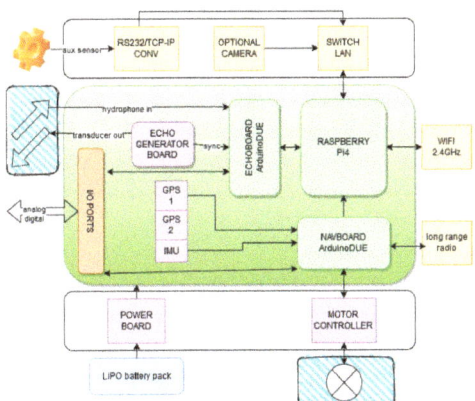

Figure 4. *Cont.*

Figure 4. Block-diagram (**top**) and photo (**bottom**) of the OpenSWAP mainboard.

3.3. Software Architecture

The software routines which control OpenSWAP are divided into two main groups, those for navigation and those for data acquisition. A block-model of the software architecture is displayed in Figure 5, and described in detail below.

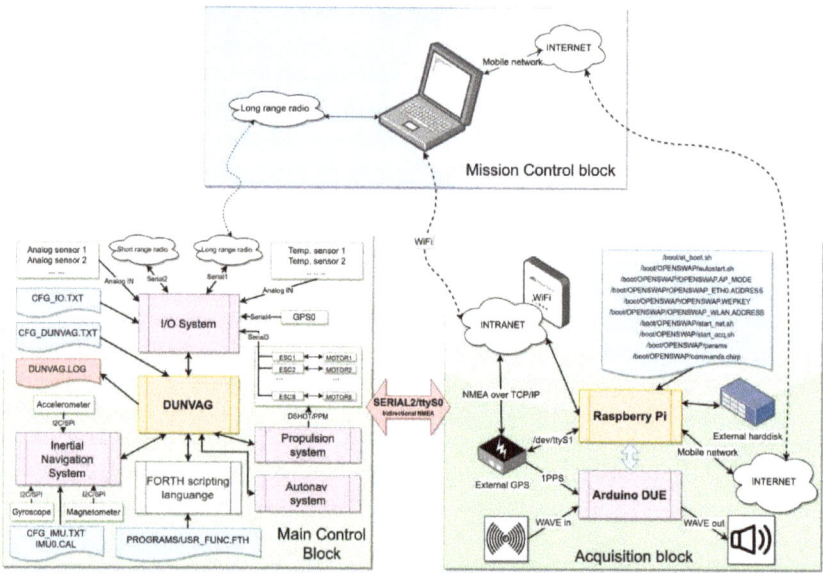

Figure 5. Block diagram showing OpenSWAP hardware/software architecture.

3.3.1. DUNVAG, the Autonomous Navigation Firmware

DUNVAG represents the "heart" of the system. It is a multiplatform firmware that runs on several different commercially available micro-controllers (i.e., ArduinoDue®, Teensy, STM32, etc.), and implements several functions, including: full control over propulsion system by using ESC (electronic speed control) boards; inertial navigation (INS), by means of MEMS (micro electro-mechanical systems) gyroscopes, accelerometers, magnetometers and single or double GPS antenna (with or without RTK correction); several algorithms for autonomous navigation and path following; wireless communication for telemetry and remote control, both for long range point-to-point connections, and by means of wide area links using 3G/4G/5G terminals. DUNVAG is configured through the editing of three text files stored inside a μSD card at the root level: CFG_NAV.TXT for configuring navigation, CFG_IMU.TXT, for the inertial system, and CFG_IO.TXT for I/O ports. A detailed description of all parameters is reported in Supplementary Tables S1–S3.

The module controlling propulsion and direction changes of the vehicle is implemented through PWM (pulse width modulation) or DSHOT1200/2400 protocols, driving two (or more) underwater electrical motors connected to STM32 based ESCs. The system is able to process feedbacks from the ESCs (when DSHOT protocol is used) to inspect the status of active motors, such as: -current absorption; -applied voltage; spinning velocity (RPM); and ESC temperature. The module is also responsible for controlling direction changes during navigation by modulating the power distributed between port and starboard. Typically, the OpenSWAP vehicles mount two motors per side, a redundancy which results critical in case of failure, but also useful for enhancing propulsion under difficult environmental conditions (strong currents, waves, wind, etc.). Since the propulsion module supports up to eight motors, and because four propellers are generally used, the spare slots are available for customizations, including implementation of different types of propulsion. An example, that was specifically tested for wetlands and swamps, is implementing an airboat by installing air propellers on top of the vehicle to avoid aquatic plants and reduce the environmental impact. One specific feature implemented in this module, is the real-time acquisition of the propulsion devices telemetry, allowing for intervention in the event of partial or total failure. Through the analysis of the ESCs status, the propulsion module can monitor at a rate of 10^2 samples per second propellers and controllers, and use such data to manage emergencies. A typical example is case of a propeller failure overcome by distributing more power to the other motors in the attempt of continue the navigation under the predefined constraints. This allows the vehicle to attempt accomplishing the mission in case of minor failures and/or re-entering the base in case o major. The configuration file including all parameters controlling the propulsion module is CFG_NAV.TXT (Supplementary Table S1).

The *I/O system* allows internal and external devices to be connected to the vehicle. It is based on several input ports, including eight RS232 serial lines and eight 0–5 V analogue inputs. In the default configuration, serial lines are configured to be connected to basic devices used for normal operations. They include:

(1) Serial-1, connected to the 433 MHz high-power, long-range, wireless module used for telemetry and remote controls;

(2) Serial-2, connected to the SBE module, is used to pass to the echosounder computer (a *Raspberry*™ PI) all the NMEA (National Marine Electronics Association) encoded sentences coming from the installed GPS, plus additional information in NMEA custom formatted strings, such as all vehicle status parameters, water temperature if temperature sensor is installed and any other readings coming from additional analog sensor eventually installed; asynchronously, *Raspberry*™ returns to the DUNVAG software the water depth computed in real time by analyzing the received echograms, all the NMEA sentences of any serial sensor connected at its ports and eventually the control-packets coming from the WAN (wide area network) detailed later on;

(3) Serial-3, connected in shared mode with every ESCs for telemetry;

(4) Serial-4, connected to the internal double RTK GPS receiver for high precision headings;

The remaining two serial lines are available for additional devices and can be accessed by the users through the internal FORTH scripting language.

Eight lines are also available for external analogue devices, accepting any input in the 0–5 V range, one already assigned to a temperature sensor embedded in the echosounder. Configuration parameters can be set to apply bias and/or scaling corrections if needed. Analogue lines could be also accessed by means of FORTH language commands, as well as through custom $IIXDR sentences sent to Serial-2. The sequence of such sentences is stored in a log file by the *Raspberry*™ echosounder module and/or eventually sent over the INTERNET when a connection is available.

The Inertial Navigation System is composed by three MEMS devices and one or two GPS receivers (see scheme in Figure 5). It is a key part of the system as it implements the HW/SW layer responsible for updating in real-time the vehicle position and its heading, also providing attitude information (pitch, roll and heave). It makes use of a MARG sensor fusion and Kalman 1D filtering algorithm [5–9] receiving data from gyroscopes, accelerometers and magnetometers, and integrating them with GPS heading and position. The inertial navigation system is configured through a text file, the CFG_IMU.TXT (Supplementary Materials), containing parameters for accelerometer and gyroscope bias, linear correction on temperature changes, and for hard and soft iron perturbation induced on magnetometer by external sources such as ferromagnetic material. Additionally, for more accurate calibrations over temperature, an IMU0.CAL file can be added, containing a list of corrections to be applied to gyroscope and accelerometer for a given temperature range (−50/100 °C). While a standard calibration is provided, in the event of calibration loss or unsatisfactory behaviors, an additional JAVA® application named MUVIMANT is available, to perform calibration of magnetometer and accelerometer/gyroscope couple.

Once calibrated, the system is able to integrate and correct the heading with using data from the GPS receiver, injecting the heading available through the IMU (Inertial Measurement System), prioritizing on most precise NMEA sentences (i.e., HDG before VTG). This provides a quick updating of the heading, since MEMS are characterized by a very fast and relatively accurate heading/positioning, and the GPSs, although slower, would provide best the accuracy in the presence of a good satellites signal.

Autonomous navigation system. Once accurate positioning and direction, as well as waypoints of the predetermined path are available to the system, they are sent to the autonomous navigation module (AIVAG), responsible for computing actual direction and power to be distributed to the propulsion module, which translate such indications into a direct control of the propellers by distributing the electrical power to the motors.

FORTH scripting language: Although it is well known that open standards and open applications are far better than the closed ones, there is still very few cases of real open applications giving to the users the possibility of "reprogramming the program", which would extend to unpredictable fields the application itself. This is mostly the case for marine oriented application, where the high software development costs do not leave additional economic resources to add extensions needed to open the application to the end-users. In most cases, not opening the software packages is among the politics of the software companies, operating in a small, very competitive market. However, in the field of Earth Sciences and more specifically in Marine Science there is a long-standing tradition for developing open software packages with high performances. Interesting examples of such packages presently diffused worldwide and used by a large audience of scientists are the Generic Mapping Tool [10] for maps and spatial data processing, MB-System [11], for processing MBES data, Seismic Unix [12] for seismic data processing, and many others. The OpenSWAP project would fall on the track of such scientific tools, and was developed assuming that all scientific applications developed by academic institutions should be, if not Open Source, at least open, giving the opportunity to adapt their functionalities to the scientific issue to be addressed. In this respect, DUNVAG implements a FORTH scripting language that gives the full control on almost every aspects of the application. It is based on a FORTH-83 implementation made available on the public domain by John Walker (https://www.fourmilab.ch/atlast/) and has been ported to DUNVAG to have access to most of the variables and parameters of the DUNVAG application. For manual of the language please refer to the above given URL. The built-in FORTH language is interfaced with DUNVAG through the execution of predefined FORTH commands, a block diagram of the DUNVAG program is shown in Figure 6, along with the FORTH entry points. Several variables of DUNVAG and some useful functions are available for full controlling the vehicle. Supplementary Tables S4 and S5 include a list of FORTH functions and variables.

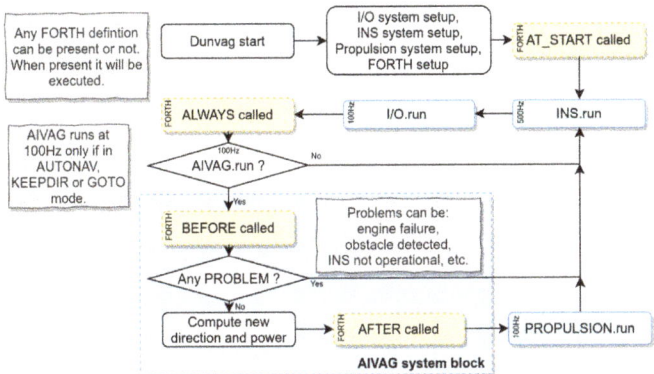

Figure 6. Block model of *DUNVAG*, the hearth of the system which implement autonomous navigation algorithms.

Since execution of the FORTH routines is placed inside the main loop, it is very important that each call to those routines will not block the normal flow for too long. For this reason, they will be interrupted after 2 ms. of CPU time. Examples of FORTH programming are given into a download area, which also includes additional contents. The reader could find into those examples, amongst others, a definition that implements a virtual GPS through reading a file with lat, lon, pitch, roll, heading and speed values, allowing for a mission replay and/or simulation.

While the DUNVAG firmware is not yet open source (it will be in the near future), the FORTH scripting capabilities opens up the SWAP autonomous class of vehicles to user extensions, hence the name OpenSWAP.

3.3.2. The Embedded SBE/SBP Instruments

The block-diagram of Figure 5, shows how the SBE is implemented. It can be both integrated into the navigation board or available as a separate module, while the chirped SBP is available only as separate board. Both instruments are installed inside the water-proof central box, together with other electronics/sensors eventually added to the payload.

SBE and SBP follows the same design, based on a *Raspberry*™ PI driving an ArduinoDue® microcontroller. The ArduinoDue® is loaded with two small pieces of software, µECHO and µCHIRP, which use the microcontroller ADC (analog to digital) and DAC (digital to analog) ports, as well as some transmitting/receiving circuits/sensors, to emit and receive acoustic signals into the water in a given time-window. The Raspberry™ PI, on the others side, runs CIAPCIAP, the software responsible for geo-referencing the data coming from the ArduinoDue® and for its storing into the µSD installed card or into an external support (hard-disk). The SBE/SBP pair can work in parallel, leading for synchronous acquisition of stratigraphic and bathymetric data. It would be possible for example, to allow the two systems modify some acquisition parameters, such as pulse length, pulse frequency, acquisition time window, etc, based on analysis of incoming data, in an AI (artificial intelligence) scheme, relating their values to actual bottom depth or other different variable environment parameters.

µECHO/µCHIRP Both SBE and SBP systems, we name µECHO and µCHIRP, respectively, are based on an ArduinoDue® microcontroller, which emits analog signals through a DAC port sampled at a very high rate (<1 µs). This performance, together with the possibility of programming the microcontroller, allow for a complete control on the emitted signal, an important characteristic to optimize signal generation in relation to specific survey, and to optimize the deconvolution of the signal. All acquisition parameters are controlled by a set of commands sent to an USB port of the micro-controller, while a second port is used for data acquisition. The software module CIAPCIAP, running on a Raspberry™ PI, is responsible for managing consistently these two ports, to implement a synchronized data acquisition. Supplementary Table S6 and S7 include the entire set of commands along with a brief description.

The μCHIRP SBP uses as emitter transducers lightweight electromagnetic resonators directly applied to the catamaran hulls, that show interesting performances in shallow water environments, and are easily installed on the vehicle. The system is composed by: (1) a digital generator of frequency-modulated signals based on an ArduinoDue® board; (2) a 600 W RMS power amplifier; (3) an array of waterproof magnetic resonators composed of four 4 Ω *MONACOR* transducers; (4) an acquisition system based on a hydrophone array, a signal amplifier, and an *ArduinoDue*® board used as analog-to-digital-converter (ADC). A Raspberry™ PI is employed to store the digital data in SEG-Y format [13] on a SD memory card.

The CIAPCIAP software module. CIAPCIAP runs on a Raspberry™ PI and is interfaced with DUNVAG (Figure 5) through a serial port to gather current position and orientation from the navigation board. On the other side, it passes to DUNVAG additional parameters such, as NMEA sentences from a GPS eventually connected to the geophysical sensors. CIAPCIAP makes also possible to connect to itself through a Wi-Fi hotspot client, or to implement an internal hotspot eventually used by external clients. It is also able to open a channel between the Mission Control Software (see next section) and DUNVAG by means of an INTERNET connection, eventually opened through available mobile networks. In such a case, a ssh tunnel with an external sshd server can be opened, allowing for connecting to the INTERNET from virtually everywhere. The operating system chosen for CIAPCIAP is Linux®, running from the installed μSD card in R/O (read only) mode for reliability reasons, and it is configurable by editing the files contained in the/boot/partitions (see Supplementary Table S8). The acquisition parameters are set by commands/parameters specified by the params, commands.echo and commands.chirp files, whose syntax is reported in the Supplementary Materials (Tables S6 and S7).

CIAPCIAP starts by reading the params file and subsequently opens the commands.echo or the commands.chirp files (the specified within params). It then sends all lines of the opened file to a proper USB port connected to the ArduinoDue® which set the emission and the acquisition parameters of the transmit/receive cycles. Such parameters are those controlling the cycle, such as the acquisition time window, the emitted waveform shape and length, the emission rate etc. Supplementary Tables S6 and S7, and could be modified by the user. This is obtained by editing a "commands file" containing initialization parameters for both, emitter and receiver. Special commands such as "WAVEOUT:" or "CHIRP:" (they are equivalent) are sent to the emitting or receiving processes. A configuration command sequence should begin with the receiver (sampler) configuration, followed by the emitter configuration. Once a WAVEOUT:MKSIGNAL command is sent, the emitting process adds a waveform with current parameters to a waveform buffer, along with current sampling parameters, allowing for the creation of waveforms of different length, duration and sweep. All waveforms are stored into an internal buffer and subsequently used to de-convolve the received signal. When the command WAVEOUT: GO is entered, the emission process starts, by generating the first available waveform since an emission timeout is reached. At this point, a second waveform (if present) could be generated, until the last waveform is reproduced. Then, the emission process starts again using the first available waveform. Synchronized with emission, the receiving procedure starts, digitally sampling the analog signal detected by the transducer/amplifier chain, according to parameters set for each configured waveform.

To start the whole emitting and receiving processes, the commands "WAVEOUT: GO" and "GO" should be entered in this order. After that, CIAPCIAP starts signal emission and the acquisition, and stops only if a "STOP" command and a "WAVEOUT: STOP" is sent.

An example of a commands file is given in Supplementary Tables S6 and S7, where a multi-chirp acquisition is activated with a start frequency of 24 Hz, and a 6 Hz frequency increasing every 4 generated waveforms. Each sweep generation is followed by data acquisition according to predefined acquisition parameters.

The acquisition process is not working separately by the other software modules. In fact, it is parallelized with several ancillary tasks dealing with: position and orientation updates (received from an external GPS and/or from the internal NAVBOARD); managing connection to a remote monitoring window; receiving commands from remote controlling applications; checking periodically whether an INTERNET connection is available. In this case, CIAPCIAP connects to a proprietary server (openswap.it) opening a bidirectional channel that allows for a remote telemetry from DUNVAG and a basic control of the vehicle. Further details on such functionality are given in the next session.

3.4. OpenSWAPNav, the Mission Control Software

The Mission Control Software, OpenSWAPNav (OSN), written in Java®, was developed to allow for interactive route planning and vehicle remote control. It runs a small C++ [14] app, (SwapCONTROLLER) that communicates through a wireless long-range (433MHz) device with the vehicle, receives the telemetry at a rate of six times per seconds, and send to the vehicle a set user-command. It can receive inputs from the user by means of a joystick or through the user interface, easily allowing manual control of the vehicle. The choice to have a small app doing the basic jobs of communicating to and controlling the remote vehicle is due to the need of a quick start/restart control system in case of problems. The SwapCONTROLLER application allows to fully control the remote vehicle in manual or auto modes.

OSN is connected with SwapCONTROLLER through a socket, for vehicle telemetry and other information. This allows the user to easily edit the routes and send the waypoints to the vehicle. As seen previously, the vehicle needs that some configuration files are present into its μSD (CFG_IMU.TXT, CFG_IO.TXT, CFG_IMU.TXT etc.), and OSN provides download and upload such files to the vehicle to update all parameters. OSN can communicate with the vehicle also through the *INTERNET* (if the vehicle is configured for the purpose), can receive telemetry also through such a connection from the vehicle and send to commands, such as the next point to reach or where to stop navigation, or even maintain the actual position. This is particularly useful when the vehicle exits the coverage of internal 433 MHz telemetry module.

A Screenshot of OSN and SwapCONTROLLER during a survey is shown in Figure 7.

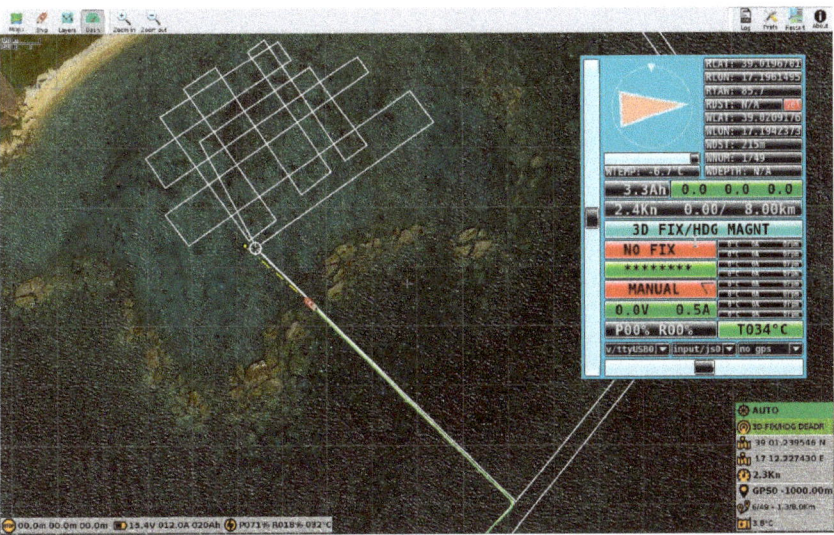

Figure 7. Screenshots of OSN and SwapController (**inset**) during a survey. Planned lines are marked in white, while navigation performed is indicated by green lines.

4. Performances of the Vehicles and Data Acquisition Examples

Performances and functionalities of the OpenSWAP prototypes were tested in several environments, under different weather and sea conditions, and different instrument payloads. Below, we report some examples of data acquisitions carried out in areas not, or very difficultly accessed by conventional surveys.

4.1. Navigation Accuracy

The accuracy in following pre-determined paths was first tested in "controlled" environments, such as small lakes, ponds and easily accessible coastal areas. Figure 8 shows an example of navigation test carried out in shallow waters along the Emilia Romagna coast, were two runs over the same planned lines were carried out. It includes a statistical evaluation of the navigation performances, i.e., of the errors that occurred in the following predetermined routes, which indicates that almost 90% of the route waypoints are within 0.30 m of polar distance from those planned. In worst cases, repeatability of the planned paths is within a mean error of about 0.40–0.50 m, reaching up to ~0.10 m in the best-case scenarios (calm weather and sea conditions). The test reported in Figure 8 consisted of navigating along a 1 km straight line towards the offshore during two different days. In this case, we used a double-antenna, double-frequency Trimble SPS461 GPS receiver, with an RTK cm accuracy. The mean error between the two path is 0.19 m, with a maximum over 99% of the points within 0.70 m.

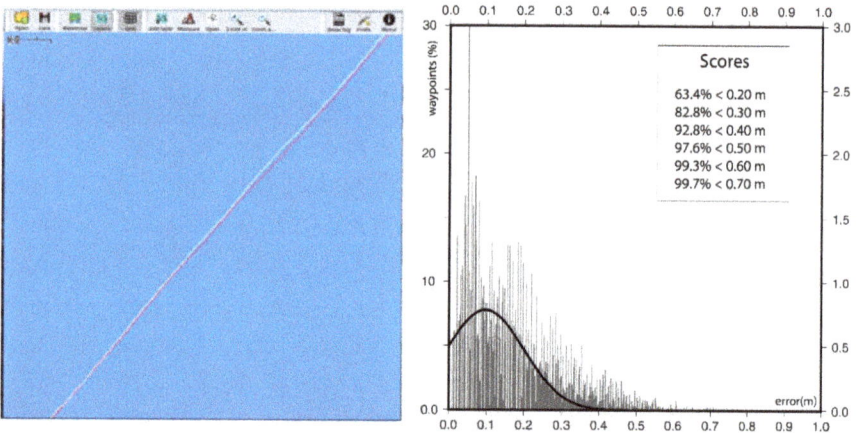

Figure 8. Navigation performance test along a 1 km straight line. **Left**: screenshot of OSN with first (white) and second (red) navigation lines. **Right**: distribution of the error modules between the two paths.

4.2. SBE

During the ASV developments, we considered the as SBE a "basic" sensor, since knowing water depth is crucial for safely navigating in shallow waters. Thus, we integrated a custom version of such instrument in the main electronic board of OpenSWAP. Our echosounder (µEcho) integrates a vertical incidence ultrasonic pinger, operating at a frequency of 200 kHz, a narrow (8°, conical) beam width, a short time-gate pulse length (350 µsec). The ultrasonic signal, is digitally sampled in a selected time-window by an ArduinoDue® board, and data (the echograms) are stored in SEGY-format files. Moreover, we implemented the procedure by Haynes et al., [14] to track the bottom reflection. First, an amplitude envelope of the echogram is computed by convolving the squared values of the original data trace with a rectangular window as wide as the source pulse:

$$a(t) = \int_{-\infty}^{\infty} [x(\tau)]^2 w(t-\tau) d\tau \qquad (1)$$

where $a(t)$ is the amplitude envelope trace, $x(t)$ is the recorded signal, is the rectangular window, and L is the window size.

$$w(t) = \begin{cases} 1 & t \in [0,L] \\ 0 & \text{elsewhere} \end{cases}$$

Subsequently, a simple threshold-time delay algorithm was used to achieve the bottom reflected signal from the amplitude envelope trace. Conversion of travel-times into water-depth could be performed once the sound-speed is determined through specific estimates.

Acquisition of the entire echosounder sweep at each sounding point, rather than the simple depth value generally provided by echosounders, give us the opportunity of estimating the relative reflectivity of the sediment-water interface. Using the SeisPrho function RCCM [3], under the vertical incidence case and neglecting the effect of energy scattering due to the bottom roughness, we obtain an estimate of the relative reflection coefficient (R) by using:

$$R = (A_r/A_s) z \qquad (2)$$

where A_r and A_s are the amplitudes of the source and reflected signals, respectively, and z is the water depth.

In order to obtain an estimate of R from our finite-length echo-sounder pulse, we used, for A_s and A_r, the values:

$$A_s = \sum_{i=0}^{W} |x_i| \qquad (3)$$

$$A_r = \sum_{i=B}^{B+W} |x_i| \qquad (4)$$

where $x(i)$ is the digital sampled signal, W is the width of the source pulse, and B is the bottom detection time. Bottom reflectivity data obtained by these echograms are used to compile reflectivity maps, that could be diagnostic of geological processes.

Propagation and scattering of high frequency acoustic sound at or near the bottom is controlled by a number of factors, including biological, geological, biogeochemical and hydrodynamic processes operating at the benthic boundary layer [15]. However, experimental measurements suggest that the single most important geotechnical property related to acoustic attenuation is the mean grain size of the insonified sediment [16–19]. In such a case, a combination of bathymetric and reflectivity maps could be used as an effective tool in analyzing geological processes acting at the sediment-water interface. Figure 9 reports an example of such procedure carried out along a river stream. We note a main erosional trough formed close to a bridge pillar, also marked by high reflectivity (red pattern in Figure 9d). We also note that areas with prevailing deposition are marked by lower reflectivity (blue pattern in Figure 9d) as along the internal part of the river meander in the northern sector of the study area. Ground-truthing reflection coefficient estimates with sediment samples is mandatory to perform more accurate bottom classifications ([20] and references therein).

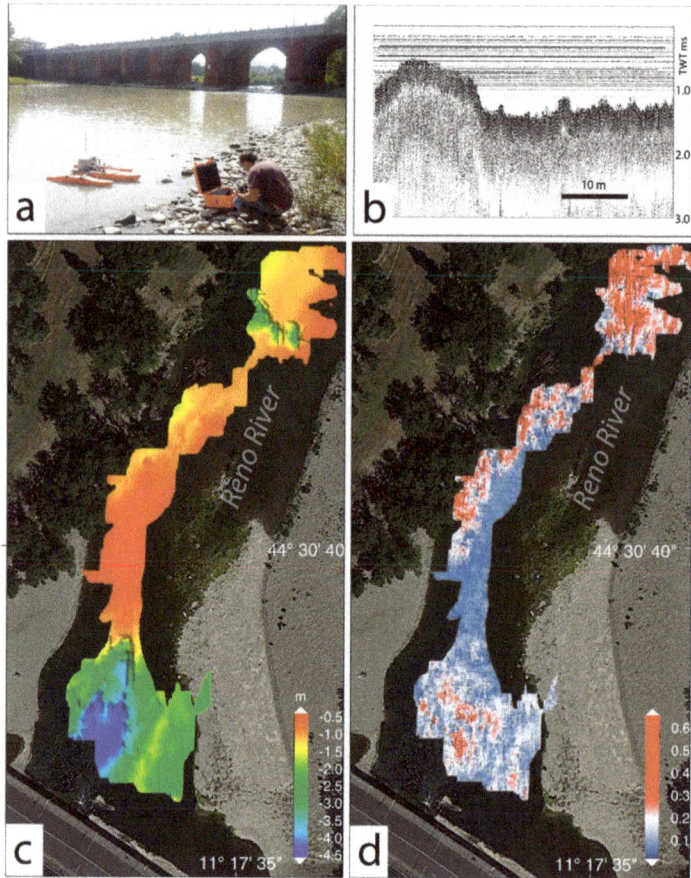

Figure 9. Echographic survey carried out along the Reno River close to an urban bridge (Bologna, Italy). (**a**) Configuration of the survey before start; (**b**) example of 200 kHz echogram in very shallow water; (**c**) morphobathymetric map highlighting the presence of a deep erosion close to the bridge pillar; (**d**) reflectivity map obtained using the same data showing sectors of prevalent erosion (red) and deposition (blue).

4.3. Side-Scan Sonar Imaging

We collected Side-Scan Sonar images along the Cavo Napoleonico artificial channel, in Northern Italy. The channel, which connects the Po and the Reno rivers in the Po plain, is oriented perpendicularly to the thrust and fold belt buried by alluvial sediments falling in the area which underwent the maximum superficial deformation during and after the Emilia 2012 earthquake sequence [21,22]. For this reason, it was chosen as an interesting site for geophysical surveys in search for co-seismic effects. Among other data collected with conventional methods, we carried out a side-scan sonar survey of this channel using a Starfish system mounted onboard of an OpenSWAP vehicle. We were searching for earthquake-related structures, such as fractures, fissures, sediment fluidization or slumps. Analysis of side-scan sonar images combined with high-resolution seismic reflection profiles (also collected with an OpenSWAP vehicle) suggests a correlation between the presence of "disturbances" at the channel floor (Figure 10) and the area o maximum deformation detected through satellite derived measures [23].

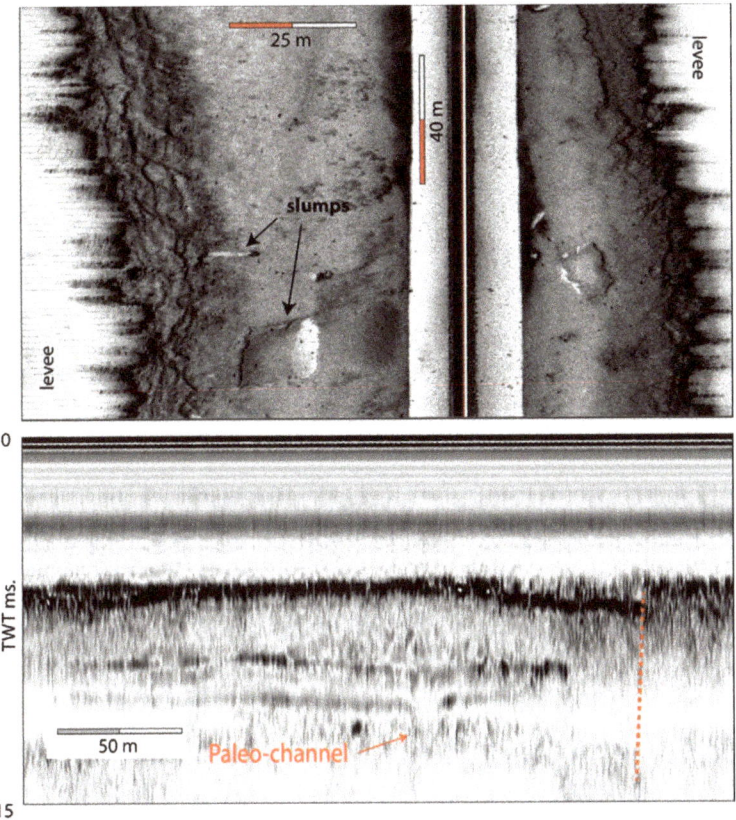

Figure 10. Top: side-scan sonar image of the Cavo Napoleonico artificial channel close to the epicenter of the Emilia 2012 earthquake, showing slumps and gravitative failures affecting the channel-floor. Bottom: SBP profile collected in the same area, penetrating the first meters of alluvial sediments, showing paleo-channels and displacements reaching up to the surface.

4.4. High-Resolution Imaging of the Subsurface Using the Embedded SBP

A typical SBP, operating with magnetostrictive transducers at hundreds to thousands of volts, is not suitable for operating on board of any OpenSWAP vehicle, either for the heavy weight and for low efficiency in converting the DC electric power of batteries to suitable high-voltages. For this reason, we developed a lightweight chirped SBP system (μCHIRP) embedded in the OpenSWAP electronics (see above). We tested the potential of our μCHIRP in Lake Trasimeno, a shallow-water tectonic lake in Central Italy. The lake was investigated through conventional systems in the frame of a geological study carried out with different geophysical methods [24,25]. However, some sectors close to the northern shore, were too shallow to be accessible using conventional systems, and were surveyed using OpenSWAP. The problem in this case was imaging at the best resolution the first tens of meters stratigraphic sequence in very shallow water, where several sources of noise affect in general the data. Prior to the survey, for a quality control of data collected by μCHIRP, we performed a comparison with an industry standard chirp-sonar system, the Benthos-Teledyne Chirp III, mounted onboard of a small boat. Results of this benchmark are reported in Figure 11, where the shallowest part (<10 ms. TWT) of the Trasimeno sedimentary sequence is imaged with the two systems, showing similar vertical resolutions and penetrations.

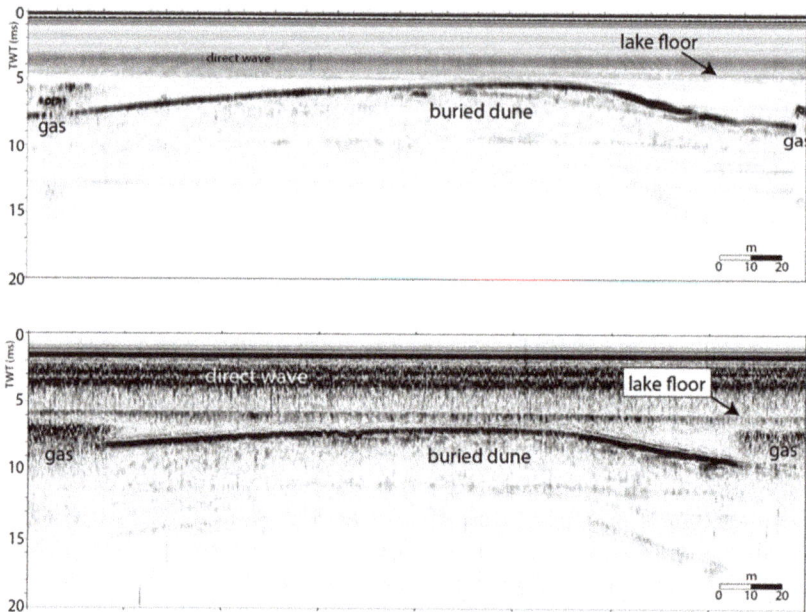

Figure 11. Example of seismic reflection profiles along the same navigation line collected with two different systems in Lake Trasimeno. **Top**: Chirp III Teledyne-Benthos, with 4 Massa transducers; **Bottom**: µCHIRP with 4 Monacor transducers. Unconsolidated sediments are penetrated down to 15–20 ms. below the lake floor, by both systems, with high vertical resolutions (tens of cm), enabling a detailed imaging the sedimentary structures.

4.5. Multibeam Echosounder Repeated Surveys

In order to test the possibility of performing MBES surveys using an OpenSWAP vehicle we used a Klein HydroChart 3500 integrated echosounder/side-scan sonar system. The HydroChart 3500 is a professional bathymetric sonar with IHO hydrographic standard for shallow water operations that integrates the characteristics of a side-scan sonar with those of an interferometric multibeam. It is a portable system that includes a motion reference unit (MRU) as well as course and sound speed sensors located in the sonar head. Each echogram includes uncertainty on the estimate of the depth and angles of the beams used for ray-tracing. In this way, the sonar propagation uncertainty model is integrated into the data processing flow, to provide uncertainty estimates for individual depth measurements that can be used by third-party bathymetric postprocessing.

We performed repeated multibeam survey offshore Calabria (Southern Italy), in the Calabrian Arc accretionary wedge, one of the most tectonically active regions in the Mediterranean Sea [26–28]. For this purpose, we surveyed repeatedly some key areas in the nearshore, where interferences between coastal sediment transport and gravitative instability in the vicinity of active faults were observed. The surveys were carried out in the same areas at different time-intervals, ranging from a few days to several months, obtaining good results. An example is reported in Figure 12, where we observe that over 95% of bathymetric measures are coherently positioned within ±30 cm of differences (the normal accuracy gathered by navigation system), and discrepancies between the two surveys (Figure 12c) are probably related to short-term seafloor changes, such as sand ripple migration, which were captured by our 4D survey. This first test indicates that repeating bathymetric measures at regularly spaced time-intervals along the same acquisition lines could be an interesting tool to determine what type of natural process is active and what is its temporal scale. Such information would be crucial in areas highly prone to slumping, seismogenic, and tsunamigenic risks.

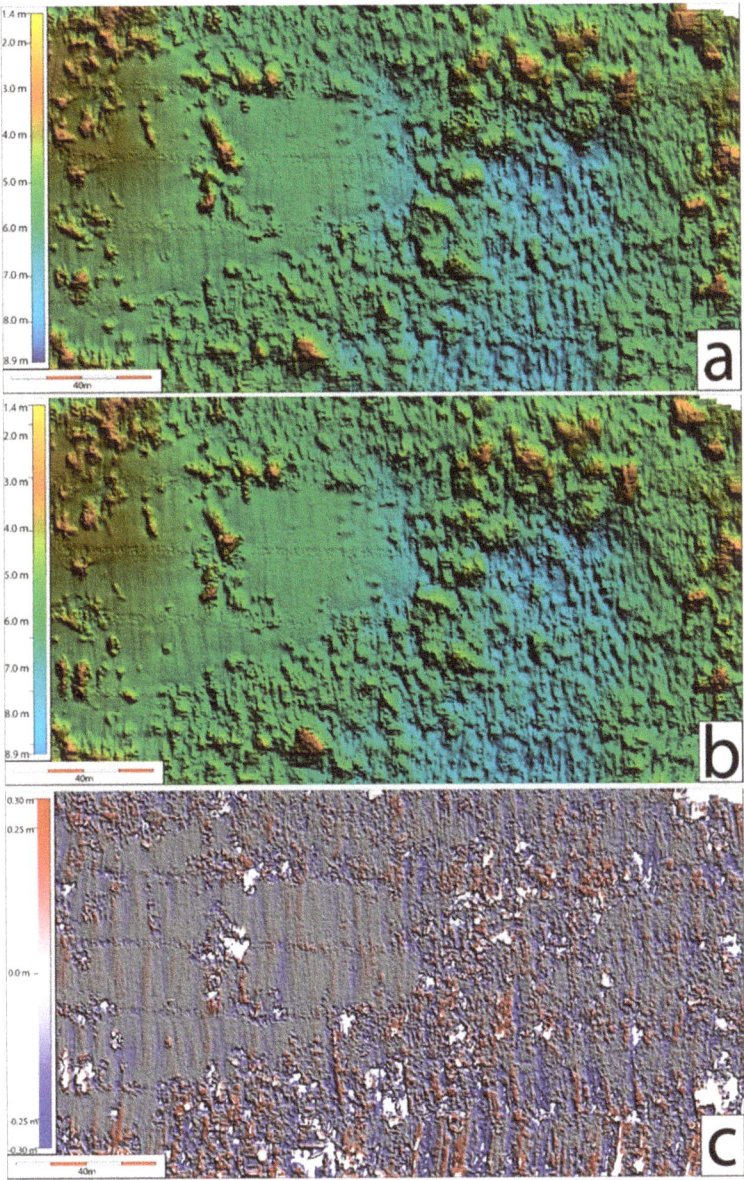

Figure 12. Bathymetric data collected using a Klein Hydrochart 3500 multibeam echosounder onboard of an OpenSWAP vehicle, including: (**a**) a first survey; (**b**) a second acquisition performed next day; and (**c**) the point-to-point difference between the two DTMs. Note that over 95% (colored dots) are within ±30 cm of difference. Red and blue undulations in c result from short-term seafloor changes, probably due to sand-ripple migration.

5. Discussion

The geophysical instruments to date installed onboard of the OpenSWAP vehicles are not exhaustive of all possible instruments that could be employed in geological, geophysical, geochemical, and oceanographic studies generally carried out in shallow water areas. In fact, the possibility of installing instruments, such as electromagnetic sensors [29,30], or instruments for physical and chemical oceanography, could represent a dramatic improvement of their potential. The peculiar design of the hull is suitable for installing ADCPs (Acoustic Doppler Current-meters Profiler), that find a wide range of application in hydraulic and physical oceanographic studies. Furthermore, some basic chemical sensors, such as oxygen, nitrates, PH, etc., now available in small and light vessels, could represent interesting payloads to carry out rapid assessment of the surface water quality. We believe that the main characteristic of the OpenSWAP vehicles, i.e., low cost, high accuracy in performing predefined routes, embedded geophysical sensors, etc. could be already interesting for different scientific purposes. We have been able to verify that the high accuracy of the navigation algorithms would open the door to 4D surveys, i.e., surveys performed more than once in any given area, with the possibility of quantitatively comparing the results within acceptable error bars. Examples could be time-dependent analysis of morphological and stratigraphic changes of the sediment/water interface and shallow substratum eventually caused by the sediment dynamics (erosion vs. deposition), slumps and gravitative failures, earthquakes (slip along seismogenic faults and secondary effects of seismic shaking), tsunamis, etc.

Since the cost of the vehicles is maintained as low relative to other commercial systems, it would be possible to survey hazardous areas, accepting the high risks of damaging or losing the vehicles. Examples could be surveys carried out at glacier fronts, always prone to unpredictable collapses, or in environments characterized by strong currents and waves, such as rivers or coastlines, as well as in highly polluted environments, not safely accessed by manned vehicles or boats. The low cost would also enhance the use of small fleets of vehicles performing cooperative and adaptive surveys.

6. Conclusions

We tested the performances of OpenSWAP, an innovative class of autonomous surface vehicles (ASV) designed to carry out geophysical surveys in different shallow water environments. Such vehicles are based on open software and hardware architectures, to comply for low cost, high performances and easy customization. The possibility of carrying out surveys in shallow-water areas quickly, and at a fraction of costs relative to conventional methods, could disclose the use of these techniques to a wide range of users, allowing for the execution of repeated (4D) and cooperative missions, giving a time-variant perspective to the study of natural process in rapidly evolving environments. We note that most of the presented case studies were not feasible using conventional methods, and probably many others are waiting for such technologies to be tested. We hope that the widest availability of these vehicles would serve to promote the study of the coastal areas, and more generally the variety of shallow water environments, complex, rapidly evolving, and so important for their economic and naturalistic value.

Supplementary Materials: The following are available online at http://www.mdpi.com/2072-4292/12/16/2575/s1, Tables S1–S8.

Author Contributions: Conceptualization, L.G. and G.S.; methodology, L.G., G.S., and F.D.B.; software, G.S.; validation, L.G. and G.S.; formal analysis, G.S. and L.G.; investigation, L.G., G.S., and F.D.B.; resources, L.G.; data curation, L.G., G.S., and F.D.B.; writing—original draft preparation, L.G. and G.S.; writing—review and editing, L.G.; visualization, L.G.; supervision, L.G.; project administration, L.G.; funding acquisition, L.G. All authors have read and agreed to the published version of the manuscript.

Funding: This research was funded by POR-FESR Emilia Romagna Project NAIADI (P.I. Luca Gasperini).

Acknowledgments: The authors are grateful to all people who helped in different ways developing OpenSWAP in the frame of NAIADI and other projects. The list is long, but should include Alexandro Palmieri, Antonella Poggi, Federica Pasini, Gianni Biasini, Alessandro Giordano, Alfredo Liverani, Eugenio Nisini, Nicolò Marinelli, Francesco Suriano, Piero Zucchini, Valentina Ferrante and Alina Polonia.

Conflicts of Interest: The authors declare no conflict of interest.

References

1. Gasperini, L.; Del Bianco, F.; Stanghellini, G.; Priore, F. Acquisition of geophysical data in shallow-water environments using autonomous vehicles: State of the art, perspectives and case histories. *GNGTS* **2014**, *1*, 10–11.
2. Dal Forno, G.; Gasperini, L. Chircor: A new tool for generating synthetic chirp-sonar seismograms. *Comput. Geosci.* **2008**, *34*, 103–114. [CrossRef]
3. Gasperini, L.; Stanghellini, G. SeisPrho: An interactive computer program for processing and interpretation of high-resolution seismic reflection profiles. *Comput. Geosci.* **2009**, *35*, 1497–1504. [CrossRef]
4. Stanghellini, G.; Carrara, G. Segy-change: The swiss army knife for the SEG-Y files. *Software-X* **2017**, *6*, 42–47. [CrossRef]
5. Kalman, R.E. A new approach to linear filtering and prediction problems. *J. Basic Eng.* **1960**, *82*, 35–45. [CrossRef]
6. Foxlin, E. Inertial head-tracker sensor fusion by a complementary separate-bias kalman filter. In Proceedings of the Virtual Reality Annual International Symposium the IEEE, Santa Clara, CA, USA, 30 March–3 April 1996; pp. 185–194.
7. Luinge, H.J.; Veltink, P.H.; Baten, C.T.M. Estimation of orientation with gyroscopes and accelerometers. In Proceedings of the First Joint [Engineering in Medicine and Biology 21st Annual Conf and the 1999 Annual Fall Meeting of the Biomedical Engineering Soc.] BMES/EMBS Conference, Atlanta, GA, USA, 13–16 October 1999; Volume 2, p. 844.
8. Sabatini, A.M. Quaternion-based extended kalman filter for determining orientation by inertial and magnetic sensing. *IEEE Trans. Biomed. Eng.* **2006**, *53*, 1346–1356. [CrossRef]
9. Madgwick, S.O.H.; Harrison, A.J.L.; Vaidyanathan, R. Estimation of IMU and MARG orientation using a gradient descent algorithm. In Proceedings of the IEEE International Conference on Rehabilitation Robotics, Zurich, Switzerland, 29 June–1 July 2011; pp. 1–7. [CrossRef]
10. Wessel, P.; Smith, W.H.F.; Scharroo, R.; Luis, J.; Wobbe, F. Generic mapping tools: Improved version released. *Eos Trans. Am. Geophys. Union* **2013**, *94*, 409–410. [CrossRef]
11. Caress, D.W.; Chayes, D.N. Improved processing of Hydrosweep DS multibeam data on the R/V Maurice Ewing. *Mar. Geophys. Res.* **1996**, *18*, 631–650. [CrossRef]
12. Stockwell, J.W., Jr. Free software in education: A case study of CWP/SU: Seismic Unix. *Lead. Edge* **1997**, *16*, 1045–1049. [CrossRef]
13. Barry, K.M.; Cavers, D.A.; Kneale, C.W. Recommended standards for digital tape formats. *Geophysics* **1975**, *40*, 244–352. [CrossRef]
14. Haynes, R.; Huws, D.G.; Davis, A.M.; Bennell, J.D. Geophysical sea-floor sensing in a carbonate sediment regime. *Geo-Mar. Lett.* **1997**, *17*, 253–259. [CrossRef]
15. Richardson, M.D.; Briggs, K.B. In situ and laboratory geoacoustic measurements in soft mud and hard-packed sand sediments: Implications for high-frequency acoustic propagation and scattering. *Geo-Mar. Lett.* **1996**, *16*, 196–203. [CrossRef]
16. Shumway, G. Sound speed and absorption studies of marine sediments by a resonance method. *Geophysics* **1960**, *25*, 451–467. [CrossRef]
17. McCann, C.; McCann, D.M. The attenuation of compressional waves in marine sediments. *Geophysics* **1969**, *34*, 882–892. [CrossRef]
18. McCann, C.; McCann, D.M. A theory of compressional wave attenuation in non-cohesive sediments. *Geophysics* **1985**, *50*, 1311–1317. [CrossRef]
19. Dunlop, J.I. Measurement of acoustic attenuation in marine sediments by impedance tube. *J. Acoust. Soc. Am.* **1992**, *91*, 460–469. [CrossRef]
20. Gasperini, L. Extremely Shallow-Water Morphobathymetric Surveys: The Valle Fattibello (Comacchio, Italy) Test Case. *Mar. Geophys. Res.* **2005**, *26*, 97–107. [CrossRef]
21. Galli, P.; Castenetto, S.; Peronace, E. May 2012 Emilia earthquakes (Mw 6, Northern Italy): Macroseismic effects distribution and seismotectonic implications. *Alp. Mediterr. Quat.* **2012**, *25*, 105–123.
22. Salvi, S.; Tolomei, C.; Boncori, J.P.M.; Pezzo, G.; Atzori, S.; Antonioli, A.; Trasatti, E.; Giulini, R.; Zoffoli, S.; Coletta, A. Activation of the SIGRIS monitoring system for ground deformation mapping during the Emilia 2012 seismic sequence, using COSMO-SkyMed InSAR data. *Ann. Geophys.* **2012**, *55*, 797–802.

23. Tizzani, P.; Castaldo, R.; Solaro, G.; Pepe, A.; Bonano, M.; Casu, F.; Manunta, M.; Manzo, M.; Samsonov, S.V.; Lanari, R.; et al. New insights into the 2012 Emilia (Italy) seismic sequence through advanced numerical modeling of ground deformation InSAR measurements. *Geophys. Res. Lett.* **2013**, *40*, 1971–1977. [CrossRef]
24. Gasperini, L.; Barchi, M.R.; Bellucci, L.G.; Bortoluzzi, G.; Ligi, M.; Pauselli, C. Tectonostratigraphy of Lake Trasimeno (Italy) and the geological evolution of the Northern Apennines. *Tectonophysics* **2010**, *492*, 164–174. [CrossRef]
25. Borgia, L.; Colombero, C.; Comina, C.; Del Bianco, F.; Gasperini, L.; Priore, F.; Sambuelli, L.; Stanghellini, G.; Trippetti, S. First results of a waterborne geophysical survey around Malpasso site (Tuoro sul Trasimeno, Italy) for geological and archeological characterization. *GNGTS* **2013**, *3*, 186–189.
26. Polonia, A.; Torelli, L.; Mussoni, P.; Gasperini, L.; Artoni, A.; Klaeschen, D. The Calabrian Arc subduction complex in the Ionian Sea: Regional architecture, active deformation, and seismic hazard. *Tectonics* **2011**, *30*, TC5018. [CrossRef]
27. Polonia, A.E.; Bonatti, E.; Camerlenghi, A.; Lucchi, R.G.; Panieri, G.; Gasperini, L. Mediterranean megaturbidite triggered by the AD 365 Crete earthquake and tsunami. *Sci. Rep.* **2013**, *3*, 1285. [CrossRef] [PubMed]
28. Bortoluzzi, G.; Polonia, A.; Torelli, L.; Artoni, A.; Carlini, M.; Carone, S.; Carrara, G.; Cuffaro, M.; Del Bianco, F.; D'Oriano, F.; et al. Styles and rates of deformation in the frontal accretionary wedge of the Calabrian Arc (Ionian Sea): Controls exerted by the structure of the lower African plate. *Ital. J. Geosci.* **2017**, *136*, 347–364. [CrossRef]
29. Forte, E.; Pipan, M. Review of multi-offset GPR applications: Data acquisition, processing and analysis. *Signal Process.* **2017**, *132*, 210–220. [CrossRef]
30. Sambuelli, L.; Fiorucci, A.; Dabove, P.; Pascal, I.; Colombero, C.; Comina, C. Case history: A 5 km long waterborne geophysical surveyalong the Po river within the city of Turin (northwest Italy). *Geophysics* **2017**, *82*, 1–11. [CrossRef]

© 2020 by the authors. Licensee MDPI, Basel, Switzerland. This article is an open access article distributed under the terms and conditions of the Creative Commons Attribution (CC BY) license (http://creativecommons.org/licenses/by/4.0/).

Article

Geophysical and Sedimentological Investigations Integrate Remote-Sensing Data to Depict Geometry of Fluvial Sedimentary Bodies: An Example from Holocene Point-Bar Deposits of the Venetian Plain (Italy)

Giorgio Cassiani *, Elena Bellizia, Alessandro Fontana, Jacopo Boaga, Andrea D'Alpaos and Massimiliano Ghinassi

Department of Geosciences, University of Padova, Via G. Gradenigo 6, IT-35131 Padova, Italy; elena.bellizia@phd.unipd.it (E.B.); alessandro.fontana@unipd.it (A.F.); jacopo.boaga@unipd.it (J.B.); andrea.dalpaos@unipd.it (A.D.); massimiliano.ghinassi@unipd.it (M.G.)
* Correspondence: giorgio.cassiani@unipd.it

Received: 30 June 2020; Accepted: 7 August 2020; Published: 10 August 2020

Abstract: Over the past few millennia, meandering fluvial channels drained coastal landscapes accumulating sedimentary successions that today are permeable pathways. Propagation of pollutants, agricultural exploitation and sand liquefaction are the main processes of environmental interest affecting these sedimentary bodies. The characterization of these bodies is thus of utmost general interest. In this study, we particularly highlight the contribution of noninvasive (remote and ground-based) investigation techniques, and the case study focuses on a late Holocene meander bend of the southern Venetian Plain (Northeast Italy). Electromagnetic induction (EMI) investigations, conducted with great care in terms of sonde stability and positioning, allowed the reconstruction of the electrical conductivity 3D structure of the shallow subsurface, revealing that the paleochannel ranges in depth between 0.8 and 5.4 m, and defines an almost 260 m-wide point bar. The electrical conductivity maps derived from EMI at different depths define an arcuate morphology indicating that bar accretion started from an already sinuous channel. Sedimentary cores ensure local ground-truth and help define the evolution of the channel bend. This paper shows that the combination of well-conceived and carefully performed inverted geophysical surveys, remote sensing and direct investigations provides evidence of the evolution of recent shallow sedimentary structures with unprecedented detail.

Keywords: electromagnetic induction; depth inversion; sedimentary processes

1. Introduction

Modern coastal landscapes are widely shaped by meandering fluvial, fluvio-tidal and tidal channels, which over the late Holocene accumulated complex and extensive sedimentary bodies. These bodies today often define subsoil permeable systems [1] that are often exploited as water reserves for agricultural, industrial and civil uses [2], and are extremely sensitive to saltwater intrusion [3,4] as well as to contamination [5,6]. These channelized bodies commonly consist of clean and poorly consolidated sand, which can also be affected by liquefaction processes [7]. The 2012 earthquake that occurred in the northeastern portion of the Po Plain (Italy) was the cause of sand eruptions that occurred along Holocene paleochannels and crevasse splay deposits down to an 8 m depth [8,9].

Aerial photographs and satellite images are excellent tools to identify and map late Holocene coastal channel networks since the former provide aerial data from the 1950s to present [10–13] and the latter provide multispectral analysis to highlight surficial paleochannel configurations [14–16].

Excellent examples of remote-sensing applications for these types of geomorphological studies can be found in the recent literature [17–21]. Despite the advantages, particularly in locating the position of these surficial bodies, and possibly distinguishing between tide- and fluvial-generated meanders [22], remote sensing alone is of course not capable of providing information at depth, thus remaining essentially a qualitative tool for the characterization of 3D geological structures. On the other hand, direct field surveys [23,24] and microrelief analysis [10,23,25] can provide further information either locally at depth or extensively at the surface. Regardless, a 3D reconstruction is still difficult with these means only [26], if not as a result of interpolation of scarce scattered data.

Addressing these issues requires that extensive and high-resolution data are available to map large areas and, at the same time, investigate the subsoil to a certain depth. This calls for geophysical methods (as they are designed to collect data informative about the subsoil, unlike remote sensing sensu strictu) that can also be deployed rapidly with limited or no ground contact so that large areas can be investigated. The most suitable methods for these purposes are those based on electromagnetic processes. In particular, approaches based on electromagnetic induction (EMI) allow for noncontact subsurface investigation, with no intrinsic limitations as posed, for instance, to wave-propagation EM methods such as ground-penetrating radar (GPR) that can be strongly limited in their depth propagation by the ground electrical conductivity.

EMI is a well-established technique that dates back nearly one century [27] and is based on Faraday's law of electromagnetic induction. The technique is articulated in a variety of specific instrument designs and investigation strategies [28] ranging in investigation depth from very shallow (the first meter or so) to tens of kilometers. For shallow applications [29,30], EMI has had widespread use in hydrological and hydrogeological characterizations [31–33], hazardous waste characterization studies [34,35], precision-agriculture application locations [36–38], archaeological surveys [39,40], geotechnical investigations [41] and unexploded ordnance (UXO) detection [42]. EMI measurements at small scale are typically conducted in the frequency domain (frequency domain electromagnetics or FDEM), and the results are classically expressed as apparent electrical conductivities (ECa) [43] using the so-called low-induction number approximation [44]. In addition to ECa mapping, the development of multifrequency and multicoil instruments has recently enabled the possibility of inversion of EMI measurements to provide quantitative models of depth-dependent electrical conductivity (EC), as the different acquisition configurations either in terms of coil geometry or frequency allow for multiple independent data to be acquired in sufficient number to warrant inversion. The majority of inversion algorithms use a 1D forward model based on either the linear cumulative sensitivity (CS) forward model proposed by [44] or nonlinear full solution (FS) forward models based on Maxwell's equations (e.g., [45,46]). As with EMI mapping, applications using inverted EMI data have also been diverse (e.g., [47–52]). Applications typically focus on using an inversion based on either the CS or an FS forward model to produce regularized, smoothly varying models of EC with fixed depths or sharply varying models of EC where layer depths are also a parameter. In the most advanced cases, a full 3D model of electrical conductivity can be reconstructed over a relatively large area, similar to what can be obtained at a larger scale by using, e.g., time-domain airborne EMI systems (e.g., [53]). The use of small FDEM measurement systems, with rapid response and easy integration into mobile platforms, is the key factor in the success of EMI techniques for near-surface investigations in these fields, as they allow dense surveying and real-time conductivity mapping over large areas in a cost-effective manner. However, sufficient control on the acquisition geometry is often needed, as the instrument response has a strong dependence also on the elevation above ground and the relative height of the primary and secondary coils [47].

The purpose of this paper is to show how the integrated use of remote sensing, EMI and direct stratigraphic investigations can provide an effective and comprehensive 3D view of the geometry of a fluvial sedimentary body. Results from the present study highlight the importance of an integrated approach to understand subsurface deposits.

2. Materials and Methods

2.1. Geological Setting and Study Area

The Venetian Plain is located at the northeastern end of the Po Plain, the largest Italian alluvial plain, and was generated during Holocene transgression by aggradation of fluvial meandering channels [23,24]. Specifically, the study area is located at the boundary between the Venetian Plain and the Po River Delta, in a zone which is characterized by a dense network of alluvial ridges and sand bodies that are the geomorphological products of the complex interaction between the Adige and Po Rivers during the late Holocene (Figure 1a) [25,54]. These sedimentary bodies currently host a multilayered system of phreatic and confined aquifers that are affected by saltwater contamination [4,55] and intensive water exploitation. The fluvial sedimentation occurred in an aggrading setting related to the marine highstand, and meander belts often correspond to fluvial ridges slightly elevated over the floodplain (i.e., 2 to 5 m above sea level (asl)) [54]. The present surface is a typical lowland landscape, which developed in the last 5000 years by the avulsions of the Adige and Po rivers [25].

The investigated site is located near the village of Anguillara Veneta (Figure 1b), about 1 km north of the current channel of the Adige River, in an area with surface elevations ranging between 0.7 and 2.0 m asl, where traces of abandoned meanders are visible in several aerial images and could be followed for about 7 km, from Stanghella to Anguillara Veneta. These paleohydrographic traces run nearly parallel to the present Adige River, even if they are slightly out of the natural levee deposits connected to the fluvial ridge of Adige. The river activated its present direction since the early Middle Age, while before it used to flow along the meander belt running from Este to Monselice to Chioggia (from west to east) [56]. Near Anguillara Veneta, the present course of the Adige River cuts the so-called fluvial ridge of Rovigo–Saline–Cona, which was formed by the Po River between 4500 and 3500 years BP [25].

The area experienced strong anthropogenic activity since the Roman period, when extensive field systems were settled in the whole Venetian Plain and parts of the Po Delta [25]. A major phase of reclamation started in the 16th century, when the Venetian Republic started the strong management of the river network, leading to the construction of the dense network of dikes, canals and ditches that still characterizes the landscape. During the same period, the Gorzone Canal was also cut, which represents the northern boundary of the study area, to convey the water discharge of the Agno–Guà–Frassine–Santa Caterina river system towards the Adriatic Sea. During the first part of the 20th century, the reclamation was extended to the coastal plain, where large portions of swamps and lagoon landscapes were drained through the excavation of canals and the use of pumping stations. These interventions made it possible to artificially lower the groundwater table below the surface and to cultivate seasonal crops (e.g., corn, wheat, bit and soya bean) and vegetables. In the last decades, several strong leveling interventions were carried out to improve the efficiency of the new draining system, as in the field investigated for this study. Besides the positive results, unfortunately, the reclamation also induced fast land subsidence caused by groundwater withdrawal, compaction of the drained soil and degradation of the organic matter formerly present in the marsh sediments [57]. From the downlift rate of the natural subsidence, ranging around 1 mm/y [58], in the last century the velocity strongly increased up to average values of 2 to 5 mm/y with large sectors up to 10 mm/y [59].

A large number of historical photos and maps are available for the Venetian Plain, along with freely distributed satellite images. Study sites are easily accessible for geophysical investigation and sedimentary cores. For these reasons, this study area represents an ideal site to develop and test the proposed integrated approach.

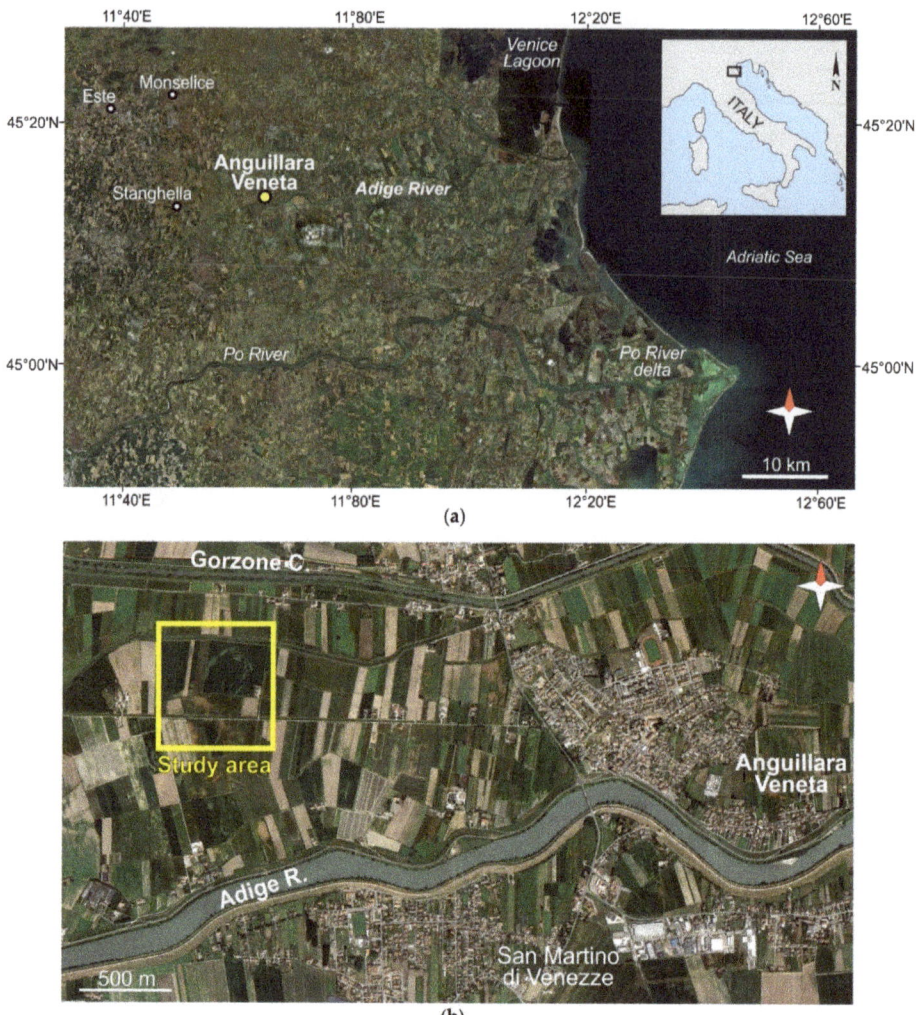

Figure 1. The study site: (**a**) location of Anguillara Veneta town in the southern Venetian Plain, at the boundary with the Po Plain sensu strictu; (**b**) satellite image (2013) of the study area (yellow box) in Anguillara Veneta town, PD (Italy).

2.2. Remote Sensing

The study site was selected after the identification of the paleomeanders in the aerial and satellite images. In particular, the first characterization was carried out on some satellite images available from Google Earth, which generally are pansharpened images of true-color composite bands of the Digital Globe company (Westminster, CO, USA) (e.g., Ikonos, QuickBird, WorldView and GeoEye missions). We selected the images with a detailed spatial resolution between 1.0–0.5 m (i.e., images 31/07/2004; 23/06/2007; 06/06/2010; 21/04/2012; 06/08/2013; 16/08/2013; 28/03/2015; 22/06/2017; 18/07/2018) and imported them into the GIS software ArcMap (version 10.7.1) [60] and QGIS 3.10 [61] for image processing and comparison with the images available as basemap reference in ESRI. Moreover, we also considered several zenithal conventional aerial pictures available from the cartographic service of

Veneto Region [62], consisting of scanned versions of black/white and color pictures from 1955 to 2008, with scales from 1:33,000 to 1:5000.

To investigate the spectral characteristics of the field surface, we analyzed some images from the satellite Sentinel-2, obtaining the normalized difference vegetation index (NDVI) and the normal difference moisture index (NDMI) [63–65]. These indices have a geometric resolution of 10 m and, being sensitive to plant health and hydraulic stress, respectively [66,67], were used to improve the identification of the traces of paleomeanders by linking sedimentology to vegetation health of the area. In particular, we produced the NDVI and NDMI not only from summer scenes, but also from different seasons and years by processing the multispectral bands through the semiautomatic classification plugin (SCP) [68] and raster calculator of QGIS 3.10. In fact, in the study area the cultivated plants change with seasons; in addition to the crops growing during summer, winter cultivations such as wheat, barley and different vegetables can be present.

Satellite, aerial and processed images were visualised in a properly georeferenced 3D space provided by the Move 2018.2™ [69] software.

2.3. Geophysical Investigations

The EMI surveys at the site were collected using a GF Instruments CMD-Explorer probe [70] that is a six-coil system (three coplanar pairs) operating at a single frequency equal to 10 kHz. The probe can be operated in both horizontal coplanar (HMD) and vertical coplanar (VMD) configurations [71], providing six independent measurements that are generally associated with six different apparent depths of investigation (Table 1). To acquire all six configurations for each geographical location it is, however, necessary to reoccupy with some acceptable degree of approximation the same location twice (once for each coil orientation).

Table 1. Technical specifications of the multicoil CMD Explorer FDEM.

Instrument Probe	Coil Interdistance (m)	Frequency	Nominal Exploration Depth (Horizontal Mode HMD/Vertical Mode VMD)
1	1.48	10 kHz	2.2/1.1 m
2	2.82	10 kHz	4.2/2.1 m
3	4.49	10 kHz	6.7/3.3 m

The FDEM probe was mounted on a specifically designed wooden carriage and connected to a Trimble 5800 GPS for continuous positioning, collecting data every second [72]. The wooden support was towed by a small tractor (Figure 2a). The acquisition apparatus adopted satisfies two fundamental requirements that proved extremely effective in terms of data quality [73,74]:

(a) Reoccupation of the same location is warranted by the GPS within the required precision (note that the sonde is a few meters long);
(b) The setting of the sonde is the same at all locations, with no changes of either the height above ground or the setting of the sonde that is maintained largely horizontal.

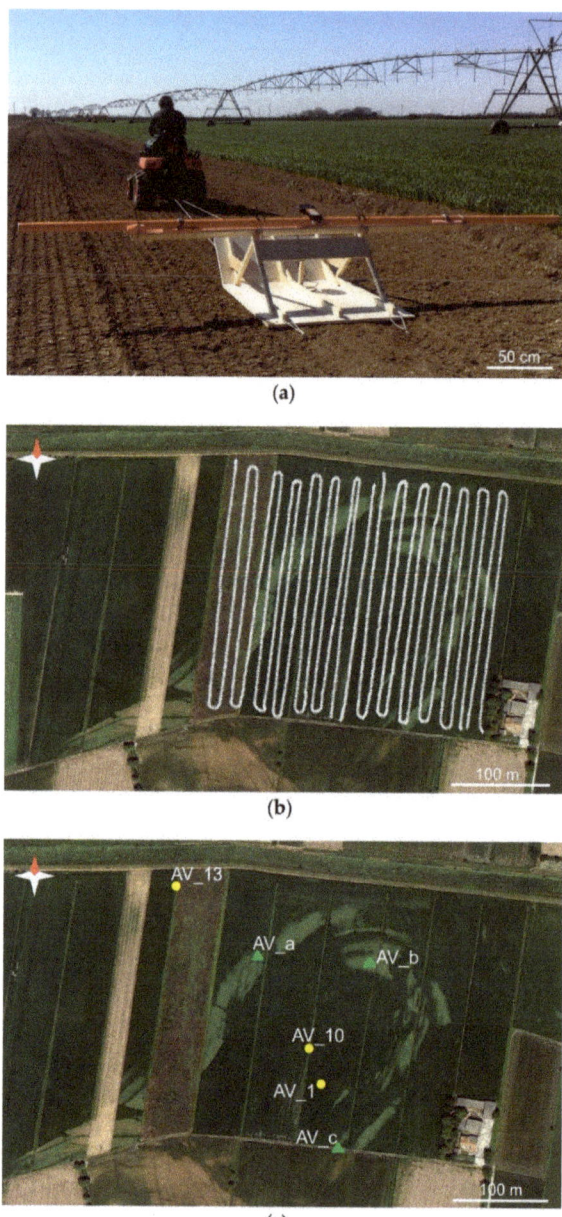

Figure 2. Methodologies. (**a**–**b**) Geophysical acquisition: (**a**) the electromagnetic tool on the wooden sledge dragged by a small tractor on the study field and (**b**) the survey path. (**c**) Position of the recovered cores. Yellow dots and green triangles indicate the hand auger core and the drilled cores with rotary technique, respectively.

In this manner, we collected about 20,000 EMI data points (Figure 2b), each in both HMD and VMD modes, with about one point every 0.5 m along the acquisition lines. The lines have a mutual distance of roughly 10 m (Figure 2b).

EMI data were then inverted to retrieve real soil conductivity values. For this purpose, we used the Interpex IX1D inversion software [75], a 1D routine based on smooth depth inversion according to the so-called Occam's approach [76]. The very dense spatial sampling allowed for the reconstruction of the subsoil practically in a 3D fashion by the juxtaposition of the 1D inverted profiles. For all locations, the same number of layers (eight) was used for the inversion, thus producing a consistent dataset. The results are presented in terms of 2D horizontal maps at several depths, that were then georeferenced using the Move 2018.2^{TM} [69] software, which also allowed creation of 3D surfaces, by the linear method, and tetravolumes.

2.4. Sedimentary Cores

Six cores were recovered at the study site to analyze sedimentary features of the study deposits and provide ground-truth for geophysical and remote-sensing data (Figure 2c). The core locations were established based on remote-sensing and geophysical data. Cores were collected by using an Eijkelkamp hand auger and a continuous drilling core sampler with rotary technique. Three cores were recovered using an Eijkelkamp hand auger, through a gouge sampler with a length of 1 m and a diameter of 30 mm, which prevented sediment compaction. Depth for these cores spanned from 2.5 to 6 m (Figure 2c). Three additional cores were recovered using a continuous drilling core sampler with rotary technique. These latter cores, which were 10 cm in diameter and reached a depth of 10 m, were located in the upstream, central and downstream part of the study bar (Figure 2c). Cores were kept humid in PVC liners and successively cut longitudinally, sampled for grain-size analysis, photographed at high resolution and preserved for making dry peels with epoxy. Each core was characterized following the basic principles of facies analysis: highlighting sediment grain size, color, oxidation, sedimentary structures and occurrence of bioturbation, plant debris and shell fragments.

The terminology used in this work is graphically summarized in Figure 3. The channel thalweg is defined as the deepest part of the active channel, where the coarsest deposits occur. Riffles and pools are situated at bend inflections and bend apexes, respectively, and correspond to the shallower and the deeper portions of the thalweg, respectively. Sinuosity is calculated as the ratio between the along-channel distance between two adjacent riffles and their linear distance (Figure 3). Straight channels are characterized by sinuosity close to 1, whereas sinuous channels reach values higher than 2.5.

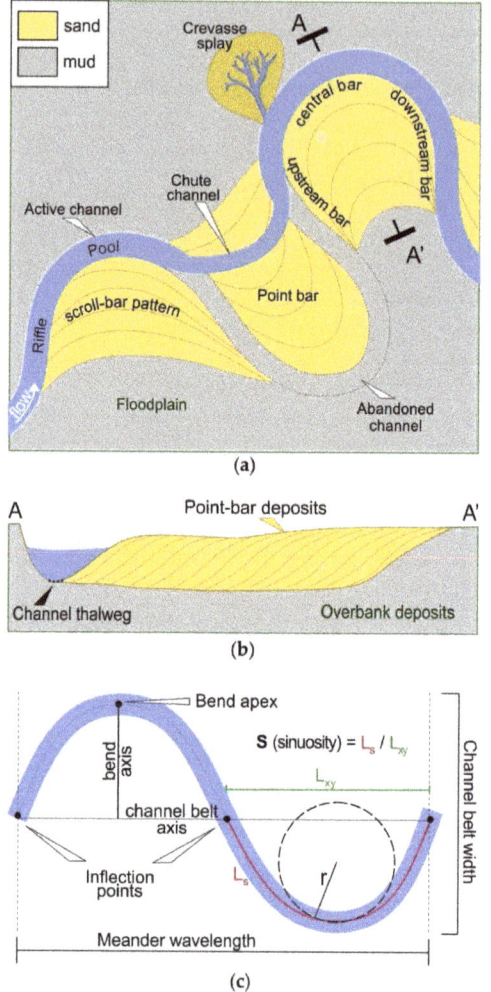

Figure 3. Descriptive terminology for fluvial meanders and related deposits: (**a**) alluvial plain morphological elements, (**b**) point-bar deposits and (**c**) meander-belt morphometry.

3. Results

3.1. Remote Sensing

The analyzed satellite images and aerial pictures give a consistent idea of the changes in the land use of the area and the visibility of the paleohydrographic traces over the last 60 years. In particular, recent high-resolution true-color composite images allow mapping these features with submetric accuracy. The traces mainly consist of cropmarks with vegetation suffering from hydraulic stress during the growing season, and dead patches in July and August. Significant changes can be observed over very short periods (compare, e.g., Figure 4d,f) during the hot season, while sensitivity is low in spring time (e.g., Figure 4e) and is lost in winter, when the fields are covered by scarce vegetation and are bare and plowed, respectively. By comparing satellite images mostly taken during the growing season (i.e., those that best show the traces) differences in vegetation colors allow the identification of buried morphologies of two distinct reaches, hereafter named Reach A and Reach B, and a crevasse splay. Reach A consists of two major bends, called Bend 1 and Bend 2 (Figure 4b). The NDVI and NDMI images derived from Sentinel-2 summer images (Figure 5c,d) clearly show the differential behavior of the paleomeanders in comparison to the external floodplain and the inner portion surrounded by Reach A; the vegetation growing on more permeable sandy soil is less healthy than the one living on finer sediments. The paleohydrographic traces are much more evidenced by NDVI and NDMI during the summer season and they clearly help in identifying the general pattern of the abandoned channels. However, the rather low resolution of the Sentinel-2 images does not contribute significantly to discriminating in detail the specific morphological and sedimentological features composing the paleohydrographic traces.

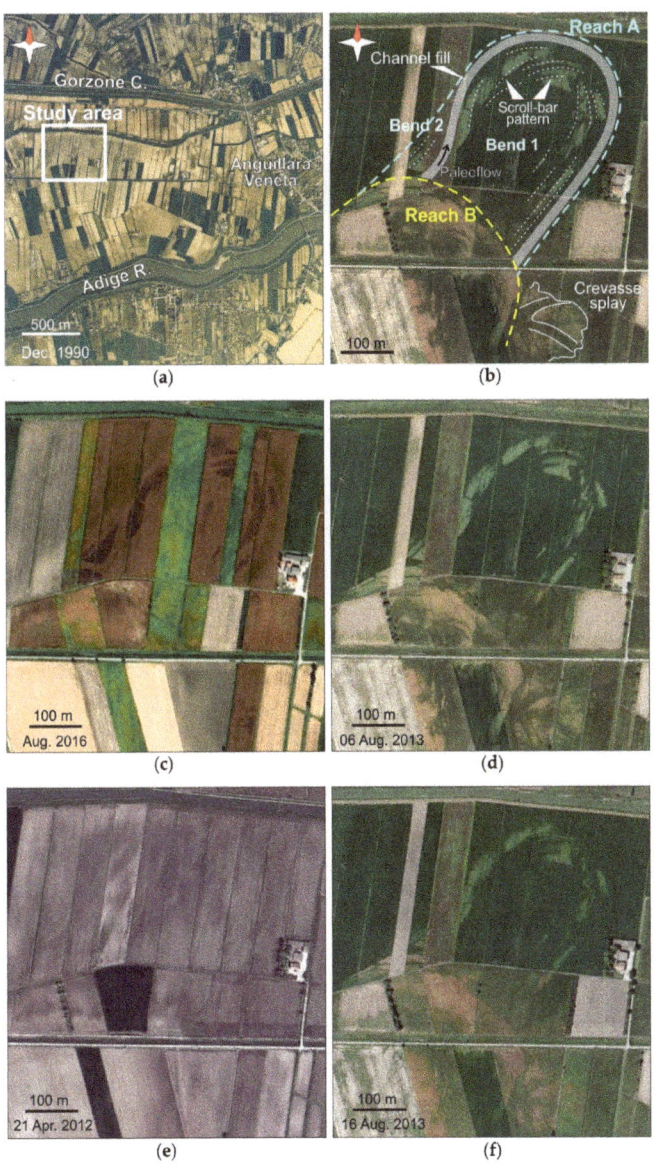

Figure 4. Remote-sensing results: (**a**) aerial photogram of the study area enhancing the poorly visible fluvial pattern during winter time; (**b**) meander belt reconstruction and fluvial morphology identification in the study area—white dashed lines highlight evidence of scroll-bar morphologies; and (**c–f**) satellite images (2016, 2013, 2012) providing information about fluvial morphology on the basis of seasons and soil use.

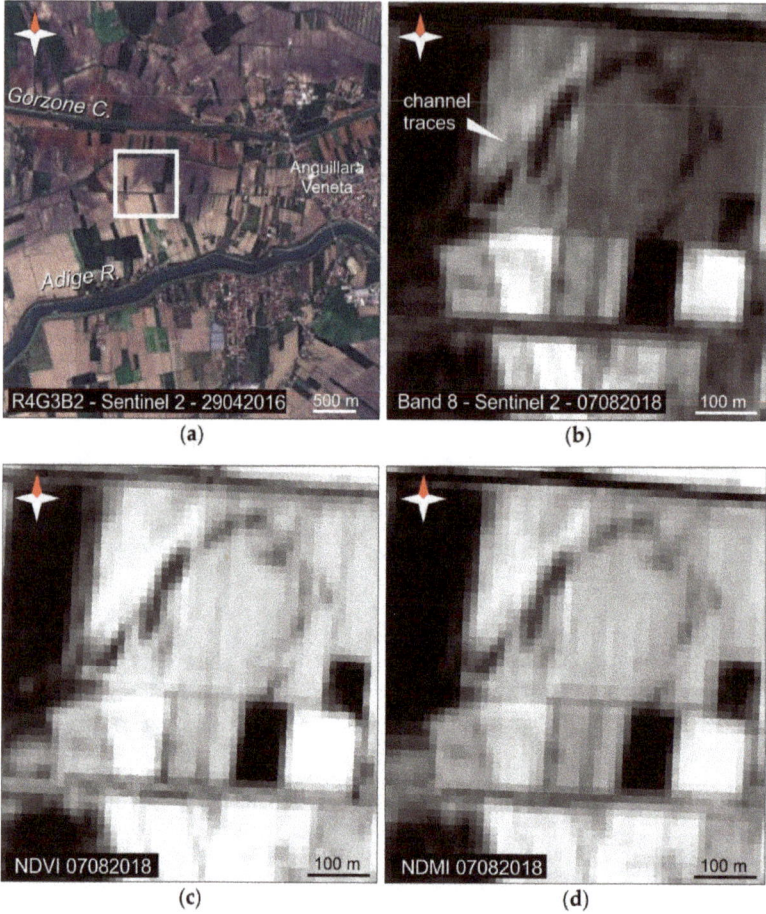

Figure 5. Processed satellite images: (**a**) RGB combination on Sentinel-2 image to highlight petrographic differences in the study area; (**b**) band 8, of Sentinel-2 image of the study area, showing paleochannel morphologies; and (**c,d**) NDVI and NDMI indexes calculated on Sentinel 2 images (2018), respectively, enhancing different plant health and hydraulic stress.

Bend 1 is an open bend, with an SSW–NNE bend axis, and is characterized by a sinuosity of about 2.2 and a radius of curvature of ca. 140 m. The scroll-bar pattern is particularly clear from cropmarks, in the northcentral portion of the bend (Figure 4c,d), showing a different signal compared to the residual channel fill (i.e., light cropmark when the others are dark, and vice-versa), testifying the progressive growth of the meander bend. The channel fill displays a width of about 15–20 m and can be better defined where bounded by opposite-trending scroll-bar patterns, like in the upstream side of the bend. The riffle-to-riffle distance on the channel fill is about 260 m.

Bend 2 is a low sinuosity bend (i.e., 1.12), with an NW–SE bend axis, characterized by an estimated radius of curvature and riffle-to-riffle distance of 135 and 230 m, respectively. Bend 2 is sited upstream from Bend 1, and shows a scroll-bar pattern that testifies a progressive expansional growth style (sensu [77]) of the bend.

Reach B forms a bend occurring south of Reach A, but is less visible from satellite images and its sinuosity cannot be defined. The radius of curvature is ca. 350 m and the axis of the bend trends ca. SW–NE, although satellite images do not show a clear bar-scroll pattern and the position of the relevant channel fill. Reach B cuts over Reach A suggesting that it developed after a chute channel that cut off Reach A [78], which was later abandoned. Additionally, along the eastern side of Reach B, (Figure 4d,f) a divergent pattern of minor channels point to a local development of a crevasse splay sourced from the downstream side of the bend. Several straight dark stripes, with a width of about 1 m, located on Reach B, can be interpreted as traces of abandoned ditches that were associated with a drainage system dating back to Renaissance times and dismissed later on (Figure 4d).

3.2. 3D Electrical Conductivity Model from EMI

The inversion of the EMI data produced a 3D volume of electrical conductivity values. The results are shown in Figure 6, where we elected to show the volume sliced horizontally at eight depth levels down to a depth of nearly 8 m below ground, corresponding each to a layer selected in the inversion approach. Note that the inversion was conducted with an Occam approach, but using a limited number of layers compatible with the information contained in the six different acquisition configurations obtainable with the CMD Explorer instrument.

An arcuate sedimentary body having low electrical conductivity (i.e., a resistive body) is clearly visible at a depth between 1 and 6 m below ground. The internal boundary of this arcuate body (see Slice 5 in Figure 6b) is fully visible in the maps, and shows a radius of curvature and a sinuosity of ca. 60 and 2.3, respectively. The external boundary (see Slice 5 in Figure 6b) slightly debouches from the maps, but its radius of curvature and sinuosity can be estimated to be ca. 135 and 2.2, respectively (Figure 6b), as also confirmed by remote-sensing results. Orientation of the outer boundary of this body fits with the orientation of meander Bend 1 of Reach A, as depicted by remote-sensing analyses, and is also consistent with the associated scroll pattern (Figure 4b), suggesting that these low-resistivity deposits represent the point-bar body associated with meander Bend 1 of Reach A. Of course, the main contribution of the EMI data is to provide continuous and extensive depth information that is not available from remote sensing. In the shallower layers (Slices 1–5), the arcuate point-bar body presents low conductivity values with $\sigma < 20$ mS/m, and its conductivity is still close to 40 mS/m at about 5–6 m below ground (Slices 6 and 7; Figure 6b). Note that the width of the most resistive part of the bar is clearly shrinking with depth, thus showing the 3D shape of the sand body. At larger depths (Slice 8—below 6.1 m) conductivity increases up to 180 mS/m, delimiting the base of the bar body. It must be noted, however, that the CMD Explorer provides, as a rule of thumb, reliable information only down to 6 m below ground and thus Slice 8 is effectively an extrapolation due to the need to have an infinite semispace at the bottom of the electrical conductivity model, and thus should be considered with care. Although the point-bar body shows a fairly homogeneous electrical resistivity, a subtle increase in resistivity values defines a 20 m narrow, NNE–SSW trending belt in the SE corner of Slices 2 to 6. The location of this belt fits with that of the abandoned channel forming the meander Bend 1 as apparent in satellite images (Figure 4b), and suggests that the higher resistivity values are linked to the coarser material of the deposits filling the abandoned channel. Deposits surrounding the low-resistivity point-bar body show conductivity values spanning from 80 to 250 mS/m, with values close to 100 mS/m down to 3 m below ground, increasing to 250 mS/m below 3 m. Comparison between geophysical data and geomorphic evidences suggest that these electrically conductive sediments represent floodplain deposits in which the Bend 1 meander was cut, thus developing the related point-bar sedimentary body.

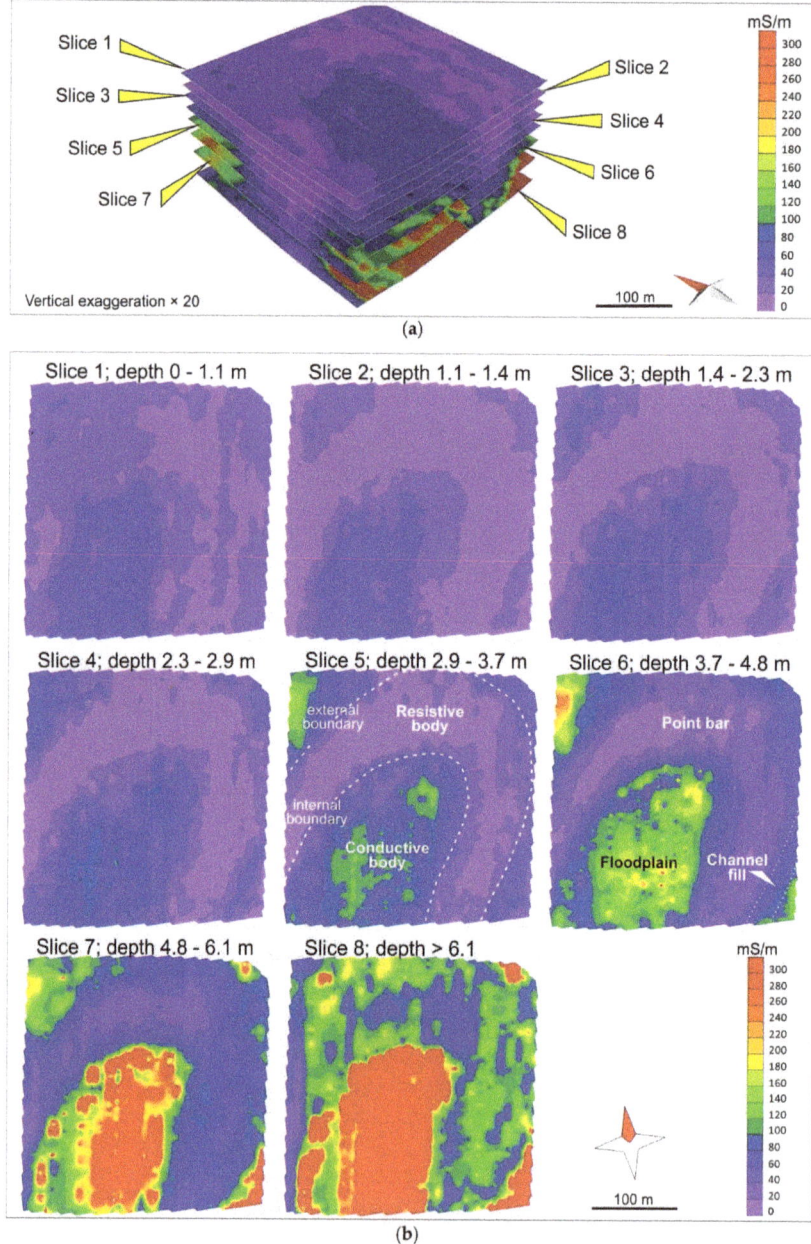

Figure 6. Geophysical results: (**a**) 3D view of the eight 2D conductivity maps; (**b**) the eight 2D maps showing difference in conductivity values and highlighting point-bar, channel-fill and floodplain morphologies.

3.3. Sedimentology

Core data provide the ground-truth information needed to calibrate/confirm geophysical data with localized information. The cores help define sedimentary features of the electrically resistive point-bar body, related channel-fill deposits and surrounding electrically conductive overbanks. All cores reveal that the deposits were intensely reworked by agricultural activities down to 80 cm below ground. Note that this fact may pose significant limitations to remote-sensing interpretation that is forcibly limited to surface images. Reworked deposits are dark brown and consist of very fine sand with a variable amount of mud. Point-bar deposits were completely cored at sites AV_a–c (Figure 7). Point-bar deposits occur from 0.8 m to a maximum of 5.4 m below ground and mainly consist of sand with a scarce percentage of mud. Cores AV_a–c reveal that bar deposits cover either organic-rich mud (core AV_b) or sandy deposits (cores AV_a and AV_c). Point-bar deposits are floored by a channel lag that consists of massive medium sand with pebble-sized mudclasts (Figure 8a). This basal lag is covered by lower bar deposits, consisting of 1–1.5 m of mud-free, well-sorted fine to medium sand, which is commonly massive or crudely plane-parallel stratified (Figure 8b). Upper bar deposits are ca. 2.5–3 m thick and consist of fine to very fine sand with subordinate mud layers. Sand is plane-parallel to ripple cross-laminated (Figure 8c) and contains mud for ca. 12%, 21% and 20% in the upstream, central and downstream zone, respectively. Mud layers (Figure 8c) range in thickness between 0.5 and 2 cm, and consist of massive or crudely laminated mud with plant debris. Lower bar deposits are ubiquitously mud free. The overall grain size of the bar deposits does not relevantly change along the bar, which appears as an almost monotonous sandy body from its upstream to downstream reach.

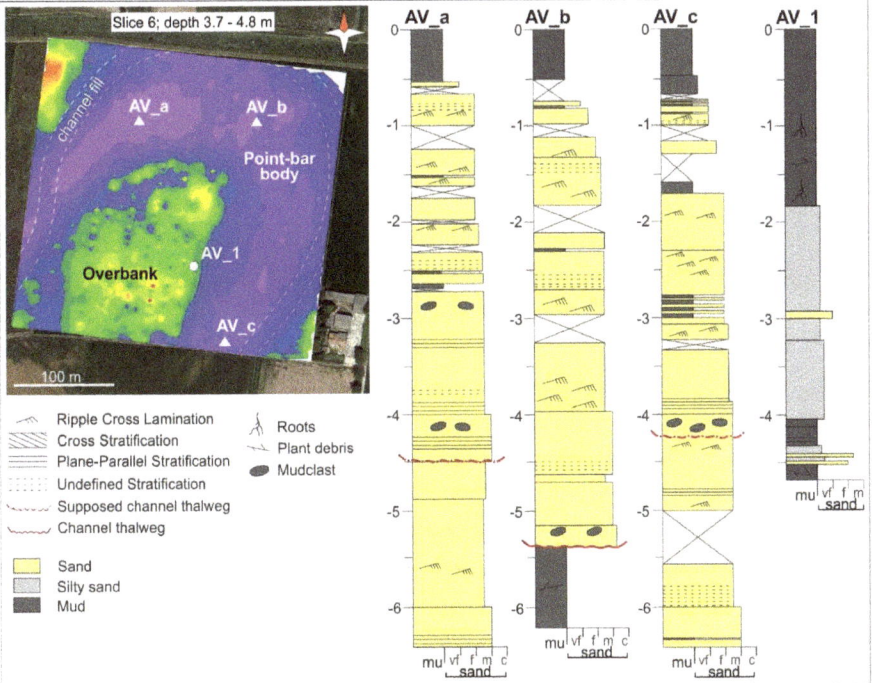

Figure 7. Sedimentary features of cores for bar and overbank deposits at the study site. Location of cores is shown with the conductivity for Slice 6.

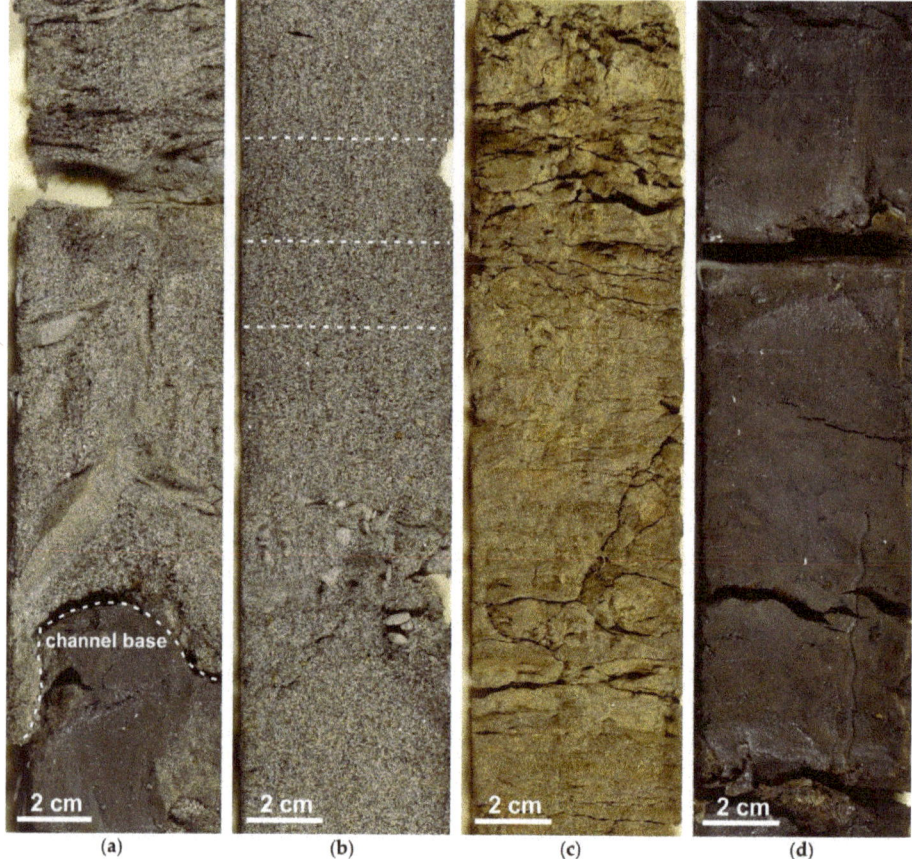

Figure 8. Cored deposits: (**a**) massive medium sand with pebble-sized mudclasts forming the channel lag; (**b**) fine to medium sands from the lower point bar; (**c**) fine to very fine sands, with cross lamination and mud layers from the upper point bar; and (**d**) massive overbank mud with moderate organic content and horizontal bedding.

The overbank deposits were cored where geophysical investigations reveal the occurrence of electrically conductive sediments. These deposits mainly consist of silt-rich mud with subordinate sandy layers with horizontal bedding. Mud is massive and can be organic-rich (Figure 8d) or slightly oxidized in the lower and upper part of the overbank succession, respectively.

4. Discussion

4.1. Implications for Noninvasive (Remote Sensing and Ground-Based Electromagnetic) Investigations

This study shows how remote sensing and ground-based geophysical data represent an ideal combination of noninvasive techniques that can guide direct investigations and, at the same time, can be integrated to provide a 3D reconstruction of the shallow subsoil once supported by the local direct evidence for verification and calibration.

Our results show that in the study area the use of aerial images is very effective in supporting the rapid recognition of geomorphological and sedimentological features with a resolution approaching 0.5 m. However, the aerial pictures dating back to before the 1980s do not provide useful paleohydrographic evidence, probably because (a) conventional zenithal pictures were originally taken for cadastral purposes and, thus, were shot during the winter season when vegetation cover was limited; and (b) widespread leveling of the fields and use of strong plowing machines was common until that time, thus causing severe erosion of the topsoil, especially in the zones where convex landforms were formerly present, with slight accumulation in the depressions. Thus, in correspondence with natural levees and sand ridges related to scroll-bar sequences, the sandy and silty sediments were exhumed, showing a lighter signal in the soilmarks.

Cropmarks appear to be, in general, the most effective indicators of shallow geomorphological features in the considered environment. This is linked to the much higher permeability of the coarser sediments, leading to greater drainage and, in absence of irrigation, water stress to crops especially during the hot season (July and August). As documented in other areas of the Venetian Plain, e.g., [79], besides the weather condition of the period before the image acquisition, the maximum detail shown by the cropmarks strongly depends on the type of cultivated plants and, in particular, it decreases with the size of the leaves and the plant spacing. The best data are generally found in zones with soya bean or hay meadow crops. Wheat, barley and corn display variable visibility, as the first two are seeded in tight rows and have small leaves, but are harvested before mid July; in contrast, corn has a larger plant size and larger distance between rows, but is harvested in September or October and thus can (usefully) experience water stress. Note that this study was carried out analyzing freely available high-resolution images, reaching a resolution of about 0.5 m. This suggests that the use of specific images, acquired through latest commercial satellite or drone-borne multispectral scanners, could easily support the recognition of features between 0.1 and 0.5 m with superior results.

Geophysical data play a critical role in our analyses. They bridge the gap between surface-extensive information provided by remote sensing (that primarily guided the identification of the study area and the location of the surveys) with the information at depth carried out through traditional drilling and sampling operations. The latter in turn were positioned on the basis of remote sensing and geophysical evidence, thus minimizing the sampling to the locations where this information was needed for verification and calibration. The geophysical data constitute the backbone of the study in that they provide ultimately the 3D information to fully reconstruct the sedimentary structure of the site. This result, however, is not trivial to achieve.

First, great care must be posed in the acquisition phase. This entails not only the choice of the instrument and the strategy developed for covering a large area—in this case a (nonconductive) sledge towed by a tractor at sufficient distance not to induce current in the metal frame of the tractor itself. In addition, care must be paid when setting the instrument to have a good control of the measurement geometry, which is in turn essential to obtaining reliable inversion results. In this case, the stability of the sledge and the positioning care allowed us to obtain, for all data points, an RMS error of less than 10% between measured and simulated apparent conductivity data at the end of the inversion process. Note that carrying the same sonde by hand, particularly on rugged terrain, can easily unbalance the instrument, with measurements thus taken with some coils much closer to the ground than others. This might induce a very large measurement error that is impossible to correct a posteriori, as the true acquisition geometry is then completely unknown.

Second, obvious outliers must be removed from the dataset. These may include negative or extremely high apparent conductivity values (that are physically implausible for the given acquisition geometry), which may be due to local metallic or magnetic features.

Third, an accurate reconstruction of positioning must be made. The availability of reliable colocated data from HMD and VMD is essential to have the six independent pieces of information necessary for depth inversion. This requires both a guided pace in the field to reoccupy roughly the same locations and proper postprocessing to assemble the data that pertain to the same reasonable surrounding of each measurement point, in this case with a radius of 1 m. Noting that the sonde size itself is of a few meters (Figure 2a), this accuracy is perfectly acceptable for the purpose at hand.

Fourth, geophysical inversion is an inherently ill-posed problem. In this case, it means that while the general pattern of the electrical conductivity variation with depth is constrained by the data, the subtler details may not be retrieved uniquely. In other words, at each measurement location, a number of different 1D models, which all, however, maintain certain key patterns, can be equivalent in terms of goodness of fit to the measured data within the data uncertainty range. In particular, for example, if conductivity increases with depth, all inverted profiles will display the same general pattern. However, the transition from lower to higher conductivity may happen continuously, or in steps, and steps may occur at slightly different depths. While this can be viewed as a weakness of geophysical methods (and EMI in this particular case), it is not without remedy. Indeed, geophysics should never be applied without some "ground-truth" coming (as in this case) from drilling investigations (local and not without their uncertainties, but still necessary). Thus, an iterative revision of the inverted vertical profiles was conducted to select, among the plausible electrical conductivity profiles those that also fit reasonably to the direct evidence where this evidence is available. The procedure was simply performed manually, particularly selecting suitable layer interfaces compatible with drilling evidence. This type of procedure should always be applied, and it is in the energy exploration where 3D seismic data are blended, e.g., with well logs coming from deep borings. For the geophysical data used in this study, the "ground-truth" is given by the local comparison between lithology from sedimentary cores (AV_1, AV_a, AV_b and AV_c) and electrical conductivity derived from EMI inversion at the same locations. A plot showing the resulting correlation, also taking into account the variability of electrical conductivity for the same lithology (shown as standard deviation error bars) is shown in Figure 9.

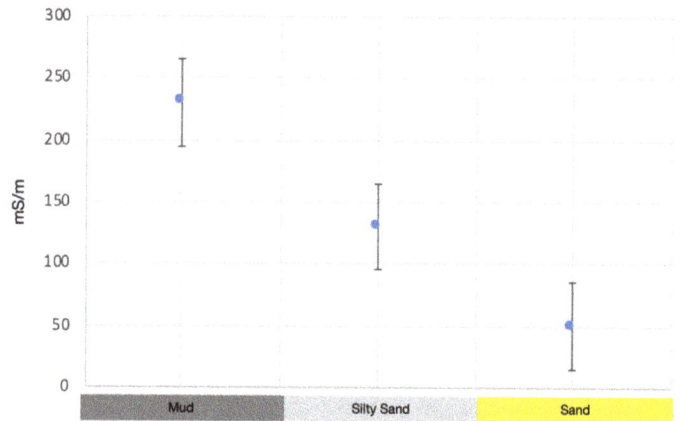

Figure 9. Plot of the interpretation of the EMI inverted conductivity intervals after the correlation with the sedimentary cores AV_1, AV_a, AV_b and AV_c.

4.2. Implications for Investigation of Meandering River Deposits

Integrated remote-sensing, geophysical and sedimentological approaches provide insights to discuss key features of investigated point-bar deposits, with specific emphasis on their genesis and internal grain-size variability.

The external boundary of this body is consistent with the morphology of a point bar that originated through lateral migration of a meandering channel [80–82]. Nevertheless, the curved profile of the inner boundary suggests that the bar accretion started from the inner bank of an arcuate channel that showed a sinuosity of ca. 2.3 (Figure 6b). This evidence contrasts the common assumption that point bars originate from a progressive increase in channel sinuosity of a relatively straight channel, which gradually migrates laterally until reaching a sinuous configuration [83–86]. The onset of accretion from a sinuous channel allows for storing a reduced volume of bar sediment in comparison with that produced by inception from a straight channel. In the study case, the 3D reconstruction of the bar body from EMI data shows that about 1.8 million m^3 of sand are stored in the arcuate point bar body (Figure 10a). If one assumes that the accretion of this bar started from a straight channel, as would be suggested by the application of classical sedimentological models to remote-sensing data, the estimated volume of the bar would have been of ca. 3.1 million m^3 of sand (Figure 10b), leading to a remarkable overestimation of the accumulated sand. The onset of point-bar bar accretion from a sinuous channel was probably driven by pre-existing floodplain morphologies, which, at the early stage of channel development, forced water to drain through the paths defined by adjacent depressed areas [87,88]. The establishment of a curved planiform morphology could have been forced by floodplain lithological and morphological heterogeneities [89,90], which are associated with the occurrence of numerous overbank subenvironments, including crevasse-splays, levees, floodplain lakes and floodbasins (cf., [91]). Different deposits and morphologies associated with these subenvironments possibly forced the newly formed watercourse to connect adjacent depressed areas and assume a sinuous geometry.

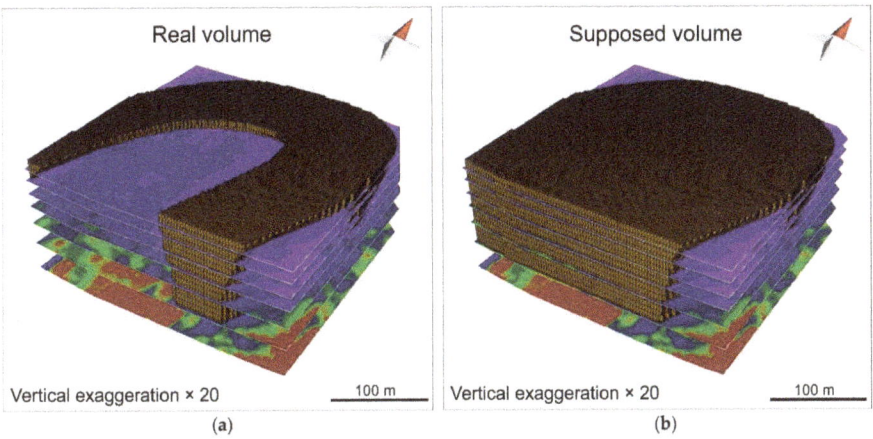

Figure 10. old9. Volume of the point-bar body (**a**) reconstructed through the present study and (**b**) inferred assuming bar growth from a straight channel, as suggested by classical sedimentological models.

The location of the sedimentary cores within the point bar provides information concerning both vertical and lateral grain-size variability within this sedimentary body, with a particular focus on the comparison between upstream and downstream bar deposits. Although a fining-upward grain size distribution has been broadly considered to be typical of point-bar deposits [85,91,92], a certain variability of vertical grain-size distribution has also been documented [93,94]. Core data show that muddy layers occur in the upper part of the bar, although mud is visibly subordinate to sand. The grain size of sand varies significantly neither vertically nor downstream, and the bar is, therefore, characterized by widespread weak vertical grain-size trends. Although the lack of a clear vertical fining in the upstream bar zone is consistent with the occurrence of high bed shear stress [95,96], the paucity of muddy deposits in the downstream part of the bar is a peculiar feature, which cannot be ascribed to the overall lack of mud in the system, being that the overbank deposits are entirely made of mud. The open morphology of the bend [97] associated with the study bar could have hindered the formation of a dead zone, which commonly forms in sharp bends [18], preventing the accumulation of mud in the downstream bar zone.

Figure 11 shows a summary of the workflow we used to extract the relevant information from each data source and blend the pieces of information towards the final conceptual 3D distributed model of the studied site.

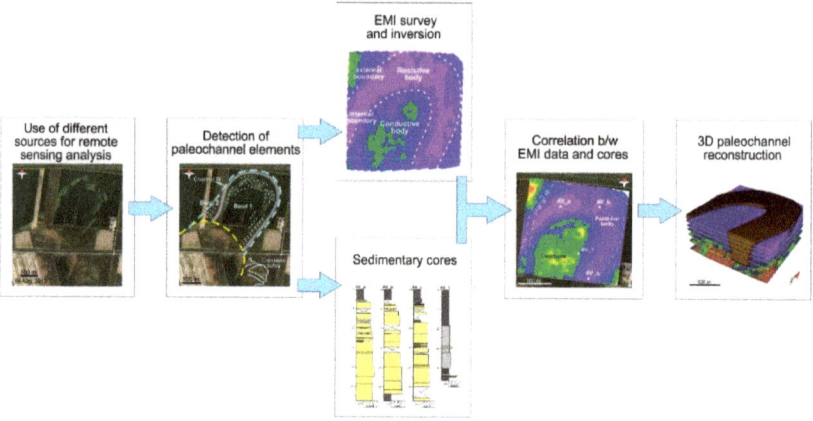

Figure 11. Workflow of the overall methodology adopted in this study.

5. Conclusions

This paper presents a successful integrated approach to analyze the distribution of sedimentary facies of a paleomeander in the Southern Venetian Plain, Northeastern Italy. The approach is based on a combination of remote-sensing (aerial and satellite) data, geophysical investigations (electromagnetic surveys) and direct sedimentary coring.

From the methodological point of view, we show that the combined use of noninvasive techniques such as remote sensing and ground-based geophysical data provides an effective method for the purpose at hand. In particular, remote sensing is quite effective for the identification of sites of interest and features at a metric scale, which is potentially linked to different subsoil structures. In the case considered here, cropmarks are the most useful features observed from the satellite images, due specifically to the water stress induced in crops by the higher permeability of sandy bodies with respect to silty sediments. However, remote sensing can only provide information on the ground surface. On the contrary, geophysical methods are specifically designed to reconstruct the subsurface structure on the basis of contrasts of geophysical parameters. In this case, we used electromagnetic induction (EMI) methods, and particularly an FDEM small-scale multicoil system. Well-designed, acquired, processed, and inverted EMI data allowed us to extend the surface information provided by remote sensing to a maximum depth exceeding 6 m below ground level, allowing the construction of a 3D model of electrical conductivity of the subsoil. Direct investigations via sedimentary core drilling were positioned on the basis of remote sensing and geophysical data, in order to confirm and calibrate the geophysical investigations, which were also partly reinverted on the basis of the new evidence. The overall cycle of investigations thus allowed us to set up a 3D stratigraphic model of the site, consistent with all available data. On the other hand, the sequence of investigation activities was designed in such a manner that the information collected at one step optimized the design of the next step, thus reducing the overall effort required to complete the task.

From the sedimentary point of view, the point bar studied shows an uncommon arcuate morphology, that contrasts the common assumption that point bars originate from a progressive sinuosity increase of a relatively straight channel that migrates laterally until reaching a sinuous configuration. This can be explained by considering the variety of alluvial subenvironments in the floodplain. These floodplain heterogeneities likely controlled water fluxes over the platform, by facilitating water drainage within traces of depressed areas, defining the sinuous shape of the study channel during the very first phases of channel formation. As far as grain size distribution is concerned, although classical facies models highlight overall trends of upward and downstream fining of grain size within point-bar deposits, the grain-size trends of the study bar do not vary significantly either vertically or laterally. The bar is, indeed, characterized by a widespread weak vertical grain-size trend, and it appears as a homogeneous body of medium to fine sand. The lower mud content in the downstream portion was probably a result of the open morphology of the bend that could have prevented the formation of the dead zone, which is commonly directly linked to mud accumulation in the downstream portion of the sharp bends.

This study provides a solid basis for developing more detailed sedimentological investigations, which could be improved including acquisition of data concerning internal stratal architecture of the alluvial deposits. GPR investigations and recovery of undisturbed sedimentary cores would provide further relevant insights to this approach, with relevant follow up in the frame of subsurface exploration or management of surficial aquifers. Detection of the distinctive morphometric and sedimentological features of Late Holocene paleochannels would allow a comparison with those of the rivers draining the area currently, and allow quantification of human impact on riverine dynamics [18].

Author Contributions: Data curation, E.B.; formal analysis, J.B. and G.C.; funding acquisition, M.G.; investigation, E.B., M.G., A.F. and J.B.; methodology, G.C., M.G. and J.B.; project administration, E.B. and M.G.; resources, M.G., J.B. and G.C.; software, G.C.; supervision, M.G.; validation, A.D.; visualization, E.B. and M.G.; writing—original draft, E.B., A.F., G.C. and J.B.; writing—review and editing, M.G., G.C. and A.D. All authors have read and agreed to the published version of the manuscript.

Funding: This work has been supported by project HYDROSEM (Progetti di Eccellenza CARIPARO 2017, Cassa di Risparmio di Padova e Rovigo): "Fluvial and tidal meanders of the Venetian-Po plain: From hydrodynamics to stratigraphy" project (PI. M. Ghinassi).

Acknowledgments: Reviewers and editors will be acknowledged. The authors thank M. Cosma for logistic support during field work and R. Bonato, the land owner of the study area.

Conflicts of Interest: The authors declare no conflict of interest.

References

1. Clement, W.P.; Barrash, W. Crosshole radar tomography in a fluvial aquifer near Boise, Idaho. *J. Environ. Eng. Geophys.* **2006**, *11*, 171–184. [CrossRef]
2. Galgaro, A.; Finzi, E.; Tosi, L. An experiment on a sand-dune environment in Southern Venetian coast based on GPR, VES and documentary evidence. *Ann. Geophys.* **2000**, *43*, 289–295. [CrossRef]
3. Nofal, E.R.; Amer, M.A.; El-Didy, S.M.; Fekry, A.M. Delineation and modeling of seawater intrusion into the Nile Delta Aquifer: A new perspective. *Water Sci.* **2015**, *29*, 156–166. [CrossRef]
4. Da Lio, C.; Carol, E.; Kruse, E.; Teatini, P.; Tosi, L. Saltwater contamination in the managed low-lying farmland of the Venice coast, Italy: An assessment of vulnerability. *Sci. Total Environ.* **2015**, *533*, 356–369. [CrossRef] [PubMed]
5. Desbarats, A.J.; Koenig, C.E.M.; Pal, T.; Mukherjee, P.K.; Beckie, R.D. Groundwater flow dynamics and arsenic source characterization in an aquifer system of West Bengal, India. *Water Resour. Res.* **2014**, *50*, 4974–5002. [CrossRef]
6. Carraro, A.; Fabbri, P.; Giaretta, A.; Peruzzo, L.; Tateo, F.; Tellini, F. Effects of redox conditions on the control of arsenic mobility in shallow alluvial aquifers on the Venetian Plain (Italy). *Sci. Total Environ.* **2015**, *532*, 581–594. [CrossRef]
7. Romeo, R.W.; Amoroso, S.; Facciorusso, J.; Lenti, L.; Madiai, C.; Martino, S.; Monaco, P.; Rinaldis, D.; Totani, F. Soil liquefaction during the Emilia, 2012 seismic sequence: Investigation and analysis. *Eng. Geol. Soc. Territ.* **2015**, *5*, 1107–1110. [CrossRef]
8. Amorosi, A.; Bruno, L.; Facciorusso, J.; Piccin, A.; Sammartino, I. Stratigraphic control on earthquake-induced liquefaction: A case study from the Central Po Plain (Italy). *Sediment. Geol.* **2016**, *345*, 42–53. [CrossRef]

9. Fontana, D.; Lugli, S.; Dori, S.M.; Caputo, R.; Stefani, M. Sedimentology and composition of sands injected during the seismic crisis of May 2012 (Emilia, Italy): Clues for source layer identification and liquefaction regime. *Sediment. Geol.* **2015**, *325*, 158–167. [CrossRef]
10. Mozzi, P. Alluvial plain formation during the Late Quaternary between the southern Alpine margin and the Lagoon of Venice (Northern Italy). *Geogr. Fis. Din. Quat.* **2005**, *7*, 219–229.
11. Ninfo, A.; Ferrarese, F.; Mozzi, P.; Fontana, A. High resolution dems for the analysis of fluvial and ancient anthropogenic landforms in the alluvial plain of Padua (Italy). *Geogr. Fis. Din. Quat.* **2011**, *34*, 95–104. [CrossRef]
12. Castigoni, G.B. Geomorphology of the Po Plain. *Earth Surf. Process. Landf. J. Br. Geomorphol. Res. Group* **1999**, *24*, 1115–1120.
13. Mehdi, S.M.; Pant, N.C.; Saini, H.S.; Mujtaba, S.A.I.; Pande, P. Identification of palaeochannel configuration in the Saraswati River basin in parts of Haryana and Rajasthan, India, through digital remote sensing and GIS. *Episodes* **2016**, *39*, 29–38. [CrossRef]
14. De Rossetti, D.F. Multiple remote sensing techniques as a tool for reconstructing late Quaternary drainage in the Amazon lowland. *Earth Surf. Process. Landf.* **2010**, *35*, 1234–1239. [CrossRef]
15. Wray, R.A.L. Palaeochannels of the Namoi River Floodplain, New South Wales, Australia: The use of multispectral Landsat imagery to highlight a Late Quaternary change in fluvial regime. *Aust. Geogr.* **2009**, *40*, 29–49. [CrossRef]
16. Entwistle, N.; Heritage, G.; Milan, D. Recent remote sensing applications for hydro and morphodynamic monitoring and modelling. *Earth Surf. Process. Landf.* **2018**, *43*, 2283–2291. [CrossRef]
17. Demarchi, L.; Bizzi, S.; Piégay, H. Regional hydromorphological characterization with continuous and automated remote sensing analysis based on VHR imagery and low-resolution LiDAR data. *Earth Surf. Process. Landf.* **2017**, *42*, 531–551. [CrossRef]
18. Piégay, H.; Arnaud, F.; Belletti, B.; Bertrand, M.; Bizzi, S.; Carbonneau, P.; Dufour, S.; Liébault, F.; Ruiz-Villanueva, V.; Slater, L. Remotely sensed rivers in the Anthropocene: State of the art and prospects. *Earth Surf. Process. Landf.* **2020**, *45*, 157–188. [CrossRef]
19. Langat, P.K.; Kumar, L.; Koech, R. Monitoring river channel dynamics using remote sensing and GIS techniques. *Geomorphology* **2019**, *325*, 92–102. [CrossRef]
20. Righini, M.; Surian, N. Remote Sensing as a Tool for Analysing Channel Dynamics and Geomorphic Effects of Floods. In *Flood Monitoring through Remote Sensing*; Springer: Cham, Switzerland, 2018; pp. 27–59.
21. Finotello, A.; Lanzoni, S.; Ghinassi, M.; Marani, M.; Rinaldo, A.; D'Alpaos, A. Field migration rates of tidal meanders recapitulate fluvial morphodynamics. *Proc. Natl. Acad. Sci. USA* **2018**, *115*, 1463–1468. [CrossRef]
22. Fontana, A.; Mozzi, P.; Bondesan, A. Alluvial megafans in the Venetian-Friulian Plain (north-eastern Italy): Evidence of sedimentary and erosive phases during Late Pleistocene and Holocene. *Quat. Int.* **2008**, *189*, 71–90. [CrossRef]
23. Fontana, A.; Mozzi, P.; Bondesan, A. Late Pleistocene evolution of the Venetian-Friulian Plain. *Rend. Lincei* **2010**, *21*, 181–196. [CrossRef]
24. Piovan, S.; Mozzi, P.; Stefani, C. Bronze age paleohydrography of the southern Venetian Plain. *Geoarchaeology* **2010**, *25*, 6–35. [CrossRef]
25. Brivio, L.; Ghinassi, M.; D'Alpaos, A.; Finotello, A.; Fontana, A.; Roner, M.; Howes, N. Aggradation and lateral migration shaping geometry of a tidal point bar: An example from salt marshes of the Northern Venice Lagoon (Italy). *Sediment. Geol.* **2016**, *343*, 141–155. [CrossRef]
26. Parasnis, D.S. *Principles of Applied Geophysics*, 5th ed.; Chapman & Hall: London, UK, 1997.
27. Telford, W.M.; Geldart, L.P.; Sheriff, R.E. *Applied Geophysics*; Cambridge University Press: Cambridge, UK, 1990.
28. Everett, M.E.; Meju, M.A. Near-Surface Controlled-Source Electromagnetic Induction: Background and Recent Advances. In *Hydrogeophysics*; Rubin, Y., Hubbard, S.S., Eds.; Springer: Berlin/Heidelberg, Germany, 2005; Volume 50, pp. 157–184. ISBN 978-1-4020-3101-4.
29. Boaga, J. The use of FDEM in hydrogeophysics: A review. *J. Appl. Geophys.* **2017**, *139*, 36–46. [CrossRef]
30. Lesch, S.M.; Strauss, D.J.; Rhoades, J.D. Spatial Prediction of Soil Salinity Using Electromagnetic Induction Techniques: 1. Statistical prediction models: A comparison of multiple linear regression and cokriging Identification and Estimation. *Water Resour. Res.* **1995**, *31*, 373–386. [CrossRef]

31. Paine, J.G. Determining salinization extent, identifying salinity sources, and estimating chloride mass using surface, borehole, and airborne electromagnetic induction methods. *Water Resour. Res.* **2003**, *39*. [CrossRef]
32. Sambuelli, L.; Leggieri, S.; Calzoni, C.; Porporato, C. Study of riverine deposits using electromagnetic methods at a low induction number. *Geophysics* **2007**, *72*, B113–B120. [CrossRef]
33. Greenhouse, J.P.; Slaine, D.D. The use of reconnaissance electromagnetic methods to map contaminant migration: These nine case studies can help determine which geophysical techniques are applicable to a given problem. *Groundw. Monit. Remediat.* **1983**, *3*, 47–59. [CrossRef]
34. Martinelli, P.; Duplaá, M.C. Laterally filtered 1D inversions of small-loop, frequency-domain EMI data from a chemical waste site. *Geophysics* **2008**, *73*, F143–F149. [CrossRef]
35. Cassiani, G.; Ursino, N.; Deiana, R.; Vignoli, G.; Boaga, J.; Rossi, M.; Perri, M.T.; Blaschek, M.; Duttmann, R.; Meyer, S.; et al. Noninvasive Monitoring of Soil Static Characteristics and Dynamic States: A Case Study Highlighting Vegetation Effects on Agricultural Land. *Vadose J.* **2012**, *11*. [CrossRef]
36. Gebbers, R.; Lück, E.; Heil, K. Depth sounding with the EM38-detection of soil layering by inversion of apparent electrical conductivity measurements. In *Precision Agricolture'07*; Stafford, J.V., Ed.; Wageningen Academic Publisher: Skiathos, Greece, 2007; pp. 95–102. ISBN 978-90-8686-024-1.
37. Yao, R.; Yang, J. Quantitative evaluation of soil salinity and its spatial distribution using electromagnetic induction method. *Agric. Water Manag.* **2010**, *97*, 1961–1970. [CrossRef]
38. Osella, A.; De la Vega, M.; Lascano, E. 3D electrical imaging of an archaeological site using electrical and electromagnetic methods. *Geophysics* **2005**, *70*, G101–G107. [CrossRef]
39. Thiesson, J.; Dabas, M.; Flageul, S. Detection of resistive features using towed Slingram electromagnetic induction instruments. *Archaeol. Prospect.* **2009**, *16*, 103–109. [CrossRef]
40. Perri, M.T.; Boaga, J.; Bersan, S.; Cassiani, G.; Cola, S.; Deiana, R.; Simonini, P.; Patti, S. River embankment characterization: The joint use of geophysical and geotechnical techniques. *J. Appl. Geophys.* **2014**, *110*, 5–22. [CrossRef]
41. Huang, H.; SanFilipo, B.; Oren, A.; Won, I.J. Coaxial coil towed EMI sensor array for UXO detection and characterization. *J. Appl. Geophys.* **2007**, *61*, 217–226. [CrossRef]
42. Corwin, D.L.; Rhoades, J.D. An Improved Technique for Determining Soil Electrical Conductivity-Depth Relations from Above-ground Electromagnetic Measurements. *Soil Sci. Soc. Am. J.* **1982**, *46*, 517–520. [CrossRef]
43. McNeill, J.D. Electromagnetic Terrain Conductivity Measurement at Low Induction Numbers. *Tech. Note* **1980**, *6*, 13.
44. Wait, J.R. *Geo-Electromagnetism*; Academic Press: New York, NY, USA, 1982.
45. Frischknecht, F.C.; Labson, V.F.; Spies, B.R.; Anderson, W.L. Profiling methods using small sources. In *Electromagnetic Methods in Applied Geophysics*; Nabighian, M.N., Ed.; Society of Exploration Geophysicists: Tulsa, OK, USA, 1991; Volume 2, pp. 105–270.
46. Deidda, G.P.; Fenu, C.; Rodriguez, G. Regularized solution of a nonlinear problem in electromagnetic sounding. *Inverse Probl.* **2014**, *30*, 125014. [CrossRef]
47. Von Hebel, C.; Rudolph, S.; Mester, A.; Huisman, J.A.; Kumbhar, P.; Vereecken, H.; van der Kruk, J. Three-dimensional imaging of subsurface structural patterns using quantitative large-scale multiconfiguration electromagnetic induction data. *Water Resour. Res. Res.* **2014**, *50*, 2732–2748. [CrossRef]
48. Saey, T.; De Smedt, P.; Delefortrie, S.; Van De Vijver, E.; Van Meirvenne, M. Comparing one- and two-dimensional EMI conductivity inverse modeling procedures for characterizing a two-layered soil. *Geoderma* **2015**, *241–242*, 12–23. [CrossRef]
49. Shanahan, P.W.; Binley, A.; Whalley, W.R.; Watts, C.W. The Use of Electromagnetic Induction to Monitor Changes in Soil Moisture Profiles beneath Different Wheat Genotypes. *Soil Sci. Soc. Am. J.* **2015**, *79*, 459–466. [CrossRef]
50. Frederiksen, R.R.; Christiansen, A.V.; Christensen, S.; Rasmussen, K.R. A direct comparison of EMI data and borehole data on a 1000 ha data set. *Geoderma* **2017**, *303*, 188–195. [CrossRef]
51. Boaga, J.; Ghinassi, M.; D'Alpaos, A.; Deidda, G.P.; Rodriguez, G.; Cassiani, G. Geophysical investigations unravel the vestiges of ancient meandering channels and their dynamics in tidal landscapes. *Sci. Rep.* **2018**, *8*, 3303905. [CrossRef]
52. Viezzoli, A.; Christensen, A.V.; Auken, E.; Sørensen, K. Quasi-3D modeling of airborne TEM data by spatially constrained inversion. *Geophysics* **2008**, *73*, F105–F113. [CrossRef]

53. Piovan, S.; Mozzi, P.; Zecchin, M. The interplay between adjacent Adige and Po alluvial systems and deltas in the late Holocene (Northern Italy). *Géomorphol. Process. Environ.* **2012**, *18*, 427–440. [CrossRef]
54. De Franco, R.; Biella, G.; Tosi, L.; Teatini, P.; Lozej, A.; Chiozzotto, B.; Giada, M.; Rizzetto, F.; Claude, C.; Mayer, A.; et al. Monitoring the saltwater intrusion by time lapse electrical resistivity tomography: The Chioggia test site (Venice Lagoon, Italy). *J. Appl. Geophys.* **2009**, *69*, 117–130. [CrossRef]
55. Mozzi, P.; Piovan, S.; Corrò, E. Long-term drivers and impacts of abrupt river changes in managed lowlands of the Adige river and northern PO delta (Northern Italy). *Quat. Int.* **2018**, *538*, 80–93. [CrossRef]
56. Teatini, P.; Tosi, L.; Strozzi, T.; Carbognin, L.; Wegmüller, U.; Rizzetto, F. Mapping regional land displacement in the Venice coastland by an integrated monitoring system. *Remote Sens. Environ.* **2005**, *98*, 403–413. [CrossRef]
57. Carminati, E.; Martinelli, G.; Severi, P. Influence of glacial cycles and tectonics on natural subsidence in the Po Plain (Northern Italy): Insights from 14C ages. *Geochem. Geophys. Geosystems* **2003**, *4*. [CrossRef]
58. Teatini, P.; Tosi, L.; Strozzi, T. Quantitative evidence that compaction of Holocene sediments drives the present land subsidence of the Po Delta, Italy. *J. Geophys. Res. Earth Surf.* **2011**, *116*. [CrossRef]
59. ArcGIS. Available online: https://www.esri.com/en-us/arcgis/about-arcgis/overview (accessed on 20 March 2020).
60. QGIS. Available online: https://www.qgis.org/it/site/ (accessed on 25 March 2020).
61. Il Geoportale Della Regione del Veneto. Available online: https://idt2.regione.veneto.it/ (accessed on 3 April 2020).
62. Recanatesi, F.; Giuliani, C.; Ripa, M.N. Monitoring mediterranean oak decline in a peri-urban protected area using the NDVI and sentinel-2 images: The Case Study of Castelporziano state natural reserve. *Sustainability* **2018**, *10*, 3308. [CrossRef]
63. Bannari, A.; Morin, D.; Bonn, F.; Huete, A.R. A review of vegetation indices. *Remote Sens. Rev.* **1995**, *13*, 95–120. [CrossRef]
64. Piedelobo, L.; Taramelli, A.; Schiavon, E.; Valentini, E.; Molina, J.L.; Xuan, A.N.; González-Aguilera, D. Assessment of green infrastructure in Riparian zones using copernicus programme. *Remote Sens.* **2019**, *11*, 2967. [CrossRef]
65. Goodwin, N.R.; Coops, N.C.; Wulder, M.A.; Gillanders, S.; Schroeder, T.A.; Nelson, T. Estimation of insect infestation dynamics using a temporal sequence of Landsat data. *Remote Sens. Environ.* **2008**, *112*, 3680–3689. [CrossRef]
66. Mather, P.M.; Koch, M. *Computer Processing of Remotely-Sensed Images: An Introduction*, 4th ed.; John Wiley & Sons: Hoboken, NJ, USA, 2011; ISBN 9780470742396.
67. Congedo, L. Semi-Automatic Classification Plugin Documentation. *Release* **2016**, *4*, 29. [CrossRef]
68. MOVE Suite-Petroleum Experts. Available online: https://www.petex.com/products/move-suite/ (accessed on 11 May 2020).
69. GF Instruments S.R.O. Available online: www.gfinstruments.cz (accessed on 10 February 2020).
70. Um, E.S.; Alumbaugh, D.L. On the physics of the marine controlled-source electromagnetic method. *Geophysics* **2007**, *72*, WA13–WA26. [CrossRef]
71. Trimble-Transforming the Way the World Works. Available online: www.trimble.com (accessed on 10 February 2020).
72. Allred, B.; Daniels, J.J.; Ehsani, R.M. *Handbook of Agricultural Geophysics*; CRC Press: Boca Raton, FL, USA, 2008; ISBN 9780849337284.
73. Delefortrie, S.; De Smedt, P.; Saey, T.; Van De Vijver, E.; Van Meirvenne, M. An efficient calibration procedure for correction of drift in EMI survey data. *J. Appl. Geophys.* **2014**, *110*, 115–125. [CrossRef]
74. Interpex Limited-Specialists in PC Based Geophysical Software. Available online: www.interpex.com (accessed on 3 February 2020).
75. Constable, S.C.; Parker, R.L.; Constable, C.G. Occam's inversion: A practical algorithm for generating smooth models from electromagnetic sounding data. *Geophysics* **1987**, *52*, 289–300. [CrossRef]
76. Ghinassi, M.; Ielpi, A.; Aldinucci, M.; Fustic, M. Downstream-migrating fluvial point bars in the rock record. *Sediment. Geol.* **2016**, *334*, 66–96. [CrossRef]
77. Ghinassi, M. Chute channels in the Holocene high-sinuosity river deposits of the Firenze plain, Tuscany, Italy. *Sedimentology* **2011**, *58*, 618–642. [CrossRef]

78. Ninfo, A.; Fontana, A.; Mozzi, P.; Ferrarese, F. The Map of Altinum, Ancestor of Venice. *Science* **2009**, *325*, 577. [CrossRef] [PubMed]
79. Bhattacharyya, P.; Bhattacharya, J.P.; Khan, S.D. Paleo-channel reconstruction and grain size variability in fluvial deposits, Ferron Sandstone, Notom Delta, Hanksville, Utah. *Sediment. Geol.* **2015**, *325*, 17–25. [CrossRef]
80. Ghinassi, M.; Nemec, W.; Aldinucci, M.; Nehyba, S.; Özaksoy, V.; Fidolini, F. Plan-form evolution of ancient meandering rivers reconstructed from longitudinal outcrop sections. *Sedimentology* **2014**, *61*, 952–977. [CrossRef]
81. Durkin, P.R.; Hubbard, S.M.; Smith, D.G.; Leckie, D.A. Predicting heterogeneity in meandering fluvial and tidal-fluvial deposits: The point bar to counter point bar transition. In *Fluvial Meanders and Their Sedimentary Products in the Rock Record*; Ghinassi, M., Colombera, L., Mountney, N.P., Reesink, J.H., Eds.; John Wiley & Sons: Hoboken, NJ, USA, 2018; Volume 48, pp. 231–250.
82. Brice, J.C. Evolution of meander loops. *Geol. Soc. Am. Bull.* **1974**, *85*, 581–586. [CrossRef]
83. Lewin, J. Initiation of bed forms and meanders in coarse-grained sediment. *Geol. Soc. Am. Bull.* **1976**, *87*, 281–285. [CrossRef]
84. Nanson, G.C.; Page, K. Lateral accretion of fine-grained concave benches on meandering rivers. In *Modern and Ancient Fluvial Systems*; Collinson, J.D., Lewin, J., Eds.; John Wiley & Sons, Inc.: Hoboken, NJ, USA, 1983; Volume 6, pp. 133–143. ISBN 0632009977.
85. Wu, C.; Bhattacharya, J.P.; Ullah, M.S. Paleohydrology and 3D facies architecture of ancient point bars, Ferron Sandstone, Notom Delta, south-central Utah, USA. *J. Sediment. Res.* **2015**, *85*, 399–418. [CrossRef]
86. Jones, H.L.; Hajek, E.A. Characterizing avulsion stratigraphy in ancient alluvial deposits. *Sediment. Geol.* **2007**, *202*, 124–137. [CrossRef]
87. Taylor, C.F.H. The role of overbank flow in governing the form of an anabranching river: The Fitzroy River, northwestern Australia. *Fluv. Sedimentol. VI Spec. Publ. Int. Assoc. Sedimentol.* **1999**, *28*, 77–91. [CrossRef]
88. Motta, D.; Abad, J.D.; Langendoen, E.J.; García, M.H. The effects of floodplain soil heterogeneity on meander planform shape. *Water Resour. Res.* **2012**, *48*, 1–17. [CrossRef]
89. Bogoni, M.; Putti, M.; Lanzoni, S. Modeling meander morphodynamics over self-formed heterogeneous floodplains. *Water Resour. Res.* **2017**, *53*, 5137–5157. [CrossRef]
90. Fidolini, F.; Ghinassi, M.; Aldinucci, M.; Billi, P.; Boaga, J.; Deiana, R.; Brivio, L. Fault-sourced alluvial fans and their interaction with axial fluvial drainage: An example from the Plio-Pleistocene Upper Valdarno Basin (Tuscany, Italy). *Sediment. Geol.* **2013**, *289*, 19–39. [CrossRef]
91. Nanson, G.C. Point bar and floodplain formation of the meandering Beatton River, northeastern British Columbia, Canada. *Sedimentology* **1980**, *27*, 3–29. [CrossRef]
92. Ielpi, A.; Ghinassi, M. Planform architecture, stratigraphic signature and morphodynamics of an exhumed Jurassic meander plain (Scalby Formation, Yorkshire, UK). *Sedimentology* **2014**, *61*, 1923–1960. [CrossRef]
93. Swan, A.; Hartley, A.J.; Owen, A.; Howell, J. Reconstruction of a sandy point-bar deposit: Implications for fluvial facies analysis. In *Fluvial Meanders and Their Sedimentary Products in the Rock Record*; Ghinassi, M., Colombera, L., Mountney, N.P., Reesink, A.J., Betaman, M., Eds.; John Wiley & Sons Ltd.: Hoboken, NJ, USA, 2018; Volume 48, pp. 445–474.
94. Frothingham, K.M.; Rhoads, B.L. Three-dimensional flow structure and channel change in an asymmetrical compound meander loop, Embarras River, Illinois. *Earth Surf. Process. Landf. J. Br. Geomorphol. Res. Gr.* **2003**, *28*, 625–644. [CrossRef]
95. Kasvi, E.; Vaaja, M.; Alho, P.; Hyyppä, H.; Hyyppä, J.; Kaartinen, H.; Kukko, A. Morphological changes on meander point bars associated with flow structure at different discharges. *Earth Surf. Process. Landf.* **2013**, *38*, 577–590. [CrossRef]
96. Finotello, A.; D'Alpaos, A.; Bogoni, M.; Ghinassi, M.; Lanzoni, S. Remotely-sensed planform morphologies reveal fluvial and tidal nature of meandering channels. *Sci. Rep.* **2020**, *10*, 54. [CrossRef]
97. Ferguson, R.I.; Parsons, D.R.; Lane, S.N.; Hardy, R.J. Flow in meander bends with recirculation at the inner bank. *Water Resour. Res.* **2003**, *39*. [CrossRef]

© 2020 by the authors. Licensee MDPI, Basel, Switzerland. This article is an open access article distributed under the terms and conditions of the Creative Commons Attribution (CC BY) license (http://creativecommons.org/licenses/by/4.0/).

Article

Mapping the Groundwater Potentiality of West Qena Area, Egypt, Using Integrated Remote Sensing and Hydro-Geophysical Techniques

Ahmed Gaber [1,*], Adel Kamel Mohamed [2], Ahmed ElGalladi [2], Mohamed Abdelkareem [3], Ahmed M. Beshr [2] and Magaly Koch [4]

1. Geology Department, Faculty of Science, Port-Said University, Port-Said 42522, Egypt
2. Geology Department, Faculty of Science, Mansoura University, Mansoura 35516, Egypt; adelkamel@mans.edu.eg (A.K.M.); galladi@mans.edu.eg (A.E.); beshr@mans.edu.eg (A.M.B.)
3. Geology Department, South Valley University, Qena 83523, Egypt; mohamed.abdelkareem@sci.svu.edu.eg
4. Center for Remote Sensing, Boston University, Boston, MA 02215, USA; mkoch@bu.edu
* Correspondence: ahmedgaber_881@hotmail.com

Received: 16 April 2020; Accepted: 11 May 2020; Published: 14 May 2020

Abstract: The integrated use of remote sensing imagery and hydro-geophysical field surveys is a well-established approach to map the hydrogeological framework, and thus explore and evaluate the groundwater potentiality of desert lands, where groundwater is considered as the main source of freshwater. This study uses such integrated approach to map the groundwater potentiality of the desert alluvial floodplain of the Nile Valley west of Qena, Egypt, as alternative water source to the River Nile. Typically ground gradient, faults and their stress field, lateral variation of rock permeability, drainage patterns, watersheds, rainfall, lithology, and soil types are the main factors believed to affect the groundwater recharge and storage from the infiltration of present-time and paleo-runoff. Following this generally accepted approach, different remote sensing data sets (SRTM DEM, Landsat-8, ALOS/PALSAR-1, Sentinel-1, and TRMM) as well as auxiliary maps (geological and soil maps) were used to identify and map these factors and prepare thematic maps portraying the different influences they exert on the groundwater recharge. These thematic maps were overlaid and integrated using weights in a GIS framework to generate the groundwater potentiality map which categorizes the different recharge capabilities into five zones. Moreover, the aeromagnetic data were processed to map the deep-seated structures and estimate the depth to basement rocks that may control the groundwater occurrence. In addition, the vertical electrical sounding (VES) measurements were applied and calibrated with the available borehole data to delineate the subsurface geological and hydrogeological setting as well as the groundwater aquifers. Different geoelectric cross-sections and hydro-geophysical maps were constructed using the borehole information and VES interpretation results to show the lateral extension of the different lithological units, groundwater-bearing zones, water table, and the saturated thickness of the aquifer. The GIS model and geophysical results show that the southwest part of Nag'a Hammadi-El-Ghoneimia stretch has very high recharge and storage potentiality and is characterized by the presence of two groundwater-bearing zones. The shallow groundwater aquifer is located at a depth of 30 m with a saturation thickness of more than 43 m. However, there are NW–SE faults crossing the study area and most likely serve as recharge conduits by connecting the shallow aquifer with the deeper ones. Such aquifers connection has been confirmed by investigating the chemical and isotopic composition of their groundwater.

Keywords: remote sensing and GIS; field geophysics; groundwater potentiality; West Qena; Egypt

1. Introduction

Water is the most important natural resource for human life; it is the basis for enabling developments (agricultural, industrial, and urban). Meanwhile, Egypt lies within an arid desert zone in the northeastern corner of Africa, where rainfall is very low and irregular. Surface freshwater resources are scarce and limited to the River Nile crossing from south to north. Most of the cultivated lands are located surrounding the Nile and its Delta, while the majority of the country is desert lands. The total cultivated area covers about 4% of Egypt and is overpopulated with about 95% of the people [1]. Since ancient times, the Nile covers all the water needs for all developmental activities. However, with the rapid and continuous increase in population, the demand for water, food, and housing is increasing. Consequently, the water of the River Nile became insufficient causing a significant fresh-water shortage. Thus, in order to maintain the current flow in the River Nile new groundwater resources need to be discovered and explored. This will allow establishing new agriculture land, industrial and urban communities to accommodate the demands of a growing population.

The old alluvial floodplain of the Nile Valley west of Qena Governorate represents one of the promising desert areas for land reclamation based on its groundwater resources. It is a part of a land reclamation mega project called "1.5 Million Acres" that has been proposed by the Egyptian Government to create new urban-agricultural communities away from the River Nile. The study area is located at the southern stretch of Qena-Nag'a Hammadi-El-Ghoneimia and extends between latitudes 25°47′N and 26°9′N and longitudes 31°51′E and 32°44′E (Figure 1). It covers a vast area of about 1432 km². It is bounded from the north and east by the cultivated young alluvial floodplain of the Nile and from the west and south by the elevated limestone plateau of the Western Desert. There are no surface water resources in the study area and the groundwater represents the only freshwater resource. The study area is characterized by a hot arid desert climate with a total annual rainfall of about 3.2 mm/year [2]. Therefore, the contribution of recent rainfall events to groundwater recharge is very low and most of the groundwater stored in the subsurface aquifers are attributed to the paleo-torrential rains and flash floods that took place during the past pluvial periods [3].

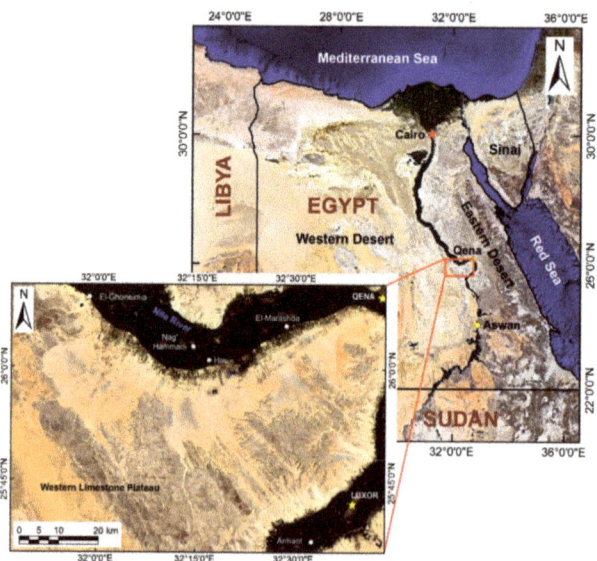

Figure 1. Location map of Egypt and West Qena area.

Satellite remote sensing is considered an effective tool for mapping the Earth's surface features. Although it does not directly observe the subsurface, it has been effectively used in groundwater exploration [4]. Analysis and interpretation of the remote sensing images enable mapping and studying the surface factors that control runoff infiltration such as geological settings, soil cover, drainage networks, and climate. These features indirectly reflect the hidden hydrogeological characteristics of the subsurface, such as the capability to recharge, accumulate, and store groundwater. Studying the influences of these factors on water infiltration capacity in an integrated manner using the Geographic Information System (GIS) facilitates a more effective delineation of zones with different groundwater potentialities [5–7]. Subsequent field geophysical surveys can this way be targeted to promising zones with high groundwater potentiality. Many studies have been attempted to integrate the groundwater controlling parameters derived from remote sensing data using GIS, such as Shaban et al., 2006; Amarsaikhan et al., 2009; Abdalla, 2012; Madani and Niyazi, 2015 [8–11].

Since satellite remote sensing observes mainly at surface (optical sensors) and near surface (radar sensors) features, its integration with other complementary data sources such as elevation data and field geophysical measurements is considered a very effective tool to characterize the hydrogeological setting of any area and particularly poorly mapped arid areas. In this context, the magnetic method is one of the oldest and widely used geophysical techniques for exploring the Earth's subsurface [12]. The magnetic method is used in hydrogeological studies to delineate the thick sedimentary basins suitable for groundwater accumulation, map the tectonic structures, and estimate the depth to basement rocks [13,14].

Moreover, over the last decades geoelectric resistivity measurements using the vertical electrical sounding (VES) technique have proven their efficiency in groundwater exploration, providing useful information about the subsurface hydrogeological framework [15]. The VES aims to measure the resistivity variation with depth. It is used to delineate and estimate the thickness and depth of the groundwater-bearing zones [16,17].

Finally, a strategy for sustainable development in arid lands requires intensive scientific studies to assess the groundwater storage and availability in order to predict the long-term impact of abstraction to decide the proper developmental plans that should be considered. The present study employs an integrated methodology based on a combination of remote sensing and geophysical techniques (aeromagnetics and VES) to explore the groundwater potentiality and subsurface hydrogeological setting of West Qena area in Egypt to support decision makers in their future developmental plans.

2. The Study Area

The study area is a part of the Nile Valley in Qena region and follows the Nile's tectonic, geological, and hydrogeological setting. The Nile Valley cuts its course through the Early Eocene limestone rocks and divides Egypt into the Eastern Desert and the Western Desert. Many authors consider the valley of tectonic origin [3,18–21]. According to these studies, the Nile Valley developed from a number of successive rivers along a path determined by structural features formed in response to the Oligo-Miocene tectonic activities in the Gulf of Suez and Red Sea region. This is supported by the fault scarps bordering the cliffs of the valley and the numerous faults observed along the valley flanks [22]. The valley north of the study area (Nag'a Hammadi) is oriented in NW–SE direction parallel to the Gulf of Suez, whereas its trend changes to NE–SW east of Nag'a Hammadi and intersects from the east with the Qena-Safaga Shear Zone. These trends were observed on the surface of the limestone plateau [23] and also recorded in the subsurface by geophysical data [24].

Geologically, the surface of the study area is covered by different alluvial sediments ranging in age from Pliocene to Quaternary [23]. These sediments were studied by Said (1981) and Omer (1996) [3,25], who divided them into a number of stratigraphic units, which are from older to younger: Muneiha Formation, Armant Formation (older fanglomerates), Issawia Formation, Qena Formation, fanglomerates, and Wadi deposits (Figure 2). The bounding plateau represents the oldest exposed rocks and consists of a succession ranging in age from Late Cretaceous to Early Eocene. The exposed rocks of the plateau were studied by Faris et al. (1985) and Issawi et al. (2009) [26,27], who divided it into four units, which are from older to younger: Dakhla Formation, Tarawan Formation, Esna Formation, and Thebes Formation.

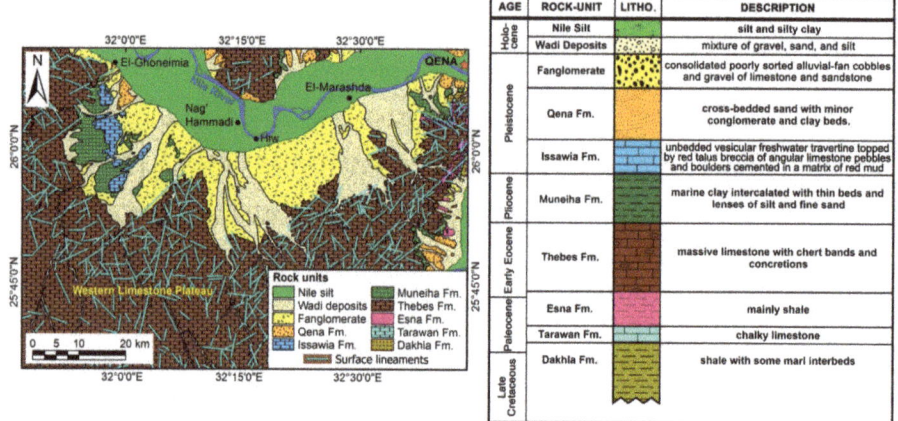

Figure 2. Geological map of the study area (modified after CONOCO and EGPC, 1987) and a composite stratigraphic column of the Nile Valley (Said, 1981; Omer, 1996; Issawi et al., 2009). The solid green color surrounding the River Nile represents the vegetation cover.

Hydrogeologically, the Quaternary aquifer represents the main groundwater resource in the Nile Valley. It consists mainly of the Pleistocene graded sand and gravel intercalated with clay lenses (i.e., Qena Formation). It is underlain by an impermeable layer of Pliocene clays (i.e., Muneiha Formation) that prevents its connection with the deeper aquifers. It is covered by a permeable layer of Wadi deposits at the old alluvial floodplain, which means that groundwater occurs under unconfined conditions. On the other hand, it occurs under semi-confined conditions at the cultivated lands where it is capped by the agricultural silt-clay layer. The thickness of the aquifer varies from about 200 m at the center of the cultivated floodplain to about 80 m at the desert fringes. It is recharged continuously from the excess irrigation water and occasionally from infrequent rainfalls [28–31].

3. Data and Methods

3.1. Remote Sensing Data

Remote sensing images and GIS analysis techniques were used to delineate the different groundwater potential zones in the old alluvial floodplain of the Nile Valley west of Qena City. The performed steps were (1) identifying and mapping the different surface factors that are believed to control the infiltration of the surface runoff to recharge the underlying unconfined aquifers; (2) reclassification of each factor with respect to its influence on infiltration/recharge (thematic layers); (3) giving each thematic layer a rank and weight based on assessing its relative importance to groundwater recharge; and finally (4) integrating all the thematic maps in a GIS model using the weighted overlay method to generate the groundwater potentiality map of the study area.

In this study, eight factors were mapped from a combination of optical, radar, and DEM data, as well as geological and soil maps. These factors include the surface topographic elevation, slope, lineament density, watershed areas, drainage density, rainfall, lithology, and soil cover. The SRTM DEM data with 30 m spatial resolution was used to represent the surface elevation and generate the slope, and delineate the different watersheds and their associated drainage networks using the ArcMap software of ESRI (2016) [32]. The drainage and lineaments densities were calculated using the spatial analyst tool in ArcGIS (Figure 3). Drainage networks were extracted using the D-8 algorithm [33,34] and applying a thresholds value of 1000 of the calculated flow accumulations. The relation between the extracted drainage patterns and the substrate soil textures was interpreted based on Ayele et al., 2017 [35]. Whereas, the lineament densities were extracted using the ordinary kriging of ArcGIS to show the spatial distribution of lineaments and highlight areas where they concentrate [36].

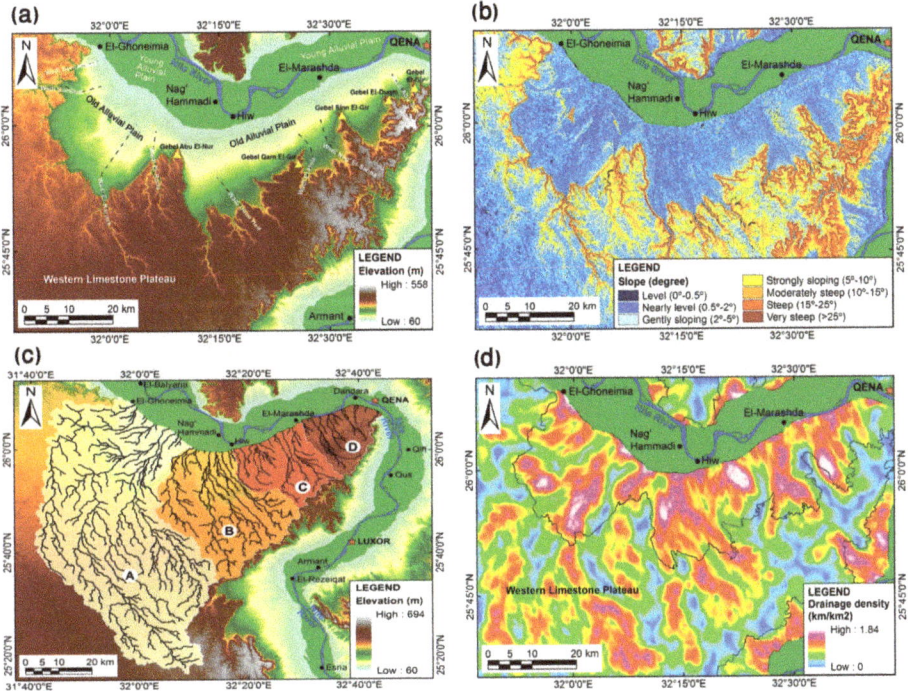

Figure 3. (**a**) SRTM digital elevation model (DEM); (**b**) surface slope; (**c**) watersheds and drainage channels; (**d**) drainage density. The solid green color surrounding the River Nile represents the vegetation cover.

The surface geological lineaments were manually extracted from different optical and radar images that included Landsat-8 OLI (https://earthexplorer.usgs.gov/) [37], ALOS/PALSAR-1 (JAXA), and Sentinel-1 (https://scihub.copernicus.eu/) [38] with the aid of the several generated hillshades with different illumination angles and geologic map. The process of lineament extraction begins with image enhancement and filtering to highlight the linear features of interest to facilitate identifying and tracing them [39]. In this context, several shaded relief maps were derived from the DEM by applying the hillshade algorithm in ArcMap using different light azimuth angles (45°, 90°, 135°, 180°, 225°, 270°, 315°, 360°) (Figure 4a). The Sobel filter [40] was applied to the Landsat-8 panchromatic band with 15 m spatial resolution to increase the contrast of edges or linear features to make them stand out from the background (Figure 4b) using the ENVI software [41]. The Sobel filter effectively enhances directional linear features by convolving a 3 × 3 window size over the image in the four principle directions: N–S, NE–SW, E–W, and NW–SE [42,43]. In this research work, the lineaments are considered as geological structures, i.e., faults, fractures, and joints that might serve as passage ways for groundwater flow along the study area [44,45]. In addition, a high spatial resolution Google Earth image overlay was used to avoid tracing any of the non-geological or man-made linear features, which are not related to structural lineaments, such as roads, tracks, and field boundaries.

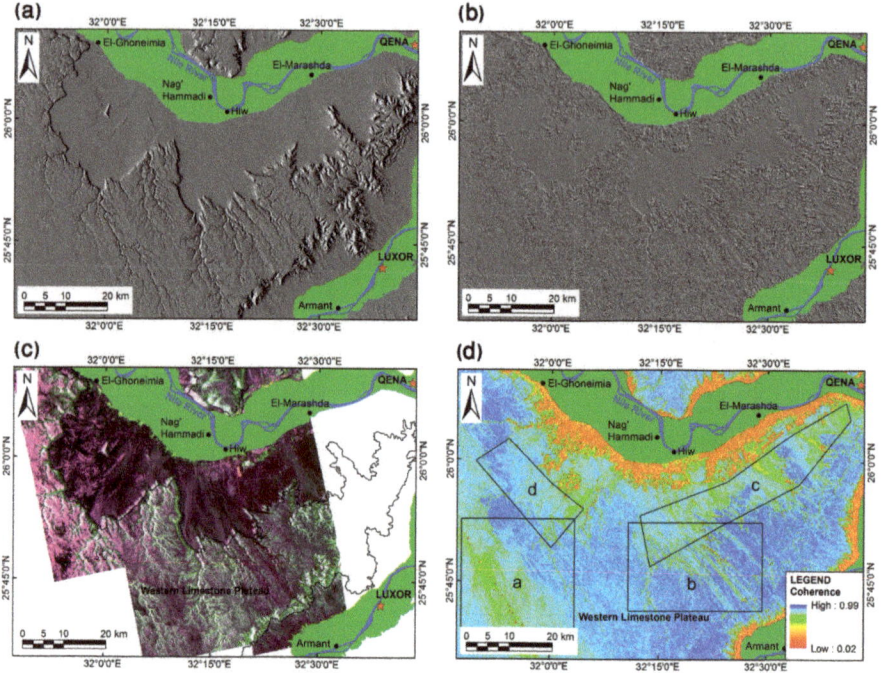

Figure 4. (a) Hillshaded DEM with an illumination angle of 270°; (b) Sobel-filtered Landsat-8 panchromatic band highlighting the NW–SE direction; (c) Pauli RGB image of the full-polarimetric ALOS/PALSAR-1 image; (d) Sentinel-1 InSAR coherence image.

Synthetic aperture radar (SAR) sensors operate with longer electromagnetic waves than optical sensors, thus they have the capability to penetrate the dry and loose sand cover to image any hidden or buried linear features [46]. In this work, both the ALOS/PALSAR L-band radar image (24 cm wavelength) and the Sentinel-1 C-band radar images (5.6 cm wavelength) were processed to map any hidden lineaments that were not revealed by the optical images along the flat alluvium floodplain of the study area. The full-polarimetric ALOS/PALSAR-1 image was processed by applying the Pauli RGB classification (Figure 4c) using SNAP software of ESA (2017) [47]. Two Sentinel-1 radar images acquired on 7 August 2015 and 13 August 2016 with Vertical-Vertical (VV) polarization were used to produce the interferometric synthetic aperture radar (InSAR) coherence image of the study area using SNAP software (Figure 4c). The InSAR coherence image represents the difference in the phase and intensity of the pixels of two radar images acquired at different times. It enables the detection of the movement and changes of mobile sands that are controlled by subsurface structures as well as the land surface erosion due to the aeolian and fluvial activities, especially in arid deserts. Such dynamic processes cause changes in their respective radar intensity and backscattered response as shown in a study by Gaber et al., 2018 [48]. The coherence intensity ranges from 0 to 1. The high coherence reflects little or no changes in the surface whereas the low coherence is indicative of major changes in the surface. Figure 5a shows the extracted surface lineaments. The lineament density map was calculated using the line density tool in ArcMap to show the spatial distribution pattern of lineaments along the study area (Figure 5b).

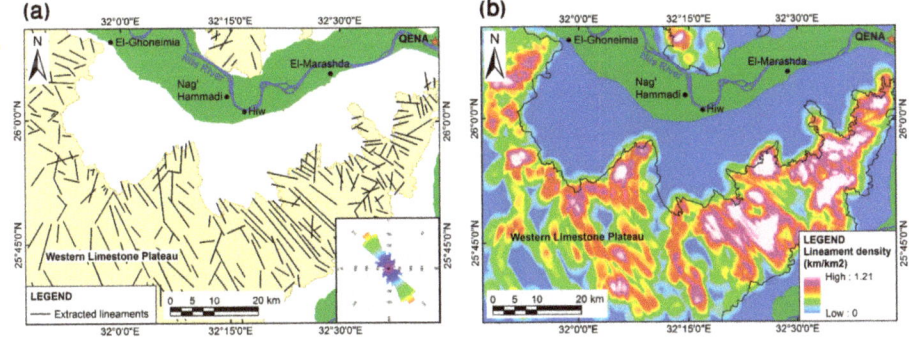

Figure 5. (a) Extracted surface lineaments from remote sensing data; (b) lineament density. The solid green color surrounding the River Nile represents the vegetation cover.

The mean annual rainfall (mm/year) in the study area was obtained from the Tropical Rainfall Measuring Mission (TRMM) rainfall data covering the period from 1998 to 2009. No rainfall has been recorded in the study area between 2009 and 2015 when the TRMM instruments were turned off in 2015 after 17 years (https://gpm.nasa.gov/trmm). The distribution of the different lithological units in the study area were obtained from the geological map of CONOCO and EGPC (1987) [23]. The surface soils were identified from the soil map produced by FAO (Figure 6) [49,50].

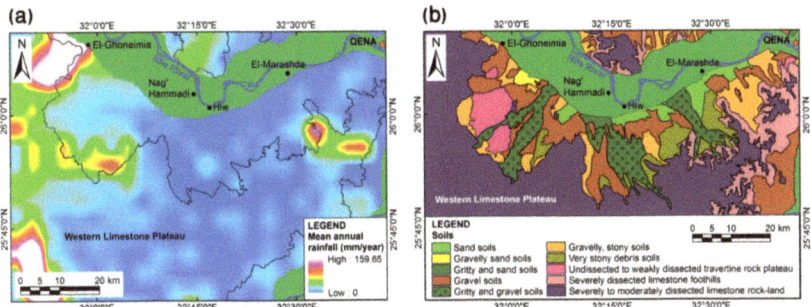

Figure 6. (a) TRMM rainfall in the period (1998–2009); (b) surface soils. The solid green color surrounding the River Nile represents the vegetation cover.

3.2. Aeromagnetic Data

The present study uses the aeromagnetic data that have been collected by Aero-Service (1984) for the Egyptian General Petroleum Cooperation (EGPC) (Figure 7a) [51]. The total magnetic intensity (TMI) data have been acquired at a barometric altitude of about 914.4 m. The TMI data were corrected for the diurnal variations in the Earth's magnetic field; a constant background value of 41,400 nT was subtracted for simplification; and were gridded with a cell size of 500 m using the minimum curvature algorithm. The values of the IGRF inclination (I) and declination (D) of the Earth's magnetic field in the study area during the survey period are about 36.7° and 1.9°, respectively.

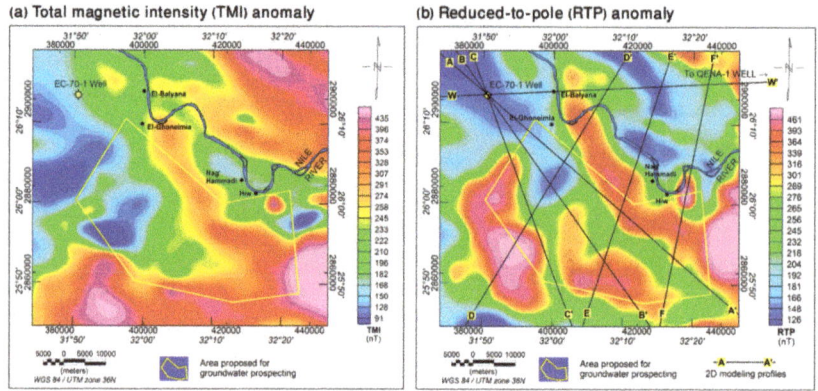

Figure 7. (a) Total magnetic intensity (TMI) anomaly map; (b) reduced-to-pole (RTP) anomaly map.

The TMI data were processed in the wavenumber domain using the Oasis Montaj software to enhance the magnetic anomalies and aid their qualitative and quantitative interpretation [12,52]. The reduction-to-pole (RTP) filter was applied to the TMI data to reduce the distortion in shape and shift in location of the magnetic anomalies relative to their subsurface basement sources due to the dipolar nature of the Earth's magnetic field and inclined magnetization (Figure 7b). The second vertical derivative (SVD) filter was applied to the RTP anomaly to enhance the short-wavelength anomalies derived from shallow sources (Figure 8a). The SVD is usually used to delineate the location of the rock contacts and faults by relating these boundaries to the inflection point of the anomaly marginal gradient. On the SVD contour map, the position of the boundaries can be identified by zero contours [53].

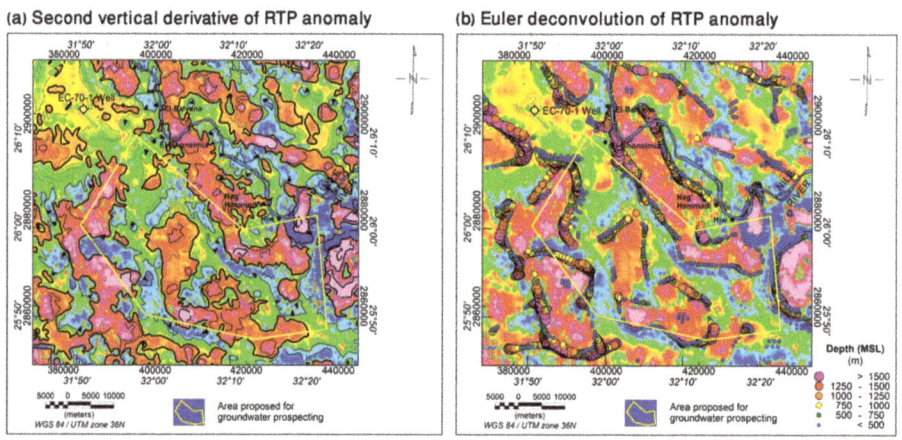

Figure 8. (a) Second vertical derivative (SVD) of the RTP anomaly; (b) Euler deconvolution solutions plotted on the SVD map.

The Euler deconvolution method was applied on the RTP aeromagnetic anomaly using a structural index (N = 0) to estimate the locations and depths of contacts and faults [54,55]. Figure 8b shows a plot of Euler deconvolution solutions on the SVD map. The solutions show tight linear clusters along the zero SVD contours that correspond to the position of contacts. The estimated Euler depths range from less than 500 m to more than 1500 m.

3.3. Vertical Electrical Sounding (VES)

A total of 13 vertical electrical sounding (VES) were carried out at pre-selected stations from remote sensing and aeromagnetic data using the Schlumberger electrode configuration with a maximum half-current electrode spacing (AB/2) of about 300–500 m depending on the availability (Figure 9). The vertical lithological successions of six productive groundwater wells distributed along the study area were obtained from Barsom (2016) [17], in order to calibrate the field collected VES and generate geological cross sections to map the structures and lateral variation in lithology along the study area (Figure 9). Some of these electrical soundings were acquired near productive groundwater wells with known lithology to calibrate the data. The IRIS Syscal Pro instrument was used to measure the apparent resistivity of the ground. The IPI2Win software of Bobachev (2002) was used to display, process, and interpret the measured VES field curves [56]. The curves were carefully processed for discontinuities, cusps, and sharp peaks to produce continuous smooth curves that can be compared with theoretical ones for 1D modeling [57]. Figure 10 shows selective examples of the processed field curves.

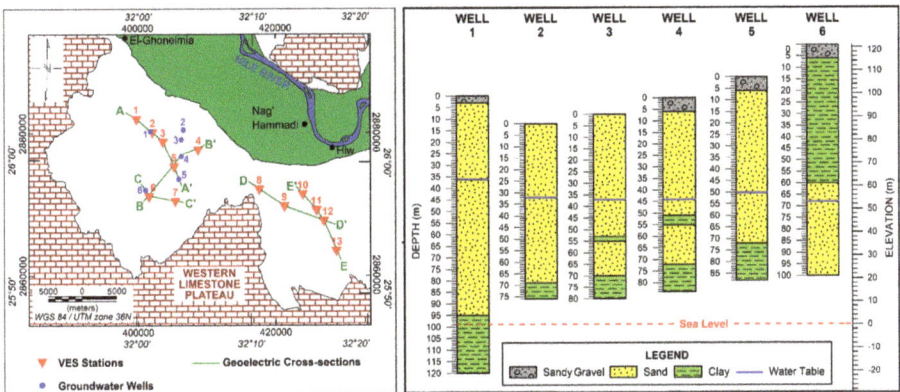

Figure 9. Location map of the vertical electrical sounding (VES) measurements and some groundwater wells in the study area.

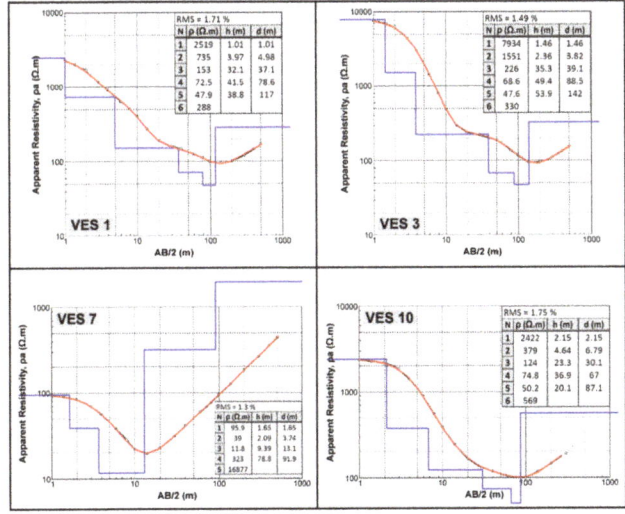

Figure 10. Selective examples of the processed VES field curves.

4. Results and Discussion

4.1. Groundwater Recharge Potentiality Mapping

The DEM shows that the elevation decreases from 558 m at the plateau to reach about 60 m at the old alluvial floodplain (Figure 3a). This variation in land surface helps in directing the runoff from the elevated plateau toward the floodplain and makes the floodplain more favorable for groundwater recharge. In addition, the floodplain is characterized by gentle slopes (Figure 3b), which slow down the runoff and hold the rainfall allowing high infiltration rates aided by the structural settings and sediment characteristics [58].

There are four delineated watersheds (A, B, C, and D) of different sizes occupied by a large network of channels and wadis that drain from the plateau toward their outlets at the alluvial floodplain (Figure 3c). Watershed (A) covers a large area of the plateau running southwards into the desert floodplain southwest of Nag'a Hammadi-El Ghoneimia stretch before reaching the Nile. Watershed (B) covers a smaller area toward the desert plain south of Hiw and Nag'a Hammadi. Watersheds (C) and (D) both cover small areas running toward the desert plain south of Qena-Hiw stretch and most of their drainages are presently reaching the Nile. Thus, it is expected that the outlet of watershed (A) receives a plausible amount of precipitation during the sporadic rainfalls on the plateau compared to the other outlets of watersheds (B), (C), and (D).

The delineated stream-networks in the study area show a dendritic pattern, which is usually developed on top of homogeneous substrate with gentle slopes [59]. On top of the plateau, the drainage channels are sub-parallel, linear, very long with several kilometers in length, and oriented along the NW–SE structural trend. The drainage density ranges from 0 to 1.84 km/km^2 (Figure 3d). The drainage density is inversely proportional to the infiltration rate that depends on the bedrock permeability. The permeable deposits are characterized by high infiltration rates causing low runoff unable of flowing for long distances or cutting further channels, which in turn show low drainage density contrary to the impermeable deposits [39].

On the other hand, all the extracted visible surface lineaments are restricted to the plateau area, which consists of hard limestone resistant to erosion, while the alluvium floodplain deposits hide all surface expression of the lineaments and no outcrops exist that would allow mapping such structures (Figure 5a). This is the reason why neither the geologic map nor the remote sensing data show any structures in the alluvial floodplain area. Consequently, several ground penetrating radar (GPR) profiles were acquired using the 100 MHz shielded antennas in an effort to map any hidden near-surface structures (shallow lineaments) along the alluvial floodplain (Figure 11). However, the GPR profiles do not reveal any near-surface geological structures. Thus, the magnetic results were the main source of mapping the structural settings along the features-less floodplain area. A total of 249 lineaments were extracted from the processed satellite images. The coordinates of the start and end of the lineaments were used to compute their lengths and bearing azimuths (0–360°), which are statistically analyzed using the RockWorks software [60] in the form of a rose diagram to determine the dominant trends in the study area. They have a total length of about 1259 km. The generated rose diagram reveals the dominance of the NW–SE trend that is parallel to the main channel of the Nile River between Qena and Asyut Governorates. Such trend is clearly consistent with the formation of the Nile Valley as a result of the tectonically uplifted Gulf of Suez-Red Sea Hills. The NE–SW trend that is parallel to the Qena Bend and Qena-Safaga Shear Zone is less prevalent. The lineament density ranges from 0 to 1.21 km/km^2 (Figure 5b). The zones where the lineaments intersect and occur with high density show the highest infiltration rates [45,61]. The lineament density is high on the eastern part of the plateau and decreases toward the west. Therefore, the old floodplain south of Qena-Hiw stretch probably receives less amount of surface water than the floodplain southwest of Hiw-El-Ghoneimia stretch, due to the possible high loss in drainage water through the dense fractures on the eastern part of the plateau. This is because lineaments may control the infiltration process of surface water and may serve as conduits of the Nile water to the plain and also help the movements of groundwater to surface through fractures and faults [62,63].

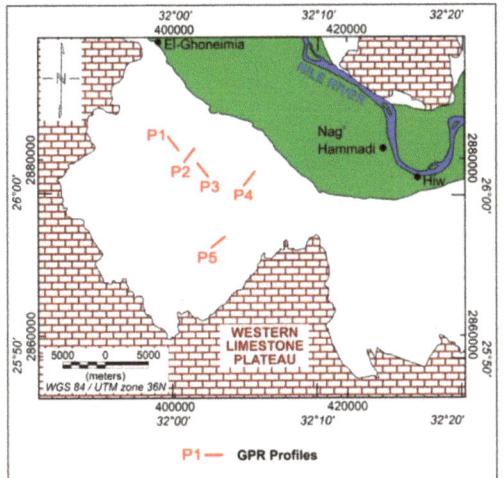

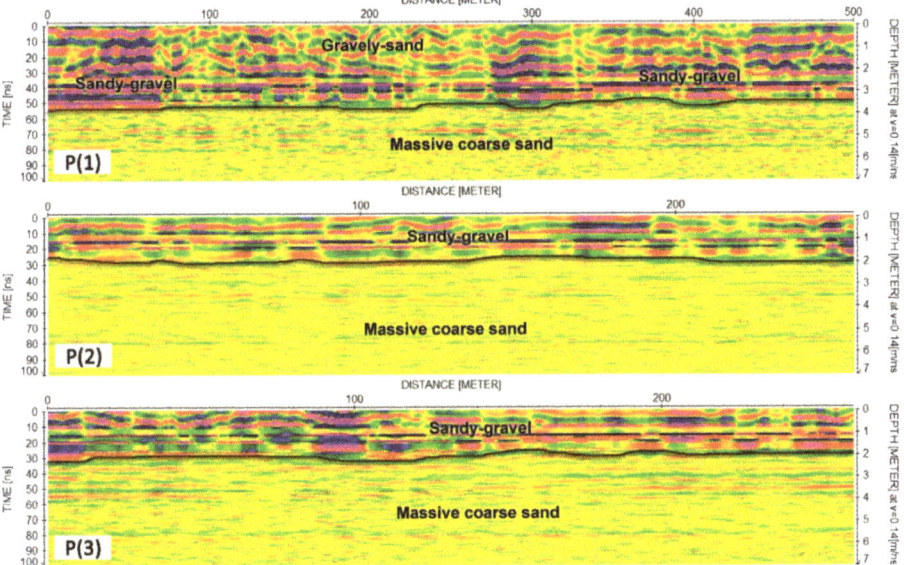

Figure 11. GPR profiles collected in the alluvial floodplain of the study area to confirm the remote sensing results and mapping any shallow lineaments if present. No lineaments were detected.

In addition, the mean annual rainfall in the available period (1998–2009) ranges between 0 and 159.65 mm/year (Figure 6a). It is high on the far west and very low in the remaining study area. The recharge increases as the rainfall increases [64]. However, the contribution of the recent rainfall in the recharge is assumed to be very low and most of the stored water in the subsurface aquifers is attributed to the past pluvial periods and the sporadic flashfloods. According to Said (1981) [3], the study area witnessed strong torrential rainfalls during the Pleistocene pluvial intervals that led to the deposition of the freshwater-travertine rocks opposite Nag'a Hammadi and the fanglomerates that cover the surface. This is also confirmed by the chemical and isotopic composition of the Nile Valley aquifer system [28]. In addition, the study area is classified as a high-risk flashflood zone in cases of sporadic showers during scarce storms [65].

Moreover, lithology plays an important role in groundwater storage (Figure 6b). The capability to infiltrate water is based on the type of rock-unit, clay content, stratigraphic facies and relationships, and the role of the secondary porosity and permeability, which might be generated due to tectonic or weathering processes. The sandy formations are characterized by high infiltration rates and good groundwater storage capacity compared to the clayey formations. The infiltration is also a function of the soil texture. The fine-grained soils of low permeability permit less infiltration rate and more runoff, while the coarse-grained soils of high permeability allow high infiltration rate [58]. The surface of the study area is covered by soils of different textures. They include gravel soils, gritty and sandy soils, gritty and gravelly soils, sandy soils, gravelly sandy soils, gravelly stony soils, stony debris soils, and un-dissected to weakly dissected travertine rocks (Figure 6). Noteworthy, the study area is covered by thick intercalation of sand and gravel deposits, which reach up to 10 m thickness based on well-logging data [17,66]. These deposit characteristics promote the high infiltration capacity to the underneath aquifers.

Accordingly, the different maps obtained from the remote sensing and auxiliary data (elevation, slope, lineament density, drainage density, watershed outlets, rainfall, aeromagnetic, lithology, and soil cover) were reclassified with respect to their influence on recharge capability and converted to thematic layers (Figure 12). Each thematic layer consists of five classes of different recharge potentiality (very high, high – moderate – low – very low) ranked on a numerical scale ranging from 5 to 1, respectively. The very high class (5) is the most preferable location for groundwater recharge contrary to the very low class (1).

GIS Integration and Modeling

Since each factor has only a partial effect on the water infiltration, it is necessary to integrate the influences of the different factors together using the GIS technique in order to get a complete picture of the groundwater recharge potentiality. This is performed by overlaying all the thematic layers using the weighted overlay method [67,68]. In this context, the thematic maps are ranked on a scale of 1–5 based on assessing the relative influences of the factors on the infiltration. Their weighted percentages (%) are then determined by dividing each map rank by the total summation of all map ranks and then multiplying by 100. Finally, the thematic maps were multiplied by their weighted percentages, and the results were added together (overlaid) using the weighted overlay analysis tool in ArcMap to produce the groundwater recharge potentiality map.

The final groundwater recharge potentiality map of the study area is classified into five zones of different recharge capabilities (Figure 13). The old alluvial floodplain is characterized by moderate to very high recharge potentialities. The desert zone located south and southwest of Hiw-El-Ghoneimia stretch shows very high recharge potentiality, whereas the area lying south of Qena-Hiw stretch shows high recharge potentiality. The plateau is characterized by low and very low recharge potentialities. Accordingly, the desert area lying south and southwest of Hiw-El-Ghoneimia stretch was identified for groundwater exploration and verifying the GIS model using the VES.

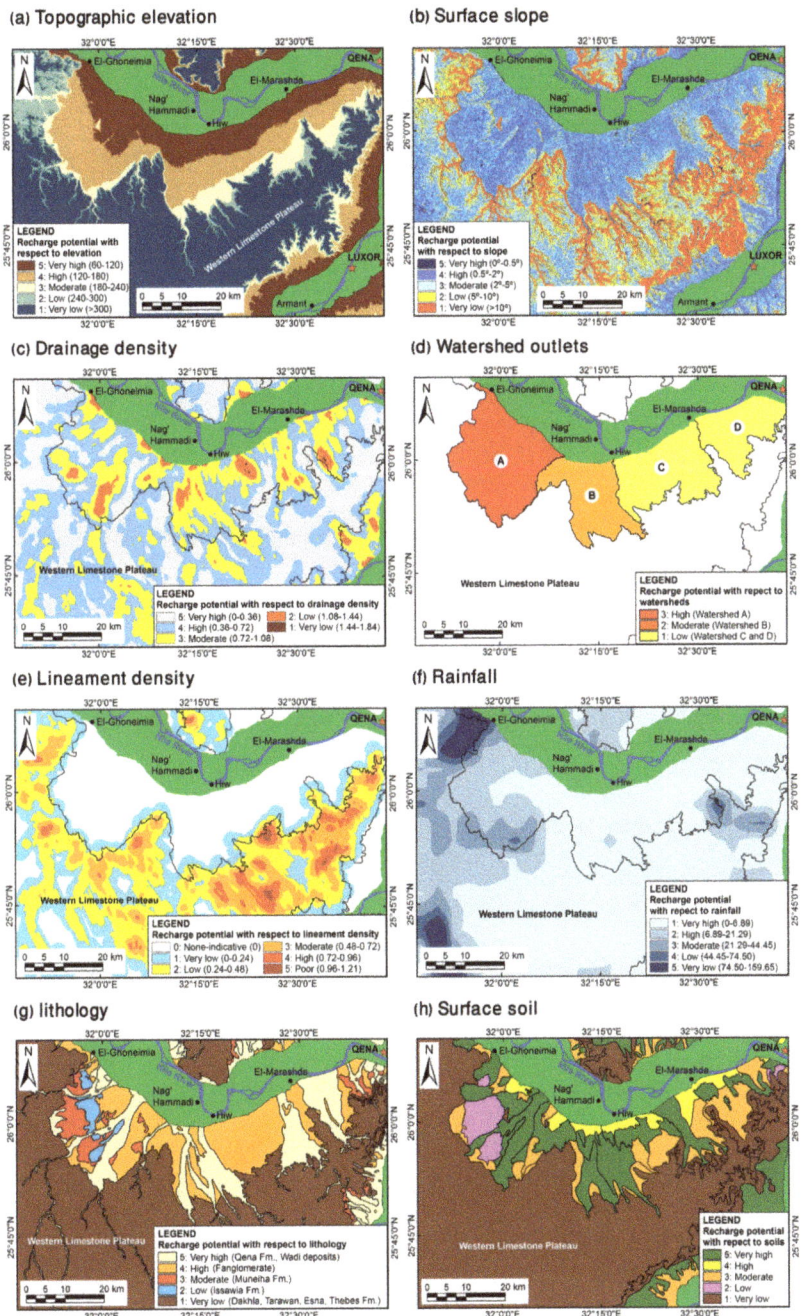

Figure 12. Thematic layers of the different factors influencing groundwater recharge potentiality.

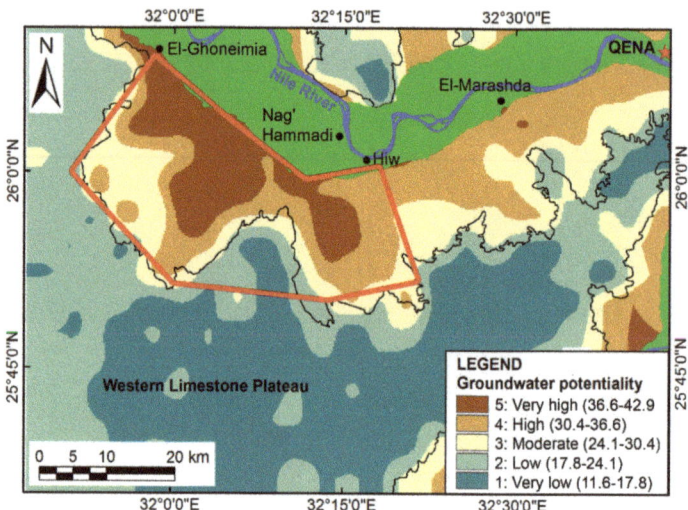

Figure 13. Groundwater recharge potentiality map of the study area (red boundary) inferred from remote sensing data and GIS modeling.

4.2. The Structural Framework of the Study Area

The deep-seated structures affecting the basement rocks along the study area were inferred from the RTP, SVD, and Euler deconvolution maps (Figure 14). These basement rocks are dissected by faults trending NW–SE and NE–SW. The corresponding rose diagram reveals the dominance of the NW–SE trend of the Gulf of Suez-Red Sea, while the NE–SW trend that is parallel to the Qena-Safaga Shear Zone is less prevalent. According to the crosscutting relationship of extracted faults from the processed aeromagnetic maps and supported by the previous works [3,69], the NW–SE trend is older than the NE–SW trend. These faults divide the basement into parallel basins (G1 and G2) bounded by basement uplifts (H1 and H2) oriented generally in the NW–SE direction. The studied old alluvial floodplain occupies the basin (G1) that takes approximately the shape and orientation of the valley, supporting that the Nile Valley has been developed on a tectonically controlled basement alignment. In addition, the surface lineaments extracted from the remote sensing data and plateau cliffs are superimposed on the basement structures and uplift at the southern and western parts, which reflects that the deep-seated faults might extend upwards to affect the overlying sedimentary succession and form vertical conduits between the deep and shallow aquifers. Studies of the chemical and isotopic composition of the groundwater of the Nile Valley aquifer system show that the shallow aquifer is recharged from the water leakage from the adjacent fractured limestone aquifer and the deep Nubian sandstone aquifer through faults, especially near the plateau [28]. The Euler deconvolution solutions show that the fault depths are shallow (less than 500 m) at the southeast and increase to more than 1500 m at the northwest.

Seven 2D magnetic models (W-W', A-A', B-B', C-C', D-D', E-E', and F-F') were constructed along profiles dissecting the study area in the NW–SE and NE–SW directions in order to display the basement topography using the GM-SYS software of NGA (2004) (Figure 7b) [70]. The basement surface was calibrated by the basement depth in El-Balyana (EC-70-1) borehole (1.737 km) at the northwest and Qena-1 borehole (0.54 km) at the northeast near the study area [71,72]. The basement rocks were considered of granitic composition and the sedimentary cover was assumed to have a minimum magnetic susceptibility of about 0.000041 cgs according to the boreholes and geophysical studies at adjacent areas (e.g., ElGalladi, 2007) [24]. The estimated average magnetic susceptibility of the basement rocks from the modeling is 0.017 cgs. This value agrees with the reported range for granitic rocks in Telford et al. (1990) [73]. Lateral heterogeneity in the basement granitic composition may cause a slight difference in the magnetic susceptibility. Figure 15 shows selective examples of the 2D magnetic models of the basement. Moreover, the basement surface was digitized from the models to construct a depth-to-basement map of the study area, which reflects the lateral variation in the thickness of the sedimentary cover (Figure 15).

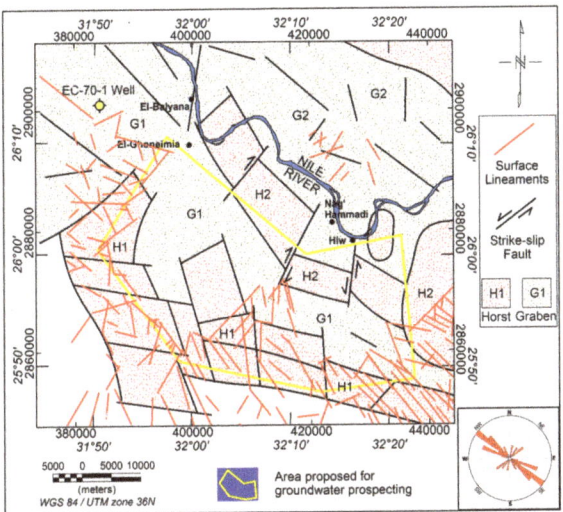

Figure 14. Basement structural map inferred from the interpretation of aeromagnetic data.

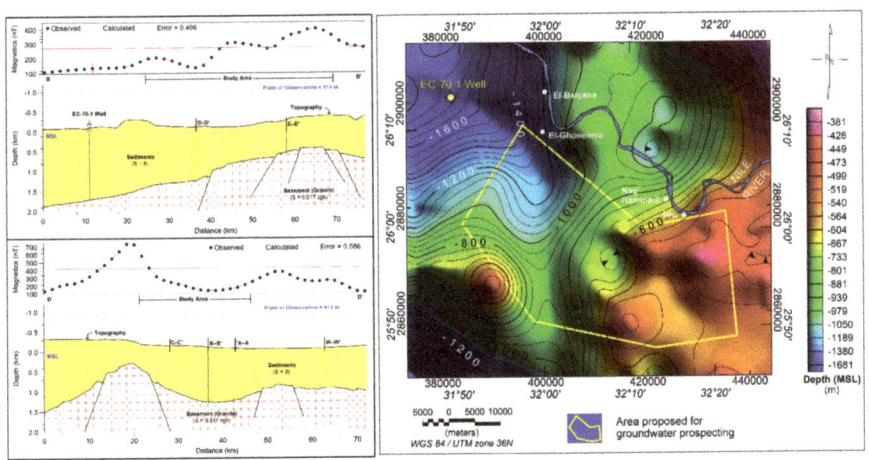

Figure 15. Selective examples of the basement 2D models (**left**) and depth-to-basement map of the study area (**right**).

The eastern part of the basin (south of Nag'a Hammadi and Hiw village) is affected by a major uplift, where the basement occurs at a shallow depth of about 0.5 km. The basin dips and increases in width toward the northwest to about 1.3 km southwest of Nag'a Hammadi-El-Ghoneimia stretch and extends further northwestwards with a large increase in width and depth to more than 1.8 km west of El-Balyana, Sohag. The increase in basement depth reflects the increase in groundwater potentiality due to the probable increase in the thickness of the groundwater-bearing sediment (e.g., the sands of Nubia Formation and Qena Formation). In general, the uplift of the basement complex of the Eastern

Desert during the Oligo-Miocene induced a tensile stress in the surrounding crust, creating the NW–SE fractures and cracks along which the valley was subsequently developed by the fluvial activities [18,63]. Such tensile stress has produced secondary porosity and increased the hydraulic conductivity of the study area.

4.3. Analysis and Interpretation of VES Data

The interpretation of VES data aims at calculating the number, resistivity, and thickness of the geoelectric layers (sedimentary layers) occurring directly under each VES. The VES data are usually correlated with the borehole data to give them a geological and hydrogeological meaning. In this context, the lithology log of Well-1 was used to calibrate the 1D model of VES-2 (Figure 16). The calibration revealed that VES-2 location consists of six geoelectric layers. The results of the calibration were used as a guide in the quantitative modeling of the other VES field curves taking into consideration geological and geophysical information from previous works. The geoelectric succession was therefore categorized into five principle layers (A, B, C, D, and E) equivalent to a stratigraphic sequence ranging in age from Eocene to Quaternary (Table 1).

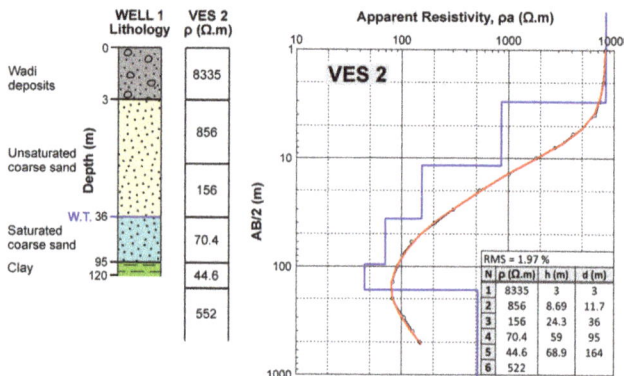

Figure 16. Calibration of the 1D model at VES-2 with the lithological log of Well-1.

Table 1. Summary of geoelectric layers and their inferred lithology and hydrogeology.

Layer		Resistivity, ρ (Ω.m)	Thickness, h (m)	Equivalent Lithology and Hydrogeology
A		95.9–9880	1.07–7.25	Surface wadi deposits made up mainly of sand and gravel (dry); The low values attribute to agricultural soil that made up of a mixture of moist sand, silt, and clay.
B	1	108–321	15.6–40.5	Coarse sand (dry)
	2	62.9–78.6	6.23–59	Coarse sand (water-bearing)
C		16.7–54	11.5–58.9	Clay
D		288–863	Unknown	Coarse sand (dry)
E		8967–16,877	Unknown	Massive limestone

Five geoelectric cross-sections (A-A′, B-B′, C-C′, D-D′, and E-E′) (Figure 9) were constructed along profiles oriented in NW–SE and NE–SW directions by combining the geoelectric succession at each VES with the lithological units from the field collected groundwater wells in order to delineate the lateral and vertical extensions of the different geological and hydrogeological units. Figure 17 illustrates selective examples of the geoelectric cross-sections. The geoelectric layers (A, B, C, D) have a large lateral extension with varying thickness, especially near the cultivated floodplain. Some of the geoelectric layers were missed toward the plateau due to the possible effect of normal faulting and the presence of the faulted Eocene limestone blocks (layer E) that has unusual high resistivity values varying from 8967 to 16,877 Ω.m (e.g., Mahmoud and Tawfik, 2015) [74].

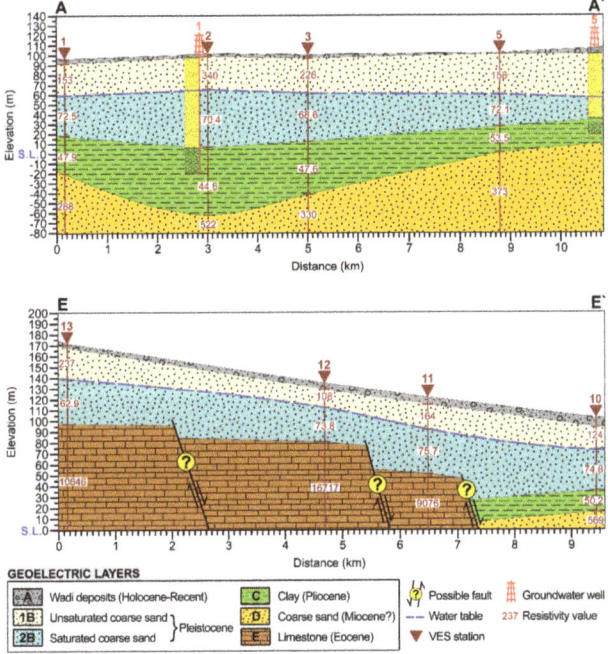

Figure 17. Selective cross-sections show the lateral distribution of the different geoelectric layers.

Importantly, the geoelectric layer (B) is interpreted as coarse sand equivalent to the shallow Quaternary aquifer that forms the main water-bearing unit in the Nile Valley. This layer is divided according to hydrogeological importance into two sub-layers: 1B and 2B. The upper sub-layer (1B) forms the dry aeration zone above the water table with resistivity values ranging from 108 Ω.m at VES-12 to 321 Ω.m at VES-8 and a thickness ranging from 15.6 m at VES-8 to 40.5 m at VES-5. The lower sub-layer (2B) is correlated to the saturation zone with low resistivity values varying slightly between 62.9 Ω.m at VES-13 and 78.6 Ω.m at VES-4 and VES-8 and its thickness ranges from 6.23 m at VES-6 to 59 m at VES-2. It occurs under unconfined condition due to its permeable cover of Wadi deposits (layer A). Moreover, the Quaternary aquifer is confined from the base by a layer of Pliocene clays (layer C) that is characterized by low resistivity values ranging from 16.7 Ω.m at VES-7 to 53.5 Ω.m at VES-5 and its thickness ranges between 11.5 m at VES-7 to 58.9 m at VES-2.

The geoelectric layer (D) has resistivity values ranging between 288 Ω.m at VES-1 and 863 Ω.m at VES-4. It is inferred as dry coarse sands, which refers to the possible existence of a second, deep sandy aquifer. A similar thick sand section dated to the Miocene is recorded below the Pliocene clays in the boreholes drilled for oil exploration in the Nile Valley at El-Balyana and Kom-Ombo [27,75]. The thickness is not detected at all in the soundings except at VES-6 and VES-7, where its thickness reaches about 97.4 and 78.8 m, respectively. It is separated from the overlying Quaternary aquifer by the Pliocene clays (layer C).

4.4. Hydro-Geophysical Maps and GIS Model Validation

Based on the above VES results, three hydro-geophysical maps were generated using the most common ordinary kriging algorithm [76]. These maps show the variation in the water-table depth, saturated thickness, and water-table elevation of the Quaternary aquifer. The groundwater occurs at shallow depth about 20 m close to the cultivated lands and the depth increases toward the plateau to reach more than 70 m (Figure 18). The saturated thickness of the aquifer increases toward the cultivated lands to reach about 35–40 m in the eastern part and more than 43 m in the western part (Figure 19). The aquifer becomes thinner toward the plateau until it disappears. The water-table elevation decreases toward the north and west from more than 85 m to less than 55 m (Figure 20). This indicates that groundwater flows toward the Nile River following the land slope. Such obtained result is consistent with Barsom (2016) [17].

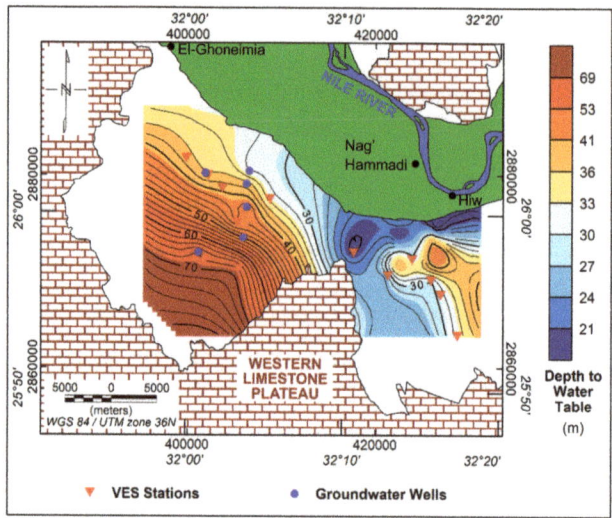

Figure 18. Depth to water contour map inferred from VES data.

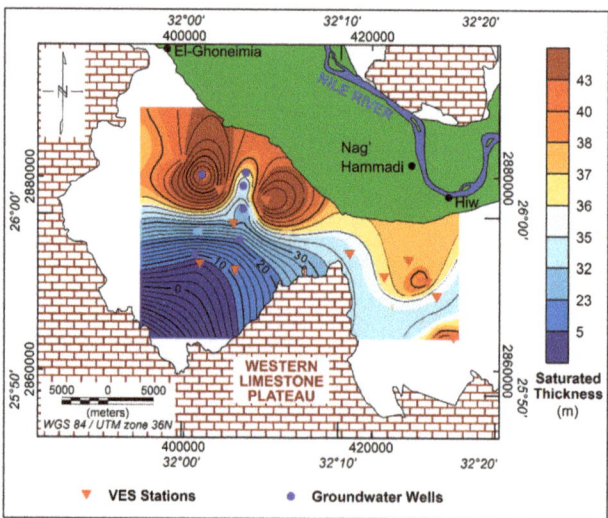

Figure 19. Saturated thickness of the Quaternary aquifer (Layer B) inferred from VES data.

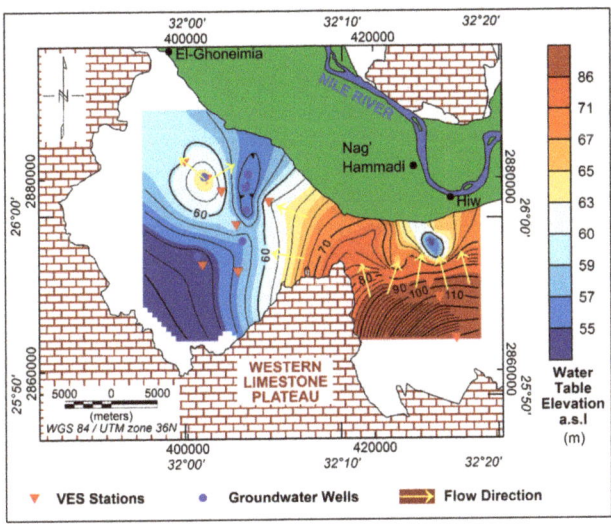

Figure 20. Water-table elevation inferred from VES data.

The desert zone located southwest of Nag'a Hammadi-El-Ghoneimia is the most promising for groundwater occurrence. It is characterized by very high groundwater recharge potentiality. Its potentiality is confirmed by the presence of a large saturated thickness of the Quaternary aquifer (more than 43 m), shallow water table (nearly 30 m), and this zone occupies a basin with a thick sedimentary succession (1.3 km).

5. Conclusions

The present study demonstrates the integration of remote sensing with the geoelectric resistivity and the aeromagnetic techniques as effective tools for exploring the groundwater potentiality of west Qena area. The study reveals that the old alluvial floodplain of the Nile Valley west of Qena is characterized by moderate to very high groundwater recharge potentiality from present-day and paleo-runoff in contrast to the plateau that shows poor and very poor potentialities. The alluvial desert plain zone located south and southwest of Hiw-El-Ghoneimia is characterized by very high potentiality. The old alluvial floodplain occupies a basin trending in the NW–SE direction and bounded by basement uplifts and plateau scarp. The eastern part of the basin south of Hiw Village is characterized by a major basement uplift at a depth of 0.5 km, whereas the basement depth increases toward the northwest to reach about 1.3 km west of Nag'a Hammadi and reaches more than 1.8 km west of Sohag. This reflects the increase in the groundwater potentiality toward the northwest due to the possible increase in the thickness of the groundwater-bearing sediments. The subsurface basement and surface of the plateau are affected by faults trending mainly in NW–SE direction parallel to the Nile Valley section between Qena and Asyut and the trend of the Gulf of Suez-Red Sea. The NE–SW trend, which is consistent with the Qena-Safaga Shear Zone, is less prevalent. These fault and fracture systems serve as preferential conduits for groundwater movement, recharge, and discharge. Based on the results of remote sensing together with the geoelectric resistivity and the aeromagnetic techniques, there is a good correlation between the surface and subsurface faults, especially near the plateau cliffs. They probably form conduits for groundwater movement and recharge.

There are two sandy aquifers in the study area: the shallow Quaternary and the deeper Miocene aquifer. The two aquifers are separated by a thick layer of Pliocene clays reaching about 59 m. The Quaternary aquifer is the main aquifer and has a large lateral extension throughout the study area. Its water table occurs at shallow depth about 20–30 m close to the cultivated lands and the depth increases toward the plateau to reach more than 70 m. The saturated thickness is more than 43 m near the cultivated lands and thins toward the plateau. The groundwater flow direction is toward the Nile as the water level of the Nile River is lower than the groundwater table of the Quaternary aquifer. Moreover, the aquifer occurs under unconfined condition due to its permeable Wadi deposits cover.

The desert zone located southwest of Nag'a Hammadi-El-Ghoneimia stretch is the most promising area for drilling groundwater wells. It is characterized by very high recharge potentiality and occupies a thick basin of about 1.3 km. In addition, it is characterized by a shallow water table at about 30 m depth and a saturated thickness of more than 43 m. Such zone is clearly controlled by fault systems and works as a trap for porous sediments accumulation and thus makes it a promising zone in terms of groundwater potentiality.

Author Contributions: Methodology, A.G. and A.E.; Resources, M.A.; Software, A.M.B.; Supervision, A.K.M.; Writing—review and editing, M.K. All authors have read and agreed to the published version of the manuscript.

Funding: This research received no external funding.

Acknowledgments: The authors would like to thank the USGS, JAXA, and ESA for providing the Landsat-8, ALOS/PALSAR-1 (through ALOS-2 6th RA PI-3131 agreement) and Sentinel-1 satellite images as well as SNAP software, respectively, free of charge. Finally, the authors are very grateful for the very helpful suggestions made by the three reviewers.

Conflicts of Interest: The authors declare no conflict of interest.

References

1. FAO. Egypt AQUASTAT-FAO's Information System on Water and Agriculture. 2016. Available online: http://www.fao.org/nr/water/aquastat/countries_regions/EGY/index.stm (accessed on 1 August 2019).
2. El-Khawaga, A. Promoting irrigation water utilization efficiency in superior vineyards. *Asian J. Crop. Sci.* **2013**, *5*, 125–138. [CrossRef]
3. Said, R. *The Geological Evolution of the River Nile*; Springer: New York, NY, USA, 1981.
4. Patra, H.P.; Adhikari, S.K.; Kunar, S. *Groundwater Prospecting and Management*; Springer: Singapore, 2016.
5. Saraf, A.K.; Choudhury, P.R. Integrated remote sensing and GIS for groundwater exploration and identifcation of artifcial recharge sites. *Int. J. Remote Sens.* **1998**, *19*, 1825–1841. [CrossRef]
6. Shahid, S.; Nath, S.; Roy, J. Groundwater potential modelling in a sof rock area using a GIS. *Int. J. Remote Sens.* **2000**, *21*, 1919–1924. [CrossRef]
7. Lee, S.; Kim, Y.-S.; Oh, H.-J. Application of a weights-of-evidence method and GIS to regional groundwater productivity potential mapping. *J. Environ. Manag.* **2012**, *96*, 91–105. [CrossRef] [PubMed]
8. Shaban, A.; Khawlie, M.; Abdallah, C. Use of remote sensing and GIS to determine recharge potential zones: The case of Occidental Lebanon. *Hydrogeol. J.* **2006**, *14*, 433–443. [CrossRef]
9. Amarsaikhan, D.; Blotevogel, H.H.; Ganzorig, M.; Moon, T.-H. Applications of remote sensing and geographic information systems for urban land-cover change studies in Mongolia. *Geocarto Int.* **2009**, *24*, 257–271. [CrossRef]
10. Abdalla, F. Mapping of groundwater prospective zones using remote sensing and GIS techniques: A case study from the Central Eastern Desert, Egypt. *J. Afr. Earth Sci.* **2012**, *70*, 8–17. [CrossRef]
11. Madani, A.; Niyazi, B. Groundwater potential mapping using remote sensing techniques and weights of evidence GIS model: A case study from Wadi Yalamlam basin, Makkah Province, Western Saudi Arabia. *Environ. Earth Sci.* **2015**, *74*, 5129–5142. [CrossRef]
12. Hinze, W.J.; Frese, R.R.B.V.; Saad, A.H. *Gravity and Magnetic Exploration: Principles, Practices, and Applications*; Cambridge University Press: New York, NY, USA, 2013.
13. Helaly, A.S. Assessment of groundwater potentiality using geophysical techniques in Wadi Allaqi basin, Eastern Desert, Egypt–Case study. *NRIAG J. Astron. Geophys.* **2017**, *6*, 408–421. [CrossRef]
14. Ibrahim, E.; Ghazala, H.; Elawadi, E.; Alfaifi, H.; Abdelrahman, K. Structural depocenters control the Nubian sandstone aquifer, Southwestern Desert, Egypt: Inferences from aeromagnetic data. *Arab. J. Geosci.* **2019**, *12*, 335. [CrossRef]
15. Mohamed, A.K. Application of DC resistivity method for groundwater investigation, case study at West Nile Delta, Egypt. *Arab. J. Geosci.* **2015**, *9*, 11. [CrossRef]

16. Reynolds, J.M. *An Introduction to Applied and Environmental Geophysics*, 2nd ed.; Wiley-Blackwell: Hoboken, NJ, USA, 2011.
17. Barsom, N. Remote sensing and groundwater data investigation in plain west of Nile, Qena Governorate, Egypt. In *Geoelectric Methods: Theory and Applications*; Bhattacharya, B., Shalivahan, S., Eds.; McGraw Hill Education: New Delhi, India, 2016; p. 147.
18. Issawi, B.; McCauley, J. The Cenozoic rivers of Egypt: The Nile problem. In *The Followers of Horus: Studies Dedicated to Michael Allen Hoffman*; Egyptian Studies Association Publication, no. 2, Oxbow Monograph 20; Friedman, R., Adams, B., Eds.; Oxbow Books, Park End Place: Oxford, England, 1992; pp. 121–146.
19. Youssef, M. Structural setting of central and south Egypt: An overview. *Micropaleontology* **2003**, *49* (Suppl. 1), 1–13. [CrossRef]
20. Abdelkareem, M.; El-Baz, F. Mode of formation of the Nile Gorge in northern Egypt: A study by DEM-SRTM data and GIS analysis. *Geol. J.* **2016**, *51*, 760–778. [CrossRef]
21. Hamimi, Z.; El-Barkooky, A.; Martínez Frías, J.; Fritz, H.; Abd El-Rahman, Y. *The Geology of Egypt*; Springer: Berlin/Heidelberg, Germany, 2020; p. 711. ISBN 978-3-030-15265-9.
22. El-Gamili, M.M. A geological interpretation of a part of the Nile Valley based on gravity data. *Egypt J. Geol. Spec. Vol. Part* **1982**, *2*, 101–120.
23. Conoco. *Geological Map of Egypt, Scale 1:500,000, Sheet NG36NE Quseir, NG36NW Asyut, NG36SE Gebel Hamata, and NG36SW Luxor, Egypt*; The Egyptian General Petroleum Corporation: Cairo, Egypt, 1987.
24. ElGalladi, A. Magneto-tectonic studies of the area west of Luxor, Upper Egypt. *Mansoura J. Geol. Geophys.* **2007**, *34*, 57–83.
25. Omer, A.A. Geological, Mineralogical and Geochemical Studies on the Neogene and Quaternary Nile Basin Deposits, Qena-Assiut Stretch, Egypt. Ph.D. Thesis, South Valley University, Sohag, Egypt, 1996.
26. Faris, M.; Allam, A.; Marzuk, A.M. Biostratigraphy of the Late Cretaceous-Early Tertiary rocks in the Nile Valley (Qena region), Egypt. *Ann. Geol. Surv. Egypt.* **1985**, *XV*, 287–300.
27. Issawi, B.; Francis, M.H.; Youssef, E.-S.A.A.; Osman, R.A. *The Phanerozoic Geology of Egypt: A Geodynamic Approach*, 2nd ed.; Egyptian Mineral Resources Authority (EMRA): Cairo, Egypt, 2009.
28. Awad, M.A.; El Arabi, N.E.; Hamza, M.S. Use of solute chemistry and isotopes to identify sources of ground-water recharge in the Nile aquifer system, Upper Egypt. *Groundwater* **1997**, *35*, 223–228. [CrossRef]
29. RIGW (Cartographer). *Hydrogeological Map of Egypt (Luxor Sheet)*. Scale 1:500,000; ASRT: Cairo, Egypt, 1997.
30. RIGW (Cartographer). *Hydrogeological Map of Egypt (Asyut Sheet)*. Scale 1:500,000; ASRT: Cairo, Egypt, 1997.
31. El Tahlawi, M.R.; Farrag, A.A.; Ahmed, S.S. Groundwater of Egypt: "an environmental overview". *Environ. Geol.* **2008**, *55*, 639–652. [CrossRef]
32. ESRI. ArcGIS Desktop 10.5 Help (Includes ArcMap, ArcCatalog, ArcGlobe, and ArcScene Softwares). 2016. Available online: https://www.arcgis.com/ (accessed on 1 June 2018).
33. O'Callaghan, J.; Mark, D. The extraction of drainage networks from digital elevation data. *Comput. Vis. Graph. Image Process.* **1984**, *28*, 323–344. [CrossRef]
34. Jenson, S.; Domingue, J. Extracting topographic structure from digital elevation data for geographic information system analysis. *Photogram. Engng Remote Sens.* **1988**, *54*, 1593–1600.
35. Fenta, A.A.; Yasuda, H.; Shimizu, K.; Haregeweyn, N.; Woldearegay, K. Quantitative analysis and implications of drainage morphometry of the Agula watershed in the semi-arid northern Ethiopia. *Appl. Water Sci.* **2017**, *7*, 3825–3840. [CrossRef]
36. Hardcastle, K.C. Photolineament Factor: A New Computer—Aided Method for Remotely Sensing the Degree to Which Bedrock Is Fractured. *Photogramm. Eng. Remote Sens.* **1995**, *61*, 739–747.
37. ESA. Copernicus Website for Downloading Free Sentinel Data. 2019. Available online: https://scihub.copernicus.eu/dhus/#/home (accessed on 9 January 2019).
38. USGS. EarthExplorer Website for Downloading Satellite Remote Sensing Data. 2019. Available online: https://earthexplorer.usgs.gov/ (accessed on 9 January 2019).
39. Meijerink, A.M.J.; Bannert, D.; Batelaan, O.; Lubczynski, M.; Pointet, T. *Remote Sensing Applications to Groundwater; IHP-VI Series on Groundwater*; United Nations Educational Scientific and Cultural Organization (UNESCO): Paris, France, 2007; Volume 16.
40. Deng, G.; Pinoli, J.-C. Differentiation-Based Edge Detection Using the Logarithmic Image Processing Model. *J. Math. Imaging Vis.* **1998**, *8*, 161–180. [CrossRef]

41. Exelis. ENVI Classic 5.3 Help. 2015. Available online: https://www.harrisgeospatial.com/ (accessed on 1 January 2019).
42. Suzen, M.L.; Toprak, V. Filtering of satellite images in geological lineament analyses: An application to a fault zone in Central Turkey. *Int. J. Remote. Sens.* **1998**, *19*, 1101–1114. [CrossRef]
43. Chang, N.-B.; Bai, K. *Multisensor Data Fusion and Machine Learning for Environmental Remote Sensing*; CRC Press: Boca Raton, FL, USA, 2018.
44. Wise, D.U.; Funiciello, R.; Parotto, M.; Salvini, F. Topographic lineament swarms: Clues to their origin from domain analysis of Italy. *GSA Bull.* **1985**, *96*, 952. [CrossRef]
45. Sander, P. Lineaments in groundwater exploration: A review of applications and limitations. *Hydrogeol. J.* **2007**, *15*, 71–74. [CrossRef]
46. Gaber, A.; Koch, M.; Griesh, M.H.; Sato, M. SAR Remote Sensing of Buried Faults: Implications for Groundwater Exploration in the Western Desert of Egypt. *Sens. Imaging: Int. J.* **2011**, *12*, 133–151. [CrossRef]
47. ESA. SNAP Software (Version 5.0): Sentinel Application Platform. 2017. Available online: http://step.esa.int/main/toolboxes/snap/ (accessed on 1 May 2018).
48. Gaber, A.; Abdelkareem, M.; Abdelsadek, I.S.; Koch, M.; El-Baz, F. Using InSAR Coherence for Investigating the Interplay of Fluvial and Aeolian Features in Arid Lands: Implications for Groundwater Potential in Egypt. *Remote. Sens.* **2018**, *10*, 832. [CrossRef]
49. FAO (Cartographer). Reconnaissance soil map of the Isna-Nag Hammadi area, Egypt. Scale 1:200,000. 1961. ESDAC. Available online: https://esdac.jrc.ec.europa.eu (accessed on 13 May 2020).
50. FAO (Cartographer). Reconnaissance soil map of the Nag Hammadi-Abu Tig area, Egypt. Scale 1:200,000. 1962; ESDAC. Available online: https://esdac.jrc.ec.europa.eu (accessed on 13 May 2020).
51. Aero-Service. Final operational report of airborne magnetic/radiometric survey in the Eastern Desert, Egypt. In *Aero-Service Division*; Western Atlas International Inc.: Houston, TX, USA, 1984.
52. Geosoft. *Oasis Montaj (Version 8.4): Software for Processing and Interpretation of Potential-Field Data*; Geosoft Inc.: Toronto, ON, Canada, 2015; Available online: http://www.geosoft.com (accessed on 1 October 2018).
53. Vacquier, V.; Steenland, N.C.; Henderson, R.G.; Zietz, I. Interpretation of aeromagnetic maps. In *Geological Society of America Memoirs*; Geological Society of America: New York, NY, USA, 1951; Volume 47, pp. 1–150.
54. Thompson, D.T. EULDPH: A new technique for making computer-assisted depth estimates from magnetic data. *Geophysics* **1982**, *47*, 31–37. [CrossRef]
55. Reid, A.; Allsop, J.M.; Granser, H.; Millett, A.J.; Somerton, I.W. Magnetic interpretation in three dimensions using Euler deconvolution. *Geophysics* **1990**, *55*, 80–91. [CrossRef]
56. Bobachev, A.A. IPI2Win: A Windows Software for an Automatic Interpretation of Resistivity Sounding Data. Ph.D. Thesis, Moscow State University, Moscow, Russia, 2002.
57. Zohdy AA, R.; Eaton, G.P.; Mabey, D.R. Application of surface geophysics to ground-water investigations. In *Techniques of Water-Resources Investigations*; John W. Powell National Center, USGS: Reston, VA, USA, 1990; Volume 02-D1.
58. Healy, R.W.; Scanlon, B.R. *Estimating Groundwater Recharge*; Cambridge University Press: New York, NY, USA, 2010.
59. Singhal, B.B.S.; Gupta, R.P. *Applied Hydrogeology of Fractured Rocks*, 2nd ed.; Springer: Dordrecht, The Netherlands, 2010.
60. RockWare. RockWorks16 Help. 2018. Available online: https://www.rockware.com/ (accessed on 1 June 2019).
61. Waters, P.; Greenbaum, D.; Smart, P.L.; Osmaston, H. Applications of remote sensing to groundwater hydrology. *Remote. Sens. Rev.* **1990**, *4*, 223–264. [CrossRef]
62. Ganapuram, S.; Kumar, G.V.; Krishna, I.M.; Kahya, E.; Demirel, M.C. Mapping of groundwater potential zones in the Musi basin using remote sensing data and GIS. *Adv. Eng. Softw.* **2009**, *40*, 506–518. [CrossRef]
63. Abdelkareem, M.; El-Baz, F. Analyses of optical images and radar data reveal structural features and predict groundwater accumulations in the central Eastern Desert of Egypt. *Geoscience* **2015**, *8*, 2653–2666. [CrossRef]
64. Moubark, K.; Abdelkareem, M.; Fakhry, M.; Barsom, N. Integration of remote sensing and hydrogeological data in the west of Qena Governorate, Egypt. In *Second Young Researchers Egyptian Universities Conference (YREUC-2)*; South Valley University: Qena, Egypt, 2015; p. 77.
65. Hefny, K.; Shata, A. *Underground Water in Egypt*; Ministry of Water Supplies and Irrigation: Cairo, Egypt, 2004; p. 295.

66. Abdalla, F.; Moubark, K. Assessment of well performance criteria and aquifer characteristics using step-drawdown tests and hydrogeochemical data, west of Qena area, Egypt. *J. Afr. Earth Sci.* **2018**, *138*, 336–347. [CrossRef]
67. Voogd, H. *Multicriteria Evaluation for Urban and Regional Planning*; Pion: London, UK, 1983.
68. NGA. *GM-SYS (Version 4.9): Gravity/Magnetic Modeling Software*; Northwest Geophysical Associates Inc.: Corvallis, OR, USA, 2004; Available online: https://www.nga.com/ (accessed on 1 June 2019).
69. Sikdar, P.K.; Chakraborty, S.; Enakshi, A.; Paul, P.K. Land use/land cover changes and groundwater potential zoning in and around Raniganj coal mining area, Bardhaman District, West Bengal–a GIS and remote sensing approach. *J. Spat. Hydrol.* **2004**, *4*, 1–24.
70. El-Soghier, M. *Gravity and Remote Sensing Investigations in the Basin West of Asyut*; Egypt MSc. South Valley University: Qena Governorate, Egypt, 2017.
71. Hussien, H.M.; Kehew, A.E.; Aggour, T.; Korany, E.; Abotalib, A.Z.; Hassanein, A.; Morsy, S. An integrated approach for identification of potential aquifer zones in structurally controlled terrain: Wadi Qena basin, Egypt. *Catena* **2017**, *149*, 73–85. [CrossRef]
72. Ghazala, H.H.; Ibraheem, I.; Haggag, M.; Lamees, M. An integrated approach to evaluate the possibility of urban development around Sohag Governorate, Egypt, using potential field data. *Arab. J. Geosci.* **2018**, *11*, 194. [CrossRef]
73. Telford, W.M.; Geldart, L.P.; Sheriff, R.E. *Applied Geophysics*, 2nd ed.; Cambridge University Press: New York, NY, USA, 1990.
74. Mahmoud, H.; Tawfik, M.Z. Impact of the geologic setting on the groundwater using geoelectrical sounding in the area southwest of Sohag–Upper Egypt. *J. Afr. Earth Sci.* **2015**, *104*, 6–18. [CrossRef]
75. Issawi, B.; Sallam, E.; Zaki, S.R. Lithostratigraphic and sedimentary evolution of the Kom Ombo (Garara) sub-basin, southern Egypt. *Arab. J. Geosci.* **2016**, *9*, 420. [CrossRef]
76. Webster, R.; Oliver, M.A. *Geostatistics for Environmental Scientists*, 2nd ed.; John Wiley & Sons, Ltd.: Hoboken, NJ, USA, 2007; Volume 1. [CrossRef]

© 2020 by the authors. Licensee MDPI, Basel, Switzerland. This article is an open access article distributed under the terms and conditions of the Creative Commons Attribution (CC BY) license (http://creativecommons.org/licenses/by/4.0/).

Article

Understanding Ancient Landscapes in the Venetian Plain through an Integrated Geoarchaeological and Geophysical Approach

Alice Vacilotto [1], Rita Deiana [1,*] and Paolo Mozzi [2]

1. Department of Cultural Heritage, University of Padova, Piazza Capitaniato 7, 35139 Padova, Italy; alice.vacilotto@unipd.it
2. Department of Geosciences, University of Padova, Via Gradenigo 6, 35129 Padova, Italy; paolo.mozzi@unipd.it
* Correspondence: rita.deiana@unipd.it

Received: 29 July 2020; Accepted: 9 September 2020; Published: 12 September 2020

Abstract: This paper reports the results of the multidisciplinary study carried out in the SE area of Ceggia, in the eastern part of the Venetian Plain. The area has been characterized, since ancient times, by numerous morphological transformation, due to the presence of lagoon and marshes, and interested by repeated reclamation. Aerial and satellite images have identified many natural and anthropogenic traces. From a geophysical point of view, electrical resistivity tomography (ERT) combined with frequency-domain electromagnetic measurements (FDEM) can help to discriminate the spatial distribution of different buried structures in conductive systems. The electrical conductivity is, in fact, directly related to the soil moisture content. The multidisciplinary approach adopted in this context, with the results obtained thanks to the contribution of aerial and satellite images, historical cartography, archaeological survey, geophysical measurements, geomorphological characterization, and ^{14}C dating, allow us to suggest a possible interpretation of the different traces highlighted in the studied area. This approach suggests a potentially useful and replicable methodology to study similar evidence, such as along the North Adriatic coast and in broad sectors of the Po Valley. The key issue, in this kind of system, lies, in fact, in the possibility to date and compare traces visible on the surface by remote sensing, establishing their interest from an archaeological and geomorphological point of view using an integration of field measurements. At the end of this research, the classification of the different anomalies found in this hydraulic variable context, thanks to the multidisciplinary approach here adopted, suggest new hypotheses for reading the complex history of this understudied area.

Keywords: aerial archaeology; landscape archaeology; electrical resistivity tomography (ERT); frequency-domain electromagnetic methods (FDEM); paleochannel

1. Introduction

The study area (Figures 1 and 2) lies in the fine-grained, low-gradient alluvial plain of the Piave River [1,2], which is part of the Piave megafan that has been forming between the Last Glacial Maximum and the Late Holocene [3]. Fluvial activity by the Piave River (a major Alpine river with a catchment in the Dolomites) and the Livenza River (a minor Prealpine river) ended here before the Roman Age [4], so in this time, the area corresponded to a stable alluvial plain. Due to relative sea-level rise in the Middle Ages, the area became part of an extensive system of coastal wetlands that was completely reclaimed only between the 19th and the 20th century. The current topographic surface, artificially drained, is 2 m below the sea level.

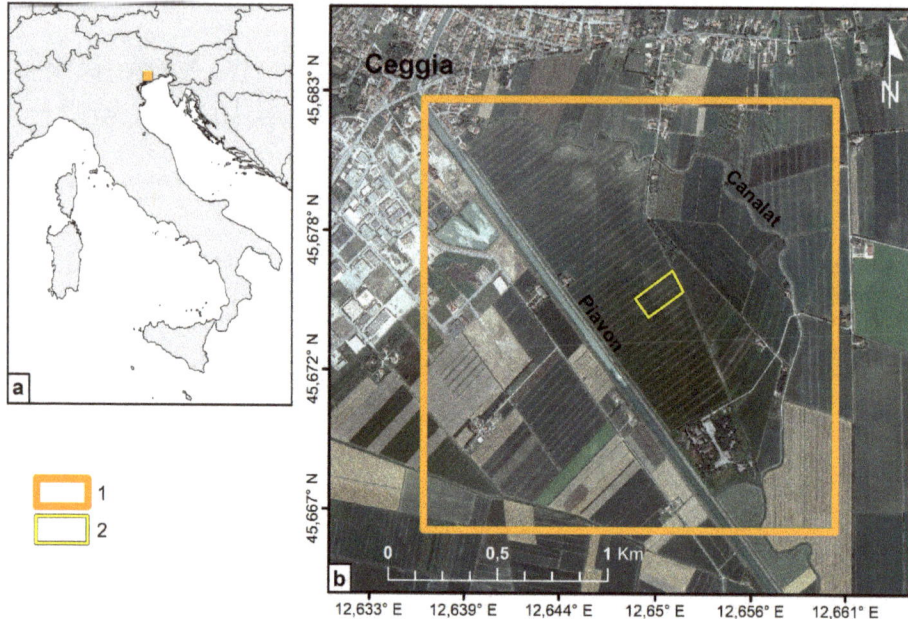

Figure 1. (**a**) Location of the study area. (**b**) Agea 2012 digital color orthophoto: (1) the study area; (2) the site investigated with geophysical prospections, core drilling, and stratigraphic analysis of the exposed sections.

In Roman times, the area of Ceggia was included in the southern part of the *"ager"* of *Opitergium* (currently Oderzo), not far from the inner edge of a sizeable lagoon-marsh extended in the vast plain between the ancient river courses of the Piave and Livenza (Figure 2a). The Roman road so-called *Via Annia* [5–7] represented the principal axis of communication on the eastern coast, marking the limit among different landscapes. The *Via Annia* crossed the territory connecting the Roman centers of *Altinum* and *Iulia Concordia* up to *Aquileia*. Different archaeological findings during the times in the area (Figure 2b) testify to its Roman occupation [8–10]. No details about the population or the layout of these landscapes are registered. The only information available is related to the so-called stream Piavon-Canalat that flows in this sector, and probably representing from ancient times one of the routes of communication and commerce [11,12] between *Opitergium*, the lagoon, and the Roman harbor to the sea, the so-called *Portus Liquentiae* (Figure 2a).

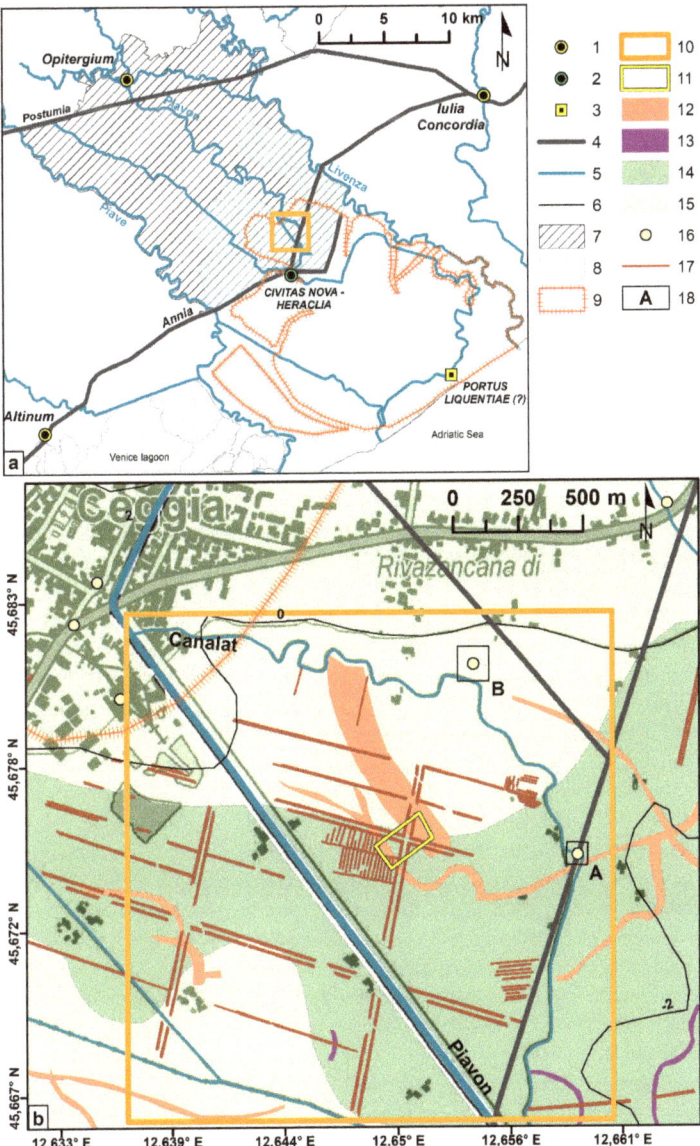

Figure 2. (**a**) Historical-topographical map; (**b**) Geoarchaeological map with traces from aerial photographs. Legend: (1) Roman cities; (2) Early Medieval towns; (3) probable site of the Roman port; (4) Roman roads; (5) main watercourses; (6) contour lines with equidistance 2 m; (7) southern countryside of *Opitergium*; (8) mechanical drainage areas; (9) extension of marshy-lagoon areas from 17th century cartography; (10) study area; (11) area studied with geophysical surveys, core drilling, and cleaning of the exposed sections; (12) paleochannels; (13) ancient lagoon canals; (14) loamy soils; (15) clayey soils; (16) archaeological findings known from the bibliography [8,10] and from data archive of the SABAP modified by the authors; (17) anthropic aerial photo lines; (18) sites of particular interest: A-Roman bridge; B-probable rustic settlement. The geomorphological elements (12), (13), (14), and (15) are taken from [1].

In the early Middle Ages, Ceggia was once again a marginal reality or rather a borderland, if we consider that it must have been on the boundary between the Byzantine territories of the coast and the Longobard hinterland, as well as between the mainland and spaces dominated by marshy and lagoon waters. Different studies conducted in the area disagree about the position of this limit, in part recognized in the route of the *Via Annia* [13], according to others, it should be searched further north [14]. Of the latter opinion were the scholars who, basing on the photo-interpretation of some aerial images, recognized in some anomalies visible in Ceggia and in the neighboring territories the result of hydraulic-agricultural interventions to be related to the nearby settlement of *Civitas Nova/Heraclia* (a center developed starting from the 7th century AD in a Byzantine lagoon environment) (Figure 2a), though not excluding a possible Renaissance origin [14,15]. The question, therefore, is now open. At the current state of knowledge, it is not possible to establish which sphere of influence falls this area or define its ancient environmental configuration.

Moving from previous considerations, we started to analyze some undefined artificial and natural evidence visible in multitemporal aerial images. Similar traces are identified in the Venetian plain [1,14], and their analysis could reveal important archeological infrastructures and sites [5,6,16,17]. More in general, these systems of anthropogenic traces are generally found in alluvial and coastal plains and often related to ancient settlements [18,19]. For this study, with the specific aim of defining the nature, chronology, and function of these undefined buried structures, in particular, we integrated multi-temporal aerial photo interpretation, geophysical measurements, core sample analysis, ^{14}C dating, analysis of historical cartography, and archaeological survey [20].

From a geophysical point of view, the specific characteristics of the current area and the open questions related to wet soils suggested the application of FDEM and ERT methods to identify both natural and anthropogenic features visible from aerial and satellite images.

Electromagnetic methods in the frequency domain, known as FDEM or electromagnetic induction methods (EMI), are widely used for soil mapping in order to obtain a quick overview of the possible heterogeneity of a given conductive system [21–27]. One of the main variables that characterize soils, determining both vertical and horizontal variability within them, beyond the grain size and composition, is undoubtedly the water content. The measurement of the electrical conductivity of soils is, in fact, closely related and dependent on their water content/moisture content. The variation in the composition of a given soil, on the other hand, determines the ability to retain the moisture content differently. In general, apart from clayey soils, whose electrical response of high conductivity is inherent like the material, the different grain size of the soil and the water content (saturated or unsaturated), determines the different electrical response of the investigated system [28–32]. In particular, soils with smaller grain size will be more conductive than those with larger ones. The possibility of detecting fast, non-direct information on these characteristics with EM methods is, therefore, at the basis of their growing popularity in contexts where these data can contribute, for example, to the optimization of cultivation practices [21,33,34].

The main advantage offered by techniques able to indirectly measure the electrical conductivity of the investigated systems, such as low-frequency EM measurements (FDEM or EMI) and electrical resistivity tomography (ERT), is inherent in the ability to establish the spatial distribution and output of bodies characterized by a different electrical conductivity (or its inverse resistivity) in an indirect way and with a different degree of detail (resolution) and at different depths. In particular, the strength of the FDEM technique lies undoubtedly in the speed of acquisition of the data of apparent conductivity of soils, made possible by the fact that direct contact with the measurement system is not necessary and therefore, the operator or the motorized vehicle carrying the instrument, connected to a GPS, can map quite quickly even large extensions giving general information on the heterogeneity of the system. The ERT method instead allows us to define in detail the exact depth and the relationship between the different buried structures in the same system. The possibility to combine these two geophysical methods through comparable measurements allow to identify targets and systems of interest. This data integration is being increasingly used in the context of geomorphological and archaeological studies.

This combination, in fact, allows to identify and describe both in a timely manner with FDEM methods natural structures related to the presence of water (e.g., paleo-environment), where the ERT method helps to better detail those and to recognize the presence of anthropogenic remains, whose exact identification requires a higher resolution than that offered by EM methods [35–43].

The more general objective of this study is to set up and suggest a workflow reproducible in other contexts to allow the proper identification, valorization, and protection of complex archaeological landscapes similar to this here analyzed.

2. Materials and Methods

2.1. Analysis of Historical Documents and Archaeological Field Data

A review of the published and unpublished material kept at the Archive of the Soprintendenza Archeologia, Belle Arti e Paesaggio (SABAP) for the metropolitan area of Venice and the provinces of Padua, Belluno, and Treviso was firstly carried out.

On the other hand, between 2014 and 2017, new data was collected in the area [20,44] through systematic archaeological surface surveys [45–47].

This data was then gathered within a GIS platform, correctly structured to manage archaeological information; and the geomorphological picture, historical cartography, and aerial image traces were analyzed.

2.2. Multitemporal Analysis of Historical Aerial Photographs

Historical aerial photographs often represent a significant resource to identify the transformations that affected the landscapes during the times, changing their original or ancient asset. The visibility of interesting potential traces on the surface linked to the past by these supports is undoubtedly conditioned by many factors as, for example, the light, the vegetation, the ground humidity, the altitude of the flight, the land use, etc. [48,49].

For our purposes, we use these materials to:

- analyze previous works to evaluate possible integration of the photo interpretations;
- obtain metric indications;
- gather the modifications in the organization of the territory, to acquire useful data to interpret the traces, eventually excluding a modern origin linked to the recent reclamations and hydraulic-agricultural interventions.

The first step of the survey consisted of the identification of the aerial photographs preserved in the archives of various entities present on the national territory (Military Geographic Institute-IGM; Centre for Cartography of the Veneto Region-ReVen).

For the above reasons and purposes, the choice of the photos (Table 1) consider the number and quality of visible anomalies, and the temporal coverage, to allow a multitemporal analysis of the landscape. The most significant images, in this sense, result in the ReVen 1983, ReVen 1990, and GAI 1954 flights (Figures 3 and 4a), while the ReVen 1999 flight returned fewer details while confirming the presence of the most relevant evidence. Even the oldest images of the IGM 1937 flight reveal the anomalies but cover only a portion of the study area.

Table 1. List of aerial photographs analyzed for the study.

Flight	Archive	Date	Type of Recording	Focal Length (mm)	Flying Height (m)	Average Frame Scale
ReVen 1999	Veneto Region	19 March 1999	Analog b/w frame (23 cm × 23 cm)	153, 31	2500	1:16,000
ReVen 1990	Veneto Region	20 March 1990	Analog color frame (23 cm × 23 cm)	152, 82	3000	1:20,000
ReVen 1983	Veneto Region	16–17 March 1983	Analog b/w frame (23 cm × 23 cm)	153, 13	2600	1:17,000
GAI 1954	IGM	11 April 1954	Analog b/w frame (23 cm × 23 cm)	153, 154	From 5000 to 10,000	From 1:30,000 to 1:62,000
IGM 1937	IGM	1937	Analog b/w frame (13 cm × 18 cm)	n.a.	From 1400 to 3400	From 1:11,000 to 1:20,000

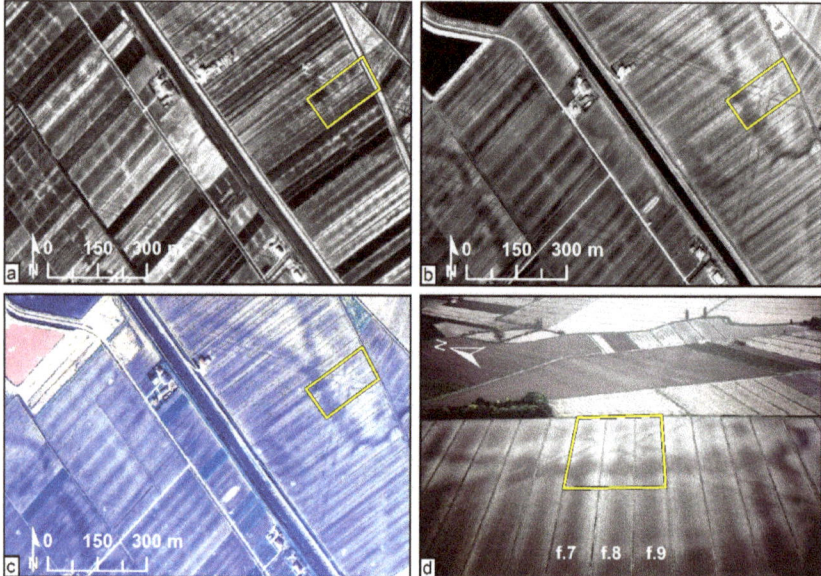

Figure 3. Area SE of Ceggia: zoom on some evident soil marks visible in historical zenithal photos of different years and a recent oblique photo. In yellow, the area subject to geophysical investigations, core drilling, and cleaning of the exposed sections. (**a**) Photo GAI 1954. (**b**) Photo ReVen 1983. (**c**) Photo ReVen 1990. (**d**) Oblique photo [5].

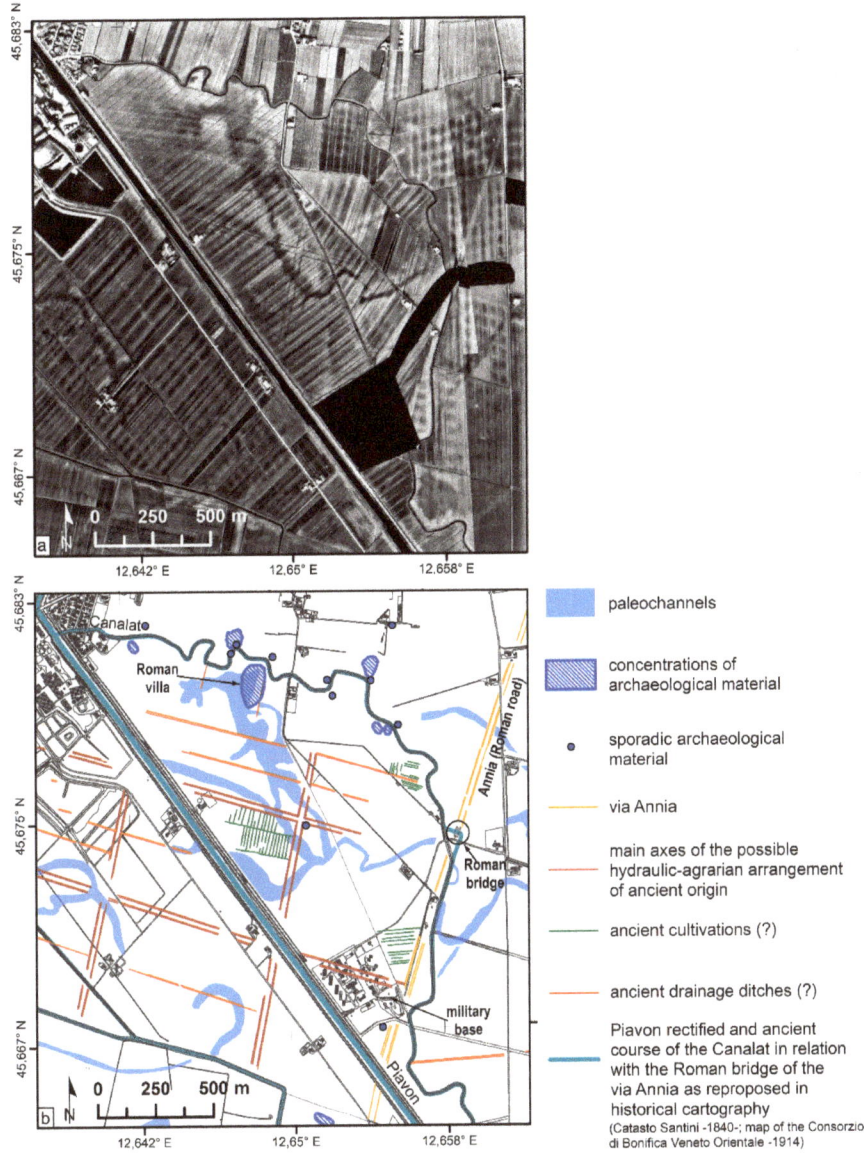

Figure 4. SE area of Ceggia. (**a**) System of traces visible in ReVen 1983 aerial frames. (**b**) Photorestitution on CTR 1:5000 of the signs detected through multitemporal analysis of historical aerial images (IGM 1937, GAI 1954, ReVen 1983 and Reven 1990) and new archaeological data deduced through surveys (2014–2017).

When digitalized, the frames of different flights have been geo-referenced in GIS in order to proceed with the identification of the anomalies and to compare with other data, (e.g., vector and raster types) available for the study area (geomorphological, archaeological, geophysical, cartographic, etc.).

The visible traces in the different temporal images were then divided into different layers, from the most recent (ReVen 1999) to the most dated (IGM 1937).

The information about the location, the source used to identify the anomaly, type of anomaly, and, when it possible, the interpretation and chronology have been associated.

2.3. Multitemporal Analysis of Historical Cartography

Historical cartography provides information on the shapes and conditions of the areas, their nature, but also on works planned realized or not realized. In our case, these instruments help to obtain information useful to interpret the traces visible in aerial images. These materials were mainly searched at the Archives of the Consorzio di Bonifica Veneto Orientale (CBVO) and the Archivio di Stato of Venice (ASVE), considering that the area of interest insists in a reclamation zone and that, since the Republic of Venice, it has been affected by operations aimed at regulating the waters. In the first case, project drawings, cadastral maps, and plans with sporadic indications about the use of defined sectors of the study area and on specific interventions concerning the hydrographic network were found [50]. The Archives of Venice, on the other hand, provide a part of the historical cartography make from the 16th century, by specialized institutes set up within the Republic of Venice to deal with the regulation of water, but also with reclamation, irrigation, and land culture. The same archive also contains maps of the Austrian Land Registry containing the description and use of the different areas. A detailed representation of the nineteenth-century natural characteristics and anthropogenic structure of the area is provided by the map of the Second Military Survey and the Kriegskarte. Other significant detailed and geometrically correct restitution of the territory between 1890–1952 has been recovered by the cartography of the IGM, that has been mapping the entire Italian territory since the second half of the 19th century.

Different cartographic products (Table 2) have then been inserted and georeferenced in a GIS project for the analysis and comparison with aerial photographs (Figure 5).

Table 2. List of historical cartography analyzed for the study.

Date	Conservation Institute	Cartographic Code	Title
1960 ca.	CBVO—(Consorzio di Bonifica Veneto Orientale)	Comune di Ceggia, FF. 2,4, 22	Piano quotato
1952	IGM	F.39 III SE e F52 IV-NE	S. Stino di Livenza e Passarella
1938	IGM	F.39 III SE e F52 IV-NE	S. Stino di Livenza e Passarella
1924	IGM	F.39 III SE e F52 IV-NE	S. Stino di Livenza e Passarella
1920 ca.	CBVO	Settore Tecnico: Progetti-Consorzio di Bonifica Bella Madonna	Comprensorio di bonifica
1914	CBVO	Settore Tecnico: Progetti-Consorzio di Bonifica Bella Madonna	Base catastale con indicazioni degli interventi da eseguire sulla rete scolante: rettifica ansa del Canalat
1890–1891	IGM	F.39 III SE e F52 IV-NE	S. Stino di Livenza e Passarella
1840 (ca.)	CBVO	Settore Catasto: Fossà	Mappa del comune censuario di Fossà con Palazzetto, distretto VII di San Donà di Piave-Catasto Santini
1807–1852	ASVE	74-Fossà con Palazzetto	Catasto Austriaco
1818–1829	Kriegsarchiv-Austrian State Archives	—	Second Military Survey
1798–1805	Kriegsarchiv-Austrian State Archives	—	Topographisch-geometrische Kriegskarte von dem Herzogthum Venedig
1789	CMC (Civico Museo Correr)	Provenienze Diverse, C 840.4	Topografia del corso delle acque del basso Trevigiano
1675	ASVE	Miscellanea Mappe, 1275	Mappa delle terre comprese tra la Livenza e il Piavon

Table 2. Cont.

Date	Conservation Institute	Cartographic Code	Title
1641	ASVE	Savi ed Esecutori alle Acque, Piave, r.106, dis.16	Piave (fiume). Il Piave da Ponte di Piave al suo sbocco a mare, con vasta zona di terre arative, pascolive, boschive, a vigneti e valli, prative e paludose ed estesa fascia litoranea da Sant'Erasmo al porto di Caorle
1639	ASVE	Savi ed Esecutori alle Acque, Piave, r.106, dis.15	Piave ed altri fiumi. Regolazione del Piave, Livenza, Sile, Zero, ed altri fiumi minori nel territorio, confluenti nel mare.
1628	ASVE	Miscellanea Mappe, 1274	Mappa di beni siti nelle adiacenze del Piavon e Magnadola
1607	ASVE	Beni Inculti Treviso-Friuli, rot.414, m.10/B, dis.3	Trevigiano (territorio). Comprensorio tra Piave e Livenza
1568	ASVE	Mensa Patriarcale, 42-55(16)	Villa de Ceia-Palu da retrazer

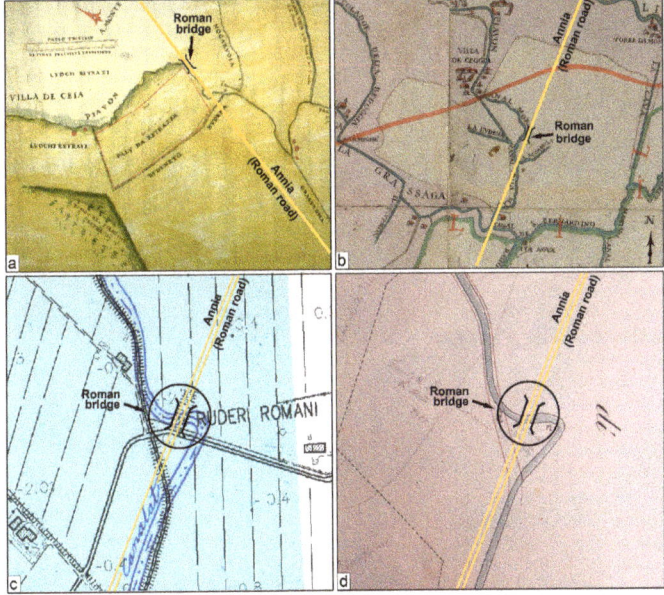

Figure 5. Historical maps of the study area (in orange), and details of the route of the *Via Annia* (in yellow) and the Roman bridge (in black). (**a**) Drawing (1568) showing the sector to be reclaimed, i.e., the "retrazer palv", south of the Piavon canal (today's Canalat)-Archivio di Stato di Venezia, MP 42-55 (16); (**b**) Map (1639) depicting the marshes (in light yellow), ploughed and inhabited land (in pink), watercourses (in blue)-Archivio di Stato di Venezia, SEA, Piave, r.106, dis.15; (**c**) Detail of the map (1840) of the Santini Cadastre (in blue), superimposed on the CTR 1:10,000 (in black), which shows the relationship of the Canalat with the *Annia* Roman bridge; (**d**) Detail of a 1914 map showing the rectification of the Canalat bend in relation to the Roman bridge-Consorzio di Bonifica Veneto Orientale. The route of the *Via Annia*, the Canalat course and the location of the Roman bridge allow to identify the study area in the figures.

We used the geographical coordinates of the IGM tablets to identify clear ground control points (GCP) useful to geo-refer the historical cartography.

At the end of this process, we produced a multi-temporal sequence of images with a window of the transformations in the area of study from the mid-16th century to the present.

2.4. Geophysical Measurements

In February 2015, we planned an FDEM field campaign to easily and quickly identify the main buried structures visible in the aerial and satellite images. Starting from these pieces of evidence, we selected a representative area, with some natural and artificial features, useful to collect field geophysical data. For the acquisition of FDEM data, we used a CMD electromagnetic conductivity meter (from GF Instruments s.r.o. Brno, Czech Republic) with a CMD-1 single-depth probe, setting high depth configuration, corresponding to −1.5 m full depth range. The resolution of the used instrument is 0.1 mS/m, where the accuracy corresponds to ±4% at 50 mS/m. The CMD-1 probe was manually moved on the ground in GPS continuous acquisition mode, connecting the instrument with a Trimble 5800 GPS receiver to register the right position of the acquisition lines. The FDEM data was collected on the field f.9 (Figure 6b) every meter, in a rectangular area, along parallel lines NE-SW oriented, for a total length of about 210 m. The entire width of the investigation area in the NW-SE direction was about 100 m. Due to the presence of small drainage channels, NE-SW oriented, the survey area has been divided into three main parts with a width of about 33 m for each (Figure 6). The conductivity values, during the processing, were converted in resistivity and then together plotted using the Surfer 10 Golden Software to obtain a map of the mean distribution of this parameter in the entire investigated area.

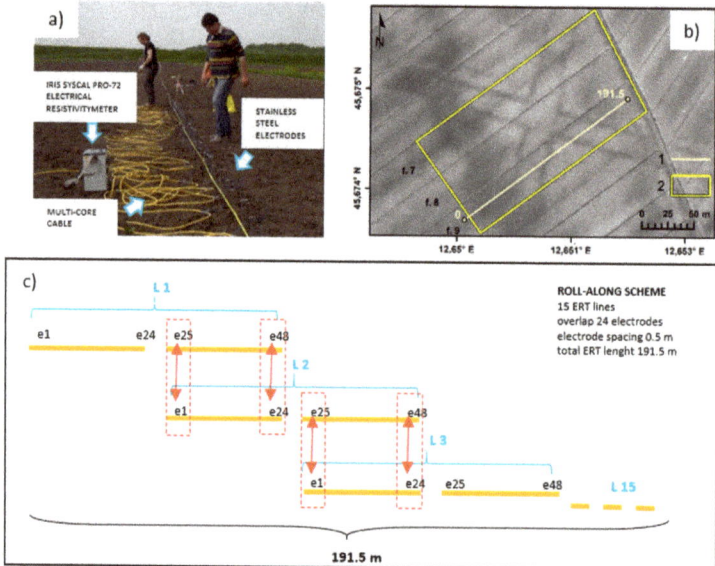

Figure 6. Geophysical measurements: (**a**) Image of an ERT line acquisition; (**b**) Localization of geophysical measurements: blue line (1) ERT, yellow area (2) FDEM; (**c**) scheme of ERT acquisition.

In the middle of the same field (f.9 in Figure 6), an electrical resistivity tomography (ERT) was acquired in the SW-NE direction. The electrical resistivity tomography (ERT) was performed by means of 15 lines, using for each line 48 electrodes spaced 0.5 m apart, overlapping line by line of 24 electrodes (roll-along) over a total length of 191.5 m. Direct and reciprocal measurements for each ERT line were acquired, by swapping current with potential electrodes, to estimate the errors in the dataset [51].

A difference between different cycle results (quality factor "Q") equal to 5% was imposed. The ERT inversion was performed using a regularized weighted least squares approach, according to the Occam's rule [52] as proposed by LaBrecque et al. [53]. The smoothness of the resistivity distribution here obtained strictly depends on the errors in each dataset.

According to Binley et al. [54], the best evaluation of errors might be obtained thanks to direct and reciprocal measurements, where these measurements shall be equal, providing the same resistance value. The possible deviation may be interpreted as an error estimate useful for the inversion. In this case study, the errors were smaller than 5%. The visualization of the inverted ERT lines was done using Surfer 10 Golden Software.

2.5. Core Sample Analysis and Radiocarbon Dating

Different core samples were carried out using an Edelman combination-type hand auger. Sediment description included grain size, sedimentary structures, Munsell Charts color, presence, size and abundance of pedologic nodules and mottles, paleontological content (shells, plant remains, wood, charcoal), archaeological content (sherds, bricks etc.). The data were compiled in stratigraphic logs and sections (Figures 7 and 8). An AMS radiocarbon date was obtained from a wood fragment (Table 3) that was collected from the undisturbed inner part of core CEG 7b and stored in aluminum foil after extraction in order to avoid contamination. The radiocarbon date was calibrated using software OxCal 4.3 [55] and applying the calibration curve IntCal13 [56].

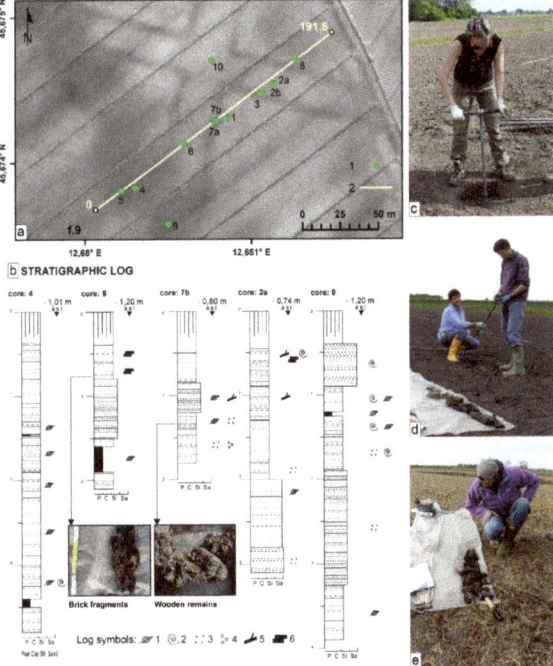

Figure 7. Drilling. (**a**) Location of the cores: (1) and ERT alignment (2). (**b**) Stratigraphic logs of core drillings performed in correspondence of paleochannels (4, 9, 8), ditches (7b), and main axes of the presumed structures of ancient origin (9, 2a). Log symbols: (1) fragments of plant remains; (2) continental mollusks; (3) carbonatic nodules; (4) nodules of iron-manganese oxides; (5) charcoal; (6) bricks fragments. On the right, from top to bottom, photos of the works: (**c**) execution of a core; (**d**) sampling of a core; (**e**) documentation of stratigraphic succession.

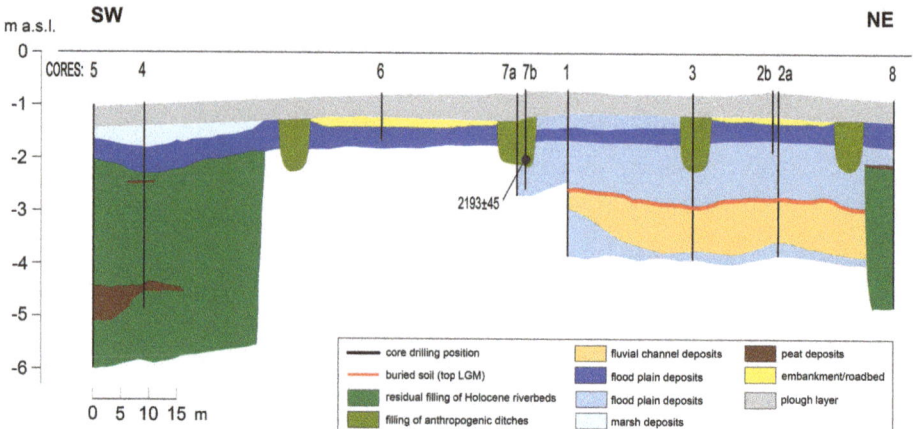

Figure 8. Section across the study site.

Table 3. Details of the ^{14}C dating of a wood sample from the core 7b (see Figure 7).

Core and Depth (cm)	LabCode	Radiocarbon Age (yr BP)	δ^{13}C (‰)	Calibrated Age B.C. [Start-End] Probability %	Material	Elevation (m a.s.l.)	Latitude	Longitude
Core 7b 135 cm	LTL1687A	2193 ± 45	−29.8 ±0.2	[384,156] 93,5% [134,115] 1,9%	wood	−0,80	45,6742° N	12,6508° E

2.6. Visual Evaluation of Exposed Section

The exposed section of the southern side of the drainage channel that limits the field f.9, where geophysical prospecting and the cores drilling were realized, allowed us to analyze the correspondent section of the traces visible in the aerial photo (Figure 9a). Thanks to this activity, we documented the stratigraphic sequence [57] of the deposits and obtained detailed information about the different soil marks revealed in the aerial photos (e.g., potential road axes, drainage ditches, and paleochannel). The position of the sections of interest were identified thanks to the georeferenced images of the area stored in the GIS platform, where their coordinates were collected using a differential GPS and positioned on the field with several material references (i.e., pickets). The section is documented by photos, drawings, and stratigraphic units, interested up to 0.9 m depth from the ground level.

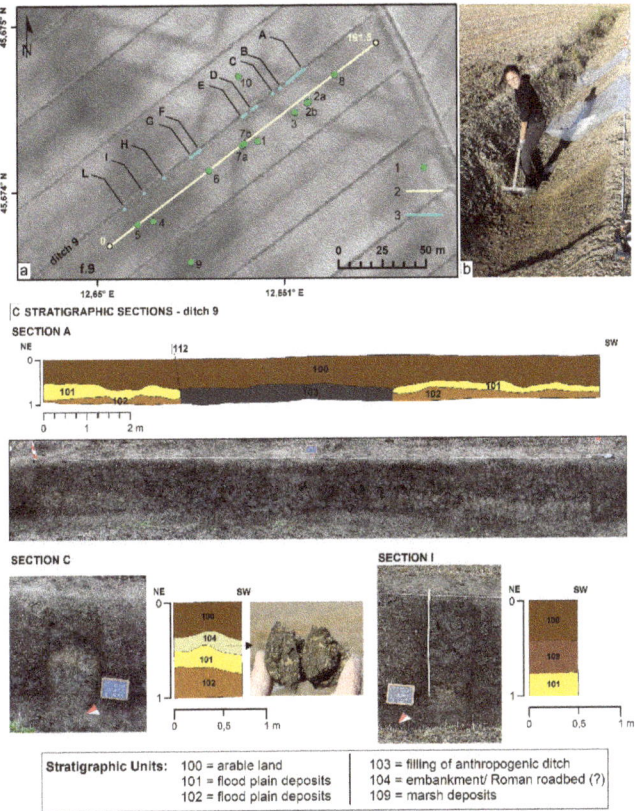

Figure 9. Cleaning of the exposed section. (**a**) Working area: (1) cores; (2) ERT line; (3) exposed sections. (**b**) Image documenting the field activity. (**c**) Stratigraphic sections detected in correspondence of some soil marks: the numbers correspond to the Stratigraphic Units (US) identified; the image shows the included angular US104 (hypothetical road/street base).

3. Results

The analysis of the pieces of evidence and documents currently available for the area, although not reported in any classical source, testifies to the Roman frequentation of the SE territory of Ceggia.

In addition to the well-known *Via Annia* (Figure 2b) and its Roman bridge (Figure 2b.A) [58] crossing a waterway active at Roman age, in the area have been discovered the remains of an apparent rural settlement, with some wall structures and a pit along the left side of the Canalat water channel (Figure 2b.B) [10], about 800 m NW to the Roman bridge. The archaeological field survey also made it possible to collect new data useful for a better knowledge of the area, never before analyzed by similar archaeological research. The survey permits the localization of the remains of the Roman road and the identification of sporadic pieces of pottery in correspondence with the linear traces visible in aerial photographs, and a significant concentration of materials of archaeological interest along the Canalat (Figure 4b). These materials are pertinent to sites from the Roman Age (1st century BC–2nd century AD), several of which were continuously used up to the Late Antiquity (IV/V AD). In some cases, it testified to the presence of Renaissance pottery (14th–16th centuries AD), while no evidence of the medieval period was registered. All the new sites identified by the survey are located near the Canalat (Figure 4b), i.e., the watercourse that crossed the study area in Roman times and that, in correspondence with the *Via Annia*, flowed under the *Annia* bridge. Among these sites, it is worth mentioning, in particular, an apparent Roman villa along the right side of the Canalat (Figure 4b), both for the valuable materials found and for its position respect to the traces visible in the aerial photos. The site is located, in fact, on the trace of the riverbed that continues towards the Roman bridge of the *Via Annia*, probably related to a minor axis of the system of artificial lines visible from aerial photos. The multitemporal analysis of the historical frames allowed the first discrimination between traces of hydrography and modern hydraulic infrastructures, as the result of interventions carried out in the second half of the twentieth century, and others of older origin found in all the aerial photos here analyzed (Figure 3). The comparison carried out was also very useful to better outline the picture of ancient traces because some images revealed particular features not visible in others, thus contributing to the formulation of a more detailed reconstruction than that proposed by previous studies. For example, in Figure 4, there is clear evidence that affects the eastern sector of the study area, represented by a white band edged by two dark lines that follow its course. This is *Via Annia*, clearly identifiable in aerial photos and on the ground, due to the dispersion on the surface of pebbles and brick fragments, as well as the presence of the remains of a bridge over a waterway. Different clear anomalies in the aerial photos document the ancient natural hydrography of the area, characteristic for their dark color, the morphology, and the meandering course further downstream. Their shape suggests the presence of a local drainage network probably linked to the channels in the marshes covering the area in the Middle Ages until the modern reclamation. Very similar traces of wetland palaeo-hydrography are reported about 10 km NE of Ceggia, beyond the Livenza river [1,4]. The central portion of the study area is crossed by one of this branched system of marshy canals, mixed with some artificial lines between the Piavon and Canalat channels, before the union in a single meander that seems to join the Canalat just upstream of the *Annia* Roman bridge. The artificial alignments just mentioned are also clearly recognizable in the whole area of investigation. By shape, these are not very different from a Roman road, appearing as light bands between dark lines, although organized in a system. These axes, about 28 m wide, including the lateral ditches, appear as a primary grid of a hydraulic-agrarian layout, which seems to insist on the *Annia* route. In the aerial photos GAI 1954 (Table 1), it is visible that one of the WNW-ESE axes extends up to the route connecting to a military base. Among the main axes of this system, some aerial photos show the presence of dark lines, probably related to minor ditches, and other very close lines that would suggest the existence of ancient cultivations. Furthermore, the bearing axes of the system are not entirely orthogonal to each other and delimit non-regular portions of land (e.g., the long side measures approximately 730 m, the short side 500 m). The contribution provided by the historical maps appeared fundamental to define the nature of the area, the modification over the centuries, and to limit the possible time range of the interesting anomalies visible in the aerial photos. The multi-temporal analysis of the cartography revealed, in fact, that the area was affected by reclamation from at least the middle of the 16th century until the end of the 19th century when an important reclamation and hydraulic-agricultural reorganization interventions profoundly transformed its asset. However, no relation appears between these recent interventions and the alignments visible in the aerial photos, placing the realization of the second ones before the presence of the marshes in the area. This evidence made by the multi-temporal analysis of the aerial photos highlights an ancient organization of the area, probably due to its hydraulic instability. The historical representations of the territory between the mid-16th and 18th centuries have also provided information on the drainage system in the area and on the interventions implemented to improve it, as well as a valuable indirect reference for the interpretation of the aerial photos. For example, the maps of the 17th and 18th centuries testify that the excavation of the straight section of Piavon is related to the burial of the same old river (today's Canalat), historically the so-called "Piavon a monito" or "Canal Morto" (Figure 5b). The comparison of these maps with a drawing of the area in 1568 (Figure 5a) dated this intervention between the end of the 16th and the beginning of the 17th century, probably during a reclamation. For our study, this information is extremely interesting because the system visible in the aerial images is crossed by the straight section of the Piavon, which suggests that it already existed at the time of the excavation of the canal (late 16th–early 17th century), also defining a timeline for the asset of the area as well as are visible in the aerial photos. The same multitemporal analysis suggests that the watercourse in the area in Roman times probably was the Canalat, a link to the *Annia* bridge at least until the beginning of the 20th century, when the interventions slightly modified its original course (Figure 5c,d).

The resistivity map obtained from the FDEM measurements made using the CMD 1 probe (Figure 10b), referring to the ground condition recorded between the surface and the maximum depth of 1.5 m, returns a pattern perfectly consistent with the anomalies visible from aerial photo. In particular, the chromatic scale that associates blue to the most conductive areas allows immediate visualization of the two paleochannels that intercept the road and the drainage ditches in the latter. Note a greater resistivity (Figure 10b) in the central part of the map corresponding to field f.10, probably linked to the greater drainage of this field compared to the two lateral ones. The result of the electrical resistivity tomography performed at the center of field f.9 (Figure 10a,c), better defines the nature and relationship between the various anomalies. It should be noted that the ERT measurement allows us to analyze the real extension of the two paleochannels, and in particular, as witnesses that the paleochannel to the east is more extended in-depth than what can be assumed by the trace visible on the surface. Another interesting fact is the extension in depth of the most resistive area of the section on which the two arms of the road axis visible from the aerial photo are set and perfectly detected even with FDEM measurements. These data testify that the road was probably crossing an area with greater resistivity than the two nearby paleochannels, perhaps indicating not only coarser materials, but also a morphologically higher area.

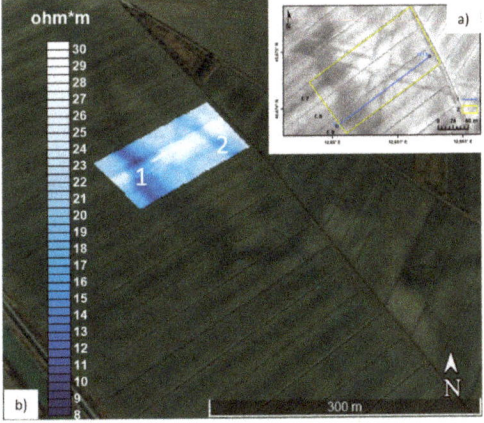

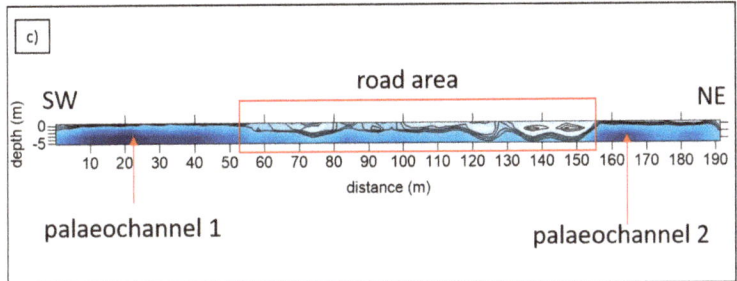

Figure 10. Results of geophysical measurements: (**a**) localization of FDEM (yellow area) and ERT line (blue); (**b**) Map of resistivity measured from the surface to −1.5 m obtained by using FDEM; (**c**) result of ERT section.

The cores, distributed along a SW-NE section across the study site, allowed to reconstruct the stratigraphy of the deposits to a maximum depth of almost 4 m (Figure 7a). In Figure 7b are reported the stratigraphic logs of the most significant cores in terms of length of the investigated succession and/or relation with the detected archaeological features. The lithostratigraphic correlation between cores is shown in Figure 8.

The alluvial succession investigated through coring is mostly alluvial silty-clay, with only a tabular body of fine silty sand with maximum thickness of 0.8 m that extends in the central part of the section, between cores 1 and 2. The top of this sand body lies at about −3 m asl and is covered by a few-decimetres-thick layer of light olive-brown (2.5Y 5/4) silty clay with abundant pedogenetic carbonate nodules (dimensions up to 2 cm, presence 10–15%) and gray mottles that evidence the occurrence of soil formation in a probably ABk horizon (after [59]). Levels rich in carbonate nodules and mottles are also present in the underlying sand body. This buried soil is covered by 1.5–2 m of olive-brown (2.5Y 4/3) silty clay up to the surface.

At the SW and NE extremes of the section, the alluvial stratigraphy is markedly different, due to the presence of two clay lenses with maximum investigated thickness of 4 m and apparently steep sides (cores 4 and 8). These soft clays have dark gray (5Y 3/1) to greenish-gray (GLEY 6/10Y) colors and contain common fragments of reeds, wood, fresh-water mollusks, and peat layers. They are buried by a laterally-continuous light yellowish brown (2.5Y 6/3) layer of silt and clay that covers the whole investigated area. Only in correspondence of the SW end, this layer is buried by about 0.5 m of very dark grey (2.5Y 3/1) clay.

Concerning the remote sensing archaeological traces, cores 3 and 7a,b were carried out in the dark linear traces. Each core shows the presence of a silty clay sedimentary body between 0.5 (just below the plough horizon) and 1.4 m depth, grayish brown (2.5Y 5/2) color with black laminae, containing common charcoal, reworked carbonate nodules, few wood fragments and just one brick sherd of few millimeters found in core 3. One wood fragment in core 7b was ^{14}C-dated at 384–156 BC (Table 3).

Cores 2a,b, 6, 9, and 10 were positioned between the dark linear traces, with the purpose of verifying the possible presence of anthropogenic strata related to a road. These cores show the presence of a 10–40 cm thick layer just below the surface plough horizon, consisting of compact silty clay with charcoals and fragments of terrestrial gastropods. This is the only stratigraphic evidence of the road, as below is an apparently natural and undisturbed alluvial succession.

The analysis of the exposed section, in the drainage channel of the field f.9, highlighted the presence of some interesting anomalies, just below the agricultural soil thickness of about 0.55 m. In particular, the section allowed us to identify the head of the drainage ditches on the sides of the visible remains of the bottom of the road. The total size of the road is about 28 m to 29 m, including the lateral drainage ditches with a variable size from about 2.8 m to 4.0 m. No evidence has been registered about the bottom of the drainage ditches as well as about the paleochannel, because of their depth, identified respectively by the sample cores at 1.35 m and 1.5 m below the surface. More in general, no structural materials (e.g., stones, wood, etc.) were identified in the section, excepting some, not relevant, small brick fragments. It should be noted that an extensive yellow clayey-loamy level, probably an alluvial deposit, of about 0.2 m in thickness (US 101 in Figure 9c) was detected. Apparently, this represents the bottom of the linear structures visible in the aerial photos, where the upper part of these structures has been probably cut by agricultural modern cultivation practices. The yellow clayey-loamy level (US 101 in Figure 9c) also covers the big paleochannel. In this position, the US 101 is deeper than in other parts of the section (0.7 m below the surface), and is covered by a grey clayey-loamy layer (US 109 in Figure 9c), of about 0.2 m of thickness, probably related to a swamping phase of the area (Figure 9c—section I) and here preserved because of its depth.

4. Discussion

The study here presented about the SE area of Ceggia, where several buried traces, potentially interesting from a geological and archaeological point of view, have been identified by aerial and satellite images, demonstrated the high value of the multidisciplinary approach to understanding, in similar contexts, the evolution of a territory.

In particular, the multitemporal combined analysis of aerial images and historical cartography of Ceggia demonstrated that the linear systems visible in the aerial photos (Figure 1) do not pertain to the modern reclamation (end of the 19th and 20th centuries), and it is not previously documented by the available historical cartography (until the middle of the 16th century), which represent only the marshes in the area. The same analysis highlighted that the artificial section of the Piavon, realized between the end of the 16th and the beginning of the 17th century, cuts this relict system, suggesting its earlier existence.

On the contrary of the hypothesis of Dorigo [15], therefore, these lines evident on the field cannot be attributable to possible interventions made during the Republic of Venice in the second half of the sixteenth century. The evidence collected in this study, thanks to the sporadic fragments of Renaissance material found along the exposed section in the analyzed drainage channel, demonstrates that the area has been undoubtedly frequented in that period but, most probably, only for hunting, fishing, and to provide natural materials.

It can, therefore, be assumed that the system visible in the photos is older than the 16th century and probably even older than the 14th century, considering two different Medieval documents [15,60] which describe the area occupied by marshes.

The preliminary information made by the multi-temporal analysis described before is then completed and validated by the new data provided by the field non-direct and direct measurements.

Starting from the buried anomalies visible in the aerial photos, the geophysical FDEM and ERT acquisitions better defined the localization and spatial relation among these different structures, driving for the sample core positioning. The geophysical data also integrated the information about the system in-depth, only partially directly investigated by the exposed section and the sample cores.

The buried soil at about 2 m depth in the central part of the stratigraphic cross-section (Figures 7 and 8) indicates the existence of a major depositional hiatus in the alluvial succession, probably related to the so-called post-Last Glacial Maximum (post-LGM) unconformity, well known in the whole Venetian-Friulian plain (e.g., [61–65]). This unconformity is associated with soil formation in the interfluves (the "caranto palaeosoil") and the development of incised valleys along with the concentrated streamflow, as a dynamic reaction of the fluvial system to the deglaciation of the mountain catchments at the end of the LGM. In this framework, the tabular sandy body below the buried soil is attributed to the LGM aggradation of the alluvial plain. The two clay lenses at the SW and NE extremes of the cross-section should represent the top portion of the infilling of post-LGM incised valleys of the Piave River, whose existence has already been reported in this distal sector of the Piave megafan (e.g., [4,66]).

The overlying tabular silty-clay layers were probably deposited by eastern branches of the Piave River before Roman times and after the 2nd millennium BC (Piavon Unit in [4]). The topmost, thin layer of gray mud deposits preserved under the plough horizon in the SW extreme of the section (i.e., US 109) probably corresponds to the so-called Ceggia unit, a thin (<2 m) sedimentary unit that was deposited in fresh-water coastal swamps during the Holocene up to modern times on wide areas of the Piave coastal plain [4]. The preservation of these deposits in the study area was probably possible thanks to the fact that its deposition took place in a depression that, in turn, was created by the higher subsidence in the organic clay fill of the incised valley in respect to the nearby interfluve (a common process in the Venetian plain, e.g., [67]). The sinuous paleochannel evident in aerial photographs is probably related to local drainage along with this elongated depression, that is used to follow the buried Holocene incised valley and debouched in the Canalat just upstream of the Roman bridge (Figure 4).

The core samples and the exposed section have, therefore, highlighted that the paleochannel visible in the aerial photos is more recent than the Roman bridge, on the contrary to as hypothesized in previous studies [68]. These data, therefore, exclude that this river branch flowed in the Roman age, at the same time making possible a hydraulic-agricultural arrangement of the area at that time. In this hypothetical scenario, the river that crossed the area flowing below the *Annia* three-arched bridge probably corresponds to the current Canalat. This hypothesis can be proved by the large amount of Roman archaeological materials found along the channel (Figure 4b) and by the remains of the Roman villa just close to it. The characteristics and geometry of sediments corresponding to the dark linear traces allow their interpretation as the infilling of couples of artificial ditches running in pairs and converging perpendicularly at the junction. The time window bracketed by the radiocarbon date spans from the beginning of the 4th century to the first half of the 2nd century BC. This confirms that the ditches are not medieval, nor modern. Considering the possibility that the dated wood pertained to a large timber of mature age, this date is not in contrast with the deforestation and installation of a pre-Roman route before the building of the *Via Annia*, attributed to the second half of the 2nd century BC, possibly as early as 153 BC [5,6]. If we take into account, moreover, that from the photos of the GAI 1954 flight, one of the axes with WNW-ESE trend delimited by these ditches seems to detach from the *Annia* at the height of the site occupied by the military base, then we have another good reason to assume that the system dates back to a phase in which the road layout was evident on the surface and possibly in use.

More scant is the subsurface evidence of the road that supposedly runs in between each pair of ditches. The relative sediments have probably been largely beheaded by ploughing and other modern agricultural activities. The US 104 is a possible remain of the lower foundation of the road; nevertheless, considering known examples of the *Via Annia* road investigated in archeological excavation (e.g., Ca' Tron Michelini; Vigoni [6]), it has to be considered that even such a major consular road outside the main cities was just a dirt road, with faint archaeological evidence. Examples of similar structures have been highlighted in Villadose in the *ager Atrianus* [69–71], but also in the territory north of Padua [72], in the Altinate [72], and in the Valli Grandi Veronesi [69,73]. If we consider, in fact, the hydraulic instability of the sector, it is likely that a system of roads on elevated embankments and ditches like the one found was the most obvious solution to be adopted to ensure the drainage of water from the fields as well as access to the area even in periods when water stagnation could still occur. Moreover, it is not to be excluded that, in Roman times, these banks were also used for seasonal grazing of sheep, in relation to the phenomenon of transhumance, which involved the wintering of flocks coming from mountain areas in the coastal plains. This practice, found in various areas of the Roman Veneto, is well documented, especially in the nearby Altinate territory [74]. Regardless of the use to which the accommodation could have been put, it should be pointed out that the mesh of the buried evidence is not perfectly orthogonal, and the surface of the plots cannot be traced back to the Roman actus. Such a finding, combined with the location of the Roman sites along the river Canalat, leads us to hypothesize that these spaces were not organized according to a centurial model, although they were equipped with hydraulic arrangements that probably guaranteed their seasonal exploitation [75,76].

As it has been pointed out for the plain of Lugo [77], during the Roman Republic, several areas exposed to hydrogeological risk were not regularly divided for the agricultural practices, probably due to the environmental constraints (e.g., presence of swamps or ponds). The conditioning by the environmental context, in terms of population and territorial structure, is also underlined by another recent study which, analyzing the southern centuriated landscape of Padua, focuses on the undefined eastern limit of the agro (coinciding with the coastline of the Roman age), assuming the presence of settlements and cultivated areas also in the area of coastal marshes mentioned by ancient authors [78].

Finally, it is not to be excluded, even if it seems less probable, that the important layout given by the Romans in the area of Ceggia, after a phase of abandonment, was restored and readapted in the Early Middle Ages, losing its original shape. If the arrangement of the sector was, therefore, to be traced back to the Roman age, as the data examined so far would seem to suggest, its progressive defunctionalisation could instead be attributed to changed environmental and land-management conditions.

There are some signs of a worsening of the state of the coastal areas of the upper-Adriatic lagoon area, already from the 3rd–4th centuries AD, when the action was taken on the *Annia* to restore some stretches of the route ruined by the marshy waters [7]. However, it is from about the middle of the VI century AD, as suggested by the chronicle of Paul the Deacon [79], that a phase of strong hydraulic instability [80], characterized by alluvial phenomena and the swamping of vast portions of territory previously inhabited [1,81], seems to have begun.

This important environmental change, which perhaps led to the deterioration of the Roman land system, had to be favored by the concomitant rarefaction of the settlement typical of the Late Antique-Early Medieval period, as well as by the lack of both maintenance and capillary water control.

On the basis of the picture outlined, it is therefore not to be excluded that the failure to find medieval material on the surface in the SE area of Ceggia is to be associated precisely with this settlement decline, due to the establishment of new socio-political and economic balances and the occurrence of climatic and environmental changes mentioned above.

If we listen to the sources, it is likely that the area of Ceggia and those close to the settlement of *Civitas Nova/Heraclia*, were already affected at the time by marshy and wooded environments and uncultivated areas for pastoral and breeding activities [82]. The natural modifications occurred during the time provide the image of a variegated landscape, without the man intervention except in an occasional and unsystematic manner. It is difficult, therefore, to think that in such a context, it has been possible to create a system like the one object of this study.

5. Conclusions

The multidisciplinary study carried out in the SE area of Ceggia (VE), in the eastern part of the Venetian Plain, provides numerous details about the high hydraulic instability that affected the area during the times, with the alternation of stable phases (land inhabited and cultivated) with unstable phases (presence of marshes).

This condition probably defined the fertility of the soils and the consequent occupation of the area for its cultivation, despite the dependence from continuous works to force and regulate the flow of water. On the other hand, it is clear that only with the beginning of modern times, with new technologies, the management of the hydraulic problems have become fully solvable. It can be easily understood, however, thanks to the data collected here, that in more ancient times, until the Roman age, the management of this instability had to be not simple and, in any case, temporary.

The survey conducted has made it possible to identify the timing of landscape transformation and to outline, albeit still with some interpretative uncertainty, the different physiognomies that the sector has been acquiring.

Starting from aerial images, it was possible to detect ancient buried traces, and thanks to the comparison with cartographic documents, ancient sources, and data acquired in the field, it was possible to formulate a hypothetical interpretation of the evidence, as well as the transformations that have affected an environment poised between land and water.

It has been highlighted that the periods of hydraulic instability were followed by periods of relative stability conditioned by the incessant anthropic intervention of regulation and maintenance of the water system. Moreover, it has been noted that in these sectors, as climatic-environmental conditions vary and in the absence of water resource management, it is the natural element that has the upper hand.

In relation to the focus of our investigation, the reconstruction of the geomorphic evolution of the study area highlights the existence of a stable alluvial plain in the 2nd century BC, when most probably the ditches were dug, as part of field systems possibly in relation with the construction of the *Via Annia* and with the occupation of the sector highlighted by a series of productive-residential settlements.

In post-Roman times, as also demonstrated by the cartography, the alluvial plain was covered by wetlands with the sedimentation of a thin veneer of organic-rich swamp deposits. Higher subsidence in the compressible soft clays and peat lowered the topographic surface along the post-LGM incised valley, allowing larger accommodation space and better preservation of the swamp deposits.

After land reclamation and areal erosion by modern agriculture, the Roman infrastructures became again visible at the surface. Locally, the remnants of the swamp persisted, producing dark-tone soil marks in aerial photographs as well as highlighting the course of low-energy, minor channels of the post-Roman swamps.

As regards the hydraulic-agricultural arrangement visible in the trace, the data collected so far would suggest a Roman dating. However, there are still several doubts linked both to the modulus of this arrangement, which cannot be traced back to Roman measurements and to the large dimensions of the main axes of the ancient arrangement. Furthermore, there is some discussion about the imperfect orthogonality of the alignments and the different inclination they present with respect to the presumed limits of the centuriation south of *Opitergium*.

Although the alignments of Ceggia can be traced in shape and width to other Roman evidence brought to light in Veneto, it should be noted that most of these comparisons refer to embankments not organized in the system. Even the most stringent comparison, which could be established with the so-called "via di Villadose", considered the *decumanus maximus* of the centuriation[n] of Adria, does not help for the moment to resolve the interpretative doubts, to which only further investigations can perhaps provide answers.

Author Contributions: A.V. contributes to the archaeological research, multitemporal analysis of aerial photos and cartography, R.D. contributes to the geophysical data acquisition, processing, and interpretation, P.M. contributes to geomorphological and geoarchaeological analysis, coring, and interpretation of the alluvial stratigraphy. All authors contributed to the conceptualization and writing of the paper. All authors have read and agreed to the published version of the manuscript.

Funding: This research received no external funding.

Acknowledgments: This paper is a part of the Ph.D. thesis of A.V. supervised by Guido Rosada (University of Padova) and Luisa Migliorati (University of Rome La Sapienza). The authors are grateful to the IGM for authorization to publish the image of Fg.39-Fight GAI 1954 str.14, n.182 (authorization number 7054, date: 27/04/2020) and to Luigi Vacilotto, Silvia Favaro and Guglielmo Strapazzon for their support during the field acquisitions. The use of the ReVen images follows the rules of "Italian Open Data License 2.0" (IODL 2.0 http://www.dati.gov.it/iodl/2.0/), which authorizes the free publication of the aerial photographs after quoting the source: "Regione del Veneto-L.R. n. 28/76 Formazione della Carta Tecnica Regionale".

Conflicts of Interest: The authors declare no conflict of interest.

References

1. Bondesan, A.; Meneghel, M. *Geomorfologia della Provincia di Venezia. Collana: Il Mito e la Storia*; Serie maggiore; Esedra: Padova, Italy, 2004.
2. Ragazzi, F.; Zamarchi, P. *Carta dei Suoli della Provincia di Venezia, Scala 1:50.000*; LAC: Firenze, Italy, 2008.
3. Carton, A.; Bondesan, A.; Fontana, A.; Meneghel, M.; Miola, A.; Mozzi, P.; Primon, S.; Surian, N. Geomorphological evolution and sediment transfer in the Piave River watershed (north-eastern Italy) since the LGM. *Géomorphologie* **2009**, *3*, 37–58.
4. Bondesan, A.; Primon, S.; Bassan, V.; Vitturi, A. *Carta Delle Unità Geologiche Della Provincia di Venezia*; Cierre: Verona, Italy, 2008.
5. Veronese, F. *Via Annia. Adria, Padova, Altino, Concordia, Aquileia. Progetto di Recupero e Valorizzazione di Un'antica Strada Romana*; Il Poligrafo: Padova, Italy, 2009.
6. Veronese, F. *Via Annia II. Adria, Padova, Altino, Concordia, Aquileia. Progetto di Recupero e Valorizzazione di Un'antica Strada Romana*; Il Poligrafo: Padova, Italy, 2011.
7. Rosada, G.; Frassine, M.; Ghiotto, A.R. *Viam Anniam Influentibus Palustribus Aquis Eververatam Tradizione, Mito, Storia e Katastrophé di una Strada Romana*; Canova: Treviso, Italy, 2010.
8. CaVe. *Carta archeologica del Veneto I*; Capuis, L., Leonardi, G., Pesavento Mattioli, S., Rosada, G., Eds.; coord. scient. L.; Bosio: Modena, Italy, 1988.

9. Croce Da Villa, P. Stele funeraria romana da Ceggia (VE). In *Studi di Archeologia Della X Regio in Ricordo di Michele Tombolani*; Scarfì, B.M., Ed.; Roma, Italy, 1994; pp. 339–343.
10. Magarotto, M. L'ambiente Antropico Delimitato Dalla SS 14 e Dagli Attuali Corsi del Piave e del Livenza Dall'epoca Romana All'affermarsi di Civitas Nova. Bachelor's Thesis, Fac. di Lettere e Filosofia, Università degli Studi di Padova, Padua, Italy, 1984–1985.
11. Giovani, E.; Rigoni, A.N. *L'agro Opitergino e i Paleoalvei alla Sinistra del Piave dai dati del Remote Sensing, in Quaderni di Archeologia del Veneto*; CEDAM: Padova, Italy, 1986; Volume II, pp. 135–139.
12. Rosada, G. *I Fiumi e i Porti Della Venetia Orientale: Osservazioni Intorno ad un Famoso Passo Pliniano*; Portus Liquentiae: I Dati e i Problemi, in Aquileia Nostra: Bollettino Dell'associazione Nazionale per Aquileia, L; Aquileia Nostra: Aquileia, Italy, 1979; pp. 174–216.
13. Bosio, L.; Rosada, G. Le presenze insediative nell'arco dell'Alto Adriatico dall'epoca romana alla nascita di Venezia. Dati e problemi topografici. In *Da Aquileia a Venezia*; Una mediazione tra l'Europa e l'Oriente dal II secolo a.C. al 6VI secolo d.C.; Libri Scheiwiller: Milano, Italy, 1989; pp. 509–567.
14. Tozzi, P.; Harari, M. *Eraclea Veneta. Immagine di una città Sepolta*; Compagnia Generale Ripreseaeree: Parma, Italy, 1984.
15. Dorigo, W. Venezie Sepolte nella Terra del Piave. Duemila Anni fra il Dolce e il Salso. 1994. Available online: https://aleph.mpg.de/F/9JH8L2FKU1AHY3N29182VTYD9GPSKK3NYJ1H3NGYBHFETLH63L-60733?func=full-set-set&set_number=116613&set_entry=000001&format=999 (accessed on 11 September 2020).
16. Mozzi, P.; Fontana, A.; Ferrarese, F.; Ninfo, A.; Campana, S.; Francese, R. The Roman city of Altinum, Venice lagoon, from remote sensing and geophysical prospection. *Archaeolog. Prospect.* **2016**, *23*, 27–44. [CrossRef]
17. Ninfo, A.; Fontana, A.; Mozzi, P.; Ferrarese, F. The map of Altinum, ancestor of Venice. *Science* **2009**, *325*, 577. [CrossRef] [PubMed]
18. Ferri, R.; Calzolari, M. *Il Contributo Dell'indagine Aerofotogrammetrica all'individuazione di Antichi Tracciati Stradali: L'esempio della Viabilità di Epoca Romana le Valli Grandi Veronesi e la Bassa Modenese, Miscellanea di Studi Archeologici e di Antichità*, 3nd ed.; Aedes Muratoriana: Modena, Italy, 1989; pp. 111–132.
19. Vermeulen, F.; Antrop, M. *Ancient Lines in the Landscape. A Geo-Archaeological Study of Protohistoric and Roman Roads and Fields Systems in: Northwestern Gaul*; Babesch. Bulletin Antieke Beschaving. Annual Papers on Classical Archaeology; Peeters Publishers: Leuven, Belgium, 2001; (Suppl. 7).
20. Vacilotto, A. Capire i segni antichi della terra per governarla. Il caso incognito dell'agro meridionale di Opitergium. In *R. PERNA, R. CARMENATI*; Giuliodori, M., Piccinini, J., Eds.; Roma ed il Mondo Adriatico. Dalla Ricerca Archeologica alla Pianificazione del Territorio, Atti del Convegno Internazionale, Macerata, 18–20 maggio 2017, Volume I; Carte archeologiche, gestione del patrimonio e parchi archeologici: Roma, Italy, 2020; pp. 33–51, in corso di stampa.
21. Zare, E.; Arshad, M.; Zhao, D.; Nachimuthu, G.; Triantafilis, J. Two-dimensional time-lapse imaging of soil wetting and drying cycle using EM38 data across a flood irrigation cotton field. *Agric. Water Manag.* **2020**, *241*. [CrossRef]
22. Boaga, J. The use of FDEM in hydrogeophysics: A review. *J. Appl. Geophys.* **2017**, *139*, 36–46. [CrossRef]
23. Anderson-Cook, C.M.; Alley, M.M.; Roygard, J.K.F.; Khosla, R.; Noble, R.B.; Doolittle, J.A. Differentiating soil types using electromagnetic conductivity and crop yield maps. *Soil Sci. Society Am. J.* **2002**, *66*, 1562–1570. [CrossRef]
24. Kitchen, N.R.; Sudduth, K.A.; Drummond, S.T. Mapping of sand deposition from 1993 midwest floods with electromagnetic induction measurements. *J. Soil Water Conserv.* **1996**, *51*, 336–340.
25. Robert, P. Characterization of soil conditions at the field level for soil specific management. *Geoderma* **1993**, *60*, 57–72. [CrossRef]
26. Saey, T.; Van Meirvenne, M.; de Smedt, P.; Neubauer, W.; Trinks, I.; Verhoeven, G.J.; Seren, S. Integrating multi-receiver electromagnetic induction measurements into the interpretation of the soil landscape around the school of gladiators at Carnuntum. *Eur. J. Soil Sci.* **2013**, *64*, 716–727. [CrossRef]
27. De Smedt, P.; Delefortrie, S.; Wyffels, F. Identifying and removing micro-drift in ground-based electromagnetic induction data. *J. Appl. Geophys.* **2016**, *131*, 14–22. [CrossRef]
28. Abdu, H.; Robinson, D.A.; Jones, S.B. Comparing bulk soil electrical conductivity determination using the DUALEM-1S and EM38-DD electromagnetic induction instruments. *Soil Sci. Soc. Am. J.* **2007**, *71*, 189–196. [CrossRef]

29. Corwin, D.L.; Rhoades, J.D. An improved technique for determining soil electrical conductivity-depth relations from above-ground electromagnetic measurements. *Soil Sci. Soc. Am. J.* **1982**, *46*, 517–520. [CrossRef]
30. Corwin, D.L.; Rhoades, J.D. Establishing soil electrical conductivity-depth relations from electromagnetic induction measurements. *Commun. Soil Sci. Plant Anal.* **1990**, *21*, 861–901. [CrossRef]
31. Friedman, S.P. Soil properties influencing apparent electrical conductivity: A review. *Comput. Electron. Agric.* **2005**, *46*, 45–70. [CrossRef]
32. Doolittle, J.A.; Brevik, E.C. The use of electromagnetic induction techniques in soils studies. *Geoderma* **2014**, *223*, 33–45. [CrossRef]
33. Allred, B.J.; Daniels, J.J.; Ehsani, M.R. *Handbook of Agricultural Geophysics*; CRC Press, Taylor and Francis Group: Boca Raton, FL, USA, 2008.
34. Koganti, T.; Van De Vijver, E.; Allred, B.; Greve, M.; Ringgaard, J.; Iversen, B. Mapping of Agricultural Subsurface Drainage Systems Using a Frequency-Domain Ground Penetrating Radar and Evaluating Its Performance Using a Single-Frequency Multi-Receiver Electromagnetic Induction Instrument. *Sensors* **2020**, *20*, 3922. [CrossRef]
35. Elmahdy, S.I.; Mohamed, M.M. Remote sensing and geophysical survey applications for delineating near-surface palaeochannels and shallow aquifer in the United Arab Emirates. *Geocarto Int.* **2015**, *30*, 723–736. [CrossRef]
36. Fitterman, D.V.; Menges, C.M.; Kamali, A.M.; Jama, F.E. Electromagnetic mapping of buried paleochannels in eastern Abu Dhabi Emirate, UAE. *Geoexploration* **1991**, *27*, 111–133. [CrossRef]
37. Kemna, A.; Binley, A.; Ramirez, A.; Daily, W. Complex resistivity tomography for environmental applications. *Chem. Eng. J.* **2000**, *77*, 11–18. [CrossRef]
38. Rudolph, S.; van der Kruk, J.; von Hebel, C.; Ali, M.; Herbst, M.; Montzka, C.; Weihermüller, L. Linking satellite derived LAI patterns with subsoil heterogeneity using large-scale ground-based electromagnetic induction measurements. *Geoderma* **2015**, *241–242*, 262–271. [CrossRef]
39. Evans, R.L.; Law, L.K.; Louis, B.S.; Cheesman, S. Buried paleo-channels on the new jersey continental margin: Channel porosity structures from electromagnetic surveying. *Mar. Geol.* **2000**, *170*, 381–394. [CrossRef]
40. Bates, M.R.; Bates, C.R.; Whittaker, J.E. Mixed method approaches to the investigation and mapping of buriedQuaternary deposits: Examples from southern England. *Archaeol. Prospect.* **2007**, *14*, 104–129. [CrossRef]
41. De Smedt, P.; Van Meirvenne, M.; Herremans, D.; De Reu, J.; Saey, T.; Meerschman, E.; Crombé, P.; De Clercq, W. The 3-D reconstruction of medieval wetland reclamation through electromagnetic inductionsurvey. *Sci. Rep.* **2013**, *3*, 1517. [CrossRef] [PubMed]
42. Deiana, R.; Bonetto, J.; Mazzariol, A. Integrated Electrical Resistivity Tomography and Ground Penetrating Radar Measurements Applied to Tomb Detection. *Surv. Geophys.* **2018**, *39*, 1081–1105. [CrossRef]
43. Tsokas, G.N.; Tsourlos, P.I.; Stampolidis, A.; Katsonopoulou, D.; Soter, S. Tracing a major Roman road in the area of Ancient Helike by resistivity tomography. *Archaeol. Prospect.* **2009**, *16*, 251–266. [CrossRef]
44. Vacilotto, A. I Segni Della Terra per la Storia Dell'assetto Agrario di Opitergium. Ph.D. Thesis, Università di Roma "La Sapienza", Roma, Italy, 2017.
45. Guaitoli, M. *Nota sulla Metodologia della Raccolta, della Elaborazione e della Presentazione dei Dati, in TARTARA P.*; Torrimpietra (IGM 149 I NO); Forma Italiae: Firenze, Italy, 1999; pp. 357–365.
46. Cambi, F.; Terrenato, N. Introduzione All'archeologia dei Paesaggi. 1998. Available online: https://www.docsity.com/it/f-cambi-n-terrenato-introduzione-all-archeologia-dei-paesaggi-roma-1994/2409147/ (accessed on 11 September 2020).
47. Corsi, C.; Slapšac, B.; Vermeulen, F. *Good Practise in Archaeological Diagnostics. Non-invasive Survey of Complex Archaeological Sites*; Springer: Berlin, Germany, 2013.
48. Piccarretta, F. *Manuale di Fotografia Aerea*; Uso Archeologico; L'Erma di Bretschneider: Roma, Italy, 1987.
49. Piccarretta, F.; Ceraudo, G. *Manuale di Aerofotografia Archeologica. Metodologia, Tecniche e Applicazioni*; Edipuglia: Bari, Italy, 2000.
50. Vacilotto, A.; Codato, D.; Novello, E. La cartografia dei Consorzi di bonifica: Recupero, analisi e valorizzazione. In Proceedings of the XXI Conferenza Nazionale ASITA, Trieste, Italy, 21–23 November 2017; pp. 1033–1040.
51. Daily, W.A.A.; Ramirez, A.; Binley, A.; LaBrecque, D. Electrical resistivity tomography. *Lead. Edge* **2004**, *23*, 438–442. [CrossRef]

52. DeGroot-Hedlin, C.; Constable, S. Occam's inversion to generate smooth, two-dimensional models from magnetotelluric data. *Geophysics* **1990**, *55*, 1613–1624. [CrossRef]
53. LaBrecque, D.J.; Morelli, G.; Daily, W.; Ramirez, A.; Lundegard, P. Occam's inversion of 3D ERT data. In *Three-Dimensional Electromagnetics*; Spies, B., Ed.; SEG: Tulsa, OK, USA, 1999; pp. 575–590.
54. Binley, A.; Ramirez, A.; Daily, W. Regularised image reconstruction of noisy electrical resistance tomography data. In Proceedings of the 4th Workshop of the European Concerted Action on Process Tomography, Bergen, Norway, 6–8 April 1995; Beck, M.S., Hoyle, B.S., Morris, M.A., Waterfall, R.C., Williams, R.A., Eds.; pp. 401–410.
55. Bronk Ramsey, C. Bayesian analysis of radiocarbon dates. *Radiocarbon* **2009**, *51*, 337–360. [CrossRef]
56. Reimer, P.J.; Bard, E.; Bayliss, A.; Beck, J.W.; Blackwell, P.G.; Bronk Ramsey, C.; Buck, C.E.; Cheng, H.; Edwards, R.L.; Friedrich, M.; et al. IntCal13 and Marine13 radiocarbon age calibration curves 0–50,000 years cal BP. *Radiocarbon* **2013**, *55*, 1869–1887. [CrossRef]
57. Harris, E.C. *Principles of Archaeological Stratigraphy*; Academic Press Limited: London, UK, 1989.
58. Galiazzo, V. *I Ponti Romani*; Catalogo Generale, Ed.; Canova: Treviso, Italy, 1995.
59. Jahn, R.; Blume, H.P.; Asio, V.B.; Spaargaren, O.; Schad, P.; Langohr, R.; Brinkman, R.; Nachtergaele, F.O.; Pavel Krasilnikov, R. *Guidelines for Soil Description*; FAO: Rome, Italy, 2006.
60. Cornaro, M. Scritture sulla laguna. In *Antichi Scrittori D'idraulica Veneta. Volume II Parte II. Scritture Sopra la Laguna di Alvise Cornaro e di Cristoforo Sabbadino, Off. grafiche C.*; Ferrari: Venezia, Italy, 1919; pp. 1442–1464.
61. Mozzi, P.; Bini, C.; Zilocchi, L.; Becattini, R.; Mariotti Lippi, M. Stratigraphy, palaeopedology and palinology of Late Pleistocene and Holocene deposits in the landward sector of the lagoon of Venice (Italy), in relation to the 'caranto' level. *Il Quat. Ital. J. Quat. Sci.* **2003**, *16*, 193–210.
62. Mozzi, P.; Ferrarese, F.; Fontana, A. Integrating digital elevation models and stratigraphic data for the reconstruction of the post-LGM unconformity in the Brenta alluvial megafan (North-Eastern Italy). *Alp. Med. Quat.* **2013**, *26*, 41–54.
63. Fontana, A.; Mozzi, P.; Marchetti, M. Alluvial fans and megafans along the southern side of the Alps. *Sedim. Geol.* **2014**, *301*, 150–171. [CrossRef]
64. Rossato, S.; Mozzi, P. Inferring LGM sedimentary and climatic changes in the southern Eastern Alps foreland through the analysis of a 14C ages database (Brenta megafan, Italy). *Quat. Sci. Rev.* **2016**, *148*, 115–127. [CrossRef]
65. Rossato, S.; Carraro, A.; Monegato, G.; Mozzi, P.; Tateo, F. Glacial dynamics in pre-Alpine narrow valleys during the Last Glacial Maximum inferred by lowland fluvial records (northeast Italy). *Earth Surf. Dyn.* **2018**, *6*, 809–828. [CrossRef]
66. Fontana, A.; Mozzi, P.; Bondesan, A. Alluvial megafans in the Venetian–Friulian Plain (north-eastern Italy): Evidence of sedimentary and erosive phases during Late Pleistocene and Holocene. *Quat. Int.* **2008**, *189*, 71–90. [CrossRef]
67. Floris, M.; Fontana, A.; Tessari, G.; Mulè, M. Subsidence zonation through satellite interferometry in coastal plain environments of NE Italy: A possible tool for geological and geomorphological mapping in urban areas. *Remote Sens.* **2019**, *11*, 165. [CrossRef]
68. Schmedt, G. La prospezione aerea nella ricerca archeologica. In Proceedings of the Convegno Internazionale Sulla Tecnica e Diritto nei Problemi Della Odierna Archeologia, Venezia, Italy, 22–24 May 1962; pp. 66–93.
69. Tozzi, P. *Memoria della Terra. Storia Dell'uomo*; La Nuova ItaliaFirenze: Firenze, Italy, 1987.
70. Masiero, E. La strada "in levada" nell'agro nord-occidentale di Adria, in JAT, IX, Roma, Italy. 1999, pp. 107–120. Available online: https://www.gcss.it/easyweb/w7044/index.php?EW_T=M1&EW_FL=w7044/ew_limiti.html&EW4_DLL=10&EW4_DLP=10&EW4_NVR=&EW4_NVT=&EW4_NMI=&EW4_CJL=1&NOICONE=1&PHPMSG=1&lang=ita&REC1MEMO=1&EW4_PY=KW=MASIERO&EW_RM=10&EW_EP=KW=MASIERO&EW_RP=2&&EW_P=LSPHP&EW_D=W7044&EW=0002562 (accessed on 11 September 2020).
71. Maragno, S. Rilievo topografico di un tratto del decumano massimo nella centuriazione fossile dell'ager Atrianus. In *La Ricerca Archeologica di Superficie in Area Padana*; Atti del workshop (Villadose, 1 ottobre 1994), a cura di E. Maragno, Stanghella (Padova); 1996; pp. 383–386. Available online: http://www.centuriazione.it/quaderni_win.asp?id=84 (accessed on 11 September 2020).
72. Bonetto, J. *Le vie Armentarie tra Patavium e la Montagna*; Zoppelli: Dosson, Italy, 1997.

73. De Guio, A. ii Progetto Alto-Medio Polesine–Basso Veronese: Sesto Rapporto, in Quaderni di Archeologia del Veneto, IX. 1993, pp. 170–186. Available online: https://culturaveneto.it/it/web/cultura/volumi-vi-x (accessed on 11 September 2020).
74. Rosada, G. Altino e la via della transumanza nella Venetia centrale, In Proceedings of the Pecus. Man and Animal in Antiquity Conference, Swedish Institute, Rome, Italy, 9–12 September 2002; Santillo Frizell: Roma, Italy, 2004; pp. 71–83.
75. Calzolari, M. Aspetti del territorio in epoca romana: Acque, bonifiche e insediamenti. In *Da Palus Maior a S. Biagio in Padule. Uomini e Ambiente nella Bassa Modenese dall'Antichità al Medioevo*; Dini: Modena, Italy, 1984; pp. 35–76.
76. Frassine, M. *Palus in Agro. Aree Umide, Bonifiche e Assetti Centuriali in Epoca Romana, Collana: Agri Centuriati. Supplementa*; Fabrizio Serra, Ed.; Pisa-Roma: Roma, Italy, 2013.
77. Franceschelli, C.; Marabini, S. *Lettura di un Territorio Sepolto. La Pianura Lughese in età Romana*; Collana "Studi e Scavi", Dipartimento di Archeologia dell'Università di Bologna; AnteQuem: Bologna, Italy, 2007; p. 224.
78. Matteazzi, M. *Il paesaggio Centuriato a sud di Padova: Una Nuova Lettura Dallo Studio Archeomorfologico del Territorio.* "Agri Centuriati"; Fabrizio Serra, Ed.; Pisa-Roma, 2014; pp. 9–29.
79. *Pauli Diaconi, Historial Langobardorum, III*; Istituto Storico Italiano: Roma, Italy, 1918; pp. 23–24.
80. Fontana, A.; Frassine, M.; Ronchi, L. Geomorphological and Geoarchaeological Evidence of the Medieval Deluge in the Tagliamento River (NE Italy). In *Fontana A. and Herget J. Palaeohydrology*; Springer: Cham, Switzerland, 2020; pp. 97–116.
81. Pinna, M. *Le Variazioni del Clima. Dall'ultima Grande Glaciazione alle Prospettive per il XXI Secolo*; Franco Angeli: Milano, Italy, 1996.
82. Pavanello, G. Di un'antica laguna scomparsa (la laguna Eracliana). In *Arch. Veneto-Tridentino*, 3nd ed.; Premiate officine grafiche C. Ferrari: Venice, Italy, 1923.

© 2020 by the authors. Licensee MDPI, Basel, Switzerland. This article is an open access article distributed under the terms and conditions of the Creative Commons Attribution (CC BY) license (http://creativecommons.org/licenses/by/4.0/).

Article

Ice Thickness Estimation from Geophysical Investigations on the Terminal Lobes of Belvedere Glacier (NW Italian Alps)

Chiara Colombero [1,*], Cesare Comina [2], Emanuele De Toma [2], Diego Franco [1] and Alberto Godio [1]

1. Department of Environment, Land and Infrastructure Engineering (DIATI), Politecnico di Torino, 10129 Torino, Italy; diego.franco@polito.it (D.F.); alberto.godio@polito.it (A.G.)
2. Department of Earth Sciences (DST), Università degli Studi di Torino, 10125 Torino, Italy; cesare.comina@unito.it (C.C.); emanuele.detoma@edu.unito.it (E.D.T.)
* Correspondence: chiara.colombero@polito.it

Received: 27 February 2019; Accepted: 2 April 2019; Published: 3 April 2019

Abstract: Alpine glaciers are key components of local and regional hydrogeological cycles and real-time indicators of climate change. Volume variations are primary targets of investigation for the understanding of ongoing modifications and the forecast of possible future scenarios. These fluctuations can be traced from time-lapse monitoring of the glacier topography. A detailed reconstruction of the ice bottom morphology is however needed to provide total volume and reliable mass balance estimations. Non-destructive geophysical techniques can support these investigations. With the aim of characterizing ice bottom depth, ground-penetrating radar (GPR) profiles and single-station passive seismic measurements were acquired on the terminal lobes of Belvedere Glacier (NW Italian Alps). The glacier is covered by blocks and debris and its rough topography is rapidly evolving in last years, with opening and relocation of crevasses and diffuse instabilities in the frontal sectors. Despite the challenging working environment, ground-based GPR surveys were performed in the period 2016–2018, using 70-MHz and 40-MHz antennas. The 3D ice bottom morphology was reconstructed for both frontal lobes and a detailed ice thickness map was obtained. GPR results also suggested some information on ice bottom properties. The glacier was found to probably lay on a thick sequence (more than 40 m) of subglacial deposits, rather than on stiff bedrock. Week deeper reflectors were identified only in the frontal portion of the northern lobe. These interfaces may indicate the bedrock presence at a depth of around 80 m from the topographic surface, rapidly deepening upstream. Single-station passive seismic measurements, processed with the horizontal-to-vertical spectral ratio (HVSR) method, pointed out the absence of sharp vertical contrast in acoustic impedance between ice and bottom materials, globally confirming the hypotheses made on GPR results. The obtained results have been compared with previous independent geophysical investigations, performed in 1961 and 1985, with the same aim of ice thickness estimation. The comparison allowed us to validate the results obtained in the different surveys, supply a reference base map for the glacier bottom morphology and potentially study ice thickness variations over time.

Keywords: Alpine glaciers; Belvedere Glacier; ice thickness estimation; ice bottom morphology and properties; ground-penetrating radar (GPR); single-station passive seismic measurements; horizontal-to-vertical spectral ratio (HVSR)

1. Introduction

Mountain glaciers are visible indicators of climate change, providing some of the clearest evidence of global warming, perturbation in the atmospheric flow pattern and consequent precipitation regime.

Due to their relatively small size and high mass turnover rates, these glacial bodies demonstrate extremely sensitive to temperature and precipitation modifications [1,2]. For the European continent, Alpine glaciers are a key component of local and regional hydrogeological cycles. The globally observed retreat of these ice masses over the last century has significant impacts on the surrounding environment, affecting both physical and socioeconomic systems [3–5]. The loss of freshwater storage and its consequent release into seas and oceans, modifications in resource availability for water consumption, irrigation and power generation, glacier-related natural hazards (e.g., landslides, avalanches and outburst of glacial lakes, causing floods and debris flows) are between the potential consequences of the general shrinkage trend.

Variations in glacier volume and mass are therefore primary targets of investigation for the understanding of ongoing modifications and the forecast of possible future scenarios. Remote sensing techniques (e.g., geodetic surveys, satellite images, GNSS or LiDAR surveys) can help to monitor these fluctuations [6–9] from accurate time-lapse reconstructions of the topographic surface of the glacier. Comprehensive knowledge of the ice bottom depth and morphology is however needed to provide a total volume estimate and reliable mass balance evaluations [10–12]. Ice thickness characterization and monitoring are also of primary importance for the modeling of future glacier dynamics [13], hydrological projections [14], glacier-related natural hazards [15] and ice core analyses [16].

Traditional glaciological measurements, i.e., local probe deployment for snow and firn/ice thickness variations or density measurements, are usually limited to the shallowest part of the glacier and do not extensively cover wide areas of investigations. Despite their relevance, ice bottom depth and morphology are often poorly known, mainly due to inadequate characterization methods and logistical issues. Geophysical methods can overcome these limitations, mitigating the operational efforts and enlarging the depth of investigation and data density of traditional techniques.

In the last decades, the ground-penetrating radar (GPR) technique has assumed an increasingly important role in the glacial exploration for the imaging of subsurface conditions [17]. The non-magnetic and low-conductive snow/ice column is indeed generally favorable for the efficient propagation of electromagnetic (EM) pulses, enabling a successful mapping of bottom morphology. In addition, GPR can also help in the evaluation of the glacier bottom properties, the search for internal water floods or underground channels and the recognition of internal glacial structures (as snow layering and crevasses detection). Several GPR applications are reported in the literature on Arctic and Antarctic ice caps and high-elevation icefields of the Tibetan Plateau [18–21]. Growing applications are also recorded in the Alpine context. Del Longo et al. [22] acquired GPR profiles on a Dolomitic glacier to identify the ice bottom depth. Binder et al. [23] investigated two glaciers in the Austrian Alps, for determination of total ice volume and ice-thickness distribution. Merz et al. [24] acquired a dense grid of helicopter-borne GPR, combined with ground-based seismic and geoelectric profiles, on a rock glacier of western Swiss Alps to reconstruct the 3D bedrock topography. Airborne GPR surveying generally overcomes problems in accessibility of the steeper glacier sectors and wave scattering on large boulders in close vicinity of the antenna, which are particularly disturbing in case of debris-cover and rock glaciers. Dossi et al. [25] performed quantitative 3D GPR analysis to estimate the total volume and water content of a glacier in Eastern Alps.

Techniques complementary to GPR surveys have also been applied. Maurer and Hauck [26] reviewed possible additional methods and proposed the joint interpretation of GPR, geoelectric and seismic surveys for analyzing the subsurface conditions of two rock glaciers in the Eastern Swiss Alps, with a special focus on the distribution of ice and water, the occurrence of shear horizons and the bedrock topography. However, geoelectric and reflection seismic methods require long arrays to reach considerable investigation depths and high-resolution imaging of the glacier subsurface, thus complicating field operations. Single-station passive seismic measurements may offer an effective and logistically affordable method to mitigate the problems of multi-channel active surveys, thanks to the use of portable and compact broadband seismometers and no need of energetic active sources. Picotti et al. [27] presented a pioneer study on the application of single-station passive measurements to determine ice thickness on five Alpine glaciers. Data were processed with the horizontal-to-vertical spectral ratio (HVSR, [28,29]) technique and validated using GPR, geoelectric and active seismic profiling methods. The HVSR method is based on the computation of the ratio between the spectra of the horizontal and vertical components of a triaxial seismic station, to retrieve the fundamental resonance frequency of a site [30,31]. Generally, the resonance originates from the trapping of seismic waves in sites with sufficiently high S-wave impedance contrast (e.g., soft sediments overlying a stiff bedrock). For these sites, a clear peak is found in the HVSR curve. The frequency of this peak is related to the depth of the interface between the two materials [32]. The origin of the HVSR peak is still debated and generally related to a superposition of Rayleigh-, Love- and/or shear-wave resonances [33].

In a glacial environment, the impedance contrast between ice and bedrock is expected to generate resonance phenomena and similar effects on the HVSR curves. Comparing HVSR data with other geophysical imaging results, Picotti et al. [27] showed that the resonance frequency depicted in the HVSR curves is correlated with the ice thickness at the site, in a wide range of ice bottom depths, from few tens of meters to more than 800 m. The reliability of the method mainly depends on the coupling of the sensor at the glacier surface and on the basal impedance contrast. However, beside the intrinsic limitation of the 1D approximation, very few Alpine glaciers lay directly on stiff bedrock for all their extent, while a basal layer of subglacial deposits and debris of significant thickness is generally present below the ablation zone [34], thus attenuating the vertical impedance contrast.

In this study, we acquired both GPR profiles and single-station passive seismic measurements on the terminal lobes of the debris-covered Belvedere Glacier (Macugnaga, NW Italian Alps), with the aim of ice thickness estimation and bottom morphology reconstruction. Very poor and conflicting reference information on ice thickness is indeed available for the glacier, from previous geophysical investigation attempts [35,36]. GPR data acquired in different seasons and with different antennas were processed and manual picking of the bottom surface was performed on the radargrams showing the clearest subsurface imaging. Retrieved ice depths were then spatially interpolated to reconstruct the 3D bottom morphology and compared with previous geophysical results [35,36]. Some single-station passive seismic tests were performed in the same area imaged with GPR profiles. Seismic recordings were processed with the HVSR method and interpreted in the light of the GPR results, for a combined evaluation attempt of the ice bottom morphology and properties.

2. Study Site

Belvedere Glacier is a debris-covered glacier located NE of the highest peaks of Monte Rosa Massif, in NW Italian Alps (Figure 1). Thanks to the debris cover and the favorable solar exposure, its frontal sectors reach considerably low altitudes and end approximately 2 km W of Macugnaga (VB). Belvedere Glacier represents the terminus of four higher-elevation glaciers (Nordend, Monte Rosa, Signal and Northern Locce Glaciers). Measurements of the Italian Glaciological Committee (CGI-CNR, http://www.glaciologia.it/i-ghiacciai-italiani) carried out in 2006 indicated a surface area of 1.46 km^2, a maximum length of 3091 m, spanning in elevation from 2397 m to 1770 m above sea level (a.s.l.), with an average slope of 8°. The terminal portion of the glacier is bilobate. Both lobes are currently exhibiting a visible retreat. Nowadays, the bigger N lobe has an average length of 650 m, reaching a minimum elevation of about 1810 m a.s.l. (i.e., 40 m higher than measurements of 2006), while the S lobe has a length of 350 m, reaching an elevation of 1840 m a.s.l. at its front. The two lobes are separated by a median morainic relief, hosting the chair lift station and the Belvedere Mountain Hut (Figure 1).

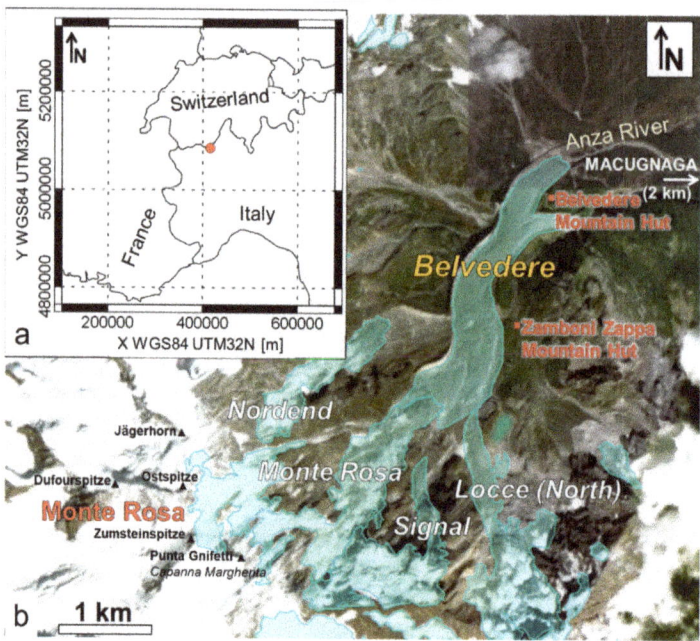

Figure 1. (a) Geographical location and (b) aerial view of the north-eastern glaciers of the Monte Rosa Massif. The main glacial masses are delimited by light blue areas (modified from the Italian Glaciological Committee (CGI-CNR), http://www.glaciologia.it/i-ghiacciai-italiani).

Despite several glaciological studies were carried out on site since the beginning of the 20th century [37,38], the first attempt towards ice thickness characterization is reported by De Visintini [35]. P-wave reflection seismic measurements were indeed carried out on the glacier in 1957 for ice bottom contouring. Explosive sources were adopted, and the active shots were recorded with a 12-channel analog seismograph. Seven local areas of the glacier were surveyed with seismic profiling, from the vicinity of the Zamboni Zappa Mountain Hut (Figure 1), down to the confluence between the two lobes. A global contour map of ice bottom depth was achieved from the interpolation of these measurements, even if some glaciers parts (e.g., the two terminal lobes) were not investigated during this survey.

A later study by VAW-ETH [36] was performed using the GPR technique, with the aim of completing ice bottom characterization in unexplored compartments. A low-frequency GPR instrumentation (USGS Monopulse-radar, with variable central frequency in the range 1–5 MHz) was adopted, with 40 sparse measurements located along 9 transverse profiles covering almost all the glacier length. For each measurement, the time delays between the transmitted EM pulse and the received echoes were recorded. The highest-amplitudes echoes were referred to ice bottom reflections, their depth was estimated from the time delay (considering a constant velocity of the EM pulse in ice) and an interpretation of the bottom morphology was retrieved from data interpolation along the 9 cross-sections.

In the upper part of the glacier (W and SW of Zamboni Zappa Mountain Hut) both radar and seismic results seemed to locate the bedrock at depths of 200–250 m. Progressing northwards, major discrepancies in ice bottom determination arose between the two techniques. Particularly, radar bottom reflectors were found to be located approximately 100 m higher than the seismic reflectors in the central part of the glacier. The two techniques, sensitive to different physical changes in the subsurface, seemed therefore to identify different interfaces. This difference in ice bottom location was interpreted as the result of the presence of a layer of subglacial deposits with relevant thickness (up to more than 100 m in the central part of the glacier) between ice and bedrock. Glacial deposits revealed transparent to the seismic reflection profiling, but were identified by the radar investigation, probably due to the significant contrast in electrical properties between these sediments and ice.

In a more recent work, Diolaiuti et al. [39] estimated the changes in ice volume between 1957 and 1991 and measured the debris thickness above the whole glacier area. The debris cover was indeed found to slow down the ablation rate of Belvedere Glacier, with respect to other similar glaciers of the same region lacking this peculiarity. A contour map of ice thickness was also presented. However, this map was based just on the results of reflection seismic [35] and did not consider the possible lower ice thicknesses, due to the presence of thick subglacial deposits, as highlighted by the later radar study [36].

In the last decades, geophysical prospection experienced greatly advanced in instrumentation, acquisition and processing techniques. Digital multi-channel recordings and high-sampling big data storage capacities allow for continuous radar profiling with respect to the study of VAW-ETH [36]. Passive seismic acquisitions require lighter instrumentation and no active sources with respect to the measurements of De Visintini [35]. At the light of these advances, new geophysical prospections at the site can therefore help in understanding the discrepancies between the previous works and provide new and more extensive information on the glacier bottom morphology and properties. Despite these considerations, the presence of blocks and debris with variable thickness and lateral distribution at the surface, the rough topography and heterogeneities inside the investigated glacial mass deeply complicate data acquisition and potentially affect the quality of the final results.

3. Methods

3.1. Ground-Penetrating Radar

Several GPR campaigns were carried out at the Belvedere Glacier between 2016 and 2018. In October 2016, 29 GPR profiles were acquired with an air-coupled 70-MHz GPR monostatic antenna (Subecho-70, Radarteam), for a total surveyed length of 2169 m (blue traces in Figure 2). The antenna was connected to an IDS K2 TR200 acquisition unit and manually transported above the ground surface. Traces were recorded for a total length of 1000 ns, with a sampling of 1024 samples/trace, and georeferenced by means of a GPS Ublox EVK-5T system.

In March 2018, 22 radar profiles were acquired with the same air-coupled 70-MHz antenna (green traces in Figure 2) transported on skis, for a total length of 2696 m of acquisitions (trace length = 2000 ns, 2048 samples/trace).

Finally, in December 2018 a lower frequency (40 MHz, Subecho-40, Radarteam) air-coupled antenna was manually moved along 15 profiles (red traces in Figure 2) focused on the terminal sector of the northern lobe of the glacier, for a total length of 2169 m of acquisitions (total recorded trace length = 1200 ns, 1024 samples/trace). The acquisition setup is shown in Figure 3a. The same acquisition unit and georeferencing instruments of October 2016 were adopted in all the later campaigns. In all surveys, the snow cover on site was limited (from absent to a few tens of centimeters on the debris cover). Unfortunately, due to the rough and rapidly evolving topography of the glacier along the years, the opening and/or relocation of crevasses and diffuse instabilities in the frontal sectors, it was not possible to follow the same profiles in each survey operating in safe conditions. In addition, due to the shape of both low-frequency antennas and the encountered topographic conditions, it was not possible to direct drag the instruments on the glacier surface to maximize the coupling. In each survey, the antenna was consequently maintained a few centimeters above the thin layer of snow partially hiding the cover of blocks and debris of the glacier (Figure 3a).

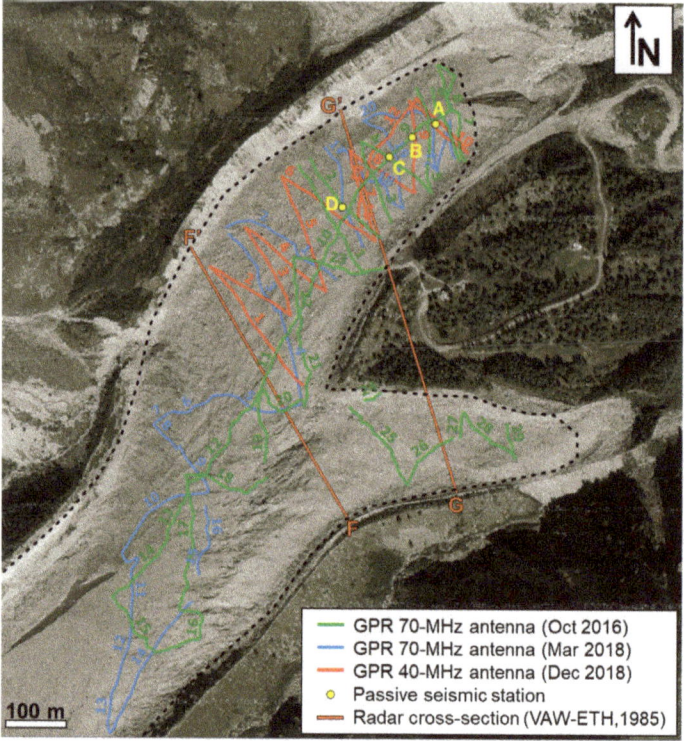

Figure 2. Geophysical surveys carried out at the Belvedere Glacier. Green lines: 70-MHz ground-penetrating radar (GPR) profiles acquired on October 2016; blue lines: 70-MHz GPR profiles acquired on March 2018; red lines: 40-MHz GPR profiles acquired on December 2018. GPR profiles are labelled with a progressive number for each survey. Yellow dots (**A–D**): Single-station passive seismic measurements (HVSR method). Orange lines (**FF′** and **GG′**): Past cross-sections of the glacier interpreted from radar measurements, from VAW-ETH [36]. The background orthophoto was acquired during summer 2015 (AGEA 2015, RGB orthophoto, WMS service at www.geoportale.piemonte.it). The black dashed line highlights the approximate glacier perimeter in October 2016 (first GPR campaign).

Data processing was performed in ReflexW software. A basic processing procedure was adopted: (i) Start time of each trace was shifted to delete samples before the main bang and obtain exact zero time; (ii) low-frequency components were removed (dewow); (iii) high-pass horizontal filtering was applied to remove horizontally coherent components (background removal); (iv) geometrical spreading correction was applied, to gain signal amplitude with depth (divergence compensation).

Manual picking of the ice bottom reflections was finally performed on the processed time sections. Due to the reduced snow cover during the surveys, the picking of the reference glacier topography (corresponding to the top of the debris cover) was neglected. A uniform ice velocity of 0.17 m/ns was considered for a time to depth conversion, disregarding the top debris cover. The latter is expected to have significant lateral and vertical variations, from a few centimeters up to a metric thickness in correspondence of boulders and blocks.

Figure 3. On-site operations for GPR and passive seismic acquisitions. (**a**) GPR instrumentation and acquisition setup. Note that the GPR antenna is not ground-coupled, but maintained a few centimeters above the ground surface during acquisition. (**b–e**) Installation of the broadband triaxial seismometer and passive seismic instrumentation details.

3.2. Single-Station Passive Seismic Measurements

Four single-station passive seismic measurements were carried out on the N lobe of the Belvedere Glacier (A to D, in Figure 2). A 3-D 1-Hz broadband seismometer (L-4-3D, Sercel Inc.) connected to a 24-bit 3-channel digitizer was adopted for the acquisitions. To obtain the best coupling between sensor and ice, the surface layers of snow, ice, and debris were removed. The 3D seismometer was placed in direct contact with compact ice and then buried with the excavated materials to minimize external noise (Figure 3b–e). At each station, noise recording lasted from 30 to 45 min, with a sampling frequency of 100 Hz.

Details on the theoretical formulation and assumption underling the application of the HVSR methods in a glacial environment can be found in Picotti et al. [27]. If the width of the glacier is considerably larger than the ice thickness, the subsurface conditions can be approximated with a 1D model for which:

$$f_0 = \frac{V_S}{4H}, \tag{1}$$

where f_0 is the resonance frequency depicted from the HVSR curve, V_S is the average S-wave velocity of ice and H is the ice thickness. If 2D effects are present at the station, f_0 value should be multiplied by correcting factors accounting for the shape of the basin [27,40]. Station locations along the axial position of the N lobe were chosen to minimize 2D lateral effects, due to the valley sides.

HVSR computation was carried out in Geopsy software. The original recordings (30–45 min) were divided into 120-s non-overlapping windows. An anti-triggering STA/LTA (short time average over long time average) algorithm was applied to each time window. This filtering procedure is based on the computation of the average signal amplitude over short (STA = 1 s) and long (LTA = 30 s) moving windows. If the STA/LTA ratio exceeds user-defined thresholds (STA/LTA < 0.2 or STA/LTA > 2.5, in this study), the window is filtered out from further computations. Otherwise, amplitude spectra of the horizontal components are combined using vector summation and the horizontal to vertical spectral ratio is computed. A smoothing function, with a 10% cosine taper and a bandwidth coefficient of 30 [32], was then applied to the resulting HVSR curve to ensure a constant number of points at low and high frequencies. This computation was repeated for all the 120-s windows passing the anti-triggering filter. The average curve and the standard deviation over the accepted HVSR curves were finally obtained. The spatial directivity of the HVSR peaks was computed using the same processing parameters, to check for the absence of 2D effects on the detected resonance frequencies.

The results were analyzed to search for HVSR peaks clearly indicating the presence of sharp subsurface contrasts in acoustic impedance and potentially estimate their embedment depth. Since no clear peaks were detected, further processing attempts, including inverse [41] or forward modeling [42,43] of the measured HVSR curves were not undertaken.

4. Results

4.1. Ground-Penetrating Radar

Representative radargrams obtained from the three GPR campaigns are shown in Figure 4. Some clear artifacts, due to electromagnetic noise are still visible in the data, even after the described processing sequence, e.g., the horizontal events at 650 ns and 700 ns in Figure 4b,c. The overall data quality is very poor if compared with the typical S/N ratio of GPR data acquired on other glaciers. High scattering is noticed along the profiles, with extremely high attenuation of the EM signals with depth. Not always clear readability in the processed profiles was observed for the 70-MHz surveys. Even increasing the total recording time (Figure 4c compared to Figure 4b), no reflections were observable in the data below 900 ns. Better results were obtained with the 40-MHz antenna (Figure 4d). In these radargrams, reflections appear more clearly and can be spatially followed and correlated between adjacent profiles. An exemplificative marked interpretation of the ice bottom morphology on the radargrams of Figure 4 is reported in Figure 5. Similar interpretations have been performed on the remaining profiles of the different campaigns for which ice bottom picking was clear and reliable.

The ice bottom appears as a discontinuous reflecting horizon in all the surveys. Layering with different orientations is noticed in some parts of the radargrams at the ice bottom location (e.g., distance = 38 m, time = 1000 ns in profile 15 of Figures 4d and 5d). A low-amplitude reflector deeper than the ice bottom is depicted in profile 12 of Figures 4d and 5d. This interface is characterized by a steeper dipping with respect to the ice bottom, in dip direction opposite to the glacier movement, reaching times higher than 1150 ns (depth > 100 m) in the center of the profile, at a distance of approximately 120 m from the glacier front.

In Figure 5c,d, the approximate locations (A' to D' and A" to D") of the single-station passive seismic measurements (A to D) are projected on the GPR sections, for further comparisons between the two geophysical methods.

The results of ice bottom picking along GPR profiles are summarized in Figure 6. All the 40-MHz radargrams provided quite clear imaging of the subsurface conditions. Conversely, only 12 radargrams acquired on October 2016 (70 MHz) supplied information on ice thickness, close to the front of the N lobe, on the S lobe and at the confluence. These areas correspond to generally lower ice thicknesses with respect to the surrounding zones. Analogously, the 70-MHz GPR campaign of March 2018 resulted in only 5 radargrams for which ice bottom picking was clearly interpretable. No information on ice thickness was recovered from the GPR data for the glacier main body, upstream the terminal bifurcation (Figure 6a). Ice thickness retrieved from ice bottom picking is shown in Figure 6b. A traditional U-shaped bottom morphology seems to be reconstructed for the N lobe, with considerable ice depths in the axial part (100–110 m) despite the proximity to the terminus. The S lobe is conversely characterized by significantly shallower ice depths and a probably flatter morphology.

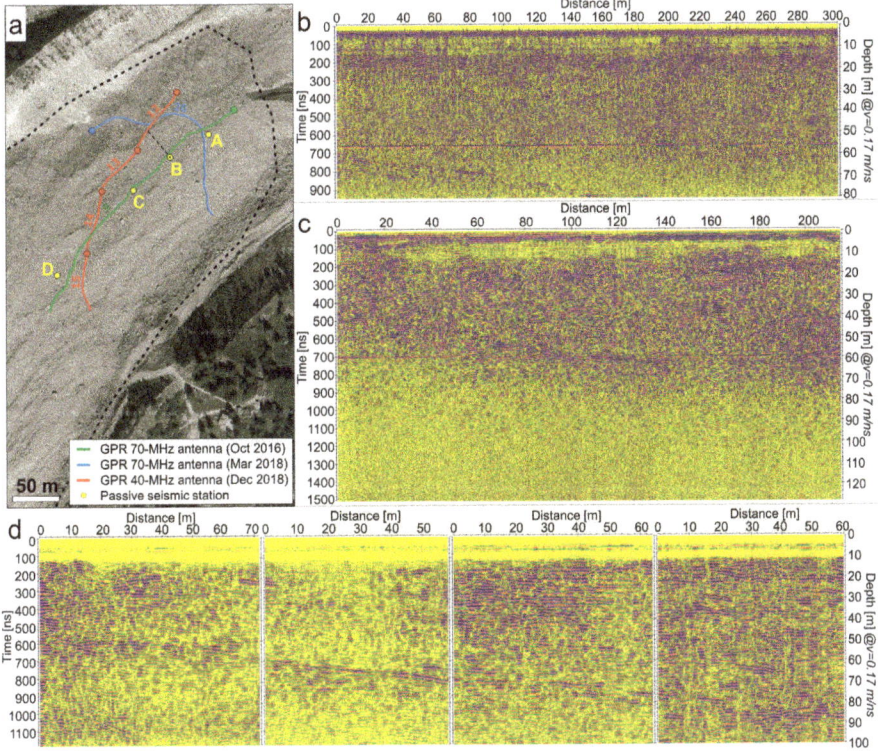

Figure 4. GPR results. (**a**) Location of the GPR profiles are shown in the right and bottom sections. The starting point of each profile is highlighted with a dot. (**b**) Profile 9 (70-MHz antenna, October 2016). (**c**) Profile 20 (70-MHz antenna, March 2018). (**d**) Profile 12 to 15 (left to right, 40-MHz antenna, December 2018). The vertical scale of some radargram is cut (no reflections for higher recorded times).

4.2. Single-Station Passive Seismic Measurements

The results of HVSR processing are shown in Figure 7. No clear single peaks are found in the results, at least two to three HVSR minor peaks can be depicted for each station. HVSR amplification is low (<2) for all the measured points. For the first three stations, the peak with the highest amplitude (f0) is located at similar frequency values, progressively decreasing from A to C. A higher frequency peak is noticed at station D.

Figure 8 shows HVSR directivity, as a function of frequency and azimuth (0° = N, 90° = E), measured at the four stations. These results clearly highlight that f0 spectral peaks spread over a wide azimuthal range, covering from 100° to 180°. These resonance frequencies seem therefore almost azimuth independent, denoting the absence of 2-D effects. The secondary peaks located at frequencies lower than 3 Hz conversely showed more focused directivities.

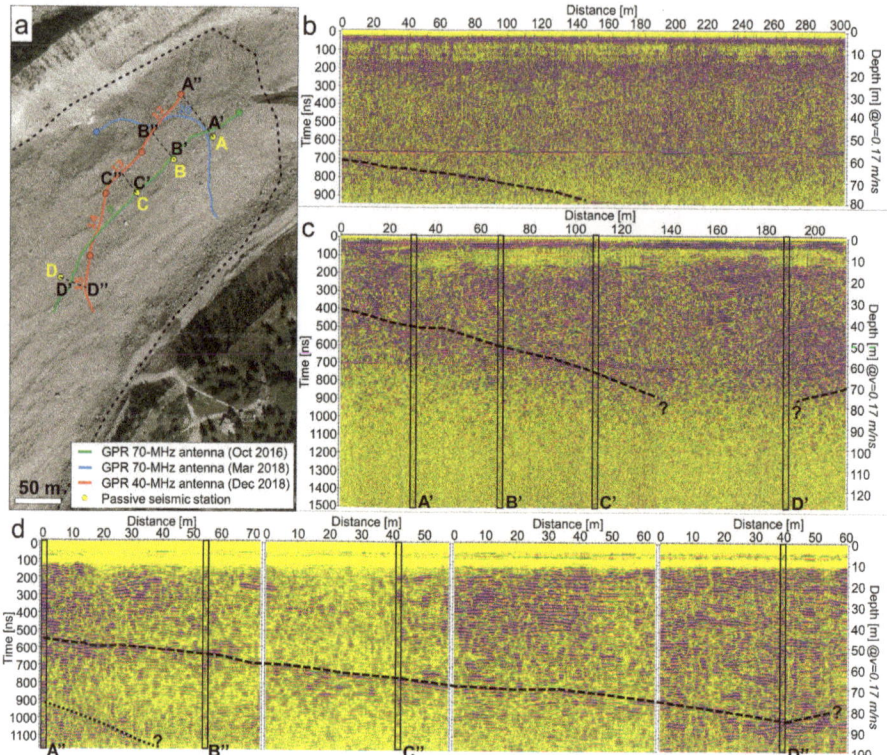

Figure 5. GPR results (same as Figure 4) with glacier bottom interpretation. The ice bottom morphology is highlighted in the radargrams with the dashed black lines. The possible bedrock top is underlined in (**d**) with the dotted black line. The approximate location of single-station passive measurements projected on the GPR profiles is shown in (**a**) with letters A′ to D′ (on 70-MHz profile 9) and A″ to D″ (on 40-MHz profiles 12 to 15), and in (**c**,**d**) with the black rectangles, for further comparison between GPR and HVSR results.

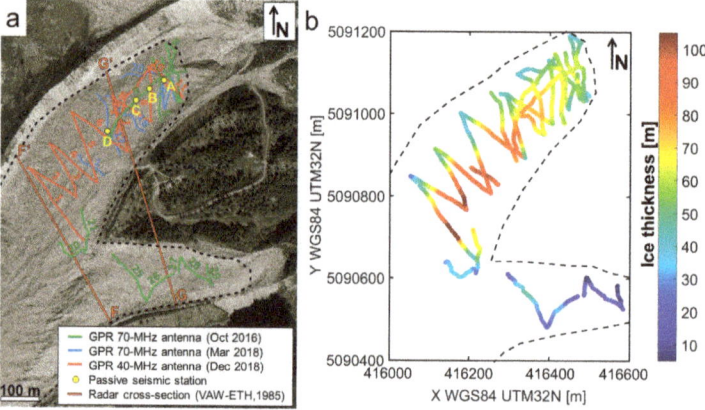

Figure 6. (**a**) Summary of the GPR profiles for which ice bottom picking was retrieved. (**b**) Map of the resulting ice bottom picks.

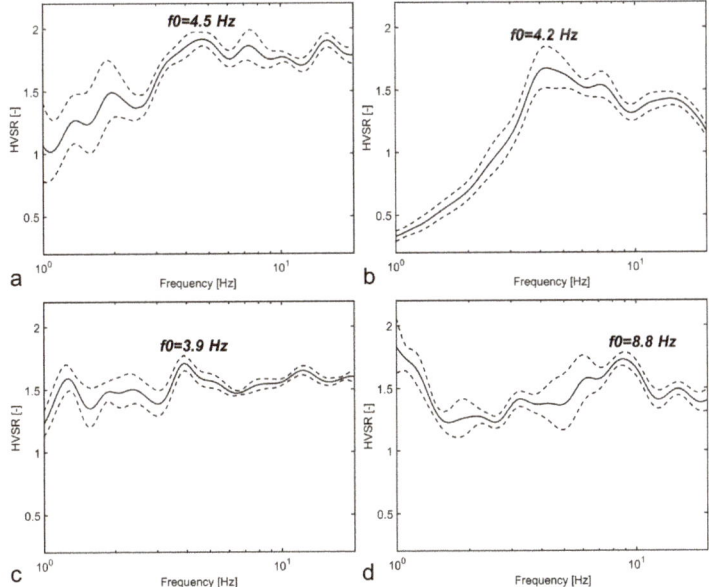

Figure 7. HVSR results on stations (**a**) A, (**b**) B, (**c**) C and (**d**) D (location in Figure 2). In each panel, the black bold and dashed curves are the average and standard deviation obtained from all the accepted 120-s window curves for each station.

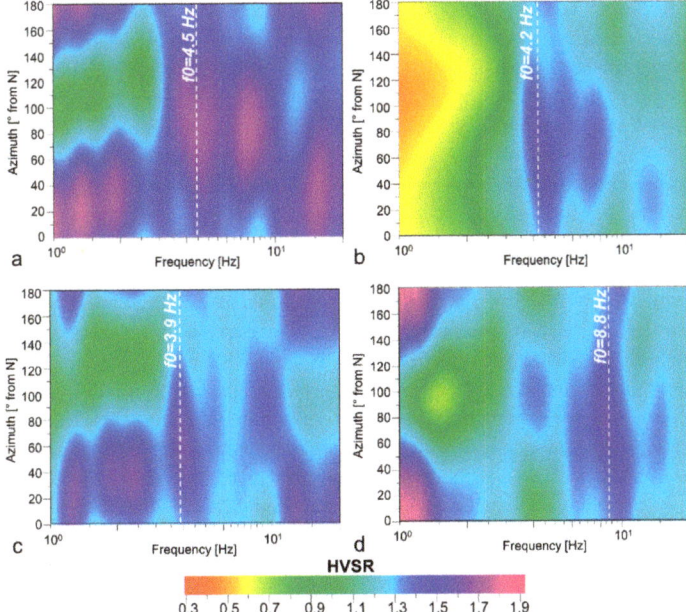

Figure 8. Directional HVSR azimuth (in degrees from N direction) on stations (**a**) A, (**b**) B, (**c**) C and (**d**) D. In each panel, the white dashed lines highlight the location of the highest spectral peaks (f0) identified in Figure 7.

5. Discussion

Geophysical characterization of rock glaciers and debris-covered glaciers is usually complicated by several factors: Logistically challenging and expensive transport of equipment and personnel on site, extremely rugged topographic conditions inhibiting antenna dragging on the surface, and inaccessibility of same areas, due to safety concerns [44]. In addition, despite the advances in geophysical prospection and instrumentation of the last decades with respect to the previous reference ice bottom investigations carried out on the Belvedere Glacier, recent surveys are not free of interpretation issues. Widespread scattering and high attenuation of the EM signals, often referred to as clutter, are noticed in all the 70-MHz GPR profiles. Slightly better results are obtained with a lower frequency of the GPR antenna (40 MHz vs. 70 MHz). No information is however retrieved for depths higher than 100–150 m below the glacier surface.

Many causes may have contributed to the poor quality of GPR imaging on this glacier. The absence of direct coupling with the ground, with possible centimetric variations in the height of the antenna, limits the amount of EM energy inserted in the subsurface, resulting in rapidly decreasing energy to image the reflectors with depth and laterally variable trace amplitudes. In addition, the presence of the top debris cover, characterized by extreme variability in grain size, thickness and lateral distribution, generates diffuse scattering and attenuation of the EM pulses transmitted to the underlying ice column, with respect to GPR prospections in glaciers lacking this surface layer. Pecci et al. [45] discussed the negative clutter effects on GPR data quality caused by the surface debris cover with a highly variable thickness of Calderone Glacier (Central Apennines, Italy). Intense clutter phenomena were observed also at a depth of few meters from the surface and were related to the presence of ice layers containing a high concentration of debris. As a consequence, the reflectors corresponding to the ice-bedrock interface could not be clearly and continuously detected.

Similar results were obtained on a debris-covered glacier of Italian Dolomites [22]. Most of the GPR scans pointed out the presence of intense clutter effects, due to the presence of heterometric debris at the surface. Authors were able to detect the ice-bedrock only with an acquisition adopting bistatic antennas in parallel-polarized modality. This arrangement was observed to be more sensitive to buried targets oriented parallel to the main axes of the antennas and relatively insensitive to depolarized scattered fields. However, moving this configuration on a rugged topography could be a logistical issue, especially when low-frequency antennas with long dipoles are adopted.

Despite these considerations, the presence of debris at the surface and the use of air-coupled antennas are common to other GPR surveys in debris-covered glaciers and rock glaciers (e.g., [24,26]) which led to satisfactory data quality. As a consequence, we hypothesize that widespread internal heterogeneities in the glacier mass may have had an additional and primary role in scattering and attenuation of the GPR signals. Beside the possible widespread presence of solid rock debris within the ice body [45], enhanced radar scattering, due to water englacial inhomogeneities is reported by several authors. Bamber [46] performed a numerical analysis to quantify the back scattering of water-filled cavities on the scale of decimeters affecting GPR results on several glaciers in Svalbard Islands. Numerical results illustrated the difficulties that may be encountered while sounding temperate glaciers possessing widespread englacial water bodies and explained the absence of bed echoes in the accumulation zone of these glaciers. GPR profiles collected by Murray et al. [47] in the ablation zone of Tsanfleuron Glacier (Swiss Alps) and Bakaninbreen Glacier (Svalbard) showed a two-layered structure, with an upper layer characterized by low returned GPR power and a lower layer of strong scattering. The thickness of these layers was observed to rapidly change along the profiles at both sites. At Tsanfleuron Glacier, the two layers were interpreted as dry ice, with a water content of 1.18%, overlapping ice containing small water bodies, up to decimeter in size, occupying 3.90% by volume. At Bakaninbreen Glacier, the upper radar layer was interpreted as cold ice with no measurable water-content and the lower layer as warm ice with a water content of 1.29%. Barrett et al. [48] modeled layers of randomly distributed scatters of decimeter-scale dimensions with an undulating upper boundary or confined to obliquely dipping planes to reproduce the scattering effects noticed on the radargrams acquired on Bakaninbreen Glacier. Numerical results supported the hypothesis that scattering originates from multiple planar sets of water-filled cavities. These features are expected to be common in glaciers surging by a thermally regulated soft bed mechanism, both at the end of the surge and into early quiescence phases. The presence of surge-type movements at the Belvedere Glacier, supporting the hypothesis of abundant water presence within the glacial mass, is well-documented in the literature [49].

As a consequence of the poor GPR data quality, ice thickness estimation was possible only in the glacier sectors characterized by the lowest ice thickness. Layering with different orientations, probably indicating the overlap of lateral morainic deposits on the bottom materials, was found to locally emphasize the bottom morphology. Only close to the terminus of the N lobe, deeper and steeper reflectors were imaged below the ice bottom. These observations suggest that the glacier bottom is not characterized by stiff bedrock, but by a thick sequence of fine-grained glacial deposits, as already hypothesize in the study of VAW-ETH [36]. If this hypothesis is valid, the glacial deposits have an average thickness of 40 m close to the glacier front and bedrock is located at a depth of approximately 80 m (e.g., rectangle A" in Figure 5d). The thickness of the bottom deposits rapidly increases upstream if the steep dipping of bedrock is constant.

Single-station passive seismic measurements confirmed the absence of sharp acoustic impedance contrasts in the glacier subsurface. The lack of a single clear peaks with high HVSR amplification values are a key piece of evidence for the absence of ice in direct contact with stiff bedrock. The occurrence of a multi-layered subsurface, with several weak contrasts in acoustic impedance at depth, can conversely help to explain the obtained HVSR results. The progressive lowering in the frequency of f0 peak from station A to C (Figure 7) is coherent with the presence of interfaces which are progressively deeper upstream, as highlighted in the GPR results (Figure 5d). HVSR complex results are however difficult to interpret with single simplified equations (e.g., Equation (1)), due to the existence of multiple interfaces whose resonance phenomena are superimposed. The stratigraphic condition appearing from all the above considerations is also at the limit of validity of the assumptions underlying further processing and interpretation of the measured curves. Despite the challenging working environment and investigated ice mass, a comparison between past and present geophysical surveys at the Belvedere Glacier is worthy of investigation. The ice-thickness map obtained from triangulation with linear interpolation of the GPR bottom peaks (Figure 6b) is shown in Figure 9a. The map confirms the previous observations of a traditional U-shaped bottom morphology for the N lobe and of shallow ice depths and flatter bottom surface for the S lobe. A digitized version of the ice thickness map of De Visintini [35], reported in Diolaiuti et al. [39], is shown for the same area of the glacier in Figure 9b. It must be noticed that the only reflection seismic measurements on this area were performed at the confluence of the two lobes (black bold lines in Figure 9b) and no data coverage was directly available on the lobes. As a consequence, this map is the result of a more approximated data interpretation with respect to the continuous GPR profiling from which Figure 9a is derived. Despite these considerations, Figure 9c shows the difference in ice thickness between the present and past maps.

A general agreement in ice thickness estimation is found along the axial position of the N lobe and on a wide area of the south lobe. Visible negative values (reduction in ice thickness) are found close to the northern front, underlying the accelerating retreat and shrinkage of the glacier in recent times. A visible decrease in ice thickness is also noticed close to the confluence between the lobes. This result can be considered quite reliable given the good data coverage of both present and previous surveys close to this point. By contrast, the unrealistic increases in ice thickness close to the glacier sides are more likely due to an erroneous approximation of ice thickness in the past study, rather than to a significant increase in ice volume in these marginal sectors.

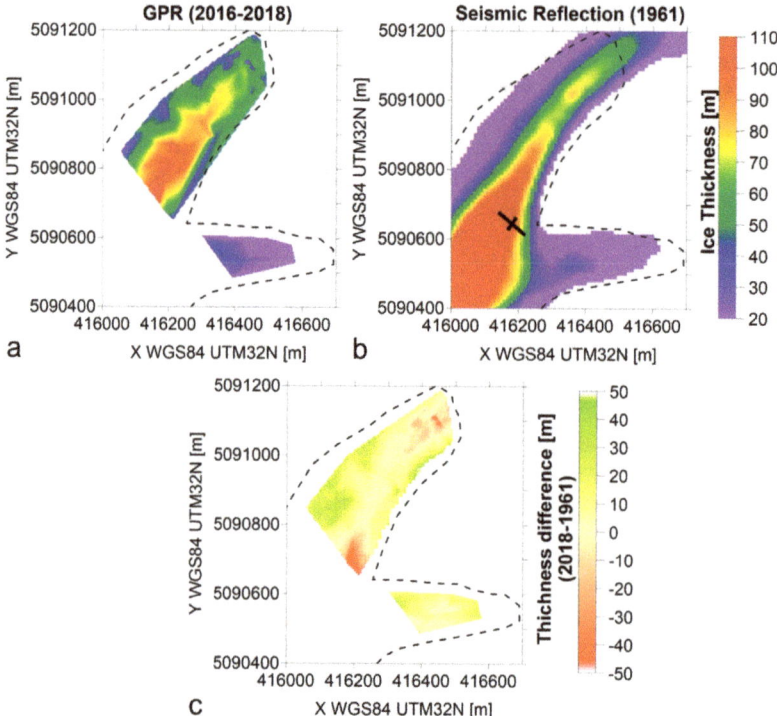

Figure 9. (**a**) Interpolated glacier bottom map from GPR results (2016–2018). (**b**) Reference glacier bottom map from De Visintini [35] and Diolaiuti et al. [39]. The black bold lines in (**b**) indicate the location of the only seismic reflection measurements in the area of interest. (**c**) Difference between (**a**) and (**b**) maps. The black dashed line highlights the approximate glacier perimeter in October 2016 (first GPR campaign).

An independent comparison between the glacier cross-sections obtained from the interpretation of the radar measurements of VAW-ETH [36] and the closest GPR profiles of the present study is presented in Figure 10. In this case, the adopted geophysical technique is the same, even if in different frequency bands. Both surveys should therefore have depicted the same interfaces. The ice bottom morphology presented in the previous work (black dashed lines in Figure 10b,c) is however the results of only two points of measure on each lobe (white triangles in Figure 10b,c), for both FF' and GG' sections [36] and not of continuous profiling.

Along the FF' cross-section, the 40-MHz profile 1 shows a depressed glacier topography with respect to the one of 1985. An acceptable discrepancy of around 15 m between the maximum ice bottom depth of the two surveys is noticed. Major differences in the morphology of the bottom surface are observed between the two campaigns. Recent GPR profiles outline a narrower and more asymmetric section of the N lobe, steeper on the side of the moraine hosting the Belvedere Mountain Hat. It is however interesting to notice that the reflection points related to the ray paths of two radar measurements performed on the N lobe well overlap with the recent bottom estimation from continuous GPR profiling. The lateral discrepancies can therefore be explained as the consequence of insufficient lateral data coverage of previous surveys. Similar results are found on GG' cross-section. Along this section, the topographic variations appear less marked on the N lobe. On the S lobe, profile 25 has higher topography than previous data, due to the fact that it is located almost 100 m upstream the reference GG' section trace. A good agreement between the present and past data is however found for the S lobe bottom. On the N lobe, the maximum depth depicted in the two surveys is the same, but the GPR profile 7 defines again a narrower ice bottom section. A wider and smoother bottom morphology was considered during the interpretation of previous radar measurements and a clear mismatch is observed between the two surveys. Differently from the original data interpretation, past radar echoes (and related ray paths) were probably originated from the deepest and narrowest axial sector of the glacial valley rather than from the lateral moraines. This comparison definitely highlights the undeniable advantages of continuous GPR profiling in maximizing data coverage and improving bottom morphology reconstruction, with respect to previous sparse and local measurements.

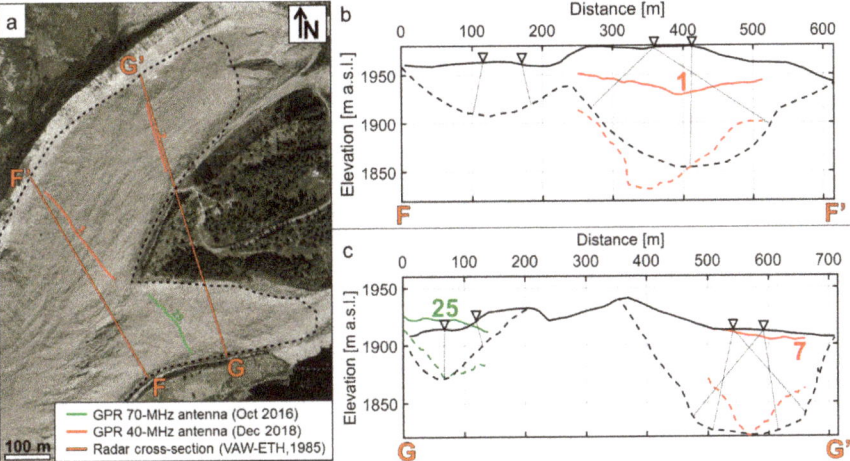

Figure 10. Comparison between the radar cross-sections of VAW-ETH [36] and the closest GPR profiles of the present study with successful glacier bottom picking. (**a**) Location of the profiles. (**b**) Section FF'. (**c**) Section GG'. The original data interpretation is shown in black (white triangles and gray rays indicate the radar measurement points and the interpreted EM ray paths as reported in [36]), while recent GPR results are shown in red for Profiles 1 and 7 (40 MHz, December 2018) and in green for Profile 25 (70 MHz, October 2016). Bold line: Topography, dashed line: Estimated ice bottom location.

6. Conclusions

Geophysical surveys, including 40-MHz and 70-MHz GPR profiling and single-station passive seismic measurements, were acquired at the Belvedere Glacier with the aim of ice thickness reconstruction. The surveys were intended to fill a knowledge gap on the glacier bottom morphology, since the existing bottom information was only retrieved from sparse and local past geophysical measurements. Despite a dense GPR survey distribution, the ice bottom surface was detected only in radargrams acquired on the frontal lobes, while noisy and uninterpretable results were obtained upstream. The globally observed low data quality was attributed to the presence of a widespread debris cover, the absence of direct coupling between antenna and glacier surface and the probable widespread presence of debris and/or small-scale water bodies within the ice column. The concurrence of all these features probably caused the observed significant scattering and attenuation of the transmitted EM pulses. The lower frequency 40-MHz survey generally showed clearer results. Ice thickness estimations based on these results were interpolated to reconstruct a detailed ice thickness map that was compared with the previous study of De Visintini [35]. Some discrepancies between the two estimations arose, mainly due to different data coverage and acceleration in the glacier retreat over the last decades. A better fitting with the results of the radar measurements of VAW-ETH [36] was obtained. Also in this case, continuous GPR profiling provided denser data coverage with respect to previous local measurements, enabling for a better definition of the ice bottom morphology. Significant ice thickness variations were detected in the upper part of the N lobe, with a transition from convex to concave glacier topography. Below the ice bottom, low-amplitude reflectors having steeper dip were identified only in the frontal portion of the N lobe. These elements may indicate the bedrock presence at a depth of around 80 m from the glacier surface close to the northern terminus and rapidly deepening upstream. Consequently, a thick layer (more than 40 m) of subglacial deposits may be present between ice and bedrock.

Single-station passive seismic measurements were processed following the HVSR method. The results showed the absence of a clear contrast in acoustic impedance (i.e., ice on bedrock) at depth, providing at least a general confirmation of the hypothesized subsurface conditions. However, without the reference GPR profiles, no information on the ice bottom could have been retrieved from HVSR curves. These tests highlighted the limitations of passive seismic measurements for glacier characterization, i.e., 1D approximation of the investigated subsurface, need to simultaneously have a dense grid of measurements but avoiding 2D effects, time-consuming acquisitions with respect to continuous GPR profiling, poor results in absence of a bottom ice-bedrock interface. Differently from GPR investigations, for which the denser data coverage of continuous profiling revealed the key point to improve the past knowledge about bottom morphology, due to the peculiar investigated conditions passive seismic methods did not succeed in improving past active seismic results, despite the lighter instrumentation, logistically easier acquisition and no need of active sources.

Future perspectives of the method may be addressed to glacier monitoring. Multi-station long-term passive seismic measurements can potentially be used for the investigation of the ongoing glacial processes (hydrogeological modifications, meltwater flow, seepage and accumulation) and of the glacier movements and stability conditions (e.g., icequakes, opening of crevasses, basal movements, serac falls and stability of the frontal compartments).

Further glaciological analyses are planned to understand the influence of the glacier subsurface conditions on the measured geophysical data. Future geophysical campaigns on site should be addressed to reach satisfactory imaging of the glacier bottom in the upstream sectors, and to monitor ice thickness variations over the investigated frontal areas. Alternative survey configurations, including the test of parallel-polarized antennas for data collection, will be tested to map the ice bottom, reducing the effect of clutter on GPR data quality.

Author Contributions: Conceptualization, C.C. (Chiara Colombero), C.C. (Cesare Comina), E.D.T., D.F. and A.G.; Data processing, C.C. (Chiara Colombero) and E.D.T.; Funding acquisition, A.G.; Investigation, D.F.; Methodology, C.C. (Chiara Colombero), C.C. (Cesare Comina) and A.G.; Supervision, C.C. (Cesare Comina) and A.G.; Validation, C.C. (Chiara Colombero); Visualization, C.C. (Chiara Colombero); Writing—original draft, C.C. (Chiara Colombero); Writing—review and editing, C.C. (Chiara Colombero), C.C. (Cesare Comina), E.D.T., D.F. and A.G.

Funding: This research was partially supported by educational funds from Alta Scuola Politecnica Milano-Torino (ASP project DREAM, DRone tEchnology for wAter resources Monitoring).

Acknowledgments: The geophysical campaigns described in this study were carried out in the framework of the Climate Change Glacier Lab of the Department of Environment, Land and Infrastructure Engineering of Politecnico di Torino. The authors are grateful to the Alpine guides that supervised and helped the field operations.

Conflicts of Interest: The authors declare no conflict of interest.

References

1. Beniston, M. Climatic Change in Mountain Regions: A Review of Possible Impacts. In *Climate Variability and Change in High Elevation Regions: Past, Present & Future*; Diaz, H.F., Beniston, M., Bradley, R.S., Eds.; Springer: Dordrecht, The Netherlands, 2003; Volume 15, pp. 5–31. [CrossRef]
2. Harris, C.; Arenson, L.U.; Christiansen, H.H.; Etzelmüller, B.; Frauenfelder, R.; Gruber, S.; Haeberli, W.; Hauck, C.; Hölzle, M.; Humlum, O.; et al. Permafrost and Climate in Europe: Monitoring and Modelling Thermal, Geomorphological and Geotechnical Responses. *Earth-Sci. Rev.* **2009**, *92*, 117–171. [CrossRef]
3. Haeberli, W.; Beniston, M. Climate Change and Its Impacts on Glaciers and Permafrost in the Alps. *Ambio* **1998**, *27*, 258–265.

4. Dyurgerov, M.B.; Meier, M.F. Twentieth Century Climate Change: Evidence from Small Glaciers. *Proc. Natl. Acad. Sci. USA* **2000**, *97*, 1406. [CrossRef]
5. Kääb, A.; Reynolds, J.M.; Haeberli, W. Glacier and Permafrost Hazards in High Mountains. In *Global Change and Mountain Regions*; Huber, U.M., Bugmann, H.K.M., Reasoner, M.A., Eds.; Springer: Dordrecht, The Netherlands, 2005; Volume 23, pp. 225–234. [CrossRef]
6. Ye, Q.; Kang, S.; Chen, F.; Wang, J. Monitoring Glacier Variations on Geladandong Mountain, Central Tibetan Plateau, from 1969 to 2002 Using Remote-Sensing and GIS Technologies. *J. Glaciol.* **2006**, *52*, 537–545. [CrossRef]
7. Bamber, J.L.; Rivera, A. A Review of Remote Sensing Methods for Glacier Mass Balance Determination. *Glob. Planet. Chang.* **2007**, *59*, 138–148. [CrossRef]
8. Kulkarni, A.; Bahuguna, I.; Rathore, B.; Singh, S.; Randhawa, S.; Sood, R.; Dhar, S. Glacial retreat in Himalaya using Indian Remote Sensing satellite data. *Curr. Sci.* **2007**, *92*, 69–74.
9. Zhou, Y.; Hu, J.; Li, Z.; Li, J.; Zhao, R.; Ding, X. Quantifying Glacier Mass Change and Its Contribution to Lake Growths in Central Kunlun during 2000–2015 from Multi-Source Remote Sensing Data. *J. Hydrol.* **2019**, *570*, 38–50. [CrossRef]
10. Farinotti, D.; Huss, M.; Bauder, A.; Funk, M. An Estimate of the Glacier Ice Volume in the Swiss Alps. *Glob. Planet. Chang.* **2009**, *68*, 225–231. [CrossRef]
11. Hagen, J.O.; Liestøl, O. Long-Term Glacier Mass-Balance Investigations in Svalbard, 1950–88. *Ann. Glaciol.* **1990**, *14*, 102–106. [CrossRef]
12. Zemp, M.; Hoelzle, M.; Haeberli, W. Six Decades of Glacier Mass-Balance Observations: A Review of the Worldwide Monitoring Network. *Ann. Glaciol.* **2009**, *50*, 101–111. [CrossRef]
13. Zekollari, H.; Fürst, J.J.; Huybrechts, P. Modelling the Evolution of Vadret Da Morteratsch, Switzerland, since the Little Ice Age and into the Future. *J. Glaciol.* **2014**, *60*, 1155–1168. [CrossRef]
14. Gabbi, J.; Farinotti, D.; Bauder, A.; Maurer, H. Ice Volume Distribution and Implications on Runoff Projections in a Glacierized Catchment. *Hydrol. Earth Syst. Sci.* **2012**, *16*, 4543–4556. [CrossRef]
15. Vincent, C.; Descloitres, M.; Garambois, S.; Legchenko, A.; Guyard, H.; Gilbert, A. Detection of a Subglacial Lake in Glacier de Tête Rousse (Mont Blanc Area, France). *J. Glaciol.* **2012**, *58*, 866–878. [CrossRef]
16. Eisen, O.; Nixdorf, U.; Keck, L.; Wagenbach, D. Alpine Ice Cores and Ground Penetrating Radar: Combined Investigations for Glaciological and Climatic Interpretations of a Cold Alpine Ice Body. *Tellus B* **2003**, *55*, 1007–1017. [CrossRef]
17. Navarro, F.; Eisen, O. Ground-penetrating radar in glaciological applications. *Remote Sens. Glaciers* **2009**, 195–229. [CrossRef]
18. Arcone, S.A. High Resolution of Glacial Ice Stratigraphy: A Ground-penetrating Radar Study of Pegasus Runway, McMurdo Station, Antarctica. *Geophysics* **1996**, *61*, 1653–1663. [CrossRef]
19. Urbini, S.; Baskaradas, J.A. GPR as an Effective Tool for Safety and Glacier Characterization: Experiences and Future Development. In Proceedings of the XIII Internarional Conference on Ground Penetrating Radar, Lecce, Italy, 21–25 June 2010; pp. 1–6. [CrossRef]
20. Navarro, F.J.; Martín-Español, A.; Lapazaran, J.J.; Grabiec, M.; Otero, J.; Vasilenko, E.V.; Puczko, D. Ice Volume Estimates from Ground-Penetrating Radar Surveys, Wedel Jarlsberg Land Glaciers, Svalbard. *Arct. Antarct. Alp. Res.* **2014**, *46*, 394–406. [CrossRef]
21. Kutuzov, S.; Thompson, L.G.; Lavrentiev, I.; Tian, L. Ice Thickness Measurements of Guliya Ice Cap, Western Kunlun Mountains (Tibetan Plateau), China. *J. Glaciol.* **2018**, *64*, 977–989. [CrossRef]
22. Del Longo, M.; Finzi, E.; Galgaro, A.; Godio, A.; Luchetta, A.; Pellegrini, G.B.; Zambrano, R. Responses of the Val d'Arcia small dolomitic glacier (Mount Pelmo, Eastern Alps) to recent climatic changes. Geomorphological and geophysical study. *Geografia Fisica e Dinamica Quaternaria* **2001**, *24*, 43–55.
23. Binder, D.; Brückl, E.; Roch, K.H.; Behm, M.; Schöner, W.; Hynek, B. Determination of Total Ice Volume and Ice-Thickness Distribution of Two Glaciers in the Hohe Tauern Region, Eastern Alps, from GPR Data. *Ann. Glaciol.* **2009**, *50*, 71–79. [CrossRef]
24. Merz, K.; Maurer, H.; Rabenstein, L.; Buchli, T.; Springman, S.M.; Zweifel, M. Multidisciplinary Geophysical Investigations over an Alpine Rock Glacier. *Geophysics* **2016**, *81*, WA147–WA157. [CrossRef]

25. Dossi, M.; Forte, E.; Pipan, M.; Colucci, R.R. Quantitative 3-D GPR Analysis to Estimate the Total Volume and Water Content of a Glacier. In Proceedings of the 2018 17th International Conference on Ground Penetrating Radar (GPR), Rapperswil, Switzerland, 18–21 June 2018; pp. 1–6. [CrossRef]
26. Maurer, H.; Hauck, C. Geophysical Imaging of Alpine Rock Glaciers. *J. Glaciol.* **2007**, *53*, 110–120. [CrossRef]
27. Picotti, S.; Francese, R.; Giorgi, M.; Pettenati, F.; Carcione, J.M. Estimation of Glacier Thicknesses and Basal Properties Using the Horizontal-to-Vertical Component Spectral Ratio (HVSR) Technique from Passive Seismic Data. *J. Glaciol.* **2017**, *63*, 229–248. [CrossRef]
28. Nogoshi, M.; Igarashi, T. On the amplitude characteristics of microtremor (part 2). *J. Seism. Soc. Jpn.* **1971**, *24*, 26–40.
29. Nakamura, Y. A method for dynamic characteristics estimation of subsurface using microtremor on the ground surface. *Quart. Rep. Railway Techn. Res. Inst.* **1989**, *30*, 25–33.
30. Lachetl, C.; Bard, P.-Y. Numerical and Theoretical Investigations on the Possibilities and Limitations of Nakamura's Technique. *J. Phys. Earth* **1994**, *42*, 377–397. [CrossRef]
31. Wohlenberg, J.; Ibs-von Seht, M. Microtremor Measurements Used to Map Thickness of Soft Sediments. *Bull. Seismol. Soc. Am.* **1999**, *89*, 250–259.
32. Konno, K.; Ohmachi, T. Ground-Motion Characteristics Estimated from Spectral Ratio between Horizontal and Vertical Components of Microtremor. *Bull. Seismol. Soc. Am.* **1998**, *88*, 228–241.
33. Bonnefoy-Claudet, S.; Köhler, A.; Cornou, C.; Wathelet, M.; Bard, P.-Y. Effects of Love Waves on Microtremor H/V RatioEffects of Love Waves on Microtremor H/V Ratio. *Bull. Seismol. Soc. Am.* **2008**, *98*, 288–300. [CrossRef]
34. Boulton, G.S.; Hindmarsh, R.C.A. Sediment Deformation beneath Glaciers: Rheology and Geological Consequences. *J. Geophys. Res. Solid Earth* **1987**, *92*, 9059–9082. [CrossRef]
35. De Visintini, G. Rilievo sismico a riflessione sul Ghiacciaio del Belvedere (Monte Rosa). *Bollettino del Comitato Glaciologico Italiano, Serie II* **1961**, *10*, 65–70.
36. VAW-ETH. Studi sul comportamento del Ghiacciaio del Belvedere, Macugnaga, Italia. *Relazione 97* **1985**, *3*, 76–109.
37. Porro, F.; Somigliana, C. Ghiacciaio di Macugnaga. *Bollettino del Comitato Glaciologico Italiano, Serie I* **1918**, *3*, 187–190.
38. Monterin, U. Il Ghiacciaio di Macugnaga dal 1870 al 1922. *Bollettino del Comitato Glaciologico Italiano, Serie I* **1922**, *5*, 12–40.
39. Diolaiuti, G.; D'Agata, C.; Smiraglia, C. Belvedere Glacier, Monte Rosa, Italian Alps: Tongue Thickness and Volume Variations in the Second Half of the 20th Century. *Arct. Antarct. Alp. Res.* **2003**, *35*, 255–263. [CrossRef]
40. Bard, P.-Y.; Bouchon, M. The Two-Dimensional Resonance of Sediment-Filled Valleys. *Bull. Seismol. Soc. Am.* **1985**, *75*, 519–541.
41. Wathelet, M.; Jongmans, D.; Ohrnberger, M. Surface wave inversion using a direct search algorithm and its application to ambient vibration measurements. *Near Surf. Geophys.* **2004**, *2*, 211–221. [CrossRef]
42. Herak, M. ModelHVSR—A Matlab®Tool to Model Horizontal-to-Vertical Spectral Ratio of Ambient Noise. *Comput. Geosci.* **2008**, *34*, 1514–1526. [CrossRef]
43. Lunedei, E.; Albarello, D. On the Seismic Noise Wavefield in a Weakly Dissipative Layered Earth. *Geophys. J. Int.* **2009**, *177*, 1001–1014. [CrossRef]
44. Malehmir, A.; Socco, L.V.; Bastani, M.; Krawczyk, C.M.; Pfaffhuber, A.A.; Miller, R.D.; Maurer, H.; Frauenfelder, R.; Suto, K.; Bazin, S.; et al. Chapter Two—Near-Surface Geophysical Characterization of Areas Prone to Natural Hazards: A Review of the Current and Perspective on the Future. In *Advances in Geophysics*; Nielsen, L., Ed.; Elsevier: Amsterdam, The Netherlands, 2016; Volume 57, pp. 51–146. [CrossRef]
45. Pecci, M.; De Sisti, G.; Marino, A.; Smiraglia, C. New radar surveys in monitoring the evolution of the Calderone glacier (Central Apennines, Italy). *Suppl. Geogr. Fis. Dinam. Quat.* **2007**, *V*, 145–150.
46. Bamber, J.L. Enhanced Radar Scattering From Water Inclusions In Ice. *J. Glaciol.* **1988**, *34*, 293–296. [CrossRef]
47. Murray, T.; Booth, A.; Rippin, D.M. Water-Content of Glacier-Ice: Limitations on Estimates from Velocity Analysis of Surface Ground-Penetrating Radar Surveys. *JEEG* **2007**, *12*, 87–99. [CrossRef]

48. Barrett, B.E.; Murray, T.; Clark, R.; Matsuoka, K. Distribution and Character of Water in a Surge-Type Glacier Revealed by Multifrequency and Multipolarization Ground-Penetrating Radar. *J. Geophys. Res. Earth Surface* **2008**, *113*. [CrossRef]
49. Haeberli, W.; Kääb, A.; Paul, F.; Chiarle, M.; Mortara, G.; Mazza, A.; Deline, P.; Richardson, S. A Surge-Type Movement at Ghiacciaio Del Belvedere and a Developing Slope Instability in the East Face of Monte Rosa, Macugnaga, Italian Alps. *Norsk Geografisk Tidsskrift—Nor. J. Geogr.* **2002**, *56*, 104–111. [CrossRef]

© 2019 by the authors. Licensee MDPI, Basel, Switzerland. This article is an open access article distributed under the terms and conditions of the Creative Commons Attribution (CC BY) license (http://creativecommons.org/licenses/by/4.0/).

Letter

The Scientific Operations of Snow Eagle 601 in Antarctica in the Past Five Austral Seasons

Xiangbin Cui [1], Jamin S. Greenbaum [2], Shinan Lang [3], Xi Zhao [4], Lin Li [1], Jingxue Guo [1,*] and Bo Sun [1]

1. Polar Research Institute of China, Shanghai 200136, China; cuixiangbin@pric.org.cn (X.C.); lilin@pric.org.cn (L.L.); sunbo@pric.org.cn (B.S.)
2. Institute for Geophysics, University of Texas at Austin, Austin, TX 78758, USA; jamin@utexas.edu
3. Faculty of Information Technology, Beijing University of Technology, Beijing 100124, China; langshinan@bjut.edu.cn
4. Chinese Antarctic Center of Surveying and Mapping, Wuhan University, Wuhan 430072, China; xi.zhao@whu.edu.cn
* Correspondence: guojingxue@pric.org.cn

Received: 3 August 2020; Accepted: 13 September 2020; Published: 15 September 2020

Abstract: The Antarctic ice sheet and the continent both play critical roles in global sea level rise and climate change but they remain poorly understood because data collection is greatly limited by the remote location and hostile conditions there. Airborne platforms have been extensively used in Antarctica due to their capabilities and flexibility and have contributed a great deal of knowledge to both the ice sheet and the continent. The Snow Eagle 601 fixed-wing airborne platform has been deployed by China for Antarctic expeditions since 2015. Scientific instruments on the airplane include an ice-penetrating radar, a gravimeter, a magnetometer, a laser altimeter, a camera and a Global Navigation Satellite System (GNSS). In the past five austral seasons, the airborne platform has been used to survey Princess Elizabeth Land, the largest data gap in Antarctica, as well as other critical areas. This paper reviews the scientific operations of Snow Eagle 601 including airborne and ground-based scientific instrumentation, aviation logistics, field data acquisition and processing and data quality control. We summarize the progress of airborne surveys to date, focusing on scientific motivations, data coverage and national and international collaborations. Finally, we discuss potential regions for applications of the airborne platform in Antarctica and developments of the airborne scientific system for future work.

Keywords: Snow Eagle 601; aerogeophysics; Princess Elizabeth Land; ice-penetrating radar; Antarctic ice sheet

1. Introduction

Antarctica is the most hostile and remote continent on Earth, possessing a large area of over 14 million km^2. Overall, 98% of the continent is covered by flowing ice sheets, causing many areas to be inaccessible from the ground [1]. The Antarctic continent and overlying ice sheet both play crucial roles in the recent sea level rise in response to global warming [1]. Crucial questions include how is Antarctica changing on regional and continental scales, what drives and controls Antarctica's changes and what is Antarctica's impact on climate and sea level changes. To answer the questions proposed by the Scientific Commission on Antarctic Research (SCAR), Antarctic and Southern Ocean Science Horizon Scan and the Antarctic Roadmap Challenges (ARC) Project [2,3], observations in Antarctica with high accuracy, resolution and reliability are urgently needed in order to improve our understanding of the continent and ice sheets and predict their changes and influences through numerical modeling.

Airborne surveys have been used to study Antarctica since the 1960s [4,5]. Multidisciplinary instruments onboard airborne platforms provide an efficient and flexible means of data collection over Antarctica, satisfying various requirements for polar exploration [6]. As a bridge between ground-based and space-borne observation, airborne surveys can acquire data with high efficiency in areas that are hard to access from the ground and achieve more accurate and higher resolution measurements relative to satellite remote sensing. Crucially, we cannot yet collect ice-penetrating radar (IPR) data from orbital platforms due primarily to interference from the ionosphere. With these advantages, airborne surveys have been continuously and extensively used in Antarctica for the past several decades. A number of large international airborne campaigns have been launched to survey Antarctica including the early SPRI-NSF-TUD (Scott Polar Research Institute, the National Science Foundation and the Technical University of Denmark) survey [7], AGAP (Antarctica's Gamburtsev Province [8]), IMAFI (Institute–Moller Antarctic Funding Initiative [9]), PolarGap [10], ICECAP (International Collaborative Exploration of the Cryosphere by Airborne Profiling [11–13]) and OIB (Operation IceBridge, https://www.nasa.gov/mission_pages/icebridge/index.html) surveys, as well as airborne surveys by German Polar 5 and Polar 6 in Dronning Maud Land and Dome F, respectively e.g., [14]. These projects have mainly focused on measuring the geometry and properties of ice sheets, subglacial topography and the geological structure of the continent. Based on airborne survey data, three important datasets, Bedmap (Antarctic Bedrock Mapping [15]), ADMAP (Antarctic Digital Magnetic Anomaly Project [16]) and AntGG (Gravity and Geoid in Antarctica [17]), have been compiled and gradually updated through international efforts. These datasets provide primary knowledge of ice thickness, bedrock topography and the geological settings of Antarctica.

China has recently become involved in Antarctic aviation. In 2015, a BT-67 airplane called Snow Eagle 601 was deployed to operate in Antarctica for Chinese National Antarctic Research Expeditions (CHINAREs [13]). An airborne IPR, a gravimeter, a magnetometer, a laser altimeter, a camera and a GNSS were configured and integrated on the airplane, offering powerful capabilities for aerogeophysical investigations. Meanwhile, as a branch of ICECAP, an international campaign of ICECAP/PEL was initiated by China to survey Princess Elizabeth Land (PEL), the largest data gap in Antarctica, using the Snow Eagle 601 airborne platform along with collaboration with the USA and the UK. In the past five austral seasons, the airborne platform has been continuously applied to survey PEL and other critical areas in East Antarctica including the Amery Ice Shelf, Ridge B, the West Ice Shelf, the Shackleton Ice Shelf and the George V Coast.

This paper reviews the scientific designs and operations of Snow Eagle 601 in Antarctica from the past five austral seasons from 2015 to 2020. First, we introduce the Snow Eagle 601 platform and ground-based scientific equipment. Next, we describe scientific operations supporting the aerogeophysical surveys including aviation support, survey design and data collection and processing. Finally, we summarize the progression of the airborne survey and discuss potential developments for airborne instruments and possible regions for additional survey in Antarctica for future years.

2. Snow Eagle 601 Airborne Surveying System

2.1. Airborne Platform

The Snow Eagle 601 airplane is a modified DC-3 aircraft, now known as a DC-3T or, more formally, a BT-67. Standard improvements include new Pratt & Whitney (East Hartford, CT, USA) turbine engines, a modern avionics system, fuel system and structural reinforcements among others. Modifications of the aircraft for polar operations include combined ski/wheel landing gear, an oxygen system, an air conditioner and a large cargo door. Ice-penetrating radar (IPR) antennas are mounted beneath the wings, a tail boom is used for a magnetometer and rolling doors beneath the fuselage are opened in flight to support a laser altimeter and a visual camera. GNSS antennas are mounted on each wing over the center of gravity (CG) and forward of the CG. Structural enhancements, electrical conduit and junction boxes to support power and data connections to science instruments are also installed on Snow Eagle 601.

Currently, the scientific instruments on the aircraft emphasize aerogeophysical investigations including an IPR made to be functionally similar to the High Capability Airborne Radar System (HiCARS) developed by the University of Texas Institute for Geophysics (UTIG) [13,18] for deep ice-penetrating capability, a GT-2A gravimeter [12], a CS-3 magnetometer, a Riegl LD90-3800-HiP laser altimeter, an Elphel NC353L downward-looking camera and a JAVAD dual frequency, four channel, carrier-phase GNSS receiver (Figure 1 and Table 1). Redundant interfaces were also reserved for the installation of more instruments in the future. More detailed descriptions of Snow Eagle 601 and the airborne scientific systems can be found in Cui et al. [13].

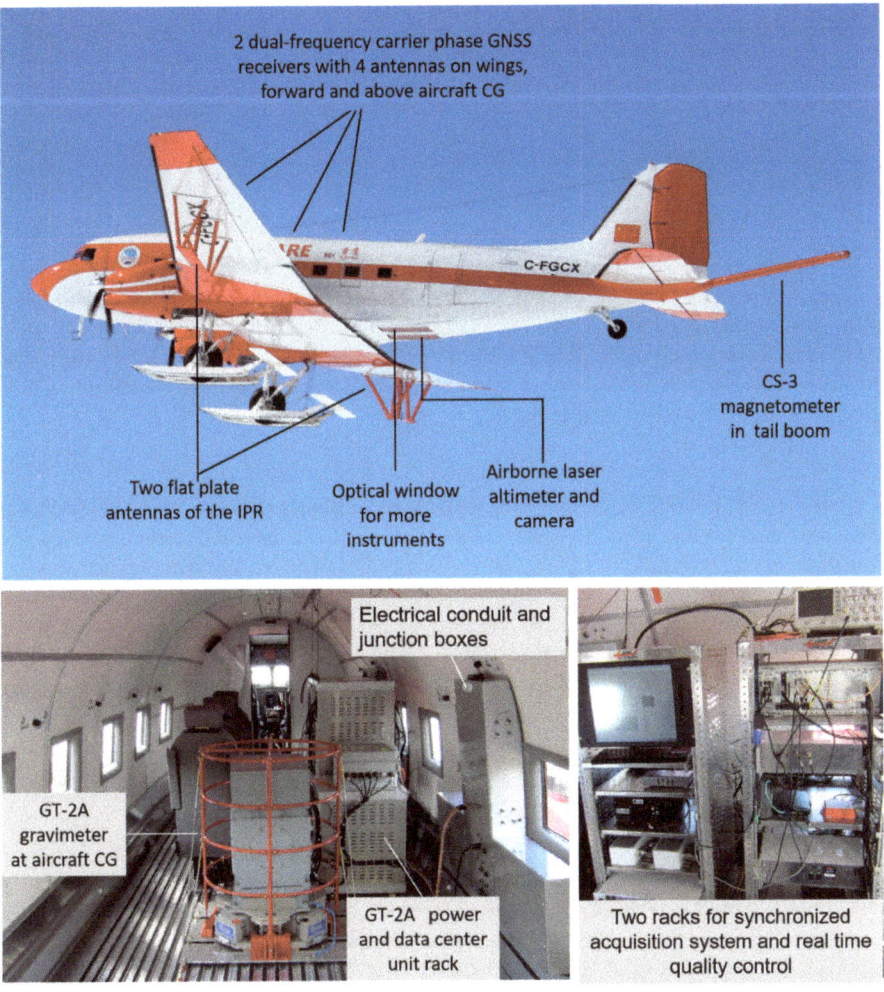

Figure 1. The Snow Eagle 601 platform and instruments.

Table 1. The Snow Eagle 601 airborne instruments and their typical performance and observing targets.

Airborne Instruments	Performance	Targets
Ice-penetrating radar (IPR)	Deep penetrating ability: >4000 m; depth resolution: 5.6 m in ice; coherent system; both low gain and high gain channels; 500–1500 m nominal surface range	Map ice sheet geometry and subglacial topography; interpret subglacial conditions
GT-2A gravimeter	Improved performance over other conventional gravimeters: dynamic range (>1000 Gal) and sensitivity (resolution: 0.02 mGal); can be used in turbulence; functions well on draped surveys	Measure gravity anomalies; infer deep geological and tectonic structures, subglacial water and sediment depth; bathymetry under ice shelves [12]
CS-3 magnetometer	High sensitivity (resolution: 0.00032 nT) and very low noise (system noise: <0.0001 nT)	Measure magnetic anomalies; infer lithology, deep geological and tectonic structures
Laser altimeter	Maximum range: 1500 m; accuracy: 15 cm	Measure accurate flight height over the snow or ice surface
Camera	-	Characterize surface features
Global Navigation Satellite System (GNSS)	Dual frequency, four-channel carrier-phase GNSS receiver	Provides aircraft position and attitude

Two to four days were needed to install the instrument suite onto the aircraft including two days for the gravimeter warm-up and auto-calibration before flights began. Ground testing of the instrument suite was required to confirm functionality before the first flight test. Normally, a short (~3–4 h) flight test would be conducted to test the operation of the suite, emphasize the high power radar amplifier and to calibrate the magnetometer. At the end of a season, airborne instruments can be de-configured from the airplane in one day.

2.2. Base Stations

As shown in Figure 2, two GPS base stations and a magnetic base station were installed beside the landing area to enable the differential processing of airborne GNSS data and to record variations in the magnetic field due to solar activity. GNSS and magnetic data from observatories in Zhongshan Station were also recorded during each flight as a backup for the instruments running at the skiway. When flights were conducted at other Antarctic stations, available GNSS and magnetic data from the nearest base station was requested for potential use. Gravity ties between the aircraft parking area and absolute gravity stations were measured at least twice during each season for airborne gravity data processing.

3. Aviation Support

3.1. Aviation Groups

Snow Eagle 601 is owned and managed by the Polar Research Institute of China (PRIC, Shanghai, China) but the airplane is registered in Canada; aviation operation and maintenance services are provided by Calgary, Alberta-based Kenn Borek Air Ltd. (KBA, Calgary, AB, Canada). KBA is a commercial company with a long history of complex polar aviation operations; only a three-person KBA crew is required for all aviation operations of the aircraft.

The PRIC determines annual missions and establishes the schedule for both the scientific and logistical operations of Snow Eagle 601. A field team from the PRIC and collaborative institutes is organized to support flight missions and carry out airborne surveys with separate scientific and logistical groups, each typically composed of six members. The scientific group is split into two sub-groups with three people managing flight operations (FOP) and the other three managing base operations (BOP). The main duty of FOP is to install and maintain airborne instruments, design flight plans and acquire data on each flight. The main duty of BOP is to conduct data processing and quality control (QC), operate and maintain base stations and prepare media and flight documents for FOP.

In a typical mission cycle, BOP initiates magnetic and GPS base station recording while FOP starts a pre-flight reference measurement on the aircraft gravimeter and starts recording GPS on the aircraft receivers at least half an hour before the aircraft moves. During the airborne survey (about seven hours), FOP will manage and monitor data acquisition on the airplane while BOP monitors the base instruments. Generally, only two of the three FOP members will fly on a given flight as each additional operator reduces survey flying time by about 15 to 30 min (80–160 km range) because the fuel load is limited by the aircraft weight. After landing, FOP and BOP require about an hour to complete a post-flight gravity reference measurement and recover all media including GPS data that are left to record for 30 min after landing. BOP collects media from the magnetometer and GPS base stations. After the post-flight data collection, both airborne data and base station data are moved to the BOP office. BOP requires about five hours to carry out data download, processing, quality control (QC), document printing and scanning and archiving. BOP and FOP discuss data QC results together to confirm that all instruments performed nominally and the collected data are of the desired quality before confirming the next flight.

3.2. Logistical Support

The scientific operations of Snow Eagle 601 mainly rely on logistical support from the Chinese Zhongshan Station and the Russian Progress Station Skiway located in the Larsemann Hills along Ingrid Christensen Coast, East Antarctica. The distance between the skiway and Zhongshan Station is about 10 km. The skiway is maintained by Progress Station during the aviation operations of Snow Eagle 601 based on collaboration between Russia and China. A temporary camp is built beside the skiway during aviation operations of each year (Figure 2). Custom-designed cabins and containers with different functions are configured for accommodation and living, power supply, field data processing and the storage of aviation accessories and scientific instruments (Figure 2). Ground support vehicles including Arctic Trucks, PistenBully snow vehicles and snowmobiles are used to transport personnel and equipment between Zhongshan and the Progress Skiway, maintain the camp and support field work.

Daily weather forecasts are provided by Zhongshan Station, Progress Station and Davis Station; the Antarctic Mesoscale Prediction System (AMPS, https://www2.mmr.ucar.edu/rt/amps) is additionally monitored regularly by the flight crew. The crew will choose from multiple flight plans based on all available weather reports.

Figure 2. Field camp for aviation and scientific operations of Snow Eagle 601 located along the Russian Progress Station Skiway.

4. Airborne Survey and Data Processing

4.1. Flight Plan Design

Flight plans are designed based on pre-season experiment designs that consider the scientific significance of each line in the context of previously completed surveying and flight limitations (e.g., maximum range and weather conditions) in target areas. Before a flight mission, two or three potential flight plans are prepared for the pilots to choose according to weather conditions. Key waypoints along a flight route are provided in the flight plans as well as surface elevation, flight range and flight altitude at each waypoint. As flight plans are created using projected coordinates but the aircraft is flown using geographic coordinates, the distance between each waypoint is kept to less than 50 km to minimize the resulting track error.

A target area can rarely be comprehensively surveyed on one flight. Systematic flight planning and survey organization allows an airborne survey to return to the target area and fill in coverage systematically and is robust against weather and equipment failures. This approach also allows elegance in analysis. We use a system of data organization developed by the University of Texas Institute for Geophysics that maps our survey plans and survey metadata into the organization of the collected data. The organizing elements are termed PSTs (project/set/transect). The project is generally defined according to the name of surveyed area or scientific campaign (e.g., PEL, representing Princess Elizabeth Land). The set indicates the survey platform used and the version of the instrument suite. For Snow Eagle 601, the first three letters in the set are taken from the final three letters of the aircraft tail number (C-FGCX); the last two numbers identify the version of the instrument suite used (e.g., GCX0g, representing the seventh version of the first acquisition system on C-FGCX). The transect represents particular survey lines within a survey area defined by the project, with orientations planned based on the target areas. Transects are typically either planned as radial lines for maximum range or as grids of perpendicular lines over an area of interest. For the latter, X lines are typically planned to run along the glacier flow and perpendicular Y lines are planned to run across the ice flow. X and Y lines are numbered sequentially from one to a number high enough to cover the survey area with a chosen line spacing; the final digit in each transect is used to denote whether the flown line is new (with an 'a' designation) and the final digit will be incremented from a to z for each repeated line or extension to the line flown on subsequent flights. PSTs divide a continuous flight into different transects by turning points and are unique transects of data collected at a discrete location and at a particular time. PSTs are imported into a data acquisition system by FOP during the flight and they are very useful for future data processing, organization and interpretation. Therefore, PSTs are clearly noted in a flight plan. The original definition of PST nomenclature can be found in UTIG's technical report database [18].

Chosen experiment designs depend on the given scientific objectives. While a radial survey plan is the most efficient way to explore large unknown areas, it also has few crossovers useful for evaluating data consistency and results in very long baselines that can result in degraded GNSS solutions. Radial lines were chosen for the first season flown by the Snow Eagle 601 airborne platform to quickly cover all of Princess Elizabeth Land. Gridded survey plans are generally chosen for investigating critical targets with a high spatial resolution requirement or where many crossovers are needed for data validation. Disadvantages to grids include longer transit lines to the survey areas, many turns and heavy logistical support required. For small outlet glaciers and ice streams, a gridded survey is the best choice as it can focus on and cover the target region in detail and also follow glaciological features.

The flight parameters cannot usually be the best for all of the instruments onboard so compromises must be made in the flight plans, often depending on the prior goals of the airborne survey. The airborne IPR should be flown at a constant height of over 500 meters above the surface to avoid saturating the low gain radar channel but preferably lower than 1500 meters to guarantee a sufficient penetrating capability. The maximum range of the laser altimeter is 1500 m, depending on surface and ambient conditions. Straight and level flight are preferred for gravity and magnetics and it is best if crossovers occur at similar elevations. During the past five austral seasons, with the highest priority of mapping the geometry and subice properties of ice sheets by airborne IPR in PEL, the Snow Eagle 601 flew at a height of ~600 m over surface most of the time, usually requiring consistent flight elevation changes to match changes in surface elevation. However, the flight height decreased to ~60 m when the signal attenuation in the ice was known to be high and penetrating capability was especially needed over areas with complex ice properties or deep subglacial valleys where the subglacial bedrock was difficult to detect because the IPR signals could be significantly attenuated or reflected away.

4.2. Data Acquisition

Except for the GT-2A gravimeter and JAVAD GNSS, the other airborne instruments are synchronously integrated by the Environment for Linked Streams Acquisition (ELSA) system [18], which was developed by UTIG. All integrated instruments can be controlled and monitored and their data are received and recorded by ELSA. ELSA also allows operators to manage data collection and check data quality in real-time on all flights. The system integration relies on an accurate time stamps (10 μs) calibrated with real-time GPS data. Data streams from different instruments with different sampling rates are received by ELSA along with the time and then written into two data files with one for radar data and the other for data from the other instruments [18]. The data files are recorded to a solid-state disk by ELSA. The gravity data are acquired and recorded separately into a USB flash disk by GT-2A's central data unit.

Flight notes are filled out, noting the times of engine start, taxi, takeoff and other major events. An instrument checklist is filled out every 30 min by FOP members and known or suspected aircraft or instrument anomalies are documented carefully. A base magnetometer and two GPS base stations are operated during each flight; at the end of each airborne survey, all media and flight notes are secured in weatherproof cases and delivered to BOP for data processing.

4.3. Data Processing

Raw data from a survey flight will be preliminarily processed by BOP before the next flight. The objectives of field data processing are to generate readable datasets of all instruments with corresponding PSTs, link all measurements to GPS positions, check the quality of the collected data, ensure all instruments functioned as designed and that data quality was sufficient to address the questions motivating the flight plan. The data processing system is set up at either the skiway camp or at Zhongshan Station and consists of two laptops for data processing, one or two QNAP (QNAP Systems, Inc.) RAIDs (redundant array of independent disks) for data storage and accessing, a printer and a tape driver for data archiving as well as a router and switch used to connect all the devices to a local network (Figure 3). Field data processing starts with downloading raw data from all media to an "orig" folder created in QNAP. Raw data are then converted into readable binary and ASCII files and separated by different instrument streams and different PSTs through a data breakout step (Figure 4). During the breakout step, each data stream from individual instruments is processed through custom code packages to improve the data quality and meet the demands of the data plotting and QC. For example, the IPR data are processed through down conversion, the removal of direct current offsets, pulse compression, coherent stacking and incoherent stacking. All measurements are linked with accurate time stamps, which can be traced to the GPS times and position with high accuracy. Data after the breakout step are saved in a "targ" folder. Finally, data from all instruments are plotted with the time and distance for each PST onto QC sheets to conveniently evaluate all data streams.

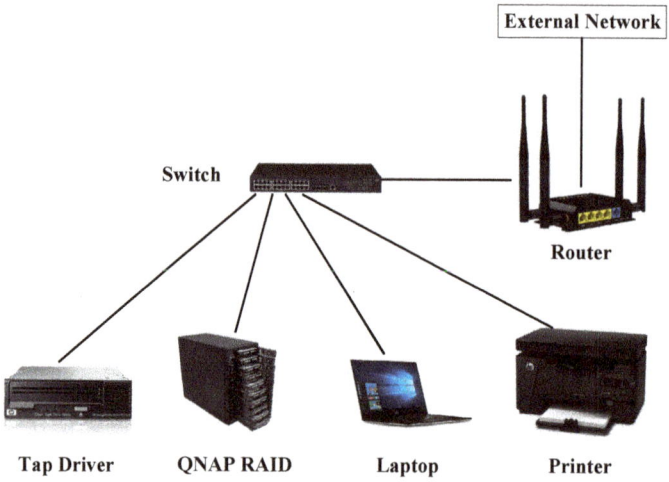

Figure 3. Field data processing, quality control and archiving system.

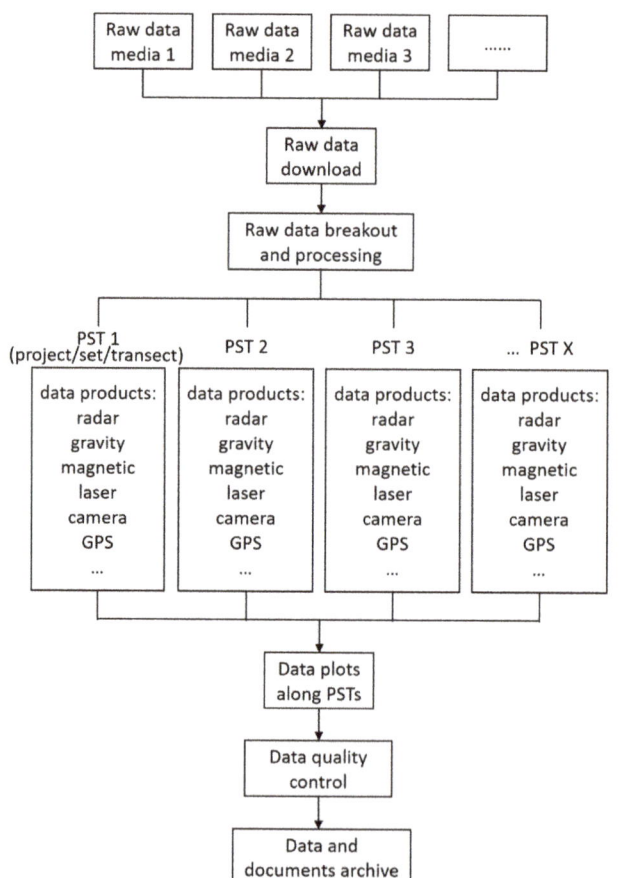

Figure 4. Flow chart of field data processing and quality control.

4.4. Data Quality Control

Data quality is checked after each flight by evaluating whether each device produced the expected data quality and, in parallel, whether the resulting data quality is sufficient to make the required interpretations of ice and subglacial geological conditions required by the motivating science questions behind the flight [18]. Device QC is checked with the data streams from four aircraft GNSS antennas, an Inertial Measurement Unit (IMU) and GNSS and magnetic base stations. Ice QC is checked with data streams from both the IPR high and low gain channels, the airborne GPS antenna at the aircraft CG and the IMU and laser altimeter are jointly analyzed. Geology QC is checked with data streams from the airborne gravimeter, magnetometer, IPR high gain channel and the airborne GNSS antenna at the CG. QC evaluations are made by BOP members based on a uniform data quality scale in four levels: good, moderate, poor and very poor (including data gaps). Joint interpretations by BOP and FOP members are made with particular emphasis on potential device problems that must be solved before another flight is possible or indications that lines must be re-flown due to poor flight conditions (e.g., excessive turbulence, clouds blocking laser data). QC records are printed in black and white and evaluations are hand-colored in four associated colors: blue = good, yellow = moderate, orange = poor, pink = very poor). A review sheet associated with the printed record is used for detailed notes during the review process.

4.5. Data and Documents Archive

All data uploaded to the "orig" and "targ" repositories including notes produced in flight and during the QC process are archived on magnetic storage tapes using a tape drive following the QC process. Four tapes are written after each flight including two redundant copies of each of the raw IPR and serial data (including all documents such as flight plans, flight notes, checklists and QC sheets). At the end of the season, the QNAP RAID and each set of tapes are shipped to PRIC using different means for added robustness against cargo delays, lost luggage or other problems [18].

5. Progress and Prospects

5.1. Progress

Scientific investigation is one of the most important tasks for Snow Eagle 601 aviation operations in Antarctica. However, the aircraft must also be available for emergency response and assisting with transportation logistics between Antarctic stations. In the past five austral seasons, Snow Eagle 601 has averaged about 120 flight hours for scientific survey out of a total of about 320 flight hours allocated for each season. Approximately 130 h are used to move the aircraft in and out of Antarctica from Calgary, Canada. Among the ~120 h of scientific flights each season, about 40 h are reserved for international collaborations. To date, China has collaborated with Australia, France, Russia, South Korea and the USA for scientific or logistical duties, sharing critical aviation capabilities in Antarctica. International collaborations improve the usability and coverage of Snow Eagle 601 in Antarctica and also benefit airborne data interpretation and scientific research. In China, scientists from many universities and institutions including Tongji University, Beijing University of Technology, Wuhan University, Beijing Normal University, Zhejiang University and the First Institute of Oceanography of the Ministry of Natural Resources (MNR) have been involved in airborne surveys in the field or through post-season data processing, interpretation and scientific research. Radioglaciological and aerogeophysical studies of Antarctica in China are significantly extended and improved through the scientific operations of Snow Eagle 601.

In its first five austral seasons, Snow Eagle 601 has acquired data along more than 175,000 line kilometers. The flight lines cover many critical areas in East Antarctica including Princess Elizabeth Land, Grove Mountains, the Amery Ice Shelf, the South Prince Charles Mountains, Ridge B, the West Ice Shelf, the Shackleton Ice Shelf, the Titan Dome, the George V Coast and the David Glacier Catchment, as shown in Figure 5.

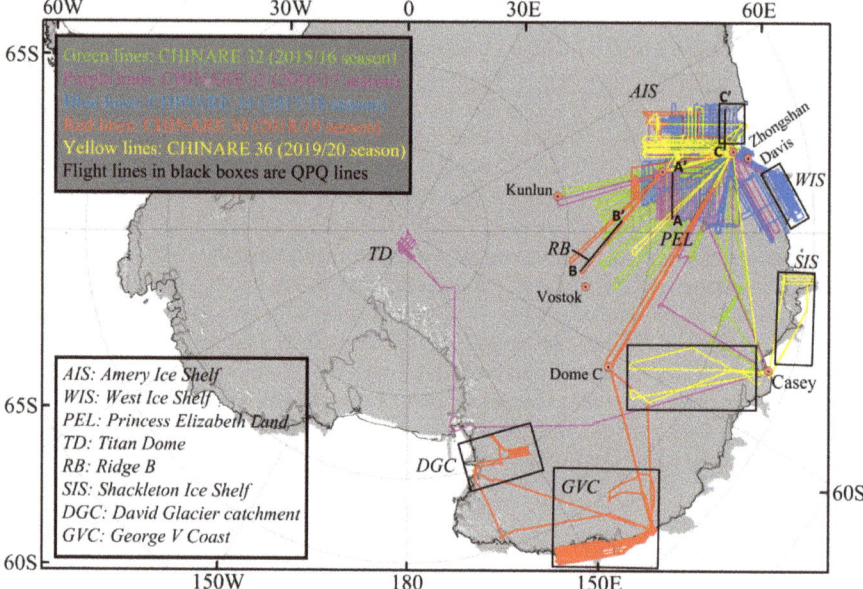

Figure 5. Flight line coverage of airborne surveys in the past five austral seasons by Snow Eagle 601. Profiles along A-A′, B-B′ and C-C′ are shown in Figure 6.

The initial scientific motivations for the scientific operations of Snow Eagle 601 were to map the ice sheet geometry, bedrock topography and infer the subglacial conditions and geological conditions of PEL, the largest data gap in Antarctica prior to these efforts. The Bedmap 2 [15], ADMAP 2 [16] and AntGG [17] datasets reflected the state of coverage from ice bottom elevation, magnetics and gravity field mapping before the beginning of Snow Eagle 601 operations, highlighting the large data gap in PEL. Moreover, ice surface features in the satellite data potentially showed a large subglacial lake in PEL and an extensive canyon system beneath the ice that may link the lake with the grounding zone of the West Ice Shelf [19]. However, these have not yet been identified and characterized by geophysical observations and their impacts on the dynamics and stability of the interior ice sheet have not been well studied, leading to our poor understanding of future mass balance in this sector and its potential contribution to the sea level. Data collected by Snow Eagle 601 over the last five years in PEL have filled this gap and ice thickness and bed topography datasets have been released by Cui et al. [20]. Airborne radar data along a transect (LSE/GCX0e/Y87a) also validate the existence of the subglacial lake (a relatively light and flat reflector in Figure 6a). Airborne survey data will continue to provide key parameters and boundary conditions for numerical models to study past and future ice sheet dynamics in PEL.

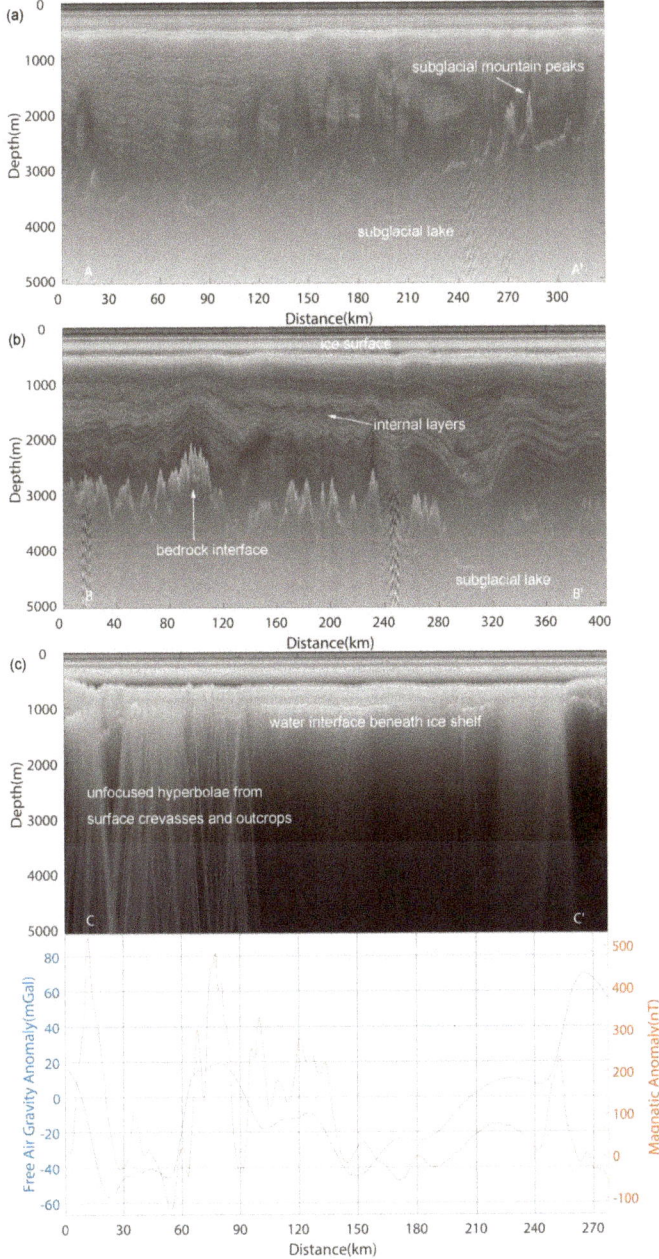

Figure 6. Data profiles along project/set/transects (PSTs) in Princess Elizabeth Land (PEL), Ridge B and the Amery Ice Shelf, respectively. (**a**) Radargram along LSE/GCX0e/Y87a, LSE = Lake of Snow Eagle; (**b**) radargram along TSH/GCX0g/R41a; (**c**) radargram, free air gravity anomaly and magnetic anomaly along AMY/GCX0e/Y184a, AMY = Amery.

Further airborne surveys have been extended to investigate the Amery Ice Shelf, the West Ice Shelf and the Shackleton Ice Shelf with the motivation of inferring bathymetry using gravity data [12], identifying properties of basal ice and grounding lines or zones and mapping ice shelf geometry with a more modern radar system relative to previous data. Along a transect near the ice front of the Amery Ice Shelf, airborne radar data have revealed a clear water interface beneath the ice shelf and unfocused hyperbolae caused by outcroppings and surface crevasses in the coastal region (Figure 6c). Gravity and magnetic anomalies along the transect show obvious relationships, which is a joint result from the deep geological structure and lithosphere (Figure 6c). Ice shelves are the most vulnerable components of ice sheets due to their direct contact with warm ocean water and play a vital role in the stability of interior ice sheets [21]. Studies show that ice shelves over most of the West Antarctic Ice Sheet (WAIS) are experiencing rapid dynamic thinning and grounding line retreat [22,23]. Ice shelves in the East Antarctic Ice Sheet (EAIS) are also beginning to earn increased attention to recently observed changes; for example, recent studies of the Totten Glacier Ice Shelf revealed rapid changes and ocean conditions similar to what was found in the coastal WAIS [12,24,25] indicating the possible instability of sectors of the EAIS. However, most ice shelves in the EAIS have not been adequately surveyed. Airborne surveys in these regions are ongoing by both the ICECAP/EAGLE (East Antarctic Grounding Lines Experiments) and ICECAP/PEL campaigns through international collaborations in Australia, China, France, Italy, South Korea, the UK and the USA. These programs have particularly emphasized coastal areas, such as the George V Coast, which is under-surveyed but is critically important for understanding the fate of the interior Wilkes Subglacial Basin.

In the austral season of 2018/19, the first effort was made to survey Ridge B. Ridge B is located in the deep interior of the EAIS. Modeling results indicated that Ridge B is one of the most likely places to host the oldest ice on Earth [26] and, therefore, the longest high resolution climate record but there are no sufficient observations there to support and verify this. Two critical flights were successfully achieved using a temporary skiway located at the Chinese Taishan Station as a transit and refueling stop. Due to the location of the interior of Antarctica, the radargram along the PST of TSH/GCX0g/R41a in Ridge B showed remarkable internal layers and continuous, clear, frequently varied bedrock topography (Figure 6b). These are essential for extracting ice thickness, bedrock elevation, past snow accumulation rate and basal conditions and have a significant value for estimating the bottom ice age in this region. The airborne data in Ridge B can improve our understanding of ice sheet geometry and improve the modeling accuracy of ice age estimation. In addition, two short surveys with similar research motivations were completed over the Titan Dome, another "old ice" target, by Snow Eagle 601 during the 2015/2016 and 2016/2017 austral seasons when the aircraft was ferrying out of Antarctica [27].

Aerogeophysical surveys in the Grove Mountains and the South Prince Charles Mountains have mainly focused on geological studies. Measurements in the regions offer a valuable extension for geological studies based on outcrops [28] and could have significant implications for the assembly and breakup of the Gondwana Supercontinent [16].

5.2. Future Work

In the past five austral seasons, Snow Eagle 601 has successfully surveyed PEL, the Amery Ice Shelf and the West Ice Shelf with moderately high spatial resolution. Data processing and publication should be accelerated to reveal the ice thickness, bedrock topography, internal layers and basal conditions of the ice sheets and ice shelves as well as analyzing the gravimetric and magnetic anomalies of the continent to an acceptable data density and accuracy. These results will significantly improve our knowledge of the ice sheet and its changes and of the geological settings of the regions. Flight lines over the South Prince Charles Mountains, Ridge B and other interior areas are still sparse. More attention should be given to these areas in future years. It is difficult to survey these areas from a coastal skiway because they require a very long transit; therefore, inland skiways such as Taishan Station, Kunlun Station, Vostok Station and South Pole Station should be considered through international collaborations and partnerships. International collaborations should also be strengthened to survey fast changing areas around the continent such as grounding zones, coastal outlet glaciers and ice shelves in the Ross Sea and Amundsen Sea sectors with powerful IPRs and aerogeophysical instruments to better characterize and infer the ice geometry, bedrock topography, basal conditions and bathymetry. Complex surface conditions and the strong attenuation of radar signals in these small-scale areas are a substantial challenge for airborne surveys. Moreover, to survey these areas, nearby landing areas are also needed. The continent–ocean transition zones in the sector of the Southern Indian Ocean are another target for the scientific operations of Snow Eagle 601 in future years. Surveys there can help to build knowledge on the connection of the geological structure between the continent and ocean there, which has significant implications for the joint interpretation of the geological evolution of Antarctica and the surrounding continents.

Updated and improved airborne instruments are also needed in the near future. The existing instruments have a strong capability for aerogeophysical investigation but are insufficient to map snow and shallow ice layers and are not sufficient to capture ice surface properties and their changes with high efficiency. Airborne shallow ice and snow radars are needed to map snow and shallow ice stratigraphy in detail, which is important to study the mass balance of Antarctica. The airborne laser altimeter and camera are not sufficient to measure surface elevation efficiently and characterize surface features in good quality. An advanced airborne optical camera or photogrammetry system as well as airborne LiDAR (light detection and ranging) and SAR (synthetic aperture radar) systems are priority choices for updating the Snow Eagle 601 airborne platform.

Snow Eagle 601 will continue operating in Antarctica for the foreseeable future. Airborne surveys remain an important part of the annual tasking; however, flight hours dedicated to scientific operations may be reduced with increased logistical requirements. Hopefully, more than 100 flight hours (~30,000 km flight lines) can be achieved in the upcoming seasons. In addition to the continuous and long-term data acquisition by Snow Eagle 601 in Antarctica, both national and international collaborations for data interpretation and research should be strengthened to extend the influence of Snow Eagle 601 scientific operations and promote scientific achievements.

6. Conclusions

In 2015, for the first time, China deployed its fixed-wing aircraft in Antarctica, which has led to substantial scientific and logistical achievements. An airborne IPR, a gravimeter, a magnetometer, a laser altimeter and a camera were simultaneously configured and integrated onto the aircraft at the same time. The airborne scientific platform has been applied to investigate many critical areas in East Antarctica in the past five austral seasons. Until now, over 175,000 line km of data have been surveyed in East Antarctica, not only filling the largest remaining data gap but also covering surrounding areas such as the Amery Ice Shelf, the West Ice shelf, Ridge B and the Titan Dome. Successful scientific operations of Snow Eagle 601 in the past five years prove the reliability and capability of the airborne platform and aviation support for airborne surveys in Antarctica.

Nationally, extensive collaboration with universities and institutes should be strengthened and deepened to carry out data interpretation and scientific research. Internationally, we have established close collaborations with the University of Texas at Austin, Imperial College London, the Australian Antarctic Division, the French Polar Institute Paul-Émile Victor and the Korea Polar Research Institute under the ICECAP framework. In the coming years, we will continue to survey the interior of the Antarctic Ice Sheet where flight lines are still sparse and launch new campaigns to survey the Ross and Amundsen Sea sectors as well as the transition zones between the continent and ocean. The resulting data will continue to contribute to international compilation efforts including the Bedmap, ADMAP and AntGG databases while improving our understanding of the ice sheet and subglacial geology of the Antarctic continent.

Author Contributions: Writing original draft, X.C.; X.Z. and S.L.; methodology and software, J.S.G.; S.L. and X.C.; investigation, J.S.G.; X.C.; J.G. and L.L.; supervision, B.S.; review and editing, all authors. All authors have read and agreed to the published version of the manuscript.

Funding: This work was supported by the National Natural Science Foundation of China (41941006, 41730102, 41776186), the National Key R&D Program of China (2019YFC1509102, 2018YFB1307504), the G. Unger Vetlesen Foundation, NSF grants PLR-1543452 and PLR-1443690, the Department for Business Innovation and Skills, the British Council and the US State Department.

Acknowledgments: We thank the Chinese Antarctic expedition teams and Kenn Borek Air Ltd. aircraft crew for their support of aviation operations as well as Laura Lindzey, Gregory Ng, Feras Habbal, Wei and other ICECAP/PEL members for their contributions to scientific operations of Snow Eagle 601. We also thank the reviewers for their constructive and helpful comments and suggestions for improving the quality of our manuscript. This is UTIG contribution #3690.

Conflicts of Interest: The authors declare no conflict of interest.

References

1. Andrew, S.; Amanda, F.H.; Louise, F.S. Trends and connections across the Antarctic cryosphere. *Nature* **2018**, *558*, 223–232.
2. Kennicutt, M.C.; Chown, S.L.; Cassano, J.J.; Liggett, D.; Massom, R.A.; Peck, L.S.; Rintoul, S.R.; Storey, J.W.V.; Vaughan, D.G.; Wilson, T.J.; et al. Polar research: Six priorities for Antarctic science. *Nature* **2014**, *512*, 23–25. [CrossRef]
3. Kennicutt ll, M.C.; Bromwich, D.; Liggett, D.; Njastad, B.; Peck, L.S.; Rintoul, S.R.; Ritz, C.; Siegert, M.J.; Aitken, A.R.A.; Brooks, C.M.; et al. Sustained Antarctic Research: A 21st Century Imperative. *One Earth* **2019**, *1*, 95–113. [CrossRef]
4. Evans, S. Radio techniques for the measurement of ice thickness. *Polar Record.* **2019**, *11*, 406–410. [CrossRef]
5. Cui, X.; Sun, B.; Zhang, X.; Zhang, D.; Li, X.; Tang, X.; Tian, G. A review of ice radar's technical development in polar ice sheet investigation. *Chin. J. Polar Res.* **2009**, *21*, 322–335.
6. Mirko, S.; Graeme, E.; Kirsty, T. Airborne platforms help answer questions in polar geosciences. *Eos* **2017**, *98*. [CrossRef]
7. Schroeder, D.M.; Dowdeswell, J.A.; Siegert, M.J.; Bingham, R.; Chu, W.; Mackie, E.J.; Siegfried, M.R.; Vega, K.I.; Emmons, J.R.; Winstein, K. Multidecadal observations of the Antarctic ice sheet from restored analog radar records. *Proc. Natl. Acad. Sci. USA* **2019**, *116*, 18867–18873. [CrossRef] [PubMed]
8. Bell, R.E.; Ferraccioli, F.; Creyts, T.T.; Braaten, D.; Corr, H.F.; Das, I.; Damaske, D.; Frearson, N.; Jordan, T.; Rose, K.; et al. Widespread Persistent Thickening of the East Antarctic Ice Sheet by Freezing from the Base. *Science* **2011**, *331*, 1592–1595. [CrossRef] [PubMed]
9. Jeofry, H.; Ross, N.; Corr, H.F.J.; Li, J.; Morlighem, M.; Gogineni, P.; Siegert, M.J. A new bed elevation model for the Weddell Sea sector of the West Antarctic Ice Sheet. *Earth Syst. Sci. Data* **2018**, *10*, 711–725. [CrossRef]
10. Jordan, T.A.; Martin, C.; Ferraccioli, F.; Matsuoka, K.; Corr, H.F.; Forsberg, R.; Olesen, A.; Siegert, M. Anomalously high geothermal flux near the South Pole. *Sci. Rep.* **2018**, *8*, 1–8. [CrossRef]
11. Young, D.A.; Wright, A.P.; Roberts, J.L.; Warner, R.C.; Young, N.W.; Greenbaum, J.S.; Schroeder, D.M.; Holt, J.W.; Sugden, D.E.; Blankenship, D.D.; et al. A dynamic early East Antarctic Ice Sheet suggested by ice-covered fjord landscapes. *Nature* **2011**, *474*, 72–75. [CrossRef] [PubMed]
12. Greenbaum, J.S.; Blankenship, D.D.; Young, D.A.; Richter, T.G.; Roberts, J.L.; Aitken, A.R.A.; Legresy, B.; Schroeder, D.M.; Warner, R.C.; van Ommen, T.D.; et al. Ocean access to a cavity beneath Totten Glacier in East Antarctica. *Nat. Geosci.* **2015**, *8*, 294–298. [CrossRef]
13. Cui, X.; Greenbaum, J.S.; Beem, L.H.; Guo, J.; Ng, G.; Li, L.; Blankenship, D.; Sun, B. The First Fixed-wing Aircraft For Chinese Antarctic Expeditions: Airframe, Modifications, Scientific Instrumentation and Applications. *J. Environ. Eng. Geophys.* **2018**, *23*, 1–13.
14. Karlsson, N.B.; Binder, T.; Eagles, G.; Helm, V.; Pattyn, F.; Van Liefferinge, B.; Eisen, O. Glaciological characteristics in the Dome Fuji region and new assessment for "Oldest Ice". *Cryosphere* **2018**, *12*, 2413–2424. [CrossRef]
15. Fretwell, P.T.; Pritchard, H.D.; Vaughan, D.G.; Bamber, J.L.; Barrand, N.E.; Bell, R.E.; Bianchi, C.; Bingham, R.G.; Blankenship, D.D.; Casassa, G.; et al. Bedmap2: Improved ice bed, surface and thickness datasets for Antarctica. *Cryosphere* **2012**, *7*, 375–393. [CrossRef]

16. Golynsky, A.V.; Ferraccioli, F.; Hong, J.; Golynsky, D.A.; Von Frese, R.R.; Young, D.A.; Blankenship, D.D.; Holt, J.W.; Ivanov, S.V.; Kiselev, A.V.; et al. New Magnetic Anomaly Map of the Antarctic. *Geophys. Res. Lett.* **2018**, *45*, 6437–6449. [CrossRef]
17. Scheinert, M.; Ferraccioli, F.; Schwabe, J.; Bell, R.E.; Studinger, M.; Damaske, D.; Jokat, W.; Aleshkova, N.; Jordan, T.; Leitchenkov, G.; et al. New Antarctic gravity anomaly grid for enhanced geodetic and geophysical studies in Antarctica. *Geophys. Res. Lett.* **2016**, *43*, 600–610. [CrossRef]
18. Ng, G.; Lindzey, L.E.; Young, D.A.; Buhl, D.P.; Kempf, S.D.; Beem, L.H.; Roberts, J.L.; Greenbaum, J.S.; Blankenship, D.D. *UTIG's Approach to Managing Polar Aerogeophysical Data in the Field: Philosophy and Examples from Fixed Wing and Helicopter Surveys*; UTIG Technical Report; University of Texas, Institute for Geophysics: Austin, TX, USA, 2020.
19. Jamieson, S.S.R.; Ross, N.; Greenbaum, J.S.; Young, D.A.; Aitken, A.R.A.; Roberts, J.L.; Blankenship, D.D.; Siegert, M.J. An extensive subglacial lake and canyon system in Princess Elizabeth Land, East Antarctica. *Geology* **2016**, *44*, 87–90. [CrossRef]
20. Cui, X.; Jeofry, H.; Greenbaum, J.S.; Guo, J.; Li, L.; Lindzey, L.E.; Habbal, F.A.; Wei, W.; Young, D.A.; Ross, N.; et al. Bed topography of Princess Elizabeth Land in East Antarctica. *Earth Syst. Sci. Data Discuss.* **2020**. in review. [CrossRef]
21. Pritchard, H.D.; Ligtenberg, S.R.M.; Fricker, H.A.; Vaughan, D.G.; Van den Broeke, M.R.; Padman, L. Antarctic ice-sheet loss driven by basal melting of ice shelves. *Nature* **2012**, *484*, 502–505. [CrossRef]
22. Paolo, F.; Fricker, H.; Padman, L. Volume loss from Antarctic ice shelves is accelerating. *Science* **2015**, *348*, 327–331. [CrossRef] [PubMed]
23. Rignot, E.; Mouginot, J.; Morlighem, M.; Seroussi, H.; Scheuchl, B. Widespread, rapid grounding line retreat of Pine Island, Thwaites, Smith, and Kohler glaciers, West Antarctica, from 1992 to 2011. *Geophys. Res. Lett.* **2014**, *41*, 3502–3509. [CrossRef]
24. Li, X.; Rignot, E.; Morlighem, M.; Mouginot, J.; Scheuchl, B. Grounding line retreat of Totten Glacier, east Antarctica, 1996 to 2013. *Geophys. Res. Lett.* **2015**, *42*, 8049–8056. [CrossRef]
25. Rintoul, S.R.; Silvano, A.; Pena-Molino, B.; van Wijk, E.; Rosenberg, M.; Greenbaum, J.S.; Blankenship, D.D. Ocean heat drives rapid basal melt of the Totten ice shelf. *Sci. Adv.* **2016**, *2*, E1601610. [CrossRef] [PubMed]
26. Van Liefferinge, B.; Pattyn, F. Using ice-flow models to evaluate potential sites of million year-old ice in Antarctica. *Clim. Past* **2013**, *9*, 2335–2345. [CrossRef]
27. Beem, L.H.; Young, D.A.; Greenbaum, J.S.; Blankenship, D.D.; Guo, J.; Bo, S. Characterization of Titan Dome, East Antarctica, and potential as an ice core target. *Cryosphere* **2020**, in review. [CrossRef]
28. Liu, X.; Zhao, Y. Geological surveys in East Antarctica by Chinese expeditions over the last 20 years: Progress and prospects. *Chin. J. Polar Res.* **2018**, *30*, 268–286.

© 2020 by the authors. Licensee MDPI, Basel, Switzerland. This article is an open access article distributed under the terms and conditions of the Creative Commons Attribution (CC BY) license (http://creativecommons.org/licenses/by/4.0/).

Article

Properties Analysis of Lunar Regolith at Chang'E-4 Landing Site Based on 3D Velocity Spectrum of Lunar Penetrating Radar

Zejun Dong [1,2], Xuan Feng [1,2,3,*], Haoqiu Zhou [1,2], Cai Liu [1,4], Zhaofa Zeng [1,4], Jing Li [1,3] and Wenjing Liang [1,2]

[1] College of Geo-Exploration Science and Technology, Jilin University, No.938 Xi MinZhu Street, Changchun 130026, China; dzj19@mails.jlu.edu.cn (Z.D.); zhouhq17@mails.jlu.edu.cn (H.Z.); liucai@jlu.edu.cn (C.L.); zengzf@jlu.edu.cn (Z.Z.); inter_lijing@jlu.edu.cn (J.L.); liangwj@jlu.edu.cn (W.L.)
[2] Science and Technology on Near-Surface Detection Laboratory, Wuxi 214035, China
[3] Institute of National Development and Security Studies, Jilin University, No.2699 Qianjin Street, Changchun 130012, China
[4] Key Laboratory of Geophysical Exploration Equipment, Ministry of Education (Jilin University), No.938 Xi MinZhu Street, Changchun 130026, China
* Correspondence: fengxuan@jlu.edu.cn; Tel.: +86-431-8502258

Received: 18 January 2020; Accepted: 12 February 2020; Published: 13 February 2020

Abstract: The Chinese Chang'E-4 mission for moon exploration has been successfully completed. The Chang'E-4 probe achieved the first-ever soft landing on the floor of Von Kármán crater (177.59°E, 45.46°S) of the South Pole-Aitken (SPA) basin on January 3, 2019. Yutu-2 rover is mounted with several scientific instruments including a lunar penetrating radar (LPR), which is an effective instrument to detect the lunar subsurface structure. During the interpretation of LPR data, subsurface velocity of electromagnetic waves is a vital parameter necessary for stratigraphic division and computing other properties. However, the methods in previous research on Chang'E-3 cannot perform velocity analysis automatically and objectively. In this paper, the 3D velocity spectrum is applied to property analysis of LPR data from Chang'E-4. The result shows that 3D velocity spectrum can automatically search for hyperbolas; the maximum value at velocity axis with a soft threshold function can provide the horizontal position, two-way reflected time and velocity of each hyperbola; the average maximum relative error of velocity is estimated to be 7.99%. Based on the estimated velocities of 30 hyperbolas, the structures of subsurface properties are obtained, including velocity, relative permittivity, density, and content of FeO and TiO_2.

Keywords: Chang'E-4; lunar penetrating radar (LPR); 3D velocity spectrum; properties analysis

1. Introduction

The Chinese Chang'E-4 mission for moon exploration has been successfully completed. The Chang'E-4 probe achieved the first-ever soft landing on the floor of Von Kármán crater (177.59°E, 45.46°S) of the South Pole-Aitken (SPA) basin on January 3, 2019 [1–3]. SPA basin is the broadest basin on the Moon. This ancient basin was born from asteroid impacts 4 billion years ago, recording the evolutionary history of the far side of the Moon, and is of great significance for researching the internal materials and structures of the Moon [4–6]. Von Kármán crater is one of the primary craters in the SPA basin, with a diameter of 186 km. Recent studies have revealed that the ejecta from adjacent craters have various contributions to the subsurface material of Von Kármán crater, which results in the complex subsurface structure at the Chang'E-4 landing site [1,3,4].

In order to patrol and investigate the lunar surface, a Yutu-2 rover is carried by Chang'E-4 probe. Yutu-2 rover is mounted with several scientific instruments containing a lunar penetrating radar (LPR)

which has been verified to be an effective device for detecting the lunar subsurface structure [7,8]. The LPR is equipped with two types of channels (CH-1 and CH-2), the center frequencies of which are 60 MHz and 500 MHz [9,10]. In addition, the CH-2 possesses one transmitting antenna and two receiving antennas of different offsets which are also known as CH-2A and CH-2B [10–12]. With different frequencies, CH-1 and CH-2 have different detecting missions. The objective of CH-1 is to detect the deep structure of Von Kármán crater [13]; CH-2 is to map the detail of near surface layers, leading to the property analysis of lunar regolith [14,15].

During the interpretation of LPR data, subsurface velocity of electromagnetic waves is a vital parameter necessary for stratigraphic division and computing other properties. However, the velocity analysis of field LPR data faces many difficulties. Complex subsurface structure and interference of noise always result in the incomplete, interlaced, and amplitude-varying hyperbolas. In previous studies of Chang'E-3, Feng et al. proposed hyperbolic fitting method for velocity analysis of LPR data [10]; Lai et al. used two-way delay method to acquire the velocity [16]; Zhang et al. applied CH-2A and CH-2B data of different offsets to estimate the velocity [17]. However, these methods need humans to select the hyperbolas, which is highly subjective. Thus, in this article, we applied 3D normalized velocity spectrum to estimate the velocity [18]. This method can automatically and objectively select hyperbolas and analyze the velocities; the normalization processing can solve the error brought by different amplitudes of different positions on hyperbolas; subsequently, during the computation, we applied a variable horizontal computation window along longitudinal direction to satisfy the field situation that rock sizes increase vertically.

This paper is organized as follows. Section 2 introduces basic theory of 3D velocity spectrum and properties analysis. In Section 3, firstly a model test is performed to verify the feasibility of 3D velocity spectrum; subsequently the method is applied to the CH-2B data analysis. Subsequently, based on the 3D velocity spectrum, we obtain the positions and velocities of the points where hyperbolas exist; then the property structures of lunar regolith are computed including velocity, relative permittivity, density, and content of FeO and TiO_2. In Section 4, we discuss the computation error of each hyperbola. Section 5 is the conclusion.

2. Methods

2.1. 3D Velocity Spectrum

The velocity spectrum analysis method is originally based on common middle point (CMP) ground penetrating radar (GPR) data [19,20], but it can also be applied to common offset data processing such as LPR data processing. In order to estimate velocity, a stacked amplitude is computed as follows:

$$S_{i,j,k} = \sum_{j=1}^{N_i} f(t_{i,j,k}, x_j) \tag{1}$$

$$i = 1, \ldots nt; \quad j = 1, \ldots nx; \quad k = 1, \ldots nv$$

where f is the LPR data in t-x domain; N_i denotes the selected horizontal computation region size at ith time; to satisfy the field situation that rock sizes increase vertically, we make N_i variations along longitudinal direction; nt, nx, and nv are the number of sampling points of each trace, number of traces, and number of velocities used in computation; x_j represents the horizontal distance between the jth point and the extreme point of the hyperbola; $t_{i,j,k}$ is the two-way time of the jth points of the hyperbola [21,22], which can be obtained using the formula below:

$$t_{i,j,k} = \left(t_i^2 + \frac{4x_j^2}{v_k^2} \right)^{1/2} \tag{2}$$

where t_i is the two-way time of extreme point of the hyperbola; v_k represents the velocity used in computation.

However, the method of (1) is not always effective in the field as the hyperbolas are incomplete, interlaced, with varying amplitude. To solve the error brought about by these situations, the normalization form of stacked amplitude is applied:

$$C_{i,j,k} = \frac{1}{N_i L} \sum_{l=i}^{L+i-1} \left[\frac{S_{i,j,k}^2}{\sum_{j=1}^{N} f^2(t_{i,j,k}, x_j)} \right] \quad (3)$$

where L is the time gate; the average in the time gate is adopted because the hyperbolas on radargram are not curves but regions with hyperbolic shapes. The normalized form can guarantee that the result $C_{i,j,k}$ is not affected by the varying amplitude. Importantly, $1/N_i$ is added in this form to compensate the energy differences caused by using different windows N_i along longitudinal direction. $C_{i,j,k}$ is 3D data, whose local maximum values indicate the time, horizontal position, and velocity of the points with hyperbolas.

2.2. Properties Computation

After we obtain the velocities from the 3D spectrum, the properties can be computed. Without considering magnetic permeability, the relative permittivity can be easily derived using the following formula:

$$\varepsilon = (c/v)^2 \quad (4)$$

where c equals 0.3 m/ns.

According to the studies of lunar regolith samples of Apollo, there is a relation between the relative permittivity and density of lunar regolith [9,17]:

$$\rho = \log_{1.919}(\varepsilon) \quad (5)$$

$$\tan \delta = 10^{0.440\rho - 2.943} \quad (6)$$

The $\tan\delta$ is the loss tangent which represents the ratio of the imaginary part of the dielectric constant to its real part. The density of lunar regolith is shown in Figure 14. Subsequently, previous studies show that the loss tangent is also related to the TiO_2 and FeO weight percentage values [9,17]:

$$\tan \delta = 10^{[0.038 \times (\omega(TiO_2) + \omega(FeO)) + 0.312\rho - 3.26]} \quad (7)$$

Combining (6) and (7), we can obtain the TiO_2 and FeO weight percentage of lunar regolith.

3. Results

3.1. Model Test

Before the processing of LPR data, a model test was performed to verify the feasibility of 3D velocity spectrum. The relative permittivity model used in test and its simulated GPR profile are shown in Figure 1a,b, respectively. In the simulation, the frequency of electromagnetic wave is 100 MHz; the time window and the sampling interval are 100 ns and 0.32 ns; the trace interval is 0.1 m; the transmitting and receiving antennas are placed on the ground surface. The radargram shows there are three pairs of hyperbolas (H1, H2, and H3); the double hyperbolas generate from the upper and bottom surface of the scatters. Subsequently, the 3D velocity spectrum is computed, whose result is shown in Figure 1c.

In fact, for a point (t_i,x_j), there is only one true velocity; this true velocity corresponds to the maximum value along velocity axis; the maximum value can help to find the position of hyperbolas on (t_i,x_j) profile. Thus, the maximum values $C_{i,j}$ along velocity axis are figured out using (8). Subsequently, in order to suppress the noise interference and obtain the accurate positions of hyperbolas, we apply a soft threshold function [23] as (9) shows. The value of ε is set to 0.3 and the result is shown in Figure 1d. The energy stacks in three rectangles indicate the positions of hyperbolas. The irregular energy stack in the ellipse is generated by the intersection of two hyperbolas.

$$C_{i,j} = \max_k C_{i,j,k} \tag{8}$$

$$C_{i,j} = \begin{cases} C_{i,j} - \varepsilon & C_{i,j} > \varepsilon \\ 0 & C_{i,j} < \varepsilon \end{cases} \tag{9}$$

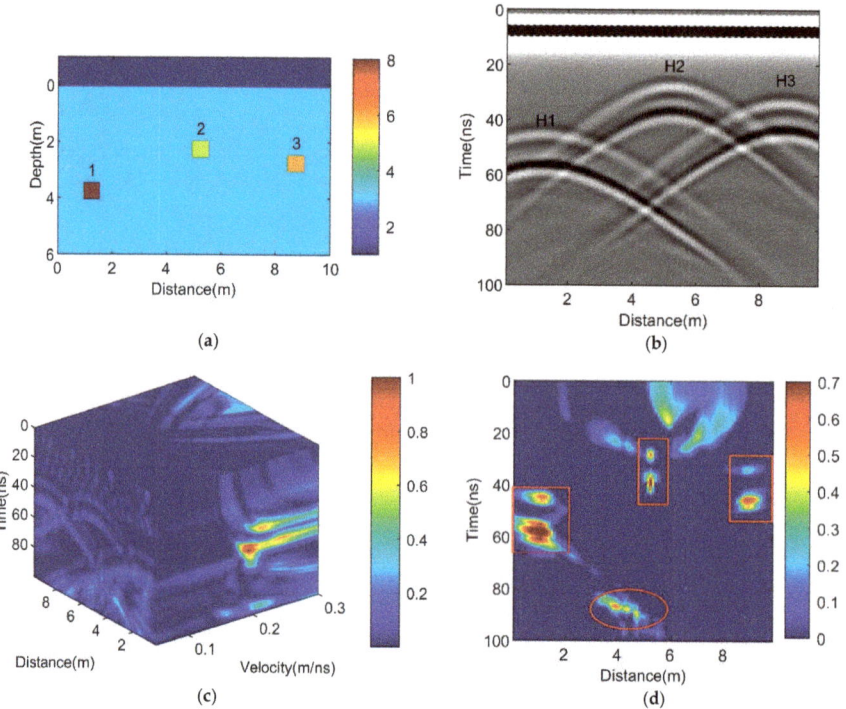

Figure 1. Model test results. (a) The model used in simulation; (b) the simulated GPR profile; (c) 3D velocity spectrum of (b); (d) maximum value along velocity axis with the soft threshold function processing.

After the x position is located, the velocity spectrum slices can be selected out to estimate the velocities. The spectrum contour slices of the three hyperbolas are shown in Figure 2. From Figure 2, the velocities can be read out. The estimated results and errors are shown in Table 1. The results show that the estimation errors of upper hyperbolas are about 1%, and the estimation errors of bottom hyperbolas are about 10%. The errors are in acceptable range; the relatively higher errors of bottom hyperbolas result from the effects of scatters; the low velocities of electromagnetic waves inside scatters reduce the estimated velocities.

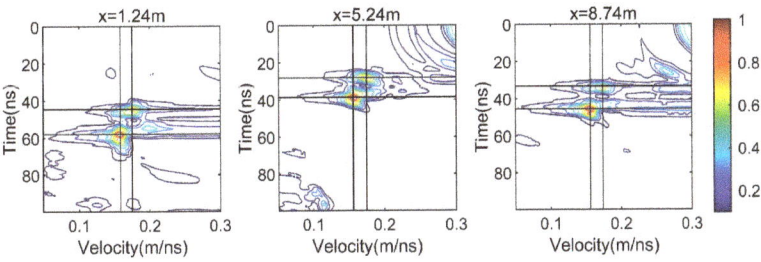

Figure 2. Velocity spectrum contour slices of three hyperbolas. The black crossed lines indicate the positions of maximum value.

Table 1. The estimated results and errors of model test.

Scatters	Upper Hyperbolas		Bottom Hyperbolas	
	Velocity (m/ns)	Error (%)	Velocity (m/ns)	Error (%)
1	0.175	1.04	0.158	8.78
2	0.175	1.04	0.156	9.93
3	0.174	0.46	0.156	9.93

3.2. Lunar Penetrating Radar (LPR) Data Processing

The walking route of Yutu-2 rover for the first day and the second day on the Moon is shown in Figure 3. The length of the path is about 105 m. Subsequently, the multi-segment data are spliced to achieve the CH-2B profile. After a series of processing (Table 2) the LPR profile is obtained, which is shown in Figure 4.

Table 2. Details of data processing.

Processing	Details
Traces amending	Adjusting the longitudinal displacement of traces, based on the phase of a strong reflection event
Traces selecting	The rover might stop at some points on the way to collect other scientific data but LPR never stops measurement, resulting in repeated acquisition of multiple traces at the same location. We average the repeated traces.
Time lag adjustment	There is a 28 ns delay for the start time of the transmitting antenna compared to the receiving antenna.
Useless data deleting	Signals after 500 ns are removed since these signals are not reliable.
Attenuation compensation Background removal	Conducting automatic gain control (AGC) to make deep information more visible Removing the average data along the rover path.
Band-pass filtering	Adopting band-pass filtering to suppress the low-frequency and high-frequency noise.

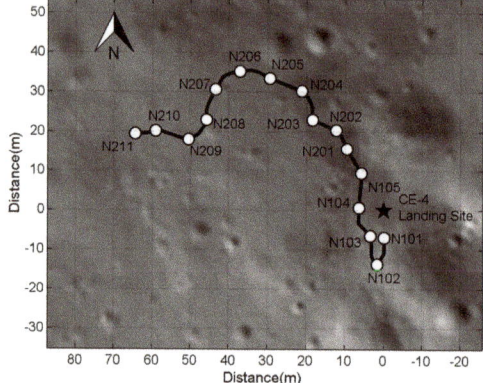

Figure 3. The walking route of Yutu-2 rover for the first day and the second day on the Moon.

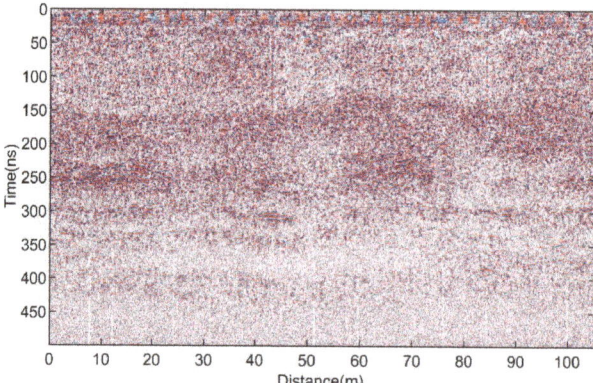

Figure 4. The CH-2B profile of lunar penetrating radar (LPR) data.

3.3. 3D Velocity Spectrum

Based on (1)–(3), the 3D velocity spectrum is computed; the range of values are limited to 0 to 1; the 3D result without compensation of $1/N_i$ is shown in Figure 5.

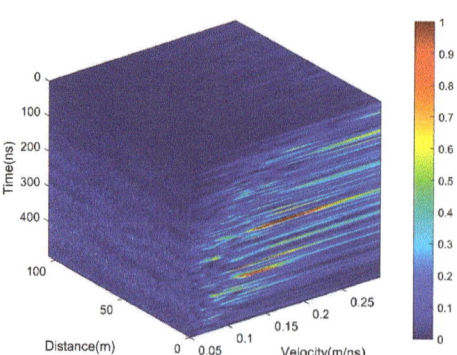

Figure 5. 3D velocity spectrum.

In order to locate the positions of hyperbolas on GPR profile, the maximum values $C_{i,j}$ along velocity axis are also figured out, which is shown in Figure 6a. Subsequently, $1/N_i$ is adopted to compensate the energy differences caused by using different windows N_i along longitudinal direction, the result is shown in Figure 6b. The result shows the energies in shallow regions are compensated.

Subsequently, the soft threshold function is also applied. The value of ε is set to 0.6. Subsequently, based on the result of soft threshold processing, the locations of local maximums are selected which are shown in Figure 7.

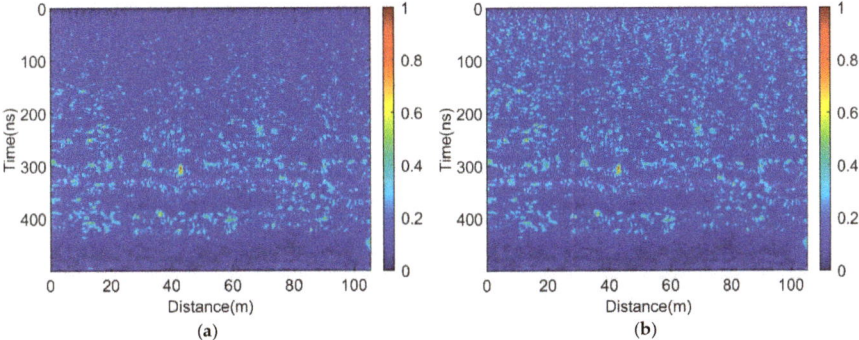

Figure 6. Maximum value along velocity axis. (a) Before compensation of $1/N_i$; (b) after compensation of $1/N_i$.

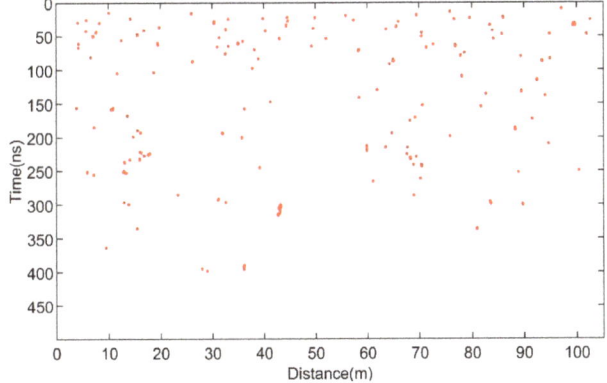

Figure 7. Initial positioning of hyperbolas based on soft threshold.

Based on Figure 7, we can search for the velocities in 3D velocity spectrum in Figure 5. However, not all the points in Figure 7 are useful, there are three reasons:

(1) For a highly flat interface, the estimated velocity will reach to 0.3 m/ns, which is obviously wrong; the lunar regolith velocity is less than 0.2 m/ns [24], so we delete the points with velocities close to 0.3 m/ns.
(2) There may be a large stacked energy even if there are no hyperbolas but interlaced regions of several hyperbolas; for this situation, we should delete the non-hyperbolic points.
(3) The hyperbolas on radargram are not curves but regions with hyperbolic shapes, so there will be several excess points at one hyperbola, which should be deleted.

After filtering using the above method, the 30 preserved hyperbolas are noted in Figure 8a; the velocity spectrum slices of these hyperbolas are shown in Figure 8b.

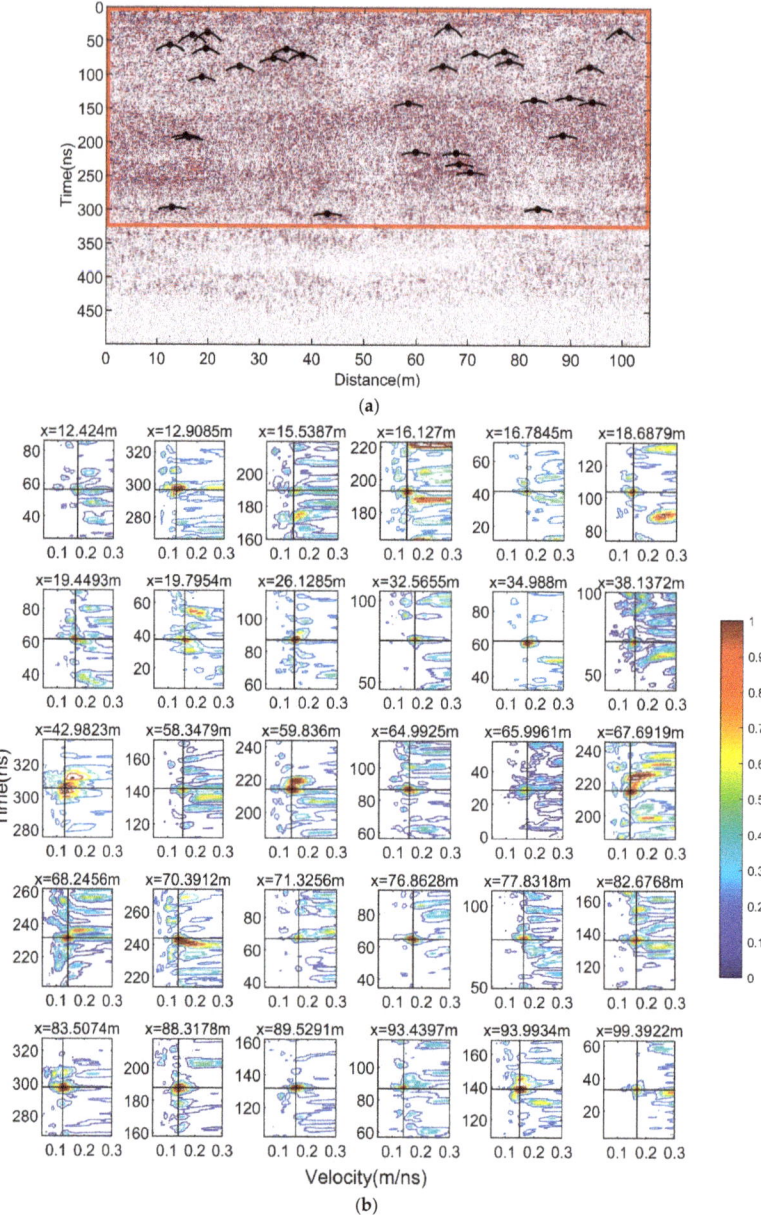

Figure 8. Result of hyperbola selection. (a) Hyperbolas on radargram; (b) velocity spectrum slices of 30 hyperbolas—the black crossed lines indicate the positions of maximum values.

3.4. Analysis of Radargram

For Figure 8a, the hyperbolas located inside of the red rectangle, the region of 320–500 ns cannot be analyzed, so we focus on the region of 0–320 ns. Based on the reflection energy, several distinct layers and special regions are noted on the radargram (Figure 9). Layer 1 is a surface layer; a distinct reflection energy difference can be seen at 25 ns. Layer 3 is the bottom layer; although its adjacent upper and lower areas show few reflections, the interface shows strong reflection energy. Layer 2 is a complex layer. The zone of 25–150 ns is relatively more uniform than the deeper zone; it contains few reflections indicating its high weathering degree. The zone of 150–300 ns is more complex possessing strong reflections; the complex signals are formed by the interlaced hyperbolas generated by the scattering of waves on rocks. Four obvious regions are selected out. The strong reflections within Regions 1 and 3 are distinct from the adjacent areas. The reflection energy in Region 2 is relatively lower than that of Regions 1 and 3, which means the material and horizontal structure changes between Regions 1 and 3. Region 4 also contains strong reflections, but its vertical size is shorter because the material and horizontal structure changes again at the position of 80 m; the material at about 250 ns in Region 3 does not extend to Region 4. Overall, the subsurface material and structure at the Chang'E-4 landing site vary both vertically and horizontally, so the simple horizontal stratigraphic division is not appropriate within the zone of 25–300 ns.

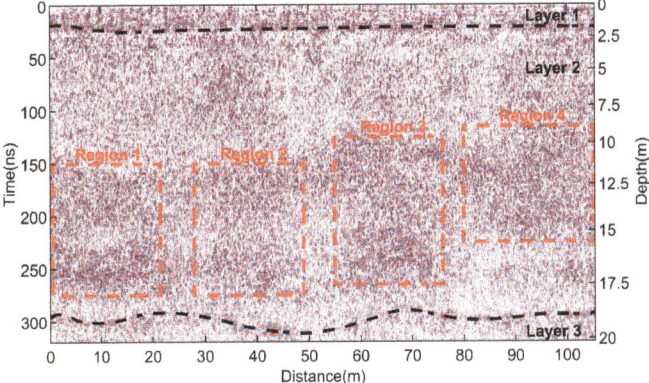

Figure 9. Analysis of radargram.

3.5. Properties Analysis

Subsequently, properties analysis is applied to LPR data. Firstly, we combine the positions and velocities of selected hyperbolas in Figure 8 and obtain a velocity scatter figure (Figure 10). The region size is the same with Figure 9, the color indicates the velocity of each point.

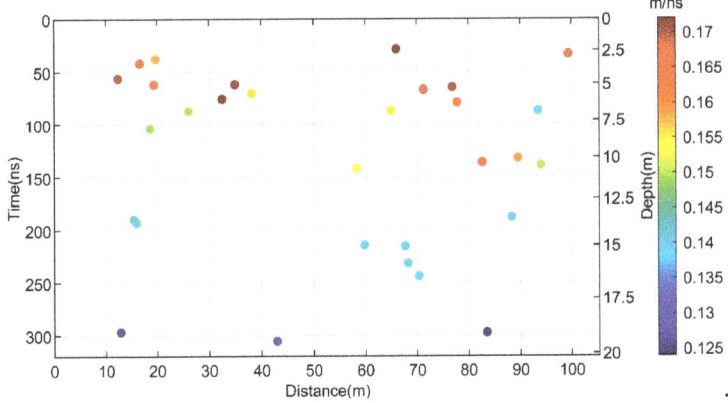

Figure 10. Velocity scatter figure.

Subsequently, the image interpolation is applied to obtain the subsurface velocity of electromagnetic wave structure, then the smoothing is performed to remove the effects of outliers. Finally, as the velocities calculated by hyperbolic fitting method are the root-mean-square velocity (RMS), the interval velocities should be derived using the Dix formula:

$$v_{int,n} = \sqrt{\frac{v_{rms,n}^2 t_n - v_{rms,n-1}^2 t_{n-1}}{t_n - t_{n-1}}} \qquad (10)$$

where $v_{int,n}$ and $v_{rms,n}$ represent the interval velocity and root-mean-square velocity of nth layer respectively, t_n is the travel time to the nth layer. By using (10), the subsurface interval velocity structure is computed, and the profile is shown in Figure 11. Subsequently, based on the velocities, the relative permittivity, density, and content of FeO and TiO_2 can be derived using (4)–(7), the results are shown in Figures 12–14, respectively.

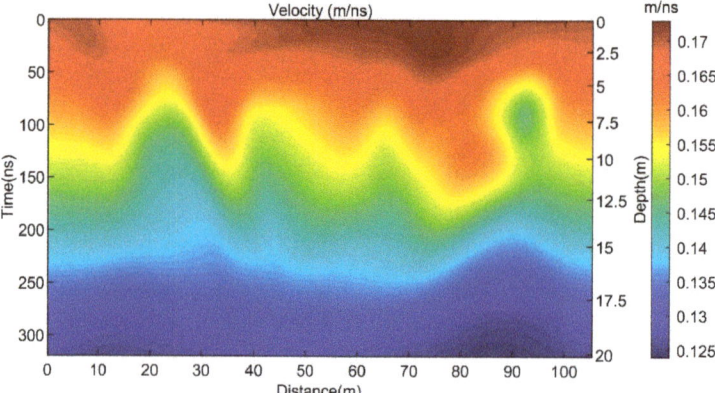

Figure 11. The subsurface electromagnetic wave velocity structure.

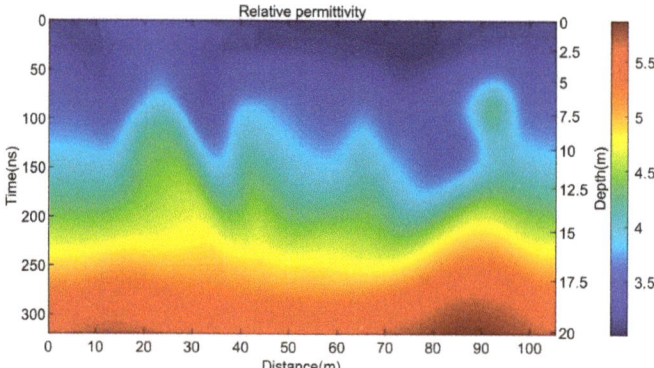

Figure 12. The subsurface relative permittivity structure.

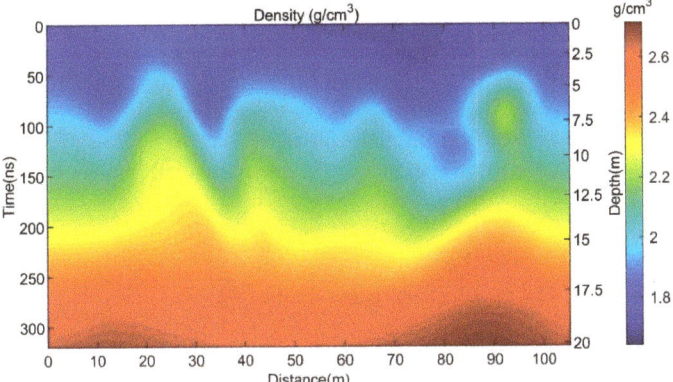

Figure 13. The density structure of lunar regolith.

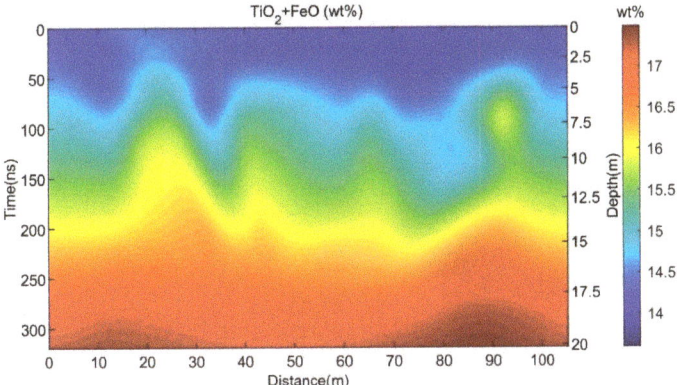

Figure 14. The TiO$_2$ and FeO content of lunar regolith.

In Figure 10, the points with different velocities distribute in different layers: 0.15 m/ns seem to be a watershed; the points whose velocities are bigger than 0.15 m/ns are mainly distributed in 0–150 ns zone; the points whose velocities are smaller than 0.15 m/ns are mainly distributed in 150–320 ns zone; this indicates the clear property difference between these two parts which can also be seen from Figures 11–14. These phenomena also correspond to the characteristic of radargram that 0–150 ns zone is of high weathering degree with few rock reflections and 150–320 ns zone is of low weathering degree with strong reflections.

Figures 11–14 show that the velocity descends as the depth increases, and the opposite varieties for velocity and relative permittivity are due to (4). In these figures, we can also find that there is a fluctuation at about 250 ns after position of 80 m, which also corresponds to the analysis in which the material and horizontal structure changes at a position of 80 m; the material about 250 ns in Region 3 does not extend to Region 4.

Previous research shows that the relative permittivity of lunar basalt is about 8 [24], so based on the velocity and relative permittivity structure we obtained, we think the region we researched (0–320 ns) is mainly the lunar regolith. The depth coordinates in Figures 9–14 are determined using average root-mean-square velocity of all traces; so the thickness of the lunar regolith is more than 20 m. In addition, the TiO$_2$ and FeO content in the bottom of Figure 14 is about 17.5 wt%, which is close to the shallow lava flow observed in [3,4], so we speculate that the surface of the basalt is within 25 m in depth.

4. Discussion

In this section, the estimation errors of two-way reflected time and velocity are discussed. Take the first slice in Figure 8b for instance, a 3D surface plot can be computed, which is shown in Figure 15a. The maximum value in this figure indicates the two-way reflected time and velocity of this hyperbola. However, some error will generate from the noise interference and incomplete hyperbola; the correct point is probably in a region; a flat at an amplitude of 0.8 is selected to split out this region. Two profiles, which are the red rectangle and black rectangle in Figure 15, are extracted out to compute the error intervals of velocity and reflected time, respectively. The two intersections of the blue curve and gray line around the maximum value are selected to compose the error limit.

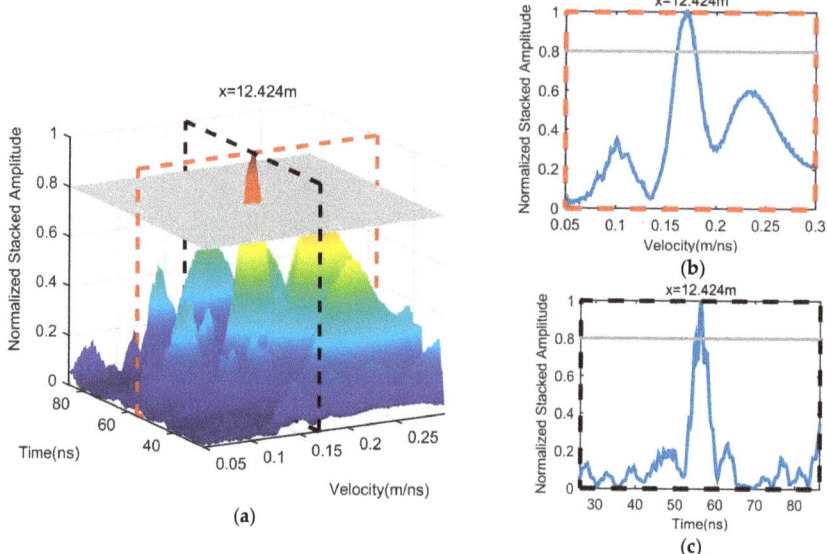

Figure 15. Error computation. (a) 3D surface plot of the first slice in Figure 8b. (b) The profile of in the red rectangle of (a). (c) The profile of in the black rectangle of (a).

In Figure 15, the error ranges of velocity and two-way reflected time are 0.161–0.179 m/ns and 55.8–56.84 ns, respectively. So, the real velocity and time may be within these two ranges. A maximum relative error (MRE) can be adopted to quantitatively describe the error:

$$\begin{cases} MRE_v = |\max(v - v_{ran})|/v \\ MRE_t = |\max(t - t_{ran})|/t \end{cases} \quad (11)$$

where v and t are the estimated values using 3D velocity spectrum, v_{ran} and t_{ran} represent the error ranges. So, the MRE_v and MRE_t in Figure 13 are 5.29% and 1.05%, respectively.

Subsequently, the errors of all hyperbolas are computed. The average MREs of time and velocity are 0.68% and 7.99%, respectively. It is clear that the error of two-way reflected time can be omitted, which is far less than the error of velocity. Furthermore, we computed the distributions of velocities and error ranges in horizontal and vertical directions which are shown in Figure 16.

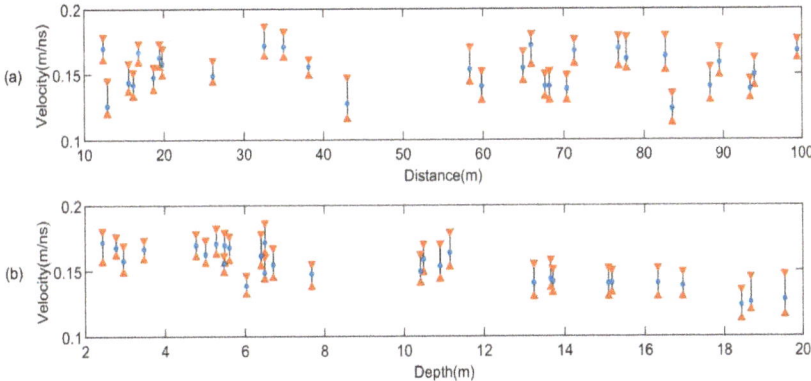

Figure 16. Distributions of velocities and error ranges. (**a**) Velocities and error ranges with distance. (**b**) Velocities and error ranges with depth. The triangular points represent the error ranges; the blue dots represent the estimated velocities.

5. Conclusions

This paper applies the 3D velocity spectrum to property analysis of lunar penetrating radar data from Chang'E-4. The result shows that 3D velocity spectrum can automatically search for hyperbolas; the maximum value along velocity axis with a soft threshold function can provide the horizontal position, two-way reflected time and velocity of each hyperbola. Based on the estimated velocities of 30 hyperbolas, the subsurface properties' structures are obtained, including velocity, relative permittivity, density, and content of FeO and TiO_2. Combining these properties and the radargram, we can conclude that 0–12 m is the fine-grained regolith; 12–20 m is the coarse regolith of different sources; the thickness of the lunar regolith is more than 20 m; the surface of the shallow basalt is within 25 m in the depth. Importantly, the subsurface material and structure at the Chang'E-4 landing site vary both vertically and horizontally, so the simple horizontal stratigraphic division is not appropriate within the zone of 25–300 ns. Finally, the estimation error is discussed, the errors of horizontal position and two-way reflected time are small which can be omitted; the average maximum relative error of velocity is 7.99%. the distributions of velocities and error ranges in horizontal and vertical directions are obtained.

Author Contributions: Conceptualization, X.F., C.L., Z.Z. and J.L.; data curation, Z.D.; formal analysis, Z.D., H.Z., C.L., Z.Z. and W.L.; funding acquisition, X.F. and W.L.; investigation, H.Z., Z.Z. and J.L.; methodology, Z.D., X.F. and H.Z.; project administration, X.F., J.L. and W.L.; resources, C.L., Z.Z., J.L. and W.L.; software, H.Z.; supervision, C.L.; validation, Z.Z.; visualization, Z.D.; writing—original draft, Z.D. and H.Z.; writing—review and editing, X.F. All authors have read and agreed to the published version of the manuscript.

Funding: This work was funded by Science and Technology on Near-Surface Detection Laboratory under Grant 6142414180911, and the Technology Development Program of Jilin Province: Suppression of Polarization Accompanying Interference for Full-Polarimetric Ground Penetrating Radar under Grant 20180101091JC, and the Fundamental Research Funds for the Central Universities.

Conflicts of Interest: The authors declare no conflict of interest.

References

1. Ling, Z.; Qiao, L.; Liu, C.; Cao, H.; Bi, X.; Lu, X.; Zhang, J.; Fu, X.; Li, B.; Liu, J. Composition, mineralogy and chronology of mare basalts and non-mare materials in Von Kármán crater: Landing site of the Chang'E4 mission. *Planet. Space Sci.* **2019**, *179*, 104741. [CrossRef]
2. Liu, J.; Ren, X.; Yan, W.; Li, C.; Zhang, H.; Jia, Y.; Zeng, X.; Chen, W.; Gao, X.; Liu, D.; et al. Descent trajectory reconstruction and landing site positioning of Chang'E-4 on the lunar farside. *Nat. Commun.* **2019**, *10*, 4229. [CrossRef]

3. Qiao, L.; Ling, Z.; Fu, X.; Li, B. Geological characterization of the Chang'e-4 landing area on the lunar farside. *Icarus* **2019**, *333*, 37–51. [CrossRef]
4. Huang, J.; Xiao, Z.; Flahaut, J.; Martiont, M.; Head, J.; Xiao, X.; Xie, M.; Xiao, L. Geological characteristics of Von Kármán crater, northwestern South Pole-Aitken Basin: Chang'E-4 landing site region. *J. Geophys. Res. Planets* **2018**, *123*, 1684–1700. [CrossRef]
5. Di, K.; Liu, Z.; Liu, B.; Wan, W.; Peng, M.; Li, J.; Xie, J.; Jia, M.; Niu, S.; Xin, X.; et al. Topographic Analysis of Chang'E-4 Landing Site Using Orbital, Descent and Ground Data. In Proceedings of the International Archives of the Photogrammetry, Remote Sensing and Spatial Information Sciences, Enschede, The Netherlands, 10–14 June 2019; pp. 1383–1387.
6. Hu, X.; Ma, P.; Yang, Y.; Zhu, M.; Jiang, T.; Lucey, P.G.; He, Z. Mineral abundances inferred from in situ reflectance measurements of Chang'E-4 landing site in South Pole-Aitken basin. *Geophys. Res. Lett.* **2019**, *46*, 9439–9447. [CrossRef]
7. Xiao, L.; Zhu, P.; Fang, G.; Xiao, Z.; Zou, Y.; Zhao, J.; Zhao, N.; Yuan, Y.; Qiao, L.; Zhang, X.; et al. A young multilayered terrane of the northern Mare Imbrium revealed by Chang'E-3 mission. *Science* **2015**, *347*, 1226–1229. [CrossRef] [PubMed]
8. Fa, W.; Zhu, M.; Liu, T.; Plescia, J.B. Regolith stratigraphy at the Chang'E-3 landing site as seen by lunar penetrating radar. *Geophys. Res. Lett.* **2015**, *42*, 10,179–10,187. [CrossRef]
9. Dong, Z.; Fang, G.; Ji, Y.; Gao, Y.; Wu, C.; Zhang, X. Parameters and structure of lunar regolith in Chang'E-3 landing area from lunar penetrating radar (LPR) data. *Icarus* **2017**, *282*, 40–46. [CrossRef]
10. Feng, J.; Su, Y.; Ding, C.; Xing, S.; Dai, S.; Zou, Y. Dielectric properties estimation of the lunar regolith at CE-3 landing site using lunar penetrating radar data. *Icarus* **2017**, *284*, 424–430. [CrossRef]
11. Su, Y.; Fang, G.; Feng, J.; Xing, S.; Jing, Y.; Zhou, B. Data processing and initial results of Chang'E-3 lunar penetrating radar. *Res. Astron. Astrophys.* **2014**, *14*, 1623–1632. [CrossRef]
12. Wang, K.; Zeng, Z.; Zhang, L.; Xia, S.; Li, J. A compressive-sensing-based approach to reconstruct regolith structure from lunar penetrating radar data at the Chang'E-3 landing site. *Remote Sens.* **2018**, *10*, 1925. [CrossRef]
13. Jia, Z.; Liu, S.; Zhang, L.; Hu, B.; Zhang, J. Weak Signal Extraction from Lunar Penetrating Radar Channel 1 Data Based on Local Correlation. *Electronics* **2019**, *8*, 573. [CrossRef]
14. Zhang, L.; Zeng, Z.; Li, J.; Lin, J.; Hu, Y.; Wang, X.; Sun, X. Simulation of the Lunar Regolith and Lunar-Penetrating Radar Data Processing. *IEEE J. Sel. Top. Appl. Earth Obs. Remote Sens.* **2018**, *11*, 655–663. [CrossRef]
15. Zhang, J.; Zeng, Z.; Zhang, L.; Lu, Q.; Wang, K. Application of Mathematical Morphological Filtering to Improve the Resolution of Chang'E-3 Lunar Penetrating Radar Data. *Remote Sens.* **2019**, *11*, 524. [CrossRef]
16. Lai, J.; Xu, Y.; Zhang, X.; Tang, Z. Structural analysis of lunar subsurface with Chang'E-3 lunar penetrating radar. *Planet. Space Sci.* **2016**, *120*, 96–102. [CrossRef]
17. Zhang, L.; Zeng, Z.; Li, J.; Huang, L.; Huo, Z.; Wang, K.; Zhang, J. Parameter Estimation of Lunar Regolith from Lunar Penetrating Radar Data. *Sensors* **2018**, *18*, 2907. [CrossRef] [PubMed]
18. Yilmaz, Ö. *Seismic Data Analysis*, 3rd ed.; Society of Exploration Geophysicists: Tulsa, Oklahoma, 2001.
19. Feng, X.; Sato, M. Pre-stack migration applied to GPR for landmine detection. *Inverse Probl.* **2004**, *20*, S99–S115. [CrossRef]
20. Fisher, E.; McMechan, G.A.; Annan, A.P. Acquisition and processing of wide-aperture ground-penetrating radar data. *Geophysics* **1992**, *57*, 495–504. [CrossRef]
21. Greaves, P.J.; Lesmes, D.P.; Lee, J.M.; Toksoz, M.N. Velocity variations and water content estimated from multi-offset, ground-penetrating radar. *Geophysics* **1996**, *61*, 683–695. [CrossRef]
22. Leparoux, D.; Gibert, D.; Côte, P. Adaptation of prestack migration to multi-offset ground-penetrating radar (GPR) data. *Geophys. Prospect.* **2001**, *49*, 374–386. [CrossRef]
23. Hu, B.; Wang, D.; Zhang, L.; Zeng, Z. Rock Location and Quantitative Analysis of Regolith at the Chang'e 3 Landing Site Based on Local Similarity Constraint. *Remote Sens.* **2019**, *11*, 530. [CrossRef]
24. Sen, P.N.; Scala, C.; Cohen, M.H. A self-similar model for sedmentary rocks with application to the dielectric constants for fused glass beads. *Geophysics* **1981**, *46*, 781–795. [CrossRef]

© 2020 by the authors. Licensee MDPI, Basel, Switzerland. This article is an open access article distributed under the terms and conditions of the Creative Commons Attribution (CC BY) license (http://creativecommons.org/licenses/by/4.0/).

Letter

Combined Study of a Significant Mine Collapse Based on Seismological and Geodetic Data— 29 January 2019, Rudna Mine, Poland

Maya Ilieva [1,*], Łukasz Rudziński [2], Kamila Pawłuszek-Filipiak [1], Grzegorz Lizurek [2], Iwona Kudłacik [1], Damian Tondaś [1] and Dorota Olszewska [2]

1. Institute of Geodesy and Geoinformatics, Wrocław University of Environmental Life and Sciences, 50-357 Wrocław, Poland; kamila.pawluszek-filipiak@upwr.edu.pl (K.P.-F.); iwona.kudlacik@upwr.edu.pl (I.K.); damian.tondas@upwr.edu.pl (D.T.)
2. Institute of Geophysics, Polish Academy of Sciences, 01-452 Warszawa, Poland; rudzin@igf.edu.pl (Ł.R.); lizurek@igf.edu.pl (G.L.); dolszewska@igf.edu.pl (D.O.)
* Correspondence: maya.ilieva@upwr.edu.pl

Received: 27 March 2020; Accepted: 12 May 2020; Published: 15 May 2020

Abstract: On 29 January 2019, the collapse of a mine roof resulted in a significant surface deformation and generated a tremor with a magnitude of 4.6 in Rudna Mine, Poland. This study combines the seismological and geodetic monitoring of the event. Data from local and regional seismological networks were used to estimate the mechanism of the source and the ground motion caused by the earthquake. Global Navigation Satellite System data, collected at 10 Hz, and processed as a long-term time-series of daily coordinates solutions and short-term high frequency oscillations, are in good agreement with the seismological outputs, having detected several more tremors. The range and dynamics of the deformed surface area were monitored using satellite radar techniques for slow and fast motion detection. The radar data revealed that a 2-km^2 area was affected in the six days after the collapse and that there was an increase in the post-event rate of subsidence.

Keywords: mine collapse; anthropogenic hazard; seismology; GNSS; InSAR

1. Introduction

Rock bursts and collapses are the most dangerous phenomena associated with seismicity in underground mines. The collapse of a mine roof can follow strong and shallow seismic events of magnitudes M > 3 [1–4]. They pose a serious threat to life for miners working in the vicinity of the collapsing rock. The strongest underground mining-induced seismic events are Volkershausen [5] and Newcastle [6], both M5.6 Although the magnitudes of such shallow seismic events are usually small in comparison with natural earthquakes, the ground shaking might be felt by local citizens and can damage surface infrastructure. Rock-burst issues have occurred in the Polish copper region known as the Legnica Głogów Copper District (LGCD). Thousands of events of M > 1.0 are recorded every year in the LGCD. Among these, several are followed by rock bursts or tunnel collapses. Both ground shaking (i.e., a seismic event) and local ground failure occur as consequences of mining. While the shaking effects are comparable to what can be felt during small to moderate natural earthquakes, the surface deformation is specific to mining areas. The subsidence is either a ground response to the closure of excavated mining panels, e.g., [7–9], or is related to strong tremors [10,11]. In this paper, we present a joint seismological and geodetic study that expands our knowledge concerning the connection between mine collapses and the ground response to seismic sources manifested by permanent ground deformations.

The analysis focused on the strongest seismic event recorded by LGCD in the last several years—namely, the 29 January 2019 M4.6 collapse in the Rudna Mine, Poland. The phenomenon was widely felt and was followed by massive collapse. While the seismological analyses were carried out using records from local and regional seismic networks, the geodetic data are comprised of Global Navigation Satellite System (GNSS) observations collected at station LES1 (Komorniki; part of the European Plate Observing System project [EPOS-PL] network) and the European Space Agency's (ESA's) Sentinel-1 radar observations. The LES1 station was aligned with a strong motion sensor (KOMR), part of the local seismic system, the Legnica–Głogow Underground Mining INduced Earthquake Observing System (LUMINEOS), located above the mine. Interferometric Synthetic Aperture Radar (InSAR) techniques were applied to analyse the co-seismic surface deformation.

2. The Collapse and Seismological Analyses

On 29 January 2019, Rudna Mine in western Poland was struck by a seismic event followed by a rock burst. It was one of a few such events in the area in recent years [4,12,13]. Using seismic signals recorded by the LUMINEOS surface local seismic network (Figure 1a) and a local velocity model [14], the epicentre was located at 51.5110°N, 16.1197°E. For our analysis, we assumed that the event depth was located at the level of the excavation; that is, 1 km below the surface. LUMINEOS consisted of 10 strong motion sensors (AC-73) and 17 five-second VE-53/BB instruments (both types manufactured by GeoSIG, Schlieren, Switzerland).

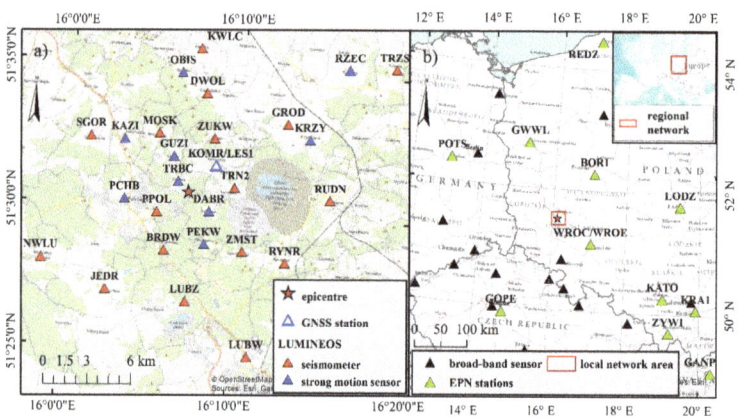

Figure 1. The seismic networks employed: (**a**) Local LUMINEOS network—strong motion sensors (blue triangles), seismometers (red triangles) and location of the KOMR GNSS station, LES1, aligned with a strong motion sensor. The epicentre is marked by a red star. (**b**) Regional network of broad-band sensors (black triangles) used to estimate the source mechanism, the closest EPN GNSS stations and the WROE reference station (green triangles). Location of these networks within Europe (inset).

The source mechanism of the event was estimated using seismic signals recorded by broad-band seismometers (STS2 or Güralp CMG3ESP; Figure 1b) at regional distances of 70 to 300 km. The signals and corresponding station responses were obtained from the ORFEUS-EIDA project page [15]. The inversion was performed using the real displacement, derived from the original signals, for 11 seismic stations located around the epicentre (Figure 1b). The mechanism (Figure 2a) was estimated in terms of the full moment tensor, which can be directly used to interpret the mechanisms of mining tremors [16]. The Pyrocko toolbox [17] and KIWI tool software [18] were used for the calculations. The software had already been tested as a tool for determining a stable point source model in mines [19], including mine collapses [4,11]. In this work, we followed the procedure previously used in [4]. The inversion was based on the full waveform displacement and spectra in the specific frequency range. The regional velocity model proposed by [20] was used. The lowest misfit was obtained for data filtered between 0.07 and 0.11 Hz. The final moment tensor solution was decomposed according to [21] into its isotropic and deviatoric parts, and ultimately into the double-couple and compensated linear vector dipole. The resulting moment tensor (m11: -4.03×10^{15} m22: -3.91×10^{15} m33: -11.9×10^{15} m12: -0.07×10^{15} m13: -0.11×10^{15} m23: -0.22×10^{15} Nm) was dominated by the implosive component (isotropic part: -55%). This kind of mechanism can be explained by a tabular cavity collapse [4,22], and it was strongly supported by the observations recorded on the strong motion sensors belonging to the LUMINEOS network(see Supporting Material: Table S1). Figure 2 shows the distribution of the peak ground acceleration around the epicentre. The clearly visible peak ground acceleration pattern corresponds very well with the collapse source model.

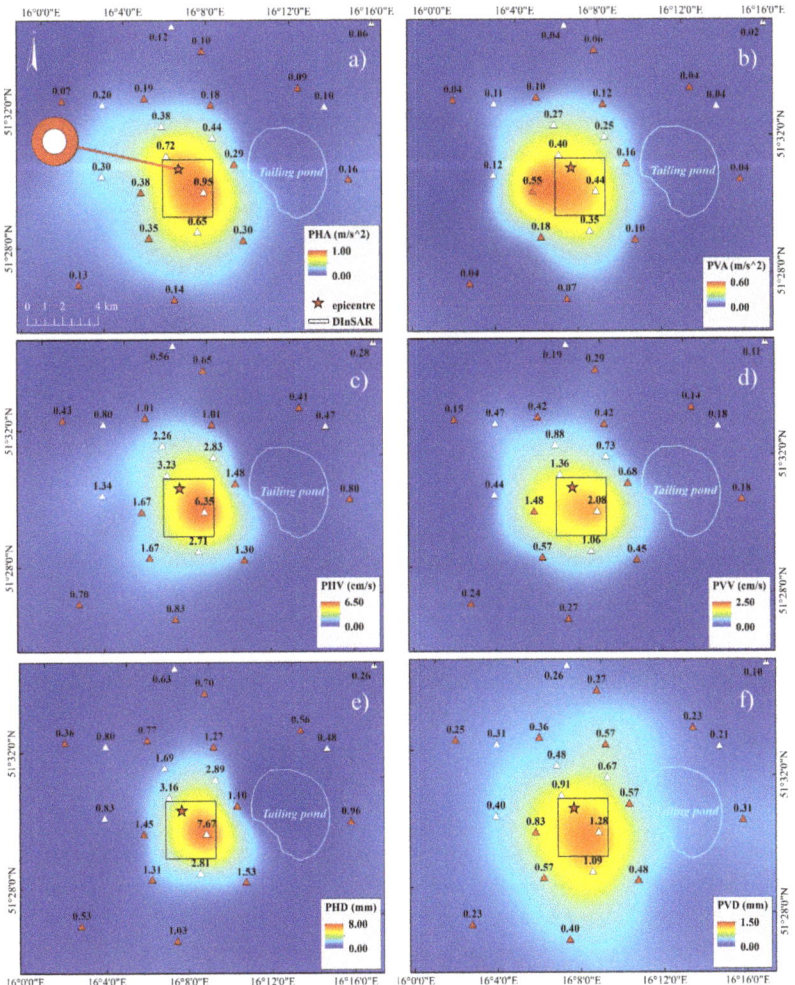

Figure 2. Peak ground acceleration: (**a**) horizontal (PHA) and (**b**) vertical (PVA); peak ground velocity: (**c**) horizontal (PHV) and (**d**) vertical (PVV); and peak ground displacement: (**e**) horizontal (PHD) and (**f**) vertical (PVD) estimated from the local LUMINEOS network record, represented as triangles. Red—seismometers. White—strong motion sensors. The location of the estimated epicentre is indicated by a star, and the beach ball represents the focal mechanism in (**a**). The black rectangle shows the range of the DInSAR plots illustrated in Section 4.2.

3. GNSS Short- and Long-Term Monitoring

The deformations of mining-induced areas might be monitored by GNSS sensors. In the literature, some research concerning the ground deformation monitoring can be found (e.g., [23–25]). However, for short-term oscillations, the papers usually are related to natural earthquake analysis (e.g., [26–29]). For the current study, the GNSS observations at station LES1, collected at 10 Hz from all visible GPS, GLONASS, Galileo and Bediou satellites, were processed in two different modes: (1) as a long-term network daily solution, producing high-quality, daily-coordinates time-series for a period of six months (three before and three after the event); and (2) short-term, high-frequency oscillations in the horizontal and vertical coordinates, observed during the event, to produce a high-frequency picture of the seismic wave passage.

3.1. Long-Term Monitoring

The daily solution was performed for station LES1 using Bernese GPS Software v.5.2 [30]. Selection of the processing parameters was based on the standard EUREF Permanent GNSS Network (EPN) processing strategy in [31], and they included the following: reference frame: ITRF 2014; orbits and clocks: CODE Final; ionosphere: CODE GLOBAL IONOSPHERE; troposphere: ZTD a priori and mapping function: Vienna Mapping Function 1; reference stations selected from EPN, independent vectors; coordinates and velocities for reference sites: EPN_A_IGS14.SNX; solution type: daily, minimum constrain on translation; estimated parameters: daily coordinates transformed into east, north, up (ENU), troposphere delays, and gradients; time span: 1 November 2018–30 April 2019; quality of estimation measured by mean Helmert transformation root-mean-square error on the daily boundaries: 2.6 mm.

The solution, based on double differences and an independent network of vectors, showed a high level of consistency between consecutive days, as confirmed by a low Helmert transformation RMS. The daily solutions were not combined into a full-period solution, so as to allow the detection of jumps related to post-seismic deformations; therefore, the impact of the plate velocities on the reference stations was removed using the EPN solution [31]. The constant velocities in each component were estimated. The analysis of six months of coordinate time-series (Figure 3a) showed that, for the north component (top panel), five significant (larger than 2σ) jump events were recorded. For the east component (middle panel), it was five events, and for the up component, it was nine events. For the investigated case (red line in Figure 3a), none of the estimated coordinates exceeded 2σ threshold. However, other events (marked as a grey dashed line in Figure 3a) coincide with significant outliers; further studies should investigate whether there is a connection between deformations and induced seismic events.

3.2. Short-Term Monitoring

The high-rate GNSS data was recorded by station LES1, which was collocated with the KOMR accelerometer, 2.5 km north-east of the epicentre. Two hours of 10-Hz GPS data were processed using GAMIT/GLOBK v.10.9 software [32] and the double-differencing method in kinematic mode. The WROE reference station was located 79 km south-east of the epicentre, outside the deformation zone. For processing, CODE final orbits and differential code biases were used. The phase and code observables were combined to remove ionospheric refraction by the Melbourne–Wübbena combination [33,34]. Additionally, the Universitat Politècnica de Catalunya's rapid Global Ionosphere Maps (UQRG) was used, which significantly reduced the noise in the resultant position time-series. The UQRG model had a 5.0 × 2.5 spatial resolution and a 15-minute temporal resolution [35].

The geocentric coordinates (XYZ) obtained were transformed into local topocentric (ENU) coordinates and displacements. To compare the GNSS-displacement time-series with the positions calculated from the accelerometric data, both time-series were reduced to the period starting 30 s before the event and finishing 90 s after (Figure 3b). The low-frequency noise of the GPS position time-series was removed using the second-order Butterworth filter and the high-pass frequency set to 0.15 Hz, based on the Fourier spectra of the seismological position time-series [27]. This reduced the standard deviations of the GPS position time-series from several millimetres to less than 1 mm. Pearson's correlation coefficient reached 0.93, 0.89 and 0.85 for the E, N and U displacement time-series, respectively. This indicates good agreement between both time series.

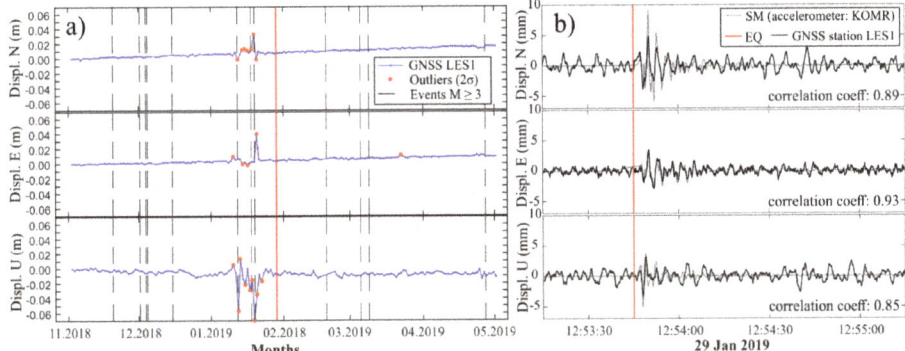

Figure 3. GNSS estimates in north, east, and up (top to bottom) components: (**a**) long-term daily coordinates estimates; (**b**) DD-GNSS and accelerometer displacement comparison. The red vertical line shows the moment of the roof collapse. The vertical dashed lines in (**a**) show that other earthquakes with a magnitude more than 3 happened in the radius of 5 km around the LES1 station.

4. InSAR Data and Analyses

Two InSAR techniques were used to obtain the areal pattern of the surface deformation: (1) a small baseline subset (SBAS) [36] for slow-rate displacements; and (2) a conventional differential InSAR (DInSAR) [37] for fast-motion monitoring around the epicentre. The selected data had the following parameters: orbits: ascending (track 73) and descending (track 22); type of images: Sentinel-1A/B (C-band), IW mode, VV polarization, SLC format; resolution: ~2.3 m in range, ~13.9 m in azimuth; time span: 28 October 2018 to 20 April 2019; revisiting time: six days, with the exception of one ascending 12-day pair (23 January–4 February 2019). In both techniques, SBAS and DInSAR, the differential interferograms were filtered with a Goldstein filter and were unwrapped using the minimum cost flow (MCF) phase unwrapping procedure [38]. The received phase differences present the surface deformation in the line-of-sight (LOS) direction to the satellite in grid images with 20-m resolution. All the interferograms were vertically translated to a common reference point, namely station LES1 (Figure 1a), and were roughly converted to vertical displacements by the division to cosine of the incidence angle. The 1 arc-second (30 m) SRTM-DEM [39] was used to remove the topographic phase component.

4.1. Slow-Rate Motion Monitoring with SBAS

This technique is typically applied for monitoring slow motions and strain accumulation. It faces some limitations regarding the capability to detect surface deformation rates above certain values depending on the wavelength of the radar signal [40]. Additionally, a bigger part of the studied area is covered by vegetation, thus it is very challenging to investigate it by using stacking-based methods (like SBAS), which depend on a long time coverage and are therefore sensitive to temporal decorrelation. Considering that SBAS has an advantage to estimate atmospheric components, and DInSAR is fully capable of detecting co-seismic displacement (see Section 4.2), these two techniques are complementary to each other and can also be evaluated with each other internally since their results are processed separately using two different softwares—SARscape [41] for SBAS and SNAP [42] for

DInSAR. Thus, in SBAS processing, a stack of multi-looked interferograms was applied according to the typical SBAS method [43,44], which is optimised for resolution cells containing distributed scatterer (DS), which are more effective in vegetated areas. Additionally, to maximize the coherence and to avoid incorrectness in the series due to a short rapid movement of the collapse, we divided our dataset for the SBAS processing into two independent datasets from descending images, from three months before (15 images) and three months after the event (14 images). The maximum critical baseline percentage was set to 2%, and the maximum temporal baseline was set to 30 days. The low-quality interferograms were removed from further processing. Finally, 36 and 21 interferograms were used for interferometric component estimation for the first (Figure 4a) and second (Figure 4b) datasets, before and after the event, respectively.

It was not possible to capture the velocity in the centre of the subsidence bowl due to the forest land cover in that area (Figure 4c), which presented an obstacle for the C-band signal. Nevertheless, the results showed a slight acceleration of the movements around the main zone of deformation in the second period, and more prominent vertical displacement, in the range of 80 cm/year, in the most affected area after the earthquake.

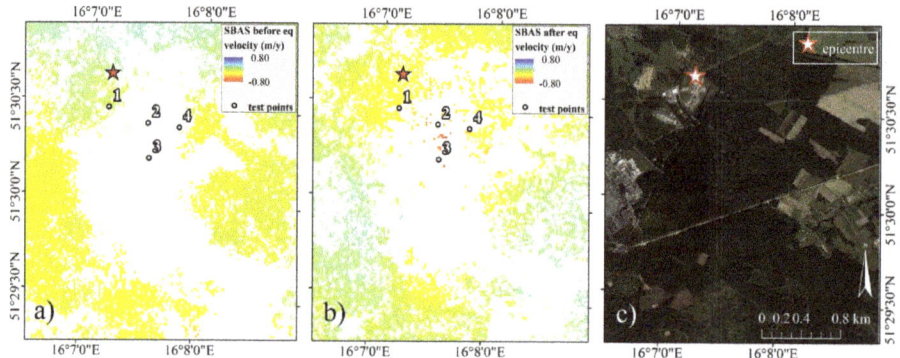

Figure 4. SBAS velocity pattern (**a**) before the event and (**b**) after the event; (**c**) land cover in the area of the subsidence (source: Google Earth).

4.2. Earthquake Surface Deformation Detected by DInSAR

For the DInSAR, the Sentinel-1 data were processed following the standard consecutive approach using the smallest temporal six-day baselines, with the exception of the one 12-day ascending pair mentioned earlier. Differential interferograms of adjacent SAR acquisitions were calculated and accumulated to provide complete time-series interferometric results. This technique uses small temporal baselines which minimize temporal decorrelation [44,45]. Vertical displacements at each pixel acquired by DInSAR estimates are presented in Figure 5.

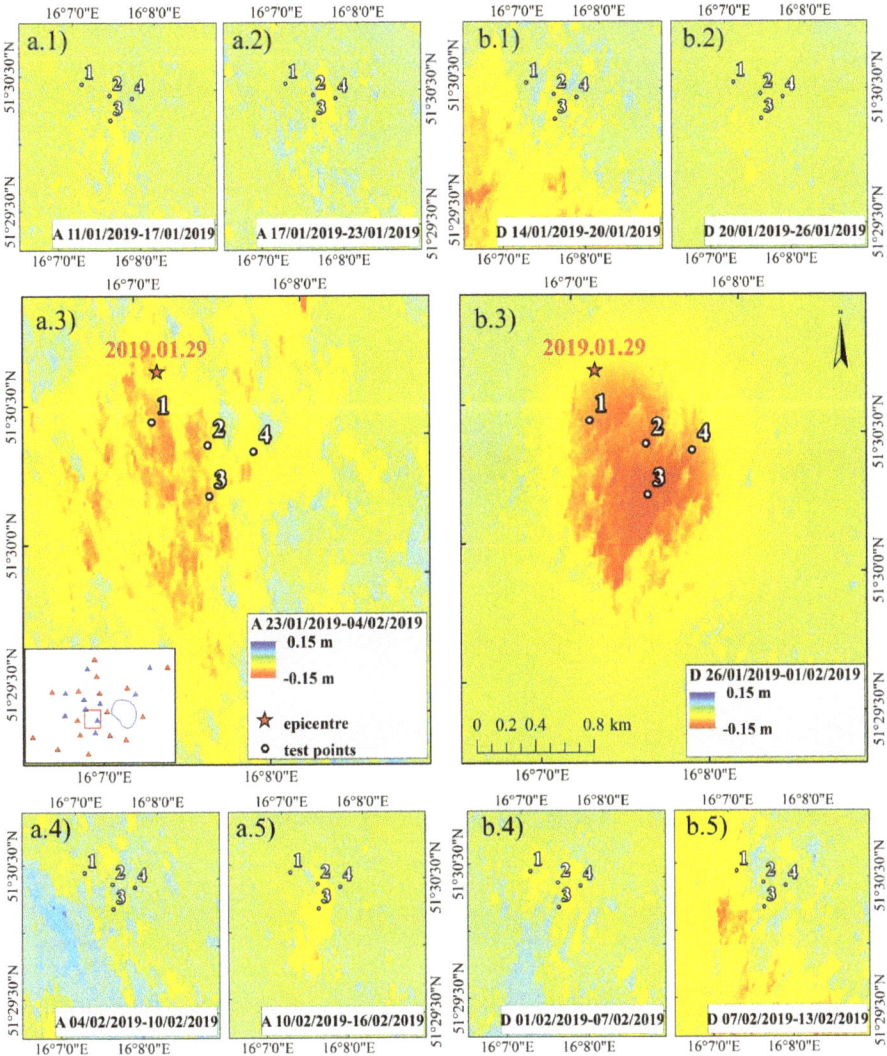

Figure 5. Vertical deformation patterns derived from the ascending (**a**) and descending (**b**) DInSAR interferograms: (**1**) and (**2**) pre-seismic; (**3**) co-seismic—(a.3) 12-day and (b.3) six-day; and (**4**) and (**5**) post-seismic. The images used for each interferogram are shown in the format DD/MM/YYYY. The red star represents the epicentre. The numbered dots are the points chosen for time-series comparison (Figure 6). The inset in (a.3) shows the footprint (red rectangle) of the DInSAR results over the local LUMINEOS network. The blue polygon represents the Żelazny Most tailings pond.

4.3. InSAR Time Series

Four test points (Figures 4 and 5) were selected in order to compare the SBAS and DInSAR vertical subsidence velocities before and after the seismic event (Figure 6a). Three of the test points (1, 2 and 4) coincided with ground objects with solid construction, such as roads and buildings, lying within highly coherent (≥0.9) pixels. Point 3 was chosen in the area of maximal detected SBAS subsidence. It was captured only in the second SBAS set.

The displacement values from the ascending and descending DInSAR deformation maps were extracted and compared by time-series (Figure 6b). The mean values of the vertical displacement in the time-series graphs were calculated by linear interpolation.

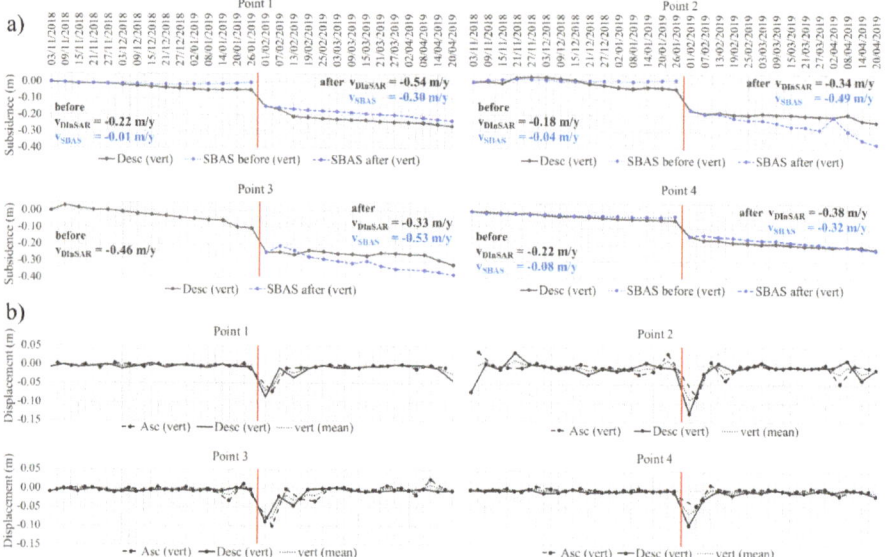

Figure 6. Time-series for selected points (Figures 4 and 5), showing: (**a**) the vertical subsidence for the period three months before and three months after the earthquake (red line), extracted from the descending SBAS (dashed blue lines) and DInSAR (solid black line); (**b**) the vertical displacement for the same period, extracted from ascending (dashed line) and descending (solid line) DInSAR data. The dotted line represents the mean value of the vertical displacements.

5. Discussion and Conclusions

Mining-induced seismicity and the associated ground deformation, measured by advanced geodetic observations, have not been studied by many authors, except in the cases of controlled salt mine cavity collapse [46–48], examples in the Upper Silesia Coal Basin [7,8,11,49,50], and in the LGCD [51,52]. Previous works concerning ground deformation studies in mining areas, however, were focused on geodetic data only (i.e., remote sensing or GNSS) ([46–48,51,53] and references therein) or used just very basic seismological parameters, usually epicentre locations provided by various seismological centres [13,50]. Some specific models of ground deformations have also been considered previously [54]. Since a mining collapse is actually an isotropic seismic source, these models seem to be in very good agreement with the seismological model of mining collapse. On the other hand, the impact of mining tremor deals with a very minor number of existing subsidence basins in mining areas. Hence, it is difficult to judge, using the geodetic data only, whether the subsidence is directly connected with a tremor or not. In all these studies, it was postulated that the increased seismicity manifested itself on the ground as a higher subsidence rate, with the measurable effects of post-seismic vertical deformation reaching up to 10 cm [52]. We applied several different techniques in order to gain a better understanding of the characteristics of the strong anthropogenic tremors induced by mining activities. These results not only support similar previous studies, but also contribute new findings.

First of all, we determined that shallow-induced seismic events with magnitudes of M4 and above can manifest not only as a relatively strong ground motion but can also provoke significant surface deformation in a very short time. Both effects have been observed in mining areas [6,11], but rarely in such a detailed way as presented here. The combination of different long- and short-time monitoring techniques provided an opportunity to track the process of evolving surface deformation in a complex regime of mining subsidence and anthropogenic seismicity in detail. Thanks to the GNSS and InSAR analysis, the rate and delineation of the subsidence was determined quite well, and we confirmed the rapid subsidence of the ground in relation to the mine roof collapse events.

High-rate GNSS observations have been applied in earthquake analysis and structural health monitoring. Previous studies of seismogeodesy have been concerned with large natural earthquakes with magnitudes exceeding M4.5 and amplitudes reaching decimetres and even meters (e.g., [27,55–57]). The event from 29th January is the first example presented of HR-GNSS application in short-term deformations caused by induced earthquakes. Studies related to induced shocks have involved long-term displacements and standard sampling frequencies of up to 1 Hz only [58]. In this study, the greater potential of the HR-GNSS for small events with centimetre-scale amplitudes was confirmed by a comparison with the seismological data. The short-term GNSS observation revealed the possibility of a wider interpretation of the source mechanism in the future. Data from the collocated GNSS and seismic sensors showed very good agreement in the displacement time-series, which indicates that the GNSS stations can be used for additional, alternative data, recorded around an epicentre in an area with sparse seismic sensors. On the other hand, the long-term GNSS study revealed the possible detection of other high-energy seismic events in the area. However, this is highly dependent on the distance between the GNSS station and the seismic sources and requires extended investigation.

The spatial range of the surface deformation derived from the InSAR studies corresponded well with the estimated location and focal mechanism of the event. The DInSAR results showed an undoubted capability to detect and delineate the scale of the surface deformations caused by such an event, as proved by the comparison between the ascending and descending data, even though the ascending co-seismic interferogram (12-day) had a lower quality. The combination of the SBAS and DInSAR results demonstrated the intensive subsidence with varying velocity associated with mining activities. Continuous subsidence occurred in the area at different velocities, depending on the dynamics of the material being extracted and also the local seismicity.

In principle, the source parameters can be derived directly or indirectly from seismograms recorded around the hypocenter. The most important and interesting in the case studied in this work is the source mechanism, which in fact is quite different in comparison with natural earthquakes that occurred on faults. In such regions, the dominance of the double couple mechanisms is observed, and no isotropic sources are expected. Concerning the mechanism, since the subsidence was completed in a very short time after the quake happened, the satellite observations can be used as an indicator for the source type determination. In our case, the geodetic studies strongly supported the full moment tensor estimated with seismological analysis. We also gained better knowledge about the size of the rupture thanks to the additional study of the seismic and geodetic data. This is quite unique in the case of seismicity related to underground mining. Our results also help in better understanding the influence of the collapse like events on surface vibration measured in terms of PGA/PGV. By geodetic techniques, we investigated the surface for possible deformations before and after the event in order to verify the proper detection of the affected area. Last, but not least, it was previously shown that seismic sources with visible rock bursts (e.g., tunnel collapses) in this area are complex (see, e.g., [4,12]). This is not obvious seismological observation and is crucial to understand the phenomena in the future. Since the subsidence is directly connected with the collapse, using geodetic analysis we would be ready to say more about the source physics.

The outcomes from the current study confirm the potential of the applied techniques for investigating ground vibrations and the surface deformation related to them. The studied event highlighted the need for a more detailed study, focused on several areas, starting with the highly complex nature of the ground deformations in areas where mining activities occur. The combination of factors provoking such ground changes, such as the very fast motions caused by moderate tremors with different magnitudes, the fast and slow nonlinear subsidence of the surface triggered by material extraction, and also the change in subsidence rates between the pre- and post-earthquake periods, the local geological structures, underground water-level changes, etc., produce a complex dynamic regime at the mine surface and in the behaviour of the cavity. The combination of different techniques provides a better understanding of the mining and its induced seismicity influence on the surface.

To increase the impact of such joint investigations, several recommendations can be made. A high density of GNSS stations, in areas with significant seismic risk also provoked by human activities, is recommended. A high density will allow a more detailed, long-term study of the horizontal and vertical trends of the surface in the area. On the other hand, the results from short-term GNSS measurements could be used to densify and verify the seismic recordings. Combined observation of long time series of seismological and geodetic data (GNSS, InSAR) could reinforce the possibility of detection of pre-seismic markers, enabling it to become a method for developing an early warning system in mining areas.

Undoubtedly, the investigation of this event has provided a new way for analysing hazardous rock bursts in underground mines. The current study has proved the high potential for obtaining a broader understanding of anthropogenic hazards by combining different Earth monitoring techniques. Joint geodetic and seismological observations can complement each other and help to overcome the individual disadvantages of the separate methods. This approach can be recommended as a complex tool, which can be used to gain a better understanding of the seismicity associated with all kinds of anthropogenic activities. The method would be especially useful in areas with sparse seismic monitoring and/or regions with already known isotropic source mechanisms, such as mining or volcanic.

Supplementary Materials: The following are available online at http://www.mdpi.com/2072-4292/12/10/1570/s1, Table S1: Ground motion parameters estimated for LUMINEOS.

Author Contributions: Conceptualization, M.I., Ł.R., G.L. and D.O.; methodology, M.I., Ł.R., K.P.-F., G.L., I.K., D.T. and D.O.; validation, M.I., Ł.R., G.L. and D.O.; formal analysis, M.I., Ł.R., K.P.-F., G.L., I.K., D.T. and D.O.; investigation, M.I., Ł.R., K.P.-F., G.L., I.K., D.T. and D.O.; resources, D.O.; data curation, M.I., Ł.R., K.P.-F., G.L., I.K., D.T. and D.O.; writing—original draft preparation, M.I., Ł.R., K.P.-F., G.L., I.K., D.T. and D.O.; writing—review and editing, M.I., Ł.R., G.L. and D.O.; visualization, M.I., Ł.R., K.P.-F. and D.O.; supervision, Ł.R.; project administration, G.L. and D.O.; funding acquisition, G.L. and D.O. All authors have read and agreed to the published version of the manuscript.

Funding: This study was conducted under the umbrella of the EPOS-PL project, POIR.04.02.00-14-A003/16, funded by the Operational Program Smart Growth 2014–2020, Priority IV: Increasing research potential, Action 4.2: Development of modern research infrastructure in the science sector. It was co-financed by the European Regional Development Fund. GNSS station LES1 was established as part of the EPOS-PL scientific geodetic infrastructure. This work was also partially financed by the Polish Ministry of Science and Higher Education in Poland as a part of its statutory activity No 3841/E-41/S/2020.

Acknowledgments: The seismic data from the local network LUMINEUS are available on the IS-EPOS platform [59] http://tcs.ah-epos.eu/#episode:LGCD, last accessed November 2019. The seismic data from the regional stations are available on the ORFEUS-EIDA project page: http://www.orfeus-eu.org/data/eida/, last accessed November 2019. The Sentinel-1 data used in the study are freely available on the ESA's Sentinel data hub: https://scihub.copernicus.eu/, last accessed November 2019. The estimated ground motion parameters for the stations from the LUMINEOS network are presented in the Supporting Information associated with this manuscript. The authors express acknowledgments to Witold Rohm and Jan Kapłon from the Institute of Geodesy and Geoinformatics at Wrocław University of Environmental and Life Sciences, for the critical reading and advices for improvement of the manuscript.

Conflicts of Interest: The authors declare no conflict of interest.

References

1. Dreger, D.S.; Ford, S.R.; Walter, W.R. Source analysis of the Crandall Canyon, Utah, mine collapse. *Science* **2008**, *321*, 217. [CrossRef] [PubMed]
2. Ford, S.R.; Dreger, D.S.; Walter, W.R. Source characterization of the 6 August 2007 Crandall canyon mine seismic event in central Utah. *Seismol. Res. Lett.* **2008**, *79*, 637–644. [CrossRef]
3. Ford, S.R.; Dreger, D.S.; Walter, W.R. Identifying isotropic events using a regional moment tensor inversion. *J. Geophys. Res. Solid Earth* **2009**, *114*. [CrossRef]

4. Rudzinski, Ł.; Cesca, S.; Lizurek, G. Complex rupture process of the 19 March 2013, Rudna mine (Poland) induced seismic event and collapse in the light of local and regional moment tensor inversion. *Seismol. Res. Lett.* **2016**, *87*, 274–284. [CrossRef]
5. Klose, C.D. Geomechanical modeling of the nucleation process of Australia's 1989 M5.6 Newcastle earthquake. *Earth Planet. Sci. Lett.* **2007**, *256*, 547–553. [CrossRef]
6. McGarr, A.; Simpson, D.; Seeber, L. Case histories of induced and triggered seismicity. In *IInternational Geophysics Series, International Handbook of Earthquake and Engineering Seismology*; Academic Press: Cambridge, MA, USA, 2002; Volume 81A, pp. 647–661, ISBN 0124406521.
7. Mirek, K.; Mirek, J. Non-parametric approximation used to analysis of psinsar^{TM} data of upper silesian coal basin, Poland. *Acta Geodyn. Geomater.* **2009**, *6*, 405–409.
8. Mirek, K.; Mirek, J. Observation of underground exploitation influence on a surface in Budryk, Sośnica, and Makoszowy Coal Mine Area. *Pol. J. Environ. Stud.* **2016**, *25*, 57–61.
9. Zheng, M.; Deng, K.; Fan, H.; Du, S. Monitoring and analysis of surface deformation in mining area based on InSAR and GRACE. *Remote Sens.* **2018**, *10*, 1392. [CrossRef]
10. Malinowska, A.A.; Witkowski, W.T.; Guzy, A.; Hejmanowski, R. Mapping ground movements caused by mining-induced earthquakes applying satellite radar interferometry. *Eng. Geol.* **2018**, *246*, 402–411. [CrossRef]
11. Rudziński, Ł.; Mirek, K.; Mirek, J. Rapid ground deformation corresponding to a mining-induced seismic event followed by a massive collapse. *Nat. Hazards* **2019**, *96*, 461–471. [CrossRef]
12. Rudziński, Ł.; Cesca, S.; Talaga, A.; Koziarz, E. Complex mechanism of rockburst observed on Polish copper mines. In Proceedings of the 9th International Symposium on Rockbursts and Seismicity in Mines—RaSiM9, Santiago, Chile, 15–17 November 2017; Vallejos, J.A., Ed.; University of Chile: Santiago, Chile, 2017; pp. 87–92.
13. Milczarek, W. Investigation of post inducted seismic deformation of the 2016 Mw4.2 Tarnowek Poland mining tremor based on DInSAR and SBAS methods. *Acta Geodyn. Geomater.* **2019**, *16*, 183–193. [CrossRef]
14. Dec, J.; Pietsch, K.; Marzec, P. Application of seismic methods to identify potential gas concentration zones at the Zechstein Limestone Level in the "Rudna" mining area, SW Poland. *Ann. Soc. Geol. Pol.* **2011**, *81*, 63–78.
15. Orfeus: European Integrated Data Archive. Available online: http://www.orfeus-eu.org/data/eida/ (accessed on 8 August 2019).
16. Hasegawa, H.S.; Wetmiller, R.J.; Gendzwill, D.J. Induced seismicity in mines in Canada—An overview. *Pure Appl. Geophys.* **1989**, *129*, 423–453. [CrossRef]
17. Heimann, S.; Kriegerowski, M.; Isken, M.; Cesca, S.; Daout, S.; Grigoli, F.; Juretzek, C.; Megies, T.; Nooshiri, N.; Steinberg, A.; et al. *Pyrocko—An Open-Source Seismology Toolbox and Library*. V. 0.3; GFZ: Potsdam, Germany, 2017.
18. Heimann, S. A Robust Method to Estimate Kinematic Earthquake Source Parameters. Ph.D. Thesis, University of Hamburg, Hamburg, Germany, 2011.
19. Sen, A.T.; Cesca, S.; Bischoff, M.; Meier, T.; Dahm, T. Automated full moment tensor inversion of coal mining-induced seismicity. *Geophys. J. Int.* **2013**, *195*, 1267–1281. [CrossRef]
20. Grad, M.; Jensen, S.L.; Keller, G.R.; Guterch, A.; Thybo, H.; Janik, T.; Tiira, T.; Yliniemi, J.; Luosto, U.; Motuza, G.; et al. Crustal structure of the Trans-European suture zone region along POLONAISE'97 seismic profile P4. *J. Geophys. Res.* **2003**, *108*, 2541. [CrossRef]
21. Jost, M.L.; Herrmann, R.B. A Student's Guide to and Review of Moment Tensors. *Seismol. Res. Lett.* **1989**, *60*, 37–57. [CrossRef]
22. Talebi, S.; Coté, M. Implosional Focal Mechanisms in a Hard-Rock Mine. In Proceedings of the Controlling Seismic Risk, 6th International Symposium on Rockburst and Seismicity in Mines, Perth, Australia, 9–11 March 2005; Potvin, Y., Hudyma, M., Eds.; Australian Centre for Geomechanics: Perth, Australia, 2005; pp. 113–121.
23. Hu, H.; Gao, J.; Yao, Y. Land deformation monitoring in mining area with PPP-AR. *Int. J. Min. Sci. Technol.* **2014**, *24*, 207–212. [CrossRef]
24. Bian, H.F.; Zhang, S.B.; Zhang, Q.Z.; Zheng, N.S. Monitoring large-area mining subsidence by GNSS based on IGS stations. *Trans. Nonferrous Met. Soc. China* **2014**, *24*, 514–519. [CrossRef]
25. Üstün, A.; Tuşat, E.; Yalvaç, S.; Özkan, İ.; Eren, Y.; Özdemir, A.; Bildirici, Ö.; Üstüntaş, T.; Kırtıloğlu, O.S.; Mesutoğlu, M.; et al. Land subsidence in Konya Closed Basin and its spatio-temporal detection by GPS and DInSAR. *Environ. Earth Sci.* **2015**, *73*, 6691–6703. [CrossRef]

26. Li, X.; Zheng, K.; Li, X.; Liu, G.; Ge, M.; Wickert, J.; Schuh, H. Real-time capturing of seismic waveforms using high-rate BDS, GPS and GLONASS observations: The 2017 Mw 6.5 Jiuzhaigou earthquake in China. *GPS Solut.* **2019**, *23*, 1–12. [CrossRef]
27. Kudlacik, I.; Kaplon, J.; Bosy, J.; Lizurek, G. Seismic phenomena in the light of high-rate GPS precise point positioning results. *Acta Geodyn. Geomater.* **2019**, *193*, 99–112. [CrossRef]
28. Michel, C.; Kelevitz, K.; Houlié, N.; Edwards, B.; Psimoulis, P.; Su, Z.; Clinton, J.; Giardini, D. The potential of high-rate GPS for strong ground motion assessment. *Bull. Seismol. Soc. Am.* **2017**, *107*, 1849–1859. [CrossRef]
29. Bock, Y.; Prawirodirdjo, L.; Melbourne, T.I. Detection of arbitrarily large dynamic ground motions with a dense high-rate GPS network. *Geophys. Res. Lett.* **2004**, *31*. [CrossRef]
30. Dach, R.; Lutz, S.; Walser, P.; Fridez, P. (Eds.) *Bernese GNSS Software Version 5.2. User Manual*; Astronomical Institute, University of Bern, Bern Open Publishing: Bern, Switzerland, 2015; ISBN 978-3-906813-05-9.
31. Bruyninx, C.; Legrand, J.; Fabian, A.; Pottiaux, E. GNSS metadata and data validation in the EUREF Permanent Network. *GPS Solut.* **2019**, *23*, 106. [CrossRef]
32. Herring, T.A.; King, R.W.; Floyd, M.A.; McClusky, S.C. *Introduction to GAMIT/GLOBK*; Release 10.8; Massachusetts Institute of Technology: Cambridge, MA, USA, 2018.
33. Melbourne, W. The Case for Ranging in GPS Based Geodetic System. In Proceedings of the 1st International Symposium on Precise Positioning with the Global Positioning System, Rockville, MD, USA, 15–19 April 1985; Goad, C., Ed.; U.S. Department of Commerce: Rockville, MD, USA, 1985.
34. Wübbena, G. Software Developments for Geodetic Positioning with GPS Using TI 4100 Code and Carrier Measurements. In Proceedings of the 1st International Symposium on Precise Positioning with the Global Positioning System, Rockville, MD, USA, 15–19 April 1985; Goad, C., Ed.; U.S. Department of Commerce: Rockville, MD, USA, 1985; pp. 403–412.
35. Orús, R.; Hernández-Pajares, M.; Juan, J.M.; Sanz, J. Improvement of global ionospheric VTEC maps by using kriging interpolation technique. *J. Atmos. Solar Terr. Phys.* **2005**, *67*, 1598–1609. [CrossRef]
36. Berardino, P.; Fornaro, G.; Lanari, R.; Sansosti, E. A new algorithm for surface deformation monitoring based on small baseline differential SAR interferograms. *IEEE Trans. Geosci. Remote Sens.* **2002**, *40*, 2375–2383. [CrossRef]
37. Massonnet, D.; Rossi, M.; Carmona, C.; Adragna, F.; Peltzer, G.; Feigl, K.; Rabaute, T. The displacement field of the Landers earthquake mapped by radar interferometry. *Nature* **1993**, *364*, 138–142. [CrossRef]
38. Chen, C.W.; Zebker, H.A. Network approaches to two-dimensional phase unwrapping: Intractability and two new algorithms. *J. Opt. Soc. Am. A* **2000**, *17*, 401–414. [CrossRef]
39. Rodriguez, E.; Morris, C.S.; Belz, J.E.; Chapin, E.C.; Martin, J.M.; Daffer, W.; Hensley, S. *An Assessment of the SRTM Topographic Products, Technical Report JPL D-31639*; Jet Propulsion Laboratory: Pasadena, CA, USA, 2005.
40. Crosetto, M.; Monserrat, O.; Cuevas-González, M.; Devanthéry, N.; Crippa, B. Persistent Scatterer Interferometry: A review. *ISPRS J. Photogramm. Remote Sens.* **2016**, *115*, 78–89. [CrossRef]
41. L3Harris Geospatial ENVI® SARscape®. Available online: https://www.harrisgeospatial.com/Software-Technology/ENVI-SARscape (accessed on 8 May 2020).
42. ESA SNAP Sentinel-1 Toolbox. Available online: https://step.esa.int/main/toolboxes/sentinel-1-toolbox/ (accessed on 8 May 2020).
43. Hooper, A.; Bekaert, D.; Spaans, K.; Arikan, M. Recent advances in SAR interferometry time series analysis for measuring crustal deformation. *Tectonophysics* **2012**, *514–517*, 1–13. [CrossRef]
44. Pawluszek-Filipiak, K.; Borkowski, A. Integration of DInSAR and SBAS techniques to determine mining-related deformations using Sentinel-1 data: The case study of Rydultowy mine in Poland. *Remote Sens.* **2020**, *12*, 242. [CrossRef]
45. Ilieva, M.; Polanin, P.; Borkowski, A.; Gruchlik, P.; Smolak, K.; Kowalski, A.; Rohm, W. Mining Deformation Life Cycle in the Light of InSAR and Deformation Models. *Remote Sens.* **2019**, *11*, 745. [CrossRef]
46. Klein, E.; Contrucci, I.; Daupley, X.; Hernandez, O.; Nadim, C.; Cauvin, L.; Pirson, M.; Klein, E.; Contrucci, I.; Daupley, X.; et al. Evolution monitoring of a solution-mining cavern in salt: Identifying and analysing early-warning signals prior to collapse. In Proceedings of the SMRI Fall Technical Conference, Austin, TX, USA, 12–14 October 2008; pp. 1–12.

47. Klein, E.; Contrucci, I.; Cao, N.T.; Bigarre, P. Mining induced seismicity—Monitoring of a large scale salt cavern collapse. In Proceedings of the 73rd European Association of Geoscientists and Engineers Conference and Exhibition 2011: Unconventional Resources and the Role of Technology. Incorporating SPE EUROPEC 2011, Vienna, Austria, 23–26 May 2011; pp. 615–619.
48. Contrucci, I.; Klein, E.; Cao, N.-T.; Daupley, X.; Bigarré, P. Multi-parameter monitoring of a solution mining cavern collapse: First insight of precursors. *Comptes Rendus Geosci.* **2011**, *343*, 1–10. [CrossRef]
49. Ćwiękała, M. Korelacja wyników monitoringu geodezyjnego z sejsmicznością indukowaną w czasie eksploatacji ściany VIb-E1 w pokładzie 703/1 w KWK ROW Ruch Rydułtowy Correlation of geodetic monitoring results with seismicity induced by exploitation of longwall VIb-E1 in. *Przegląd Górniczy* **2019**, *75*, 4–15.
50. Krawczyk, A.; Grzybek, R. An evaluation of processing InSAR Sentinel-1A/B data for correlation of mining subsidence with mining induced tremors in the Upper Silesian Coal Basin (Poland). *E3S Web Conf.* **2018**, *26*. [CrossRef]
51. Szczerbowski, Z.; Jura, J. Mining induced seismic events and surface deformations monitored by GPS permanent stations. *Acta Geodyn. Geomater.* **2015**, *12*, 237–248. [CrossRef]
52. Szczerbowski, Z. High-energy seismic events in Legnica–Głogów Copper District in light of ASG-EUPOS data. *Rep. Geod. Geoinform.* **2019**, *107*, 25–40. [CrossRef]
53. Wang, L.; Deng, K.; Zheng, M. Research on ground deformation monitoring method in mining areas using the probability integral model fusion D-InSAR, sub-band InSAR and offset-tracking. *Int. J. Appl. Earth Obs. Geoinf.* **2020**, *85*, 101981. [CrossRef]
54. Milczarek, W. Application of a Small Baseline Subset Time Series Method with Atmospheric Correction in Monitoring Results of Mining Activity on Ground Surface and in Detecting Induced Seismic Events. *Remote Sens.* **2019**, *11*, 1008. [CrossRef]
55. Psimoulis, P.; Houlié, N.; Meindl, M.; Rothacher, M. Consistency of PPP GPS and strong-motion records: Case study of Mw9.0 Tohoku-Oki 2011 earthquake. *Smart Struct. Syst.* **2015**, *16*, 347–366. [CrossRef]
56. Nie, Z.; Zhang, R.; Liu, G.; Jia, Z.; Wang, D.; Zhou, Y.; Lin, M. GNSS seismometer: Seismic phase recognition of real-time high-rate GNSS deformation waves. *J. Appl. Geophys.* **2016**, *135*, 328–337. [CrossRef]
57. Fratarcangeli, F.; Savastano, G.; D'Achille, M.C.; Mazzoni, A.; Crespi, M.; Riguzzi, F.; Devoti, R.; Pietrantonio, G. VADASE reliability and accuracy of real-time displacement estimation: Application to the Central Italy 2016 earthquakes. *Remote Sens.* **2018**, *10*, 1201. [CrossRef]
58. Xu, C.; Gong, Z.; Niu, J. Recent developments in seismological geodesy. *Geod. Geodyn.* **2016**, *7*, 157–164. [CrossRef]
59. Orlecka-Sikora, B.; Lasocki, S.; Kocot, J.; Szepieniec, T.; Robert Grasso, J.; Garcia-aristizabal, A.; Schaming, M.; Urban, P.; Jones, G.; Stimpson, I.; et al. An open data infrastructure for the study of anthropogenic hazards linked to georesource exploitation. *Sci. Data* **2020**, *7*, 1–16. [CrossRef] [PubMed]

© 2020 by the authors. Licensee MDPI, Basel, Switzerland. This article is an open access article distributed under the terms and conditions of the Creative Commons Attribution (CC BY) license (http://creativecommons.org/licenses/by/4.0/).

Article

Rheology of the Zagros Lithosphere from Post-Seismic Deformation of the 2017 Mw7.3 Kermanshah, Iraq, Earthquake

Xiaoran Lv [1,2], Falk Amelung [3], Yun Shao [1,2,4,*], Shu Ye [1,2], Ming Liu [4] and Chou Xie [1,2,4]

[1] Aerospace Information Research Institute, Chinese Academy of Sciences, Beijing 100094, China; lvxr@radi.ac.cn (X.L.); 201421480013@mail.bnu.edu.cn (S.Y.); xiechou@radi.ac.cn (C.X.)
[2] University of Chinese Academy of Sciences, Beijing 100049, China
[3] Rosenstiel School of Marine and Atmospheric Sciences, University of Miami, Miami, FL 33149, USA; famelung@rsmas.miami.edu
[4] Laboratory of Target Microwave Properties (LAMP), Zhongke Academy of Satellite Application in Deqing (DASA), Deqing 313200, China; euler@dasa.net.cn
* Correspondence: shaoyun@radi.ac.cn

Received: 26 May 2020; Accepted: 23 June 2020; Published: 24 June 2020

Abstract: We use 2018–2020 Sentinel-1 InSAR time series data to study post-seismic deformation processes following the 2017 Mw 7.3 Kermanshah, Iraq earthquake. We remove displacements caused by two large aftershock sequences from the displacement field. We find that for a six month period the response is dominated by afterslip along the up-dip extension of the coseismic rupture zone, producing up to 6 cm of radar line-of-sight displacements. The moment magnitude of afterslip is Mw 5.9 or 12% of the mainshock moment. After that period, the displacement field is best explained by viscoelastic relaxation and a lower crustal viscosity of $\eta_{lc} = 1^{+0.8}_{-0.4} \times 10^{19}$ Pas. The viscosity of the uppermost mantle is not constrained by the data, except that it is larger than 0.6×10^{19} Pas. The relatively high lower crustal and uppermost mantle viscosities are consistent with a cold and dry lithosphere of the Zagros region.

Keywords: post-seismic deformation mechanism; InSAR time series algorithm; Kermanshah earthquake; viscoelastic relaxation

1. Introduction

The Mw 7.3 Kermanshah earthquake of 12 November 2017 near the Iran/Iraq boundary occurred along a thrust fault of the Zagros Mountains. It killed more than 400 people and destroyed 12,000 buildings. There were serious economic losses caused by this event [1]. It was one of the largest earthquakes in the northwestern Zagros during the instrumental era (others were the 1909 Mw 7.3 earthquake and 1957 Mw 7.1 earthquake) [2,3]. Observations of post-seismic deformation provide an opportunity to investigate the fault and rheological properties of the Zagros lithosphere.

Post-seismic displacements can be caused by (i) afterslip along the fault, either along the patch that ruptured during the earthquake or next to it [4–7]; (ii) viscoelastic relaxation of the lower crust and/or upper mantle [8–10]; (iii) poroelastic rebound following earthquake-induced pore-fluid pressure changes [11–13]; or (iv) large aftershocks. These processes have different spatial-temporal characteristics [14]. Afterslip and poroelastic rebound cause ground displacements in the near field and last for a few months to years. In contrast, viscoelastic relaxation causes displacements in the far field and can last several decades. Understanding the origin of post-seismic deformation is important for seismic hazard because it changes the crustal stress field, loading or unloading nearby faults [15–17].

In this study we use 2.3 years (until February 2020) of Sentinel-1 data to investigate post-seismic displacements after the Kermanshah earthquake. After removing the coseismic deformation of the largest aftershocks, we investigate whether surface displacements reflect afterslip and/or viscoelastic relaxation and explore the afterslip characteristics and the lithospheric rheology of this region.

2. Geologic Setting and Aftershock Sequence

The Kermanshah earthquake is located on the southwestern edge of the northwestern Zagros fold-and-thrust belt (ZFTB), one of the most seismically active intra-continental belts and relatively young orogens in the world [8,18]. ZFTB is formed due to continental collision between the Arabian and Eurasian plates in western Asia. The convergence rate between these two plates is approximately ~2 cm/year [19] and ZFTB accommodates one third of the total collision rate, according to the NUVEL-1A plate motion model [20]. GPS data show that the collision rate decreases steadily from 9 mm/year in the southeastern section to 7 mm/year and 4 mm/year in the central and northwestern Zagros, respectively [20]. The Zagros Main Recent Fault (MRF), the High Zagros Fault (HZF) and the Zagros Mountain Front Fault (MFF) are major faults in the northwestern Zagros region (Figure 1).

Several studies using seismic and geodetic data concluded that this earthquake occurred on a shallowly east-dipping blind fault with 15° dip angle and between 13 and 20 km depth [3,18,20–24]. For the viscoelastic relaxation simulation below we use the source parameters of Vajedian [3].

The Kermanshah earthquake was followed by strong aftershock sequences. There were two sequences of aftershocks with several events with magnitude 5.0 or larger that occurred in close vicinity within a short time period. The first one was a one-hour-long sequence on 11 January 2018 with magnitude ranging from Mw 5.1 to Mw 5.5; the second one was a seven-hour-long sequence on 25 November 2018 with magnitude ranging from Mw 5.0 to Mw 6.3. These sequences (Figure 1, Table 1) are referred to as aftershock sequence 1 and 2, respectively. The third largest aftershock was an Mw 6.0 on 25 August 2018.

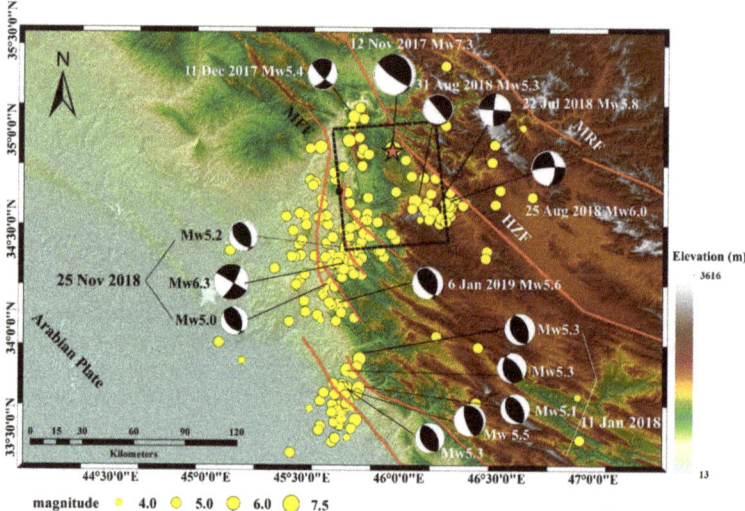

Figure 1. Map of the study area. The red star represents the location of the mainshock (from USGS earthquake Catalog). The black dashed rectangle marks the coseismic fault [3]. The black dot indicates the upper edge of fault. The red lines are nearby faults (MRF: the Zagros Main Recent Fault, HZF: the High Zagros Fault, MFF: the Zagros Mountain Front Fault). Yellow circles mark aftershocks (Mw > 3.0) up to 839 days after the mainshock (from USGS earthquake catalog: https://earthquake.usgs.gov/earthquakes/).

Table 1. Aftershock sequences [a].

Time (UTC)	Magnitude (Mw)	Type	Depth (km)
Aftershock sequence on 11 January 2018 (aftershock sequence 1)			
06:59:30	5.5	Thrust	13.5
07:00:52	5.3	No information	No information
07:14:15	5.3	Thrust	13.5
07:55:00	5.1	No information	No information
08:00:39	5.3	Thrust	17.5
Aftershock sequence on 25 November 2018 (aftershock sequence 2)			
16:37:32	6.3	Strike-slip	23.5
17:09:36	5.2	Thrust	17.5
23:00:46	5.0	No information	No information

[a] earthquake information came from USGS earthquake catalog: https://earthquake.usgs.gov/earthquakes/.

3. InSAR Data and Processing Methodology

We used a total of 134 ascending Sentinel-1 A/B data from Path 72 (from 17 November 2017 to 5 February 2020; the first image 5 days after the mainshock) and 54 descending images from Path 79 (from 18 November 2017 to 31 January 2020; the first image 6 days after the mainshock).

We used ISCE stack processor [25,26] to generate the interferograms with 21 looks in the azimuth and 7 looks in the range direction. Precise orbit data and the 3 arc-second DEM from the Shuttle Radar Topography Mission (SRTM) [27] were used to simulate and remove the phase error caused by topography and earth curvature from each interferogram. Multi-looked and filtered interferograms were registered to a master SAR image by finding offsets between single-look complex (SLC) images and the master SLC using DEM and orbit vectors. Finally, the phases of the coregistered interferograms were unwrapped using the statistical-cost network-flow algorithm (SNAPHU) [28].

For time series processing we used the routine workflow of the Miami InSAR time series software in Python (MintPy) [29]. In this workflow the network of interferograms was inverted for the raw phase time series and then corrected for the tropospheric delay (we used the ERA5 global atmospheric model) and topographic residuals to obtain the noise-reduced displacement time series from which the average line of sight (LOS) velocity was estimated pixel by pixel.

4. Modelling Approach

To estimate the aftershock and afterslip parameters, we consider uniform rectangular dislocations in a homogeneous elastic half-space [30]. The model parameters are fault location, length, width, dip angle, strike angle, depth, strike slip and dip slip. To estimate the rheological structure we consider Maxwell rheology for the lower crust and upper mantle, each characterized by two parameters, the steady state shear modulus and steady state viscosity. We use the RELAX software [31–33], which uses the Fourier–domain elastic Green's functions and the equivalent body-force representation of coseismic and post-seismic deformation processes to calculate the viscoelastic relaxation displacement. The model parameters are the lower crust and upper mantle viscosities.

We sample from the data using a downsampling method combining uniform and quadtree sampling. We use quadtree sampling in the region with significant deformation and in the far field the uniform sampling approach (see Supplemental Material for downsampling results). We calculate for each model a misfit function χ^2 defined as:

$$\chi^2 = (\mathbf{d}_{obs} - \mathbf{d}_{sim})\mathbf{C}^{-1}(\mathbf{d}_{obs} - \mathbf{d}_{sim})^{-1}$$

where $\mathbf{d}_{obs}$ and $\mathbf{d}_{sim}$ are the observed and simulated displacements, respectively, and $\mathbf{C}$ is the covariance matrix [34]. We solve the non-linear inversion problem for the elastic dislocations using a Bayesian approach with the Geodetic Bayesian Inversion Software (GBIS) [34]. We also use GBIS to estimate C. We invert for the rheological structure using a grid search approach.

5. Results

5.1. Post-Seismic Displacement Field

Figure 2 shows the line-of-sight (LOS) displacements from ascending and descending orbits after the earthquake (17 and 18 November 2017, respectively) to February 2020, where the positive value means movement towards the sensors and negative values mean movement away from the sensor. The ascending data shows a lobe of up to 8 cm of LOS decrease (yellow/red color) and the descending data shows lobes of both LOS decrease and LOS increase (up to 10 cm, blue color). The ascending and descending displacement time series for points in the epicentral area show initially rapidly and then slowly growing and decaying signals, respectively (Figure 2c). We therefore consider two time periods characterized by different displacement patterns, from 17 November 2017 to 10 June 2018 (time period 1) and from 10 June 2018 to February 2020 (time period 2). We consider four different areas (Figure 2a,b). In addition to the area used for studying viscoelastic relaxation (area 1), we consider two aftershock areas (areas 2 and 3) and an area to investigate afterslip (area 4).

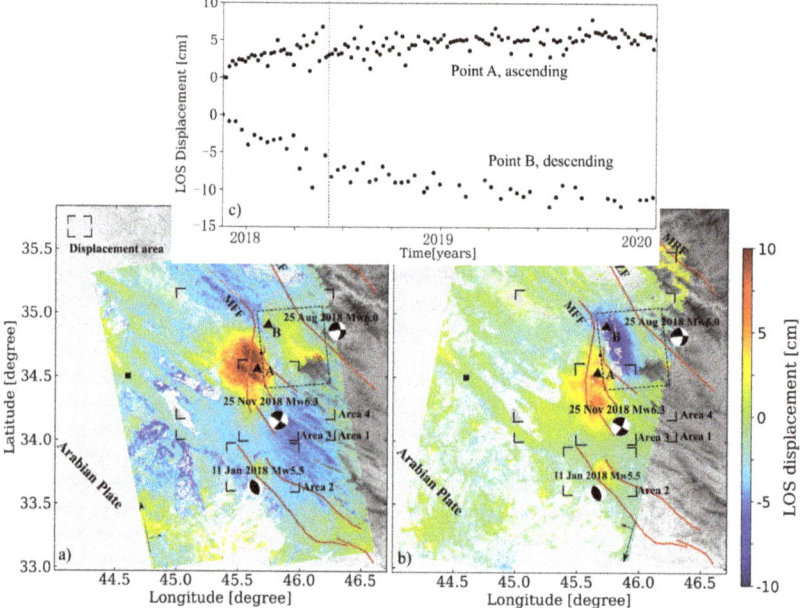

Figure 2. (a) Ascending and (b) descending LOS displacement field from November 2017 to February 2020. (c) LOS time series for two points located in the epicentral area. Black dashed rectangle in (a), (b): coseismic fault. Black corners: displacement areas. Black solid square: reference point (used through the whole paper except for aftershock sequence 1 and 2). Black dot: upper edge of fault. Black triangle: two points to show displacement time series.

5.2. Aftershocks

We consider both aftershock sequences as composite events and we obtain elastic dislocation models for aftershock sequence 1 and 2 so that their contributions can be removed from the post-seismic displacement fields. We do not consider the 25 August 2018 aftershock located east of the epicenter because there is no clear signal associated with it in ascending and descending data (Figure S1) and no geodetic focal mechanism can be estimated. For both ascending and descending orbits, we use interfergrams reconstructed from the raw time series. They are less noisy than the observed interferograms because the network inversion of interferograms filters out temporal decorrelation noise.

For aftershock sequence 1, the best-fitting dislocation has a length and width of 12.9 km and 5.4 km and 0.4 m slip. For aftershock sequence 2, the best-fitting dislocation has a length and width of 12.7 km and 8.8 km and 1.8 m slip (Table 2). The observations, modelled displacement and difference between observations and model are shown in Figure 3a,b. The red lines are nearby faults. The geodetic focal mechanisms (shown in Figure 3) are very similar to the Centroid-Moment-Tensor (CMT) solutions (shown in Figure 1).

5.3. Afterslip for Time Period 1

To test whether the surface displacements are consistent with afterslip, we use the ascending and descending data until June 2018 (Figure 4a) after removal of the contribution from the January 2018 aftershock sequence. We estimate the best-fitting uniform dislocation using a homogeneous elastic half space model. The best-fitting solution (Table 2, black solid rectangle in Figure 4a) shows that the observations are well explained by 0.8 m of uniform slip along one dislocation with a length and width of 45.5 and 6.4 km located on the near-horizontal, western, up-dip continuation of the coseismic fault.

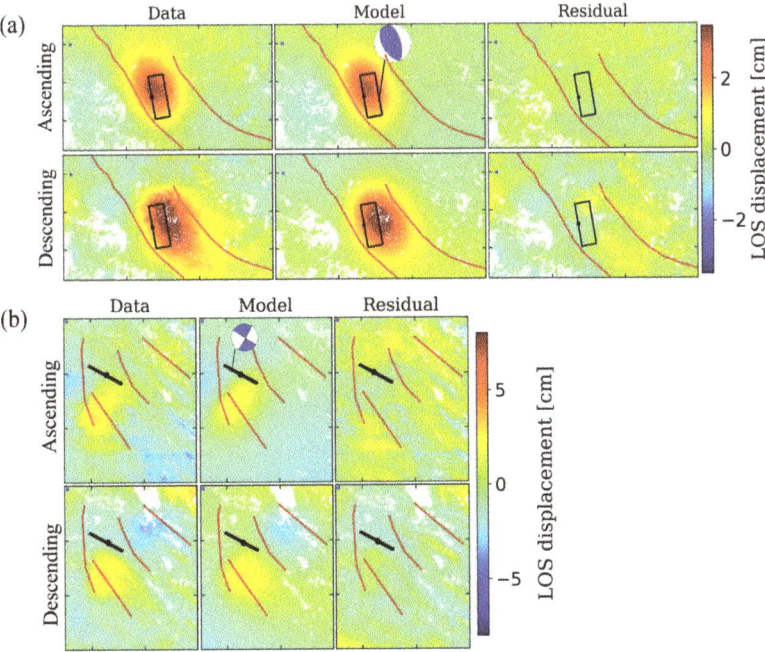

Figure 3. Uniform slip inversion result of aftershock sequence 1, aftershock sequence 2 for ascending data and descending data. (**a**) Aftershock sequence 1, ascending data from 10 January 2018 to 16 January 2018, descending data from 5 January 2018 to 17 January 2018; (**b**) aftershock sequence 2, ascending data from 18 November 2018 to 30 November 2018, descending data from 25 November 2018 to 7 December 2018. Black solid rectangle: best-fitting dislocation. Red lines: nearby faults. Black dot: upper edge of fault. The beach balls for aftershock sequence 1 and 2 are obtained based on the inverted fault parameters. Blue points for aftershock sequence 1 and 2: reference point.

Table 2. Range, optimal and uncertainties of InSAR-inversion fault slip parameters for the aftershock sequence that occurred on 11 January 2018, and the aftershock sequence that occurred on 25 November 2018 and afterslip for time period 1.

	Length (km)	Width (km)	Depth (km)	Dip (°)	Strike (°)	Dip Slip [b] (m)	Strike Slip [c] (m)	Lat [d,e]	Lon [d,e]	Mw [e]	Rms (m)
					Aftershocks 1 (Mw5.5)						
Lower	5	5	1	−90	90	0	−0.2	—	—	—	
Upper	50	35	30	90	360	6	0.2	—	—	—	
Optimal	12.9	5.4	13.0	51	344	0.39	−0.08	33.71	45.68	5.94	0.002
2.5% [a]	8.7	1.8	11.9	42	335	0.21	−0.25	—	—	—	
97.5% [a]	15.3	10.1	15.3	58	352	0.59	0.05	—	—	—	
					Aftershocks 2 (Mw6.3)						
Lower	5	5	1	50	90	−0.3	−2	—	—	—	
Upper	30	30	40	90	360	0	2	—	—	—	
Optimal	12.7	8.8	22.0	90	302	−0.29	1.80	34.36	45.70	6.52	0.008
2.5%	6.2	2.9	13.2	87	292	−0.41	0.91	—	—	—	
97.5%	21.2	14.7	32.0	90	310	−0.02	2.25	—	—	—	
					Afterslip						
Lower	5	1	5	−20	330	0	−3	—	—	—	
Upper	70	50	20	20	360	5	3	—	—	—	
Optimal	45.5	6.4	11.5	−4	346	0.69	−0.45	34.45	45.62	6.56	0.018
2.5%	38.9	1.0	10.3	−10	340	0.13	−0.76	—	—	—	
97.5%	54.6	11.5	13.6	5	350	1.56	−0.24	—	—	—	

[a] maximum posteriori probability solutions of 2.5% and 97.5% from the posterior probability density function of fault parameters. [b] The positive value means the hanging wall showed relative uplift; [c] the negative value means that the hanging wall direction of the motion was opposite to the strike; [d] longitude and latitude of the corner of upper edge; [e] calculated based on the optimal inversion parameters.

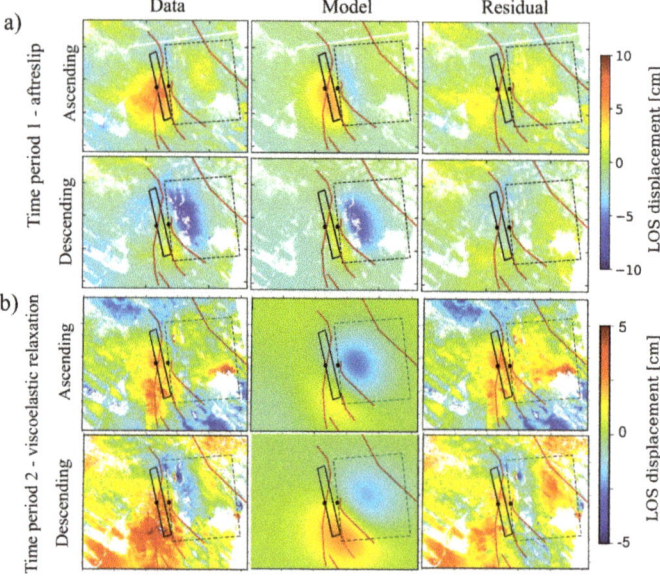

Figure 4. Best-fitting modelling results for (**a**) afterslip period (until June 2018) and (**b**) viscoelastic relaxation period (from June 2018 to February 2020). Black thin dashed rectangle: coseismic fault. Black solid rectangle: best-fitting afterslip dislocation. Black dot: upper edges of fault. Red lines: nearby faults. The reference point is shown in Figure 2.

5.4. Viscoelastic Relaxation for Time Period 2

We assume that the displacements in time period 2 are caused by viscoelastic relaxation. In order to constrain the rheological structure we use a 25 km-thick elastic upper crust and a 20 km-thick viscoelastic Maxwell lower crust over a viscoelastic Maxwell upper mantle (Figure 5a; the average depth of the regional Moho is 45 km according to the CRUST 1.0 model) [35,36]. In the grid search the viscosities of the lower crust and upper mantle range from 1×10^{17} Pas to 1×10^{21} Pas; we use a step size for the logarithm of viscosity of 0.25.

Our result (Figure 5b) yields the best estimates of the parameters $\eta_{lc} = 1^{+0.8}_{-0.4} \times 10^{19}$ Pas, where the confidence interval is calculated using F–Test [37–39]. Our data resolve the viscosity of the lower crust but not the viscosity of the uppermost mantle, except that it is larger than 0.6×10^{19} Pas.

The comparison between the observed and simulated displacement (Figure 4b) shows that the model explains the descending data well but there is a discrepancy for the ascending data. This discrepancy can be attributed to the larger tropospheric noise for the ascending data which are acquired during the day when the tropospheric variability is high (local acquisition time of 15:00 and 3:00 for ascending and descending data, respectively). What is more, the color discontinuity in the ascending data (around center-bottom of Figure 4b, top) is caused by the subtraction of coseismic displacement of aftershock sequence 2.

Furthermore, the viscoelastic relaxation signal during this 1.75 year period is relatively small (up to 4 cm). The small signal is a consequence of the fault geometry. A horizontal fault causes largely horizontal flow.

We also explored afterslip models and concluded that they can be ruled out. The fault plane of the best-fitting model locates a few kilometers above the coseismic rupture in the phanerozoic cover (Figures S2 and S3), which is geologically not plausible. Besides, the best-fitting viscoelastic relaxation model is characterized by a significantly better rms ($rms = \sqrt{\frac{\sum_{i=1}^{n} residual_i^2}{n}}$) than the best-fitting afterslip model (0.015 m versus 0.025 m). We can also rule out afterslip on the coseismic rupture or its downdip extension because of even higher rms. We did not consider combined models with both afterslip and viscoelastic relaxation because the data lack the resolution to constrain all model parameters.

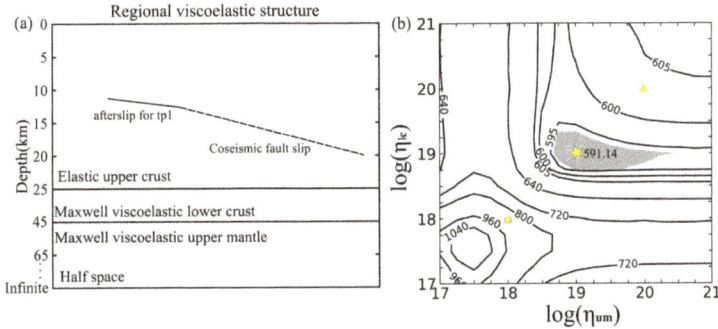

Figure 5. (a) Regional layered model. The upper crust depth is shown at 4:1 scale for visual clarity; the lower crust and upper mantle depth are shown at 1:1 scale; (b) misfit contour map in η_{lc} and η_{um} space. The yellow star is model 1 (best-fitting model) in Figure 6; the yellow square is model 2 in Figure 6; the yellow triangle is model 3 in Figure 6. The black dash line in (a) is the coseismic fault. The black solid line in (a) is the afterslip fault. The gray-shaded area in (b) denotes 95% confidence interval.

5.5. Summary of Post-Seismic Models

To visualize the fits of the post-seismic models we combine the ascending and descending displacement fields for the two time periods to the quasi-vertical and quasi-horizontal displacements in east–west direction [40] (Figure 6a,b). The north–south displacement cannot be calculated because only right-looking InSAR data are available and because polar-orbiting sensors are not sensitive to north–south displacements. We generate simulations for models with lower and higher viscosities in both the lower crust and uppermost mantle. We use $\eta_{lc} = \eta_{um} = 10^{18}$ Pas for the low viscosity model (model 2) and $\eta_{lc} = \eta_{um} = 10^{20}$ Pas for the high viscosity model (model 3), respectively.

For time period 1, the low viscosity model produces vertical displacements comparable to the observations, but the east–west displacements are different than observed (model 2, Figure 6a). This reinforces that afterslip was the dominating process during this period. For time period 2, the low and high viscosity models predict significantly more and less displacements than observed, respectively (models 2, 3 in Figure 6f,g), reinforcing that the viscosities are of the order of 10^{19} Pas (model 1, Figure 6d).

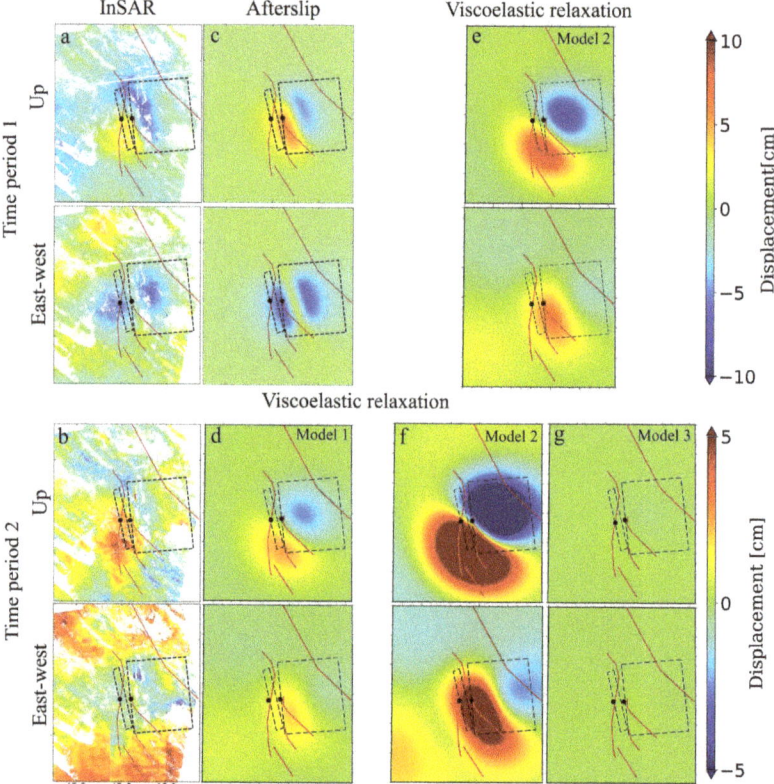

Figure 6. InSAR-derived horizontal east–west and vertical displacements with aftershocks' contribution removed and modelling results. (**a**) and (**b**) InSAR observation; (**c**) afterslip model; (**d**) viscoelastic relaxation model 1; (**e**) and (**f**) viscoelastic relaxation model 2; (**g**) viscoelastic relaxation model 3. Black dot: upper edge of the fault. Red line: nearby faults. Thin dashed rectangle: best-fitting afterslip dislocation and coseismic fault, respectively.

6. Discussion

6.1. Afterslip for Time Period 1

Our analysis shows that the post-seismic response through about June 2018 was dominated by afterslip along the up-dip continuation of the coseismic rupture, consistent with Barnhart [18] and Yang [24] who studied shorter time periods. The moment released by afterslip was 1.13×10^{19} Nm, which was about 12% of that by the coseismic slip. To put these results into perspective, we compiled the location of afterslip and the ratios between afterslip and coseismic moment release found in previous studies of earthquakes, which are listed in the supplementary Table S1.

The afterslip on the up-dip continuation of the coseismic rupture indicates a zone along the fault with velocity-strengthening frictional behavior [41] above a zone with velocity weakening behavior that ruptured during the earthquake. Up-dip afterslip was also observed for the 2003 Mw 6.9 Boumerdes [42] and the 2003 Mw6.8 Zemmouri, Algeria [43] earthquakes. For many other earthquakes, afterslip occurs on the down-dip extension of the coseismic rupture zone. Examples include the 2015 Mw 7.8 Gorkha earthquake [44], the 2011 Mw 9.0 Tohoku earthquake [45–47] and the 2008 Mw 7.9 Wenchuan earthquake [48].

The afterslip duration time of six months is comparable to that of the 2014 Napa earthquake [10] but much smaller than for example for the 2004 Mw6.0 Parkfield, California, earthquake [49] after which it continued for about 12 years. The relative amount of afterslip moment release is at the lower end compared to other earthquakes (10%, 28% and 56% for the Gorkha, Tohoku and Pakistan earthquakes, respectively).

Afterslip is promoted by clay-rich sediments/fault with clay-rich gouge, high temperature and elevated pore-fluid pressure [7]. Therefore, the low heat flow (average of 74 mW/m^2 [50]) and relatively low pore-fluid pressure (indicated by the relative low Vp/Vs ratio about 1.73 [51]) might contribute to the relative short duration time and small moment of Kermanshah afterslip compared to the mainshock moment.

6.2. Viscoelastic Relaxation for Time Period 2

Our modelling results show that the best-fitting viscosity of the lower crust is 10^{19} Pas while the viscosity of upper mantle is 10^{19} Pas or larger, i.e., the viscosity of the lower crust is equal or lower than that of the upper mantle. The inferred lower crust viscosity beneath the Kermanshah region equals to that in the regions of the Hector Mine [52], EL Mayor Cucapah [53], Iceland [54], Loma Prieta [55], Irpinia [56], Gonghe [38] and Bam [57] earthquakes (Table S2). It is lower than that of the Central Nevada Seismic Belt [58], and above that of Hebgen Lake [59], Northridge [60], Izmit [61], Chi-Chi [62], Manyi [63] and Mongolia earthquakes [64]. The inferred upper mantle viscosity range is consistent with upper mantle viscosities found for the regions of the Izmit, Northridge, Wenchuan [48] (Tibet), Bhuj [65], Gonghe, Bam, Northridge, Loma Prieta, Chi-Chi, Manyi and Kokoxili [66] earthquakes, but is higher than for the Hebgen Lake, Landers [39], Hector Mine, Denali [67], Central Nevada Seismic Belt, El Mayor Cucapah, Iceland, Parkfiled [68] and Mongolia earthquakes.

Laboratory experiments and previous studies show that lithospheric viscosity is mainly dependent on the composition, temperature and water content [69–71]. Lower temperature and less water lead to higher viscosity. The relatively high upper mantle viscosity of the Kermanshah region compared to the western US (Landers, Hector Mine, Central Nevada Seismic Belt, EL Mayor Cucapah, Heban Lake earthquakes), South central Alsaka (Denali earthquake) and Iceland is consistent with the relatively lower average heat flow (Iran: 74 mW/m^2 [50], Western US: 91 mW/m^2 [50], South central Alsaka [72]: 89 mW/m^2; Iceland: 175 mW/m^2 [73]). It is also consistent with the presence of the high-velocity anomalies in the upper mantle beneath the Zagros [74] which mark the presence of dry and cold lithospheric mantle [48]. What is more, the relatively high viscosity of the lower crust is also consistent with the relatively low average Vp/Vs value (Iran: 1.73 [51]; Chi–Chi: 1.9 [75]), which can be an indication of low fluid content [76]. Lower temperatures might also contribute to a higher viscosity in the lower crust.

There was no significant relaxation deformation in the first six months, which can be explained by the relatively high lower crust and upper mantle viscosities. Furthermore, laboratory experiments also suggest that the power–law viscous flow often occurs on the hot lithosphere rocks [77], thus the relatively cold lithosphere rocks in our study region may mean the low possibility of the power–law viscous flow occurring in the first few months, which may also contribute to this no significant relaxation deformation in the first six months phenomenon.

7. Conclusions

We obtained the post-seismic displacement field of the Kermanshah earthquake for the November 2017 to February 2020 time period, longer than previous studies (4 months for Barnhart's study, 7 months for Yang's study), from Sentinel-1 ascending and descending data. We came to the following conclusions:

For the first six months (until June 2018) the post-seismic displacement was due to the afterslip along the up-dip extension of the rupture zone. The moment released by afterslip was 12% of that released by coseismic slip. This was at the lower end of the observed range of afterslip durations and relative moment release. Possible explanations for little afterslip were the relatively low heat flow and pore-fluid pressures in the Zagros region.

For the subsequent period (we considered data until February 2020), the post-seismic displacement field was not consistent with afterslip but more consistent with viscoelastic relaxation in the lower crust. Assuming that this was the only process causing surface displacements, we found a best-fitting viscosity of $\eta_{lc} = 1^{+0.8}_{-0.4} \times 10^{19}$ Pas for lower crust. The data do not have the resolution to constrain the viscosity of the uppermost mantle but we could infer a lower bound of 0.6×10^{19} Pas. A relatively strong lower crust and upper mantle was consistent with the relatively lower heat flow, low average Vp/Vs value and the existence of the upper mantle high-velocity anomalies in the Zagros region.

Supplementary Materials: The following are available online at http://www.mdpi.com/2072-4292/12/12/2032/s1: Figure S1: coseismic displacement field caused by the aftershock of 25 August 2018; Figure S2: uniform slip inversion result for afterslip for time period 2; Figure S3: conceptual model of the spatial relationship between coseismic fault slip and afterslip for time period 2; Figure S4: sampled data used for modelling of aftershock sequence 1 and aftershock sequence 2; Figure S5: same as Figure S4 but for time period 1 and time period 2; Table S1: reported ratio of the afterslip moment release relative to the coseismic moment release and the position of afterslip relative to the coseismic slip; Table S2: rheologic structures inferred in previous studies; Table S3: range, optimal and uncertainties of InSAR-inversion fault slip parameters for the afterslip for time period 2.

Author Contributions: Funding acquisition, S.Y., M.L. and C.X.; investigation, X.L.; methodology, X.L. and F.A.; software, X.L.; supervision, F.A. and Y.S.; writing–original draft, X.L. All authors have read and agreed to the published version of the manuscript.

Funding: This research was funded by National Key Research and Development Program of China: Construction and Demonstration of Accurate Emergency Service System for aerial-space-ground based cooperative remote sensing, grant number Y6A0022010, and the National Science Foundation of China, grant number 41431174.

Acknowledgments: We would like to thank the anonymous reviewers for their helpful suggestions, which improved the quality of this manuscript.

Conflicts of Interest: The authors declare no conflicts of interest.

References

1. Zare, M.; Kamranzad, F.; Parcharidis, I. *Preliminary Report of Mw7.3 Sarpol-e Zahab, Iran Earthquake on November 12. 2017*; EMSC Report; EMSC/CSEM: Essonne, France, 2018.
2. Peyret, M.; Rolandone, F.; Dominguez, S.; Djamour, Y.; Meyer, B. Source model for the Mw 6.1, 31 March 2006, Chalan-Chulan Earthquake (Iran) from InSAR. *Terra Nova* **2008**, *20*, 126–133. [CrossRef]
3. Vajedian, S.; Motagh, M.; Mousavi, Z.; Motaghi, K.; Fielding, E.; Akbari, B.; Wetzel, H.-U.; Darabi, A. Coseismic Deformation Field of the Mw 7.3 12 November 2017 Sarpol-e Zahab (Iran) Earthquake: A Decoupling Horizon in the Northern Zagros Mountains Inferred from InSAR Observations. *Remote Sens.* **2018**, *10*, 1589. [CrossRef]
4. Shen, Z.-K.; Jackson, D.D.; Feng, Y.; Cline, M.; Kim, M.; Fang, P.; Bock, Y. Postseismic deformation following the Landers earthquake, California, 28 June 1992. *Bull. Seismol. Soc. Am.* **1994**, *84*, 780–791.
5. Freed, A.M. Earthquake triggering by static, dynamic, and postseismic stress transfer. *Annu. Rev. Earth Planet. Sci.* **2005**, *33*, 335–367. [CrossRef]

6. Wei, S.; Barbot, S.; Graves, R.; Lienkaemper, J.J.; Wang, T.; Hudnut, K.; Fu, Y.; Helmberger, D. The 2014 Mw 6.1 South Napa Earthquake: A Unilateral Rupture with Shallow Asperity and Rapid Afterslip. *Seismol. Res. Lett.* **2015**, *86*, 344–354. [CrossRef]
7. Avouac, J.-P. From Geodetic Imaging of Seismic and Aseismic Fault Slip to Dynamic Modeling of the Seismic Cycle. *Annu. Rev. Earth Planet. Sci.* **2015**, *43*, 233–271. [CrossRef]
8. Deng, J.; Gurnis, M.; Kanamori, H.; Hauksson, E. Viscoelastic Flow in the Lower Crust after the 1992 Landers, California, Earthquake. *Science* **1998**, *282*, 1689. [CrossRef]
9. Wiseman, K.; Bürgmann, R.; Freed, A.M.; Banerjee, P. Viscoelastic relaxation in a heterogeneous Earth following the 2004 Sumatra–Andaman earthquake. *Earth Planet. Sci. Lett.* **2015**, *431*, 308–317. [CrossRef]
10. Pollitz, F.F. Lithosphere and shallow asthenosphere rheology from observations of post-earthquake relaxation. *Phys. Earth Planet. Inter.* **2019**, *293*, 106271. [CrossRef]
11. Peltzer, G.; Rosen, P.; Rogez, F.; Hudnut, K. Poroelastic rebound along the Landers 1992 earthquake surface rupture. *J. Geophys. Res. Solid Earth* **1998**, *103*, 30131–30145. [CrossRef]
12. Hughes, K.L.H.; Masterlark, T.; Mooney, W.D. Poroelastic stress-triggering of the 2005 M8.7 Nias earthquake by the 2004 M9.2 Sumatra–Andaman earthquake. *Earth Planet. Sci. Lett.* **2010**, *293*, 289–299. [CrossRef]
13. Hu, Y.; Bürgmann, R.; Freymueller, J.T.; Banerjee, P.; Wang, K. Contributions of poroelastic rebound and a weak volcanic arc to the postseismic deformation of the 2011 Tohoku earthquake. *Earth Planets Space* **2014**, *66*, 106. [CrossRef]
14. Zhao, B.; Bürgmann, R.; Wang, D.; Tan, K.; Du, R.; Zhang, R. Dominant Controls of Downdip Afterslip and Viscous Relaxation on the Postseismic Displacements Following the Mw7.9 Gorkha, Nepal, Earthquake. *J. Geophys. Res. Solid Earth* **2017**, *122*, 8376–8401. [CrossRef]
15. Mikumo, T.; Yagi, Y.; Singh, S.K.; Santoyo, M.A. Coseismic and postseismic stress changes in a subducting plate: Possible stress interactions between large interplate thrust and intraplate normal-faulting earthquakes. *J. Geophys. Res. Solid Earth* **2002**, *107*, ESE 5-1–ESE 5-12. [CrossRef]
16. Cattania, C.; Hainzl, S.; Wang, L.; Enescu, B.; Roth, F. Aftershock triggering by postseismic stresses: A study based on Coulomb rate-and-state models. *J. Geophys. Res. Solid Earth* **2015**, *120*, 2388–2407. [CrossRef]
17. Verdecchia, A.; Pace, B.; Visini, F.; Scotti, O.; Peruzza, L.; Benedetti, L. The Role of Viscoelastic Stress Transfer in Long-Term Earthquake Cascades: Insights After the Central Italy 2016–2017 Seismic Sequence. *Tectonics* **2018**, *37*, 3411–3428. [CrossRef]
18. Barnhart, W.D.; Brengman, C.M.J.; Li, S.; Peterson, K.E. Ramp-flat basement structures of the Zagros Mountains inferred from co-seismic slip and afterslip of the 2017 Mw7.3 Darbandikhan, Iran/Iraq earthquake. *Earth Planet. Sci. Lett.* **2018**, *496*, 96–107. [CrossRef]
19. Vernant, P.; Nilforoushan, F.; Hatzfeld, D.; Abbassi, M.R.; Vigny, C.; Masson, F.; Nankali, H.; Martinod, J.; Ashtiani, A.; Bayer, R.; et al. Present-day crustal deformation and plate kinematics in the Middle East constrained by GPS measurements in Iran and northern Oman. *Geophys. J. Int.* **2004**, *157*, 381–398. [CrossRef]
20. Chen, K.; Xu, W.; Mai, P.M.; Gao, H.; Zhang, L.; Ding, X. The 2017 Mw 7.3 Sarpol Zahāb Earthquake, Iran: A compact blind shallow-dipping thrust event in the mountain front fault basement. *Tectonophysics* **2018**, *747–748*, 108–114. [CrossRef]
21. Feng, W.; Samsonov, S.; Almeida, R.; Yassaghi, A.; Li, J.; Qiu, Q.; Li, P.; Zheng, W. Geodetic Constraints of the 2017 Mw7.3 Sarpol Zahab, Iran Earthquake, and Its Implications on the Structure and Mechanics of the Northwest Zagros Thrust-Fold Belt. *Geophys. Res. Lett.* **2018**, *45*, 6853–6861. [CrossRef]
22. Ding, K.; He, P.; Wen, Y.; Chen, Y.; Wang, D.; Li, S.; Wang, Q. The 2017 Mw 7.3 Ezgeleh, Iran earthquake determined from InSAR measurements and teleseismic waveforms. *Geophys. J. Int.* **2018**, *215*, 1728–1738. [CrossRef]
23. Yang, Y.H.; Hu, J.C.; Yassaghi, A.; Tsai, M.C.; Zare, M.; Chen, Q.; Wang, Z.G.; Rajabi, A.M.; Kamranzad, F. Midcrustal Thrusting and Vertical Deformation Partitioning Constraint by 2017 Mw 7.3 Sarpol Zahab Earthquake in Zagros Mountain Belt, Iran. *Seismol. Res. Lett.* **2018**, *89*, 2204–2213. [CrossRef]
24. Yang, C.; Han, B.; Zhao, C.; Du, J.; Zhang, D.; Zhu, S. Co-and post-seismic Deformation Mechanisms of the MW 7.3 Iran Earthquake (2017) Revealed by Sentinel-1 InSAR Observations. *Remote Sens.* **2019**, *11*, 418. [CrossRef]
25. Rosen, P.A.; Gurrola, E.; Sacco, G.F.; Zebker, H. The InSAR scientific computing environment. In Proceedings of the EUSAR 2012; 9th European Conference on Synthetic Aperture Radar, Nuremberg, Germany, 23–26 April 2012; pp. 730–733.

26. Fattahi, H.; Agram, P.; Simons, M. A Network-Based Enhanced Spectral Diversity Approach for TOPS Time-Series Analysis. *IEEE Trans. Geosci. Remote Sens.* **2017**, *55*, 777–786. [CrossRef]
27. Farr, T.G.; Rosen, P.A.; Caro, E.; Crippen, R.; Duren, R.; Hensley, S.; Kobrick, M.; Paller, M.; Rodriguez, E.; Roth, L.; et al. The Shuttle Radar Topography Mission. *Rev. Geophys.* **2007**, *45*, RG2004. [CrossRef]
28. Chen, C.W.; Zebker, H.A. Two-dimensional phase unwrapping with use of statistical models for cost functions in nonlinear optimization. *J. Opt. Soc. Am. A* **2001**, *18*, 338–351. [CrossRef]
29. Yunjun, Z.; Fattahi, H.; Amelung, F. Small baseline InSAR time series analysis: Unwrapping error correction and noise reduction. *Comput. Geosci.* **2019**, *133*, 104331. [CrossRef]
30. Okada, Y. Internal deformation due to shear and tensile faults in a half-space. *Bull. Seismol. Soc. Am.* **1992**, *82*, 1018–1040.
31. Barbot, S.; Fialko, Y.; Sandwell, D. Three-dimensional models of elastostatic deformation in heterogeneous media, with applications to the Eastern California Shear Zone. *Geophys. J. Int.* **2009**, *179*, 500–520. [CrossRef]
32. Barbot, S.; Fialko, Y. Fourier-domain Green's function for an elastic semi-infinite solid under gravity, with applications to earthquake and volcano deformation. *Geophys. J. Int.* **2010**, *182*, 568–582. [CrossRef]
33. Barbot, S.; Fialko, Y. A unified continuum representation of post-seismic relaxation mechanisms: Semi-analytic models of afterslip, poroelastic rebound and viscoelastic flow. *Geophys. J. Int.* **2010**, *182*, 1124–1140. [CrossRef]
34. Bagnardi, M.; Hooper, A. Inversion of Surface Deformation Data for Rapid Estimates of Source Parameters and Uncertainties: A Bayesian Approach. *Geochem. Geophys. Geosyst.* **2018**, *19*, 2194–2211. [CrossRef]
35. Laske, G.; Masters, G.; Ma, Z.; Pasyanos, M. Update on CRUST1.0—A 1-degree Global Model of Earth's Crust. *Geophys. Res. Abstr.* **2013**, *15*, 2658.
36. Hatzfeld, D.; Tatar, M.; Priestley, K.; Ghafory-Ashtiany, M. Seismological constraints on the crustal structure beneath the Zagros Mountain belt (Iran). *Geophys. J. Int.* **2003**, *155*, 403–410. [CrossRef]
37. Stein, S.; Gordon, R.G. Statistical tests of additional plate boundaries from plate motion inversions. *Earth Planet. Sci. Lett.* **1984**, *69*, 401–412. [CrossRef]
38. Hao, M.; Shen, Z.-K.; Wang, Q.; Cui, D. Postseismic deformation mechanisms of the 1990 Mw 6.4 Gonghe, China earthquake constrained using leveling measurements. *Tectonophysics* **2012**, *532–535*, 205–214. [CrossRef]
39. Pollitz, F.F.; Peltzer, G.; Bürgmann, R. Mobility of continental mantle: Evidence from postseismic geodetic observations following the 1992 Landers earthquake. *J. Geophys. Res. Solid Earth* **2000**, *105*, 8035–8054. [CrossRef]
40. Wright, T.J.; Parsons, B.E.; Lu, Z. Toward mapping surface deformation in three dimensions using InSAR. *Geopyhs. Res. Lett.* **2004**, *31*, L01607. [CrossRef]
41. Marone, C.J.; Scholtz, C.H.; Bilham, R. On the mechanics of earthquake afterslip. *J. Geophys. Res. Solid Earth* **1991**, *96*, 8441–8452. [CrossRef]
42. Mahsas, A.; Lammali, K.; Yelles, K.; Calais, E.; Freed, A.M.; Briole, P. Shallow afterslip following the 2003 May 21, Mw = 6.9 Boumerdes earthquake, Algeria. *Geophys. J. Int.* **2008**, *172*, 155–166. [CrossRef]
43. Cetin, E.; Meghraoui, M.; Cakir, Z.; Akoglu, A.M.; Mimouni, O.; Chebbah, M. Seven years of postseismic deformation following the 2003 Mw = 6.8 Zemmouri earthquake (Algeria) from InSAR time series. *Geopyhs. Res. Lett.* **2012**, *39*. [CrossRef]
44. Jiang, G.; Wang, Y.; Wen, Y.; Liu, Y.; Xu, C.; Xu, C. Afterslip evolution on the crustal ramp of the Main Himalayan Thrust fault following the 2015 Mw 7.8 Gorkha (Nepal) earthquake. *Tectonophysics* **2019**, *758*, 29–43. [CrossRef]
45. Ozawa, S.; Nishimura, T.; Suito, H.; Kobayashi, T.; Tobita, M.; Imakiire, T. Coseismic and postseismic slip of the 2011 magnitude-9 Tohoku-Oki earthquake. *Nature* **2011**, *475*, 373–376. [CrossRef] [PubMed]
46. Ozawa, S.; Nishimura, T.; Munekane, H.; Suito, H.; Kobayashi, T.; Tobita, M.; Imakiire, T. Preceding, coseismic, and postseismic slips of the 2011 Tohoku earthquake, Japan. *J. Geophys. Res.Solid Earth* **2012**, *117*, 7404. [CrossRef]
47. Noda, A.; Takahama, T.; Kawasato, T.; Matsu'ura, M. Interpretation of Offshore Crustal Movements Following the 2011 Tohoku-Oki Earthquake by the Combined Effect of Afterslip and Viscoelastic Stress Relaxation. *Pure Appl. Geophys.* **2018**, *175*, 559–572. [CrossRef]
48. Diao, F.; Wang, R.; Wang, Y.; Xiong, X.; Walter, T.R. Fault behavior and lower crustal rheology inferred from the first seven years of postseismic GPS data after the 2008 Wenchuan earthquake. *Earth Planet. Sci. Lett.* **2018**, *495*, 202–212. [CrossRef]

49. Lienkaemper, J.J.; McFarland, F.S. Long-Term Afterslip of the 2004 M 6.0 Parkfield, California, Earthquake—Implications for Forecasting Amount and Duration of Afterslip on Other Major Creeping Faults. *Bull. Seismol. Soc. Am.* **2017**, *107*, 1082–1093. [CrossRef]
50. Wright, T.J.; Elliott, J.R.; Wang, H.; Ryder, I. Earthquake cycle deformation and the Moho: Implications for the rheology of continental lithosphere. *Tectonophysics* **2013**, *609*, 504–523. [CrossRef]
51. Afsari, N.; Sodoudi, F.; Taghizadeh Farahmand, F.; Ghassemi, M.R. Crustal structure of Northwest Zagros (Kermanshah) and Central Iran (Yazd and Isfahan) using teleseismic Ps converted phases. *J. Seismol.* **2011**, *15*, 341–353. [CrossRef]
52. Pollitz, F.F.; Wicks, C.; Thatcher, W. Mantle Flow Beneath a Continental Strike-Slip Fault: Postseismic Deformation After the 1999 Hector Mine Earthquake. *Science* **2001**, *293*, 1814. [CrossRef]
53. Pollitz, F.F.; Bürgmann, R.; Thatcher, W. Illumination of rheological mantle heterogeneity by the M7.2 2010 El Mayor-Cucapah earthquake. *Geochem. Geophys. Geosyst.* **2012**, *13*, Q06002. [CrossRef]
54. Jónsson, S. Importance of post-seismic viscous relaxation in southern Iceland. *Nat. Geosci.* **2008**, *1*, 136–139. [CrossRef]
55. Pollitz, F.F.; Bürgmann, R.; Segall, P. Joint estimation of afterslip rate and postseismic relaxation following the 1989 Loma Prieta earthquake. *J. Geophys. Res. Solid Earth* **1998**, *103*, 26975–26992. [CrossRef]
56. Dalla Via, G.; Sabadini, R.; De Natale, G.; Pingue, F. Lithospheric rheology in southern Italy inferred from postseismic viscoelastic relaxation following the 1980 Irpinia earthquake. *J. Geophys. Res. Solid Earth* **2005**, *110*, B06311. [CrossRef]
57. Wimpenny, S.; Copley, A.; Ingleby, T. Fault mechanics and post-seismic deformation at Bam, SE Iran. *Geophys. J. Int.* **2017**, *209*, 1018–1035. [CrossRef]
58. Gourmelen, N.; Amelung, F. Postseismic Mantle Relaxation in the Central Nevada Seismic Belt. *Science* **2005**, *310*, 1473. [CrossRef]
59. Nishimura, T.; Thatcher, W. Rheology of the lithosphere inferred from postseismic uplift following the 1959 Hebgen Lake earthquake. *J. Geophys. Res.* **2003**, *108*. [CrossRef]
60. Deng, J.; Hudnut, K.; Gurnis, M.; Hauksson, E. Stress loading from viscous flow in the lower crust and triggering of aftershocks following the 1994 Northridge, California, Earthquake. *Geopyhs. Res. Lett.* **1999**, *26*, 3209–3212. [CrossRef]
61. Wang, L.; Wang, R.; Roth, F.; Enescu, B.; Hainzl, S.; Ergintav, S. Afterslip and viscoelastic relaxation following the 1999 M 7.4 İzmit earthquake from GPS measurements. *Geophys. J. Int.* **2009**, *178*, 1220–1237. [CrossRef]
62. Sheu, S.; Shieh, C. Viscoelastic–afterslip concurrence: A possible mechanism in the early post-seismic deformation of the Mw 7.6, 1999 Chi-Chi (Taiwan) earthquake. *Geophys. J. Int.* **2004**, *159*, 1112–1124. [CrossRef]
63. Ryder, I.; Parsons, B.; Wright, T.J.; Funning, G.J. Post-seismic motion following the 1997 Manyi (Tibet) earthquake: InSAR observations and modelling. *Geophys. J. Int.* **2007**, *169*, 1009–1027. [CrossRef]
64. Vergnolle, M.; Pollitz, F.; Calais, E. Constraints on the viscosity of the continental crust and mantle from GPS measurements and postseismic deformation models in western Mongolia. *J. Geophys. Res. Solid Earth* **2003**, *108*, 2502. [CrossRef]
65. Reddy, C.D.; Sunil, P.S.; Bürgmann, R.; Chandrasekhar, D.V.; Kato, T. Postseismic relaxation due to Bhuj earthquake on January 26, 2001: Possible mechanisms and processes. *Nat. Hazards* **2013**, *65*, 1119–1134. [CrossRef]
66. Ryder, I.; Bürgmann, R.; Pollitz, F. Lower crustal relaxation beneath the Tibetan Plateau and Qaidam Basin following the 2001 Kokoxili earthquake. *Geophys. J. Int.* **2011**, *187*, 613–630. [CrossRef]
67. Johnson, K.M.; Bürgmann, R.; Freymueller, J.T. Coupled afterslip and viscoelastic flow following the 2002 Denali Fault, Alaska earthquake. *Geophys. J. Int.* **2009**, *176*, 670–682. [CrossRef]
68. Bruhat, L.; Barbot, S.; Avouac, J.-P. Evidence for postseismic deformation of the lower crust following the 2004 Mw6.0 Parkfield earthquake. *J. Geophys. Res. Solid Earth* **2011**, *116*, B08401. [CrossRef]
69. Dixon, J.E.; Dixon, T.H.; Bell, D.R.; Malservisi, R. Lateral variation in upper mantle viscosity: Role of water. *Earth Planet. Sci. Lett.* **2004**, *222*, 451–467. [CrossRef]
70. Hirth, G.; Kohlstedt, D. Rheology of the Upper Mantle and the Mantle Wedge: A View from the Experimentalists. In *Inside the Subduction Factory*; Geophysical Monograph Series; Eiler, J., Ed.; Blackwell Publishing Ltd.: Oxford, UK, 2004; pp. 83–105.

71. Bürgmann, R.; Dresen, G. Rheology of the Lower Crust and Upper Mantle: Evidence from Rock Mechanics, Geodesy, and Field Observations. *Annu. Rev. Earth Planet. Sci.* **2008**, *36*, 531–567. [CrossRef]
72. Batir, J.F.B.; David, D.; Richards, M.C. Updated Surface Heat Flow Map of Alaska. In Proceedings of the Geothermal Resources Council Transactions, 2013 Annual Meeting, Las Vegas, NV, USA, 30 September–3 October 2013.
73. Hjartarson, Á. Heat Flow in Iceland. In Proceedings of the World Geothermal Congress 2015, Melbourne, Australia, 19–25 April 2015.
74. Mahmoodabadi, M.; Yaminifard, F.; Tatar, M.; Kaviani, A.; Motaghi, K. Upper-mantle velocity structure beneath the Zagros collision zone, Central Iran and Alborz from nonlinear teleseismic tomography. *Geophys. J. Int.* **2019**, *218*, 414–428. [CrossRef]
75. Chen, C.-H.; Wang, W.-H.; Teng, T.-L. 3D Velocity Structure around the Source Area of the 1999 Chi-Chi, Taiwan, Earthquake: Before and After the Mainshock. *Bull. Seismol. Soc. Am.* **2001**, *91*, 1013–1027. [CrossRef]
76. Moreno, M.; Haberland, C.; Oncken, O.; Rietbrock, A.; Angiboust, S.; Heidbach, O. Locking of the Chile subduction zone controlled by fluid pressure before the 2010 earthquake. *Nat. Geosci.* **2014**, *7*, 292–296. [CrossRef]
77. Freed, A.M.; Bürgmann, R. Evidence of power-law flow in the Mojave desert mantle. *Nature* **2004**, *430*, 548–551. [CrossRef] [PubMed]

© 2020 by the authors. Licensee MDPI, Basel, Switzerland. This article is an open access article distributed under the terms and conditions of the Creative Commons Attribution (CC BY) license (http://creativecommons.org/licenses/by/4.0/).

Article

Shear-Wave Tomography Using Ocean Ambient Noise with Interference

Guoli Wu [1,2], Hefeng Dong [2,*], Ganpan Ke [3] and Junqiang Song [1]

1. College of Meteorology and Oceanography, National University of Defense Technology, Changsha 410073, China; wuguoli13@nudt.edu.cn (G.W.); junqiang@nudt.edu.cn (J.S.)
2. Department of Electronic Systems, Norwegian University of Science and Technology, 7491 Trondheim, Norway
3. Equinor ASA, Arkitekt Ebbells Vei 10, 7005 Trondheim, Norway; gank@equinor.com
* Correspondence: hefeng.dong@ntnu.no

Received: 18 June 2020; Accepted: 21 August 2020; Published: 11 September 2020

Abstract: Ambient noise carries abundant subsurface structure information and attracts ever-increasing attention in the past decades. However, there are lots of interference factors in the ambient noise in the real world, making the noise difficult to be utilized in seismic interferometry. The paper performs shear-wave tomography on a very short recording of ocean ambient noise with interference. An adapted eigenvalue-based filter is adopted as a pre-processing method to deal with the strong, directional interference problem. Beamforming and the noise crosscorrelation analyses show that the filter works well on the noise recorded by the array. Directional energy is significantly suppressed and the background diffuse component of the noise is relatively enhanced. The shear-wave tomography shows a 4-layer subsurface structure of the area covered by the array, with relatively homogeneous distribution of the shear-wave velocity values in the top three layers and a complicated structure in the bottom layer. Moreover, 3 high-velocity zones can be recognized in the bottom layer. The result is compared with several other tomography results using different methods and data. It demonstrates that, although the ambient noise used in this paper is very short and severely contaminated, a reasonable tomography result can be obtained by applying the adapted eigenvalue-based filter. Since it is the first application of the adapted eigenvalue-based filter in seismic tomography using ambient noise, the paper proves the effectiveness of this technique and shows the potential of the technique in ambient noise processing and passive seismic interferometry.

Keywords: passive seismic interferometry; surface wave; inversion; shear-wave velocity; ambient noise; dispersion curve

---

1. Introduction

Passive seismic interferometry (PSI) makes use of the noise crosscorrelations (NCC) recorded by receiver pairs to reconstruct the subsurface impulse response. It turns one of the receivers in the pairs into a virtual source whose hypothetical reflection/energy is imaged by the other receiver. In this way, an estimate of the Green's function between the receivers can be obtained [1,2]. The waves in the Green's function usually carry much useful information of the environment between the receivers. This technique provides geophysicists a new perspective to view noise and study the structure of subsurface with no need for an active seismic source [3]. As an environmental-friendly passive imaging technique, PSI significantly enhances the importance of naturally occurring ambient noise and makes it possible to image subsurface structure and monitor temporal subsurface changes conveniently, environmental-friendly, and economically [4–6].

Aki [7] and Claerbout [8] developed the early framework of the technique. The theoretical basis of the technique was completed in the early 2000s [2,9–13]. Later, the application of the theory spreads in multiple research fields including seismology [1,8], helioseismology [14], ultrasonics [9], and underwater acoustics [15]. Nowadays, PSI has become a popular technique in subsurface tomography. Various applications have been done at different scales and for various purposes all over the world [16–20].

One of the key tasks of PSI is the reliable approximation of Green's functions between receiver pairs. It requires that the noise field should be diffuse and equipartitioned [21,22]. Unfortunately, the noise field is usually contaminated by strong, directional sources in the real world, introducing biases into the approximated Green's functions [23–27]. Wu et al. [22] reviewed the techniques which have been proposed to attenuate the interference of strong directional sources, and developed an adapted eigenvalue-based filter to improve the quality of the estimation of the Green's function. The filter is adaptable for different data sets and can reduce the influence of strong, directional sources significantly.

The main objective of the paper is to perform shear-wave tomography on a very short recording of ocean ambient noise with interference and compare the tomographic picture obtained with other tomography results using active sources or much longer ambient noise recordings at the same field. The noise was recorded by a permanent reservoir monitoring receiver array (PRMRA) installed on the seabed of an offshore oil field in Norwegian North Sea. The adapted eigenvalue-based filter is applied as a pre-processing method to suppress the strong, directional interference in the recording and retrieve reliable approximations of Green's functions. Note that the paper focuses on the same dataset with Wu et al. [22] but Wu et al. [22] focuses on the analysis of noise recording at one sensor, while this paper focuses on the whole data array.

The paper is organized as follows: First, we describe the array geometry and perform time and frequency analysis of the measured ambient noise. Then, we summarize the most important ideas of the adapted eigenvalue-based filter and apply the adapted eigenvalue-based filter to the measured ambient noise. Conventional beamforming is applied to the filtered sample covariance matrix (SCM) to evaluate the effectiveness of the adapted eigenvalue-based filter. The NCC is also retrieved based on the filtered SCM at this time. Later, the phase-speed dispersion curves are extracted and a 3D shear-wave tomography is performed. The result is compared with several previously published tomography results using different methods. Finally, we summarize the work and draw some conclusions.

2. Array Geometry and Noise Analysis

2.1. Array Geometry

The passive noise data were acquired by the PRMRA installed by Equinor in the Norwegian North Sea. The PRMRA contains 17 cables with length from 1.5 to 12.85 km, covering an area of approximately 50 km^2. The Array Geometry is shown in Figure 1. The PRMRA is oriented in a 26–206° direction. The cable spacing is 300 m and the receiver spacing in cable is 50 m.

2.2. Noise Analysis

The noise data used here were continuously recorded by hydrophones in each receiver station for 1.02 h at a sampling rate of 500 Hz on 14 September 2015.

2.2.1. The 'Natural Ambient Noise'

Figure 2a shows the noise recording on one receiver. Relatively speaking, the noise recordings marked with blue box in Figure 2a are not affected by severe events and can be denoted it as 'natural ambient noise'. Figure 2c shows the enlarged view of the 'natural ambient noise' and Figure 2d presents the corresponding power spectral density (PSD) of it in blue. The PSD of the 'natural ambient noise' (blue) shows a peak at 0.7 Hz, which can be recognized as the peak of the microseisms in the ocean [28]. As the frequency increases from 0.7 Hz to 4.5 Hz, the PSD decreases gradually and no other peak appears. Actually, the frequency band of 0.2 to 4.5 Hz of the ocean ambient noise is usually dominated by the microseisms, which are believed to be generated by wind-related wave–wave interaction [28]. The microseisms are thought to be more evenly distributed in the ocean and the frequency band of 0.2 to 4.5 Hz is chosen to apply PSI in this paper.

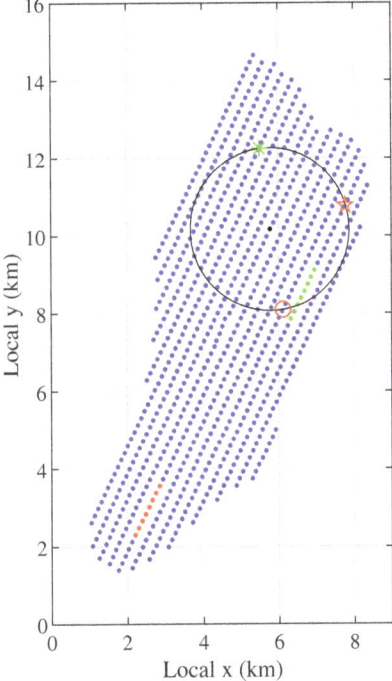

Figure 1. The PRMRA geometry. Not all the receivers are plotted in the figure for display purposes. The gather marked in red is used in the following example of beamforming. The gathers marked in green and red are used in the following examples of the NCC retrieval. The black circle and marks on it are used in the following study of events.

2.2.2. Event A

In Figure 2a, waves with large amplitude (denoted as Event A) dominate the noise in the range of 4 to 26 min and 46 to 57 min. Figure 2b shows the averaged spectrogram of all receivers. Most of the energy of Event A spreads within the frequency band larger than 5 Hz. Part of the energy also affects the frequency band of 2 to 5 Hz considerably (especially in the time band of 20 to 26 min). Figure 2d presents the corresponding PSD of part of Event A (marked by a green box in Figure 2a) in green. It shows a similar peak with 'natural ambient noise' (blue) at 0.7 Hz but also shows increased energy beyond 2.5 Hz. 2 other higher peaks appears at about 8 Hz and 15 Hz.

2.2.3. Event B

In Figure 2a, Event B (marked by red box) happened during the period of 31 to 36 min. It is not obvious in the raw data. However, it becomes apparent in the spectrogram (solid red box in Figure 2b) and shows high energy in the frequency band of 1.5 to 5 Hz. Figure 2d presents PSD of Event B in red. It shows a similar trend to 'natural ambient noise' (blue) and Event A (green) below 0.7 Hz, but shows increased energy in the frequency band of 0.7 to 20 Hz. A higher peak appears at about 3 Hz.

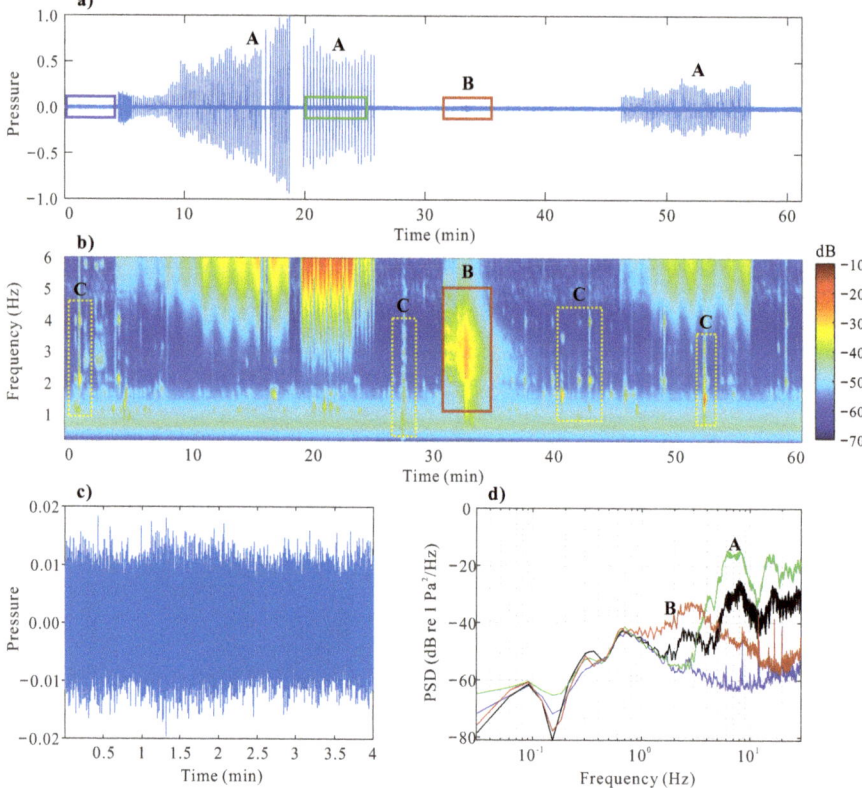

Figure 2. Temporal and spectral analysis. Letters A, B and C denote different events detected. (**a**) the pressure recorded at one receiver. The pressure is normalized by the maximum amplitude. The blue box selects a window where the noise is not affected by severe events. The green box selects a window where the noise is affected by Event A. The red box selects a window where the noise is affected by Event B; (**b**) the averaged PSD (dB re 1 Pa^2/Hz) of all receivers as a function of time and frequency (spectrogram); (**c**) the enlarged view of the noise marked by blue box in (**a**); (**d**) the PSD of the corresponding noise marked by boxes in (**a**). The curves are related to the noise by the color of the boxes in (**a**). The black curve is the averaged PSD over 60 min.

2.2.4. Events C

Some other events can be recognized in the spectrogram (denoted as Events C and marked by dashed box in Figure 2b). These events have higher energy than the background noise and are usually generated by strong, directional sources.

2.2.5. Discussion of the Events

Overall, Figure 2d indicates that the PSD are clearly contaminated by Events A and B which are generated by strong and directional sources. The black curve shows the PSD of 60-min recordings. It describes the average effect of all events. Compared with the blue line, it shows increased energy beyond 1.5 Hz and is especially influenced by Events A and B.

The existence of Events A, B, and C are common and unpredictable in the real world. It damages the diffuse and equipartition property of the noise and bring challenges to the PSI.

2.2.6. Additional Insights on the Events

Figure 2 shows that Event A has a very regular time interval. The temporal and spectral signatures of Event A indicate that it comes from artificial sources which are very like airgun shots from other seismic surveys near this field. Unfortunately, there is no exact information available on these shots. In order to further study these signals, a travel time study is employed. Thirty-one sensors on the circle in Figure 1 are selected and numbered counterclockwise (with the red five-pointed star numbered as 1). Figure 3a shows the arrival time of the airgun signal on different sensors. The 4th sensor gets roughly the earliest arrival (green) and the 25th sensor gets the latest arrival (red). The 4th sensor and the 25th sensor are marked by a green asterisk and the red small circle in Figure 1, respectively. It indicates that the angle of incidence of the airgun shots is approximately 45° (the normal direction, which is perpendicular to the cable, is taken as a reference in the paper). The velocity of the signal is about 1450 m/s, which is estimated by travel time analysis. Figure 3b shows the spectrum of the signals in Figure 3a. It shows that the main energy of the airgun is concentrated in the frequency band of 3 to 200 Hz.

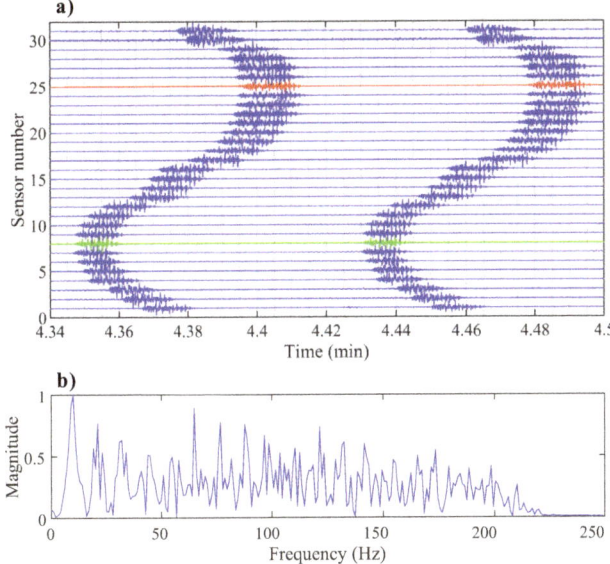

Figure 3. (a) arrival analysis on Event A. The green waveform represents the earliest arrival and the red waveform represents the latest arrival. The amplitudes of the signals are normalized by the maximum of each waveform; (b) the spectrum of Event A. Note that the magnitude is normalized to the maximum value.

Figure 2b shows that the energy of Event B concentrate in the frequency band of 1.5 to 5 Hz. It fits the spectrum of earthquake very well [29]. Thus, it is possibly a small earthquake which lasts few minutes.

3. NCC Retrieval Using the Adapted Eigenvalue-Based Filter

Events similar to A, B, and C are unavoidable and unpredictable in the real world, challenging the accuracy and stability of the PSI. The adapted eigenvalue-based filter can help to mitigate the influence of these events and obtain improved NCC [22]. We apply the adapted eigenvalue-based filter to the noise data recorded by the PRMRA. In the following subsections, we summarize the basic ideas of the adapted eigenvalue-based filter (a detailed description can be found in [22]) and retrieve the NCC from the contaminated noise data using the adapted eigenvalue-based filter.

3.1. An Overview of the Adapted Eigenvalue-Based Filter

The main operation of the adapted eigenvalue-based filter is carried out on the SCM $\hat{\mathbf{R}}(f)$. Supposing that the noise records are segmented into M segments, the SCM $\hat{\mathbf{R}}(f)$ can be computed by

$$\hat{\mathbf{R}}(f) = \frac{1}{M} \sum_{m=1}^{M} \mathbf{u}_m(f)\mathbf{u}_m(f)^H, \tag{1}$$

where $\mathbf{u}_m(f)$ is the Fourier coefficients vector of the noise records at a particular frequency f and H denotes Hermitian transpose.

The adapted eigenvalue-based filter tries to separate the SCM into different components as

$$\begin{aligned}\hat{\mathbf{R}} &= \sum_{k=1}^{K} \hat{\lambda}_k \hat{\mathbf{v}}_k \hat{\mathbf{v}}_k^H + \sum_{k=K+1}^{N'} \hat{\lambda}_k \hat{\mathbf{v}}_k \hat{\mathbf{v}}_k^H + \sum_{k=N'+1}^{N} \hat{\lambda}_k \hat{\mathbf{v}}_k \hat{\mathbf{v}}_k^H \\ &= \hat{\mathbf{R}}_s + \hat{\mathbf{R}}_d + \hat{\mathbf{R}}_i,\end{aligned} \tag{2}$$

where $\hat{\lambda}_k$ (ascending order) and $\hat{\mathbf{v}}_k$ are the eigenvalues and eigenvectors of $\hat{\mathbf{R}}$, respectively, $\hat{\mathbf{R}}_s$ is the strong, directional noise-related component, $\hat{\mathbf{R}}_d$ is the diffuse noise-related component, and $\hat{\mathbf{R}}_i$ is the uncorrelated noise-related component. The objective of the separation is to get a wavefield which is closer to be equipartitioned by suppressing $\hat{\mathbf{R}}_s$ and $\hat{\mathbf{R}}_i$ in the SCM.

The key point of the separation is the determination of values for N' and K. N' is called the cutoff number. Eigenvalues smaller than $\hat{\lambda}_{N'}$ are thought to be related to uncorrelated noise (such as electronic noise and sensor self-noise) and are filtered. N' can be determined by a theoretically derivation based on the geometry and degrees of freedom of the wavefield [22,30]. Eigenvalues larger than $\hat{\lambda}_K$ are thought to be related to strong, directional sources and need to be suppressed. A statistical hypothesis test is applied to determine the value of K [22]. By comparing the largest eigenvalues of the SCM and the statistical model of the ideal SCM (generated by diffuse, equipartitioned wavefield), all the eigenvalues larger than $\hat{\lambda}_K$ will be rejected. The weight of the hypothesis test enhances the effectiveness of the adapted eigenvalue-based filter and makes it more adaptable for different datasets.

The basic strategies to determine K and N' are provided in Appendix A. Once the values for K and N' are determined, we can obtain the filtered SCM $\hat{R}'_{ij}(f)$ by suppressing $\hat{\mathbf{R}}_s$ and $\hat{\mathbf{R}}_i$. The time-domain NCC function $C_{ij}(t)$ is defined as

$$C_{ij}(t) = \int_0^T s_i(\tau)s_j(\tau+t)d\tau, \tag{3}$$

where $s_i(t)$ and $s_j(t)$ denote the ambient-noise records obtained by receivers i and j, respectively, and T denotes the observation period. In the frequency domain, given the filtered SCM $\hat{\mathbf{R}}'(f)$, we can compute the NCC of the data as

$$\hat{C}_{ij}(t) = \mathcal{F}^{-1}[\hat{R}'_{ij}(f)], \tag{4}$$

where $\mathcal{F}^{-1}$ denotes the inverse Fourier transform, f denotes the frequency, and $\hat{R}'_{ij}(f)$ denotes the entry of $\hat{\mathbf{R}}'(f)$.

3.2. The Result

3.2.1. Beamforming

Wu et al. [22] studied one straight cable of the PRMRA and showed the good effect of the adapted eigenvalue-based filter. Since the PRMRA covers an area of 50 km^2, local environments for different sensors and the coherence of the noise sources around different sensors may be different. It is necessary to study the application of the adapted eigenvalue-based filter to the entire array further. The study is also valuable for the shear-wave tomography in the following section.

To investigate the directionality of the wavefield, conventional beamforming is applied to the data recorded by different gathers (choose 30 continuously adjacent sensors in one cable as a gather, as is marked in red in Figure 1, for example). If the wavefield is diffuse and equipartitioned, the beamforming energy should be evenly distributed in all directions. An empirical estimation of the average phase speed 900 m/s is used in the beamforming. Although the estimation is rough, it does not make a big difference and is good enough to show the distribution of incoming energy. In addition, an iterative study can be done after the dispersion curves of the surface wave are extracted if it is necessary.

Figure 4 shows the beamforming result of one gather which is marked in red in Figure 1. Figure 4a,c show the beamforming energy distribution of the original wavefield. The energy is higher in the directions of −20° to 20° in Figure 4a than the background level. The energy becomes extraordinarily high from 30 to 40 min at 0° and can be recognized as Event B considering the high consistency in time and frequency. It indicates that Event B comes from the normal direction of the array. In Figure 4c, a source dominates the beam power with an almost constant direction of about 40°. It can be recognized as Event A considering the high consistency in time. The travel time study of Event A shows the incident angle of the airgun shots is approximately 45°. The misfit is caused by the usage of averaged phase speed in the beamforming. Figure 4a,c indicate that the original wavefield is affected by several strong, directional sources at both 2 Hz and 4 Hz. Event C, however, is not visible in the beamforming results because they have relatively lower energy than other events.

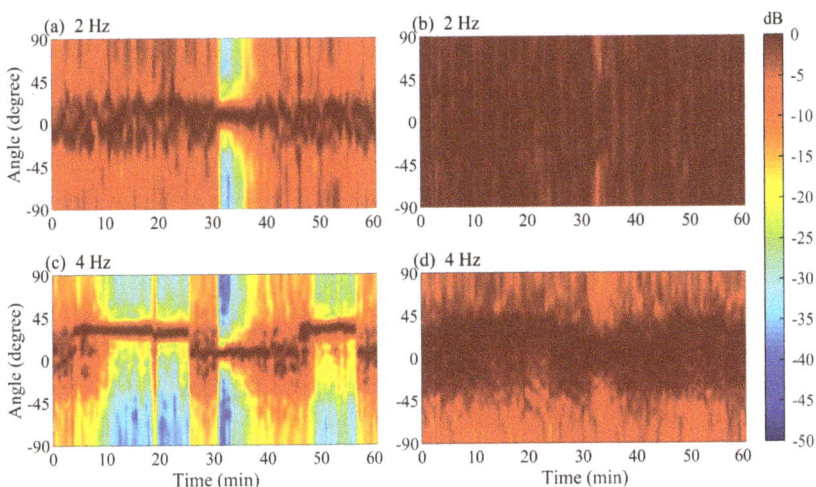

Figure 4. Beamforming results. (**a**,**b**) show the beamforming energy distribution of 2 Hz; (**c**,**d**) show the beamforming energy distribution of 4 Hz; (**a**,**c**) show the beamforming energy distribution of the original wavefield; (**b**,**d**) show the beamforming energy distribution of the filtered wavefield using the adapted eigenvalue-based filter.

Figure 4b,d illustrate the beamforming energy distribution of the filtered wavefield using the adapted eigenvalue-based filter. Figure 4b,d show relatively uniform distributed energy compared with Figure 4a,c. It indicates that the adapted eigenvalue-based filter can suppress most of the high energy and partially recover the expected diffuse and equipartitioned wavefield.

Keeping all the parameters (especially the weight) of the adapted eigenvalue-based filter the same, similar results have been obtained for other gathers. It indicates that the influence of non-microseismic events on the receiver array over 50 km^2 is more or less the same and that the adapted eigenvalue-based filter and the parameters used keep effective at the scale of 50 km^2.

3.2.2. The NCC Retrieval

Selecting 30 continuously adjacent sensors in one cable as a gather and 25 sensors as the overlap between the adjacent gathers, 573 gathers are obtained from the entire PRMRA. We have retrieved the NCC for all gathers using Equation (4).

Figure 5 displays the NCC results for two gathers marked in red (Figure 5a,b) and green (Figure 5c,d) in Figure 1. All the NCCs are bandpass filtered between 0.2 and 4.5 Hz. Figure 5a,c displays the NCC retrieved from unfiltered SCM. The waveforms are apparently contaminated by directional, energetic events. The NCCs in Figure 5a uses the same gather with beamforming in Figure 4. The narrow peaks (close to 0 s) traveling with a velocity of about 1450 m/s are nondispersive and can be recognized as Event A. However, the effect of Events B and C is not very tyical in the NCC, making it hard to recognize. The existence of these directional, energetic events shields the expected surface waves severely. Figure 5b,d displays the NCC retrieved from the filtered SCM. Most of the influence of the directional, energetic events has been removed from the NCC and the expected surface waves are significantly enhanced. Figure 5b,d also show a clear dispersion feature, which is a typical characteristic for surface waves.

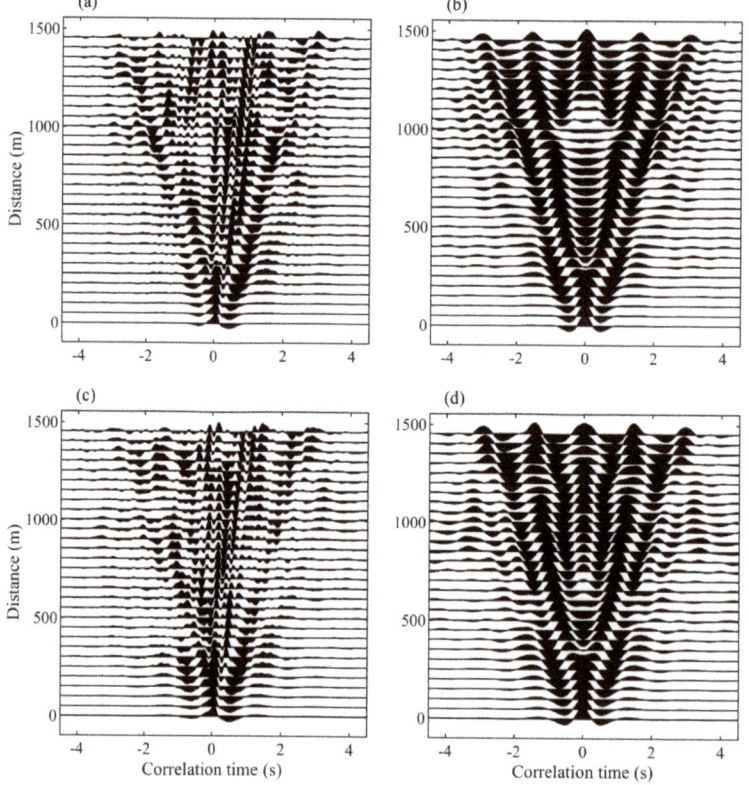

Figure 5. The NCC results. Note that the NCCs are bandpass filtered between 0.2 and 4.5 Hz. (**a**,**b**) displays the NCC of the gather marked in red in Figure 1; (**c**,**d**) displays the NCC of the gather marked in green in Figure 1; (**a**,**c**) displays the NCC retrieved from unfiltered SCM while (**b**,**d**) displays the NCC retrieved from the filtered SCM.

The values of basic parameters used in the AEF are displayed in Figures 6 and 7. The value of N' shown in Figure 6 is a function of frequency when gathers with the same shape are used and an averaged slowness of the medium 1.1 s/km is chosen. The weight $w = 0.2$ is used for this study. Note that K may be different for different gathers. Figure 7 displays the values of K of the gather marked in red in Figure 1. It shows increased values for larger frequencies because energetic events dominate higher frequencies over almost the entire time, as is indicated in Figure 4c. Between 30–40 min at around 1.5–3 Hz, Figure 7 shows larger values of K. It corresponds well with Figure 4a, where Event B happens at 30–40 min. For frequencies lower than 1.5 Hz, the values of K are very small, indicating a better equipartitioned wavefield for lower frequencies.

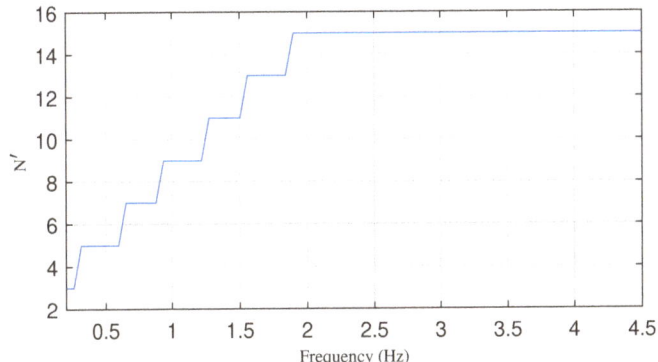

Figure 6. The value of N' as a function of frequency.

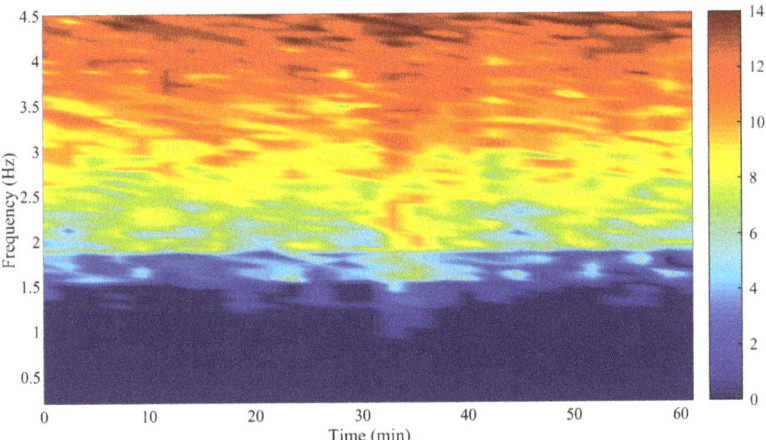

Figure 7. The value of K as a function of frequency and time. The chosen gather is marked in red in Figure 1.

4. Shear-Wave Tomography

The dispersive behavior of the retrieved surface waves carry useful information about the subsurface. The dispersion property of the retrieved surface waves is highly sensitive to shear-wave velocities at different depths and is used to perform 3D shear-wave tomography of the area.

The slowness-frequency transform (SFT) [31] is applied to the retrieved NCC of each gather. The result of the SFT can be expressed as a Gabor diagram, which shows signal energy as a function of frequency and phase velocity. Then, the phase–velocity dispersion curves can be extracted from the resulting Gabor diagram of the SFT. Examples of the retrieved dispersion curves can be found in [22].

4.1. The ASSA Inversion

A nonlinear optimization method, adaptive simplex simulated annealing algorithm (ASSA) [32], is applied to the fundamental mode of extracted dispersion curves of all gathers for estimating shear-wave velocity as a function of depth and range in the sea bottom.

The ASSA combines the simulated annealing and downhill simplex method. It operates on a simplex of M models and randomly perturbs the parameters after a downhill simplex step. The obtained trial model is subsequently accepted or rejected according to the Metropolis criteria

$$P(\Delta E) = \exp(-\Delta E/T), \tag{5}$$

where T denotes the temperature and E denotes the mismatch. The temperature T is decreased every N_p accepted perturbations using

$$T_j = \beta^j T_0, \tag{6}$$

where T_0 denotes the initial temperature and β denotes a temperature reduction factor satisfying $0 < \beta < 1$. The procedure continues until the models in the simplex satisfy the convergence condition

$$\frac{E_{high} - E_{low}}{\left(E_{high} + E_{low}\right)/2} < \epsilon, \tag{7}$$

where E_{high} and E_{low} represent the highest and lowest mismatch model in the simplex, respectively. Assuming that the measured dispersion data errors (including measurement error and theory error) are independent, Gaussian-distributed random processes, the likelihood function can be written as

$$L(\mathbf{m}) = \frac{1}{(2\pi)^{N/2} \prod_{i=1}^{N} \sigma_i} \exp\left\{-\sum_{i=1}^{N} \frac{[d_i - d_i(\mathbf{m})]^2}{2\sigma_i^2}\right\}, \tag{8}$$

where $\mathbf{d}$ denotes the dispersion data vector with N entries, $\mathbf{d}(\mathbf{m})$ denotes the modeled data vector and σ is the standard deviation of $\mathbf{d}$. The data misfit

$$E(\mathbf{m}) = \sum_{i=1}^{N} \frac{[d_i - d_i(\mathbf{m})]^2}{2\sigma_i^2} \tag{9}$$

is used as the objective function in the inversion. Therefore, the ASSA minimizes the misfit and yields an maximum-likelihood solution.

The chosen values of the mentioned parameters in this paper are listed in Table 1. Note that M is chosen based on the dimensional of the search space, E_0 is the initial misfit of the initialized model and σ_i is assumed to be the same for all entries of σ.

Table 1. Values of the parameters used in the ASSA inversion.

Parameter	Description	Value
M	Number of models in the simplex	11
T_0	Initial temperature	E_0
N_p	Temperature decrease frequency	10
β	Temperature reduction factor	0.995
ϵ	Convergence tolerance	0.001
σ_i	Standard deviation of data (km/s)	0.1

The forward problem is solved using the DISPER80 subroutines [33], which can calculate phase–velocity dispersion curves given a geoacoustic model.

4.2. Seabed Parameterizations

A 6-layer geoacoustic model (shown in Figure 8) is used in the inversion. The 6th layer is a semi-infinite space. Considering that the dispersion property of the surface wave is not sensitive to the density ρ and the compressional velocity v_P in the frequency range used (0.2 to 4.5 Hz), we use the following empirical equation [34,35]

$$v_P = 1.16 \times v_S + 1.36, \qquad (10)$$

$$\rho = 1.74 v_P^{0.25}, \qquad (11)$$

to evaluate the values of them during the inversion.

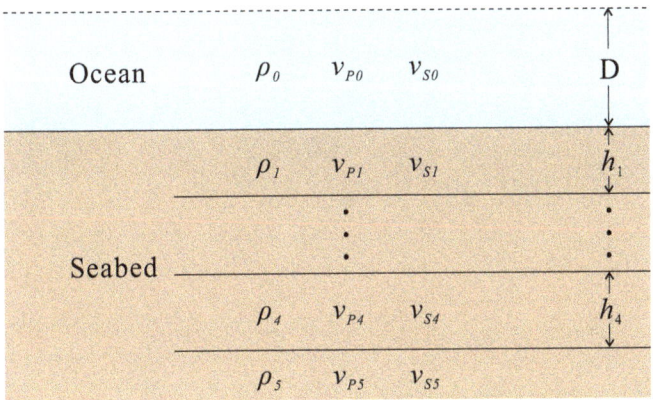

Figure 8. Geoacoustic model used in the inversion.

The shear-wave velocity v_S and the thickness of the layer h are variable in the iteration of the inversion process. Prior bounds are applied to the geoacoustic parameters for excluding physically unreasonable values. Table 2 shows the chosen bounds for different parameters.

Table 2. Search bound of seabed parameters.

Parameter	Bound (km/s)	Parameter	Bound (km)
v_{S1}	[0.1, 0.5]	h_1	[0.01, 0.10]
v_{S2}	[0.2, 1.0]	h_2	[0.05, 0.80]
v_{S3}	[0.3, 2.0]	h_3	[0.10, 0.80]
v_{S4}	[0.3, 2.0]	h_4	[0.20, 0.80]
v_{S5}	[0.3, 2.0]	h_5	semi-infinite
v_{S0}	0	D	0.125
v_{P0}	1.49		

4.3. Tomography Result

The ASSA provides the maximum-likelihood solution of the data. We have performed 100 independent ASSA inversions on each gather to form a mean shear-wave velocity profile. A 3D shear-wave structure (Figures 9a and 10) can be reconstructed based on the solution.

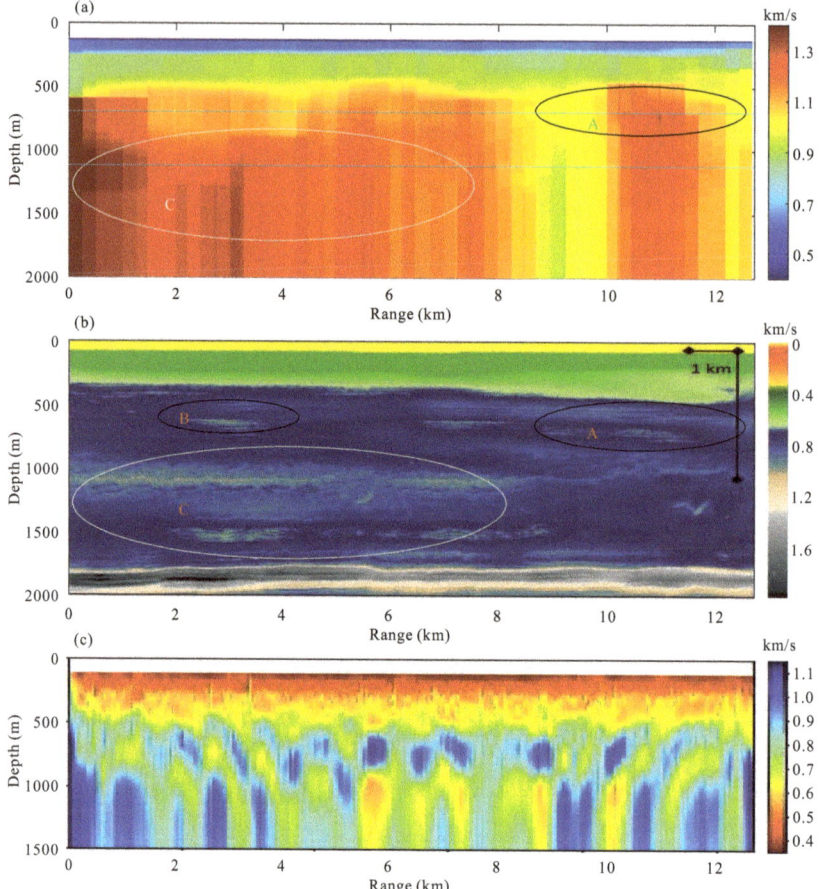

Figure 9. (**a**) a vertical slice of the resolved mean shear-wave subsurface structure under the 10th cable counted from left to right of the PRMRA in Figure 1. The top white layer represents the water layer. The dashed lines indicate the horizontal slices used in Figure 10; (**b**) a vertical slice of the shear-wave subsurface structure from active-source reflection tomography [36]; (**c**) a vertical slice of the shear-wave subsurface structure from 1D Monte Carlo ambient noise tomography [37]. The circles in (**a**,**b**) mark the high-velocity zones.

Figure 9a displays a vertical slice of the shear-wave subsurface structure under the 10th cable counted from left to right of the PRMRA in Figure 1. Four subsurface layers (excluding the water layer) are well resolved from a holistic perspective, although a 6-layer model is used in the inversion. The top white layer (0–125 m) denotes the water layer. The first layer (125–140 m) is a low-velocity layer (lower than 300 m/s) which is very thin. This layer locates near the ocean bottom and is thought to be very soft. The second layer (140–240 m) is relatively thicker and harder, and contains a velocity of about 500 to 600 m/s. As the depth goes deeper, the layers become much thicker and contain larger velocities. The third layer (240–500 m) contains a velocity of about 800 to 900 m/s. The interfaces for the top three layers below the water layer are clear and the velocity values distribute relatively homogeneous over range. The bottom layer (>500 m) is a high-velocity layer. The velocity values of it mostly span from 900 to 1300 m/s (can be even larger in several areas), indicating a more complicated structure compared with the top three layers. Note that the resolution in range of Figure 9 is 200 m, which is dependent on the chosen overlap between the adjacent gathers.

Figure 9b displays a similar slice of the shear-wave subsurface structure from active-source reflection tomography at the same field from [36]. The active-source has higher energy and the resolution of the reflection tomography is better. Thus, we consider it as a good approximation to the true model of the seabed. The top thin layer (125–140 m) of Figure 9a can not be recognized in Figure 9b. It might be caused by the regularization method which smooths the layers. However, it can be recognized easily in the P-wave velocity structure in [36]. The existence of the top thin layer in Figure 9a is reasonable because the top layer of shallow ocean is usually very soft with low shear-wave velocity. Figure 9b combines the first and second layer (125–240) of Figure 9a and shows a velocity of around 500 m/s at this layer. When the depth goes deeper in Figure 9b (240–500 m, corresponding to the third layer of Figure 9a), the velocity increases from 500 to 900 m/s. The third layer of Figure 9a can be considered as a transition layer at this place, compared with Figure 9b. In the depth range of 500 to 1700 m, although Figure 9a generally shows higher velocities than Figure 9b, they do have some similarities in the structure. At the depth of 600 m, Figure 9b shows two high-velocity zones A and B. We can find a similar zone A in Figure 9a, while zone B is not visible here possibly because of limited resolution of inversion using ambient noise. If we look further at the horizontal slice of the subsurface structure at the depth of 600 m in Figure 10c, both the high-velocity zones A and B can be recognized. At the depth of 1000 to 1600 m, Figure 9b shows another large high-velocity zone C. We can find a similar zone C in Figure 9a, although it is not as large as the zone in Figure 9b. The horizontal slice of Figure 9a at the depth of 1100 m demonstrates the distribution of this zone in Figure 10d, which is larger than it shows in Figure 9a. The high similarity in these zones of Figure 9a,b indicates that these high velocity anomalies are reliable and can be recognized as real features of the seabed. Below the depth of 1700 m, Figure 9b indicates a layer with very high velocities, which can not be recognized in Figure 9a. One possible reason is that surface wave can not penetrate to this depth, so that the inversion of Figure 9a is not sensitive to the shear-wave velocity at this depth.

Figure 9c displays a similar slice of the shear-wave subsurface structure from 1D Monte Carlo ambient noise tomography using 12 h of continuous ambient noise records at the same field from [37]. Although Zhang et al. [37] also did 2D and 3D tomography with the same data and obtained smoother results, Figure 9c is more comparable with the result of the present paper other than 2D and 3D results because they both use 1D parametrization. Figure 9c illustrates continuously increasing velocities from the depth of 125 to 500 m. The general structure corresponds well with Figure 9a,b. When the depth goes deeper than 500 m, Figure 9c shows complicated structure. These complicated features are thought to be artifacts due to the inversion strategy. Although lateral spatial constraints in 2D and 3D tomography make the structure smoother [37], the zones recognized in Figure 9a,b are not visible. One possible reason is that the authors used a narrower frequency band (0.35–1.5 Hz) compared with what we have used, and the information of the dispersion curves is limited. If more modes are included in the inversion, the frequency band needs to be wider. On the other side, people have to deal with the problem of the interferences included when a wider frequency band is used. At a depth deeper than 500 m, Figure 9c seems to get lower mean velocities than Figure 9a. The difference can be explained by several factors. One factor is the measurement errors of the dispersion data. The depth sensitivity (described in the following section) indicates that only lower frequency components can penetrate to a deeper depth so that the measurement errors at low frequencies may result in large bias to the velocities of deeper layers. The other factor is the large uncertainty at deeper layers caused by the inversion method, which can be quantified by statistical methods. The uncertainty for depth deeper than 500 m of Figure 9c is about 250 m/s [37] and the uncertainty for depth deeper than 500 m of Figure 9a is about 100 m/s (described in the following section).

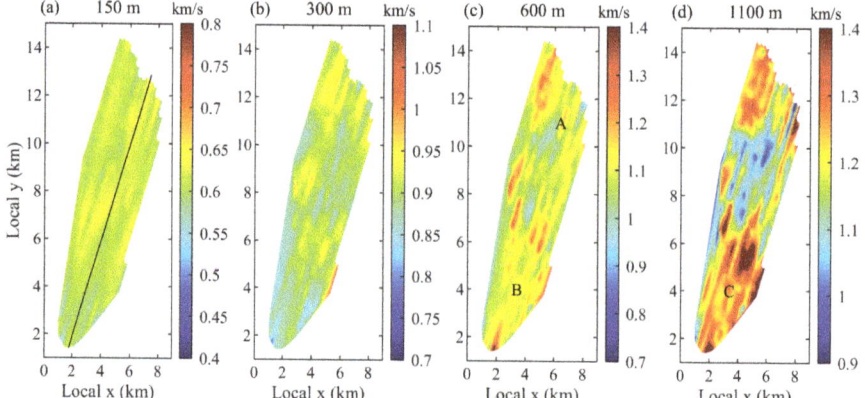

Figure 10. (**a**–**d**) show horizontal slices of the mean shear-wave subsurface structure at the depth of 150, 300, 600 and 1100 m, respectively. The black line in (**a**) shows the location of Figure 9. The letters in (**c**,**d**) correspond to the letters in Figure 9.

Figure 10 displays four horizontal slices of the shear-wave subsurface structure at different depths. The distribution is roughly 'flat' for Figure 10a,b except for some small areas with a little bit high shear-wave velocities. Figure 10c,d show the horizontal distribution of the shear-wave velocities corresponding to the straight dashed lines marked in Figure 9a. In Figure 10c, zones A and B are a little discrete. This is probably because the results slightly overfit the local data. Adding lateral constraints to the parametrization of the inversion may help to improve it. The high velocity patches in Figure 10c,d correspond to the high velocity patches in Figure 9b well. It indicates that there may be similar geological features at this field although the ground-truth is not known.

5. Discussion

5.1. Depth Sensitivity

To evaluate the depth sensitivity of the inversion, we investigate the depth sensitivity kernels of the surface waves based on the inverted shear-wave velocity model. It is constructed by computing partial derivative of the predicted phase velocity with respect to velocity perturbation at different depths.

Figure 11 shows the normalized depth sensitivity kernels for different frequencies of surface waves. It shows that lower-frequency kernels have peak sensitivity at deeper depths while higher-frequency kernels have less sensitivity to deep subsurface structure. It also indicates that deeper layers may be mostly determined by frequencies lower than 1 Hz.

5.2. Tomography Resolution

The ASSA inversions are performed on 1D gathers and then all gathers are used to reconstruct the 3D shear-wave tomographic model. Since we use 30 continuously adjacent sensors (with 50 m spacing) in one cable as a gather and 25 sensors as the overlap between the adjacent gathers, the resolution along the cable is 250 m. The resolution across the cables depends on the spacing between cables, which is 300 m. The thickness of each layer is adaptive during the inversion process. Generally speaking, shallow layers have higher resolution due to the high sensitivity of both high frequencies and low frequencies of surface waves while deeper layers (>1000 m) have lower resolution since they only have some sensitivity to very low frequencies of surface waves.

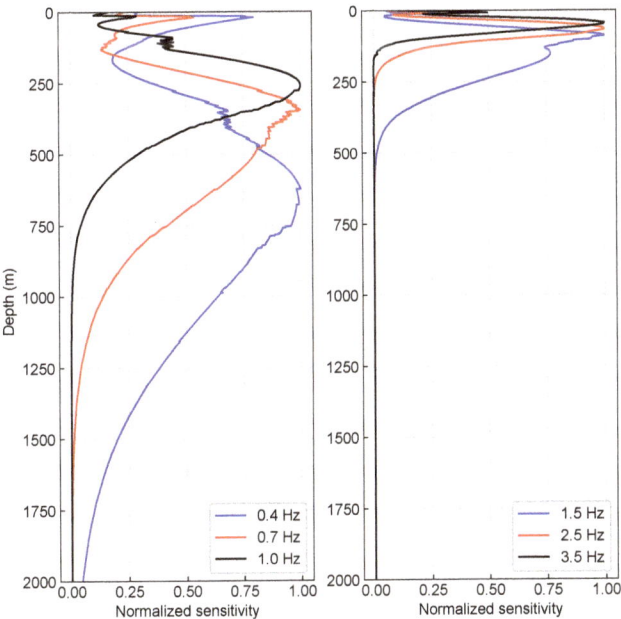

Figure 11. Normalized depth sensitivity kernels for different frequencies of surface waves.

5.3. Non-Uniqueness and Uncertainty

An important issue should be mentioned is that the non-uniqueness of the solution. Several factors can account for the non-uniqueness of the solution of the shear-wave inversion problem. One factor is that the problem is a highly nonlinear inverse problem and there is no analytical solution available. Numerical methods are used to solve such problems and the uniqueness of the solution to given data are not proven. Another factor is that the given data are usually noisy. In this paper, we assume that the dispersion data uncertainty are independent and Gaussian distributed (Equation (8)). In general, an infinite number of models fitting the noisy data can be found to an acceptable level [32]. The third factor is that, in practice, the dispersion curves are usually cut off to a certain frequency band or sometimes only part of the dispersion curves can be excited by the source, leading to some uncertainties to the solution.

To quantify the uncertainty of the inversion, we have calculated the standard deviation of 100 independent ASSA inversions. Figures 12 and 13 show a vertical and a horizontal slice of the standard deviation of the resolved shear-wave subsurface structure, respectively. Figure 12 indicates an average standard deviation of about 100 m/s. The standard deviation becomes larger at interfaces of layers and zones with a complicated structure (500 to 1500 m). This is because the thickness of each layer is adaptively determined, making slightly different layers for each independent inversion and resulting in larger standard deviation at the interfaces. Figure 13 indicates a generally homogeneous standard deviation on horizontal slices. The uncertainty for deeper layers is not as large as Zhang et al. [37] because our parameterization on number of layers and larger thickness of deeper layers, which constrains the freedom and uncertainty although limiting the resolution at the same time.

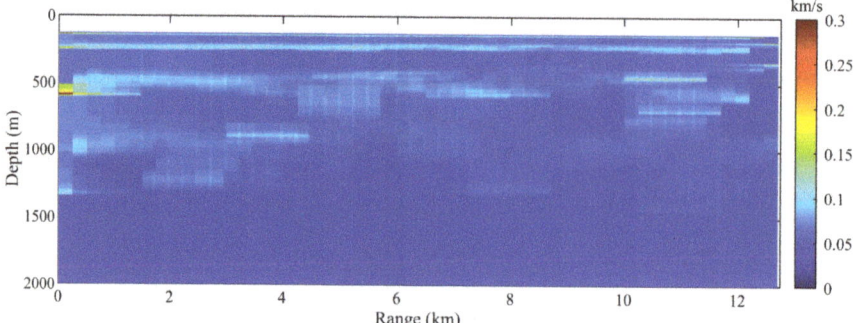

Figure 12. A vertical slice of the standard deviation of the resolved shear-wave subsurface structure under the 10th cable counted from left to right of the PRMRA in Figure 1.

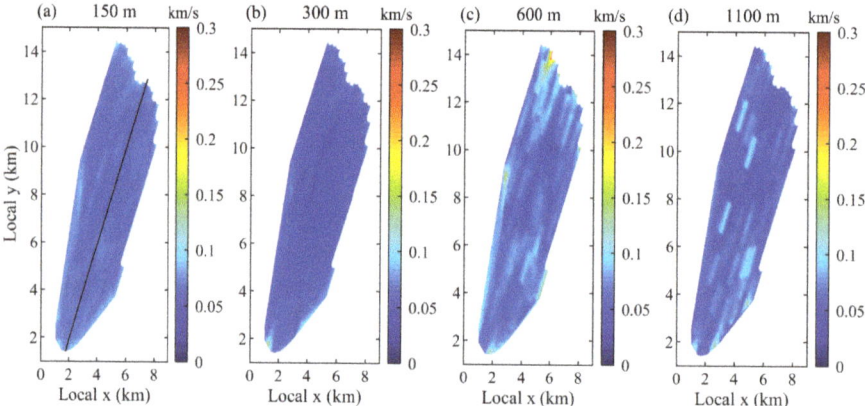

Figure 13. (a–d) show horizontal slices of the standard deviation of the shear-wave subsurface structure at the depth of 150, 300, 600 and 1100 m, respectively. The black line in (a) shows the location of Figure 12.

6. Conclusions

Shear-wave tomography is performed using only 1.02-h ambient noise recorded by a permanent reservoir monitoring receiver array installed in the Norwegian North Sea. Passive tomography needs the wavefield to be diffuse and equipartitioned. However, the wavefield can usually be contaminated by various events in the real world. Time and frequency analysis indicates that the ambient noise used in the paper is contaminated by several high energy events. The beamforming results further clarify that these events are directional and should be suppressed before the noise is cross-correlated. An adapted eigenvalue-based filter is applied to accomplish the suppression and shows good effects both in beamforming and noise crosscorrelation results. The filter makes it possible to implement shear-wave tomography on the ambient noise with interference. Finally, an adaptive simplex simulated annealing algorithm (ASSA) is applied for estimating the three-dimensional subsurface shear-wave structure. The results indicate a four-layer structure with a generally increasing velocity as the depth goes deeper. The velocity values of the first three layers distribute relatively homogeneous, while the fourth layer shows a more complicated distribution, indicating a heterogeneous structure under 500 m. The comparison of the tomography result with other methods demonstrates that the tomography presented in this paper is reasonable and encouraging, although the ambient noise is very short and contaminated.

Author Contributions: Formal analysis, G.W., H.D., G.K., and J.S.; Funding acquisition, H.D. and J.S.; Methodology, G.W. and H.D.; Resources, G.W. and G.K.; Software, G.W. and H.D.; Writing—original draft, G.W.; Writing—review and editing, H.D., G.K., and J.S. All authors have read and agreed to the published version of the manuscript.

Funding: This research was funded by the National Natural Science Foundation of China, Grant Number [41605070].

Acknowledgments: The authors would like to acknowledge Grane license partners (Equinor ASA, Petoro AS, Vår Energi AS, and ConocoPhillips Skandinavia AS) for allowing to present this work. The views and opinions expressed in this paper are those of the authors and are not necessarily shared by the license partners; H.D. thanks the Norwegian Research Council and the industry partners of the GAMES consortium at NTNU for financial support (Grant No. 294404); We would like to give special thanks to section managing editor Bommie Xiong and unknown reviewers for their help and comments that significantly improved the manuscript.

Conflicts of Interest: The authors declare no conflict of interest.

Abbreviations

The following abbreviations are used in this manuscript:

PSI	Passive seismic interferometry
NCC	Noise crosscorrelations
1D/2D/3D	1/2/3 dimensional
PRMRA	Permanent reservoir monitoring receiver array
SCM	Sample covariance matrix
ASSA	Adaptive simplex simulated annealing algorithm

Appendix A. The Strategies to Determine N' and K

This appendix provides the strategies we follow to determine the value of N' and K. The strategies are from [22] and a more complete description can be found there.

Appendix A.1. The Determination of N'

Supposing that γ is the slowness of the medium, f is the frequency, N is the number of sensors, r_{ij} is the great-circle distance between two sensors i and j, and $\bar{r}$ is the typical separation of the array which is defined as

$$\bar{r} = \frac{2}{N(N-1)} \sum_{i=1}^{N} \sum_{j>i}^{N} r_{ij}, \tag{A1}$$

we can determine the value of N' using

$$N'(f) = min\{2\lceil 2\pi f \gamma \bar{r} \rceil + 1, N/2\}, \tag{A2}$$

where $\lceil x \rceil$ is the least integer greater than or equal to x.

Equation (A2) implies that the determination of N' only relies on the frequency, the slowness of the medium, the typical separation, and the number of sensors used. Since the slowness of the medium is not known beforehand, a rough estimation of the average slowness is sufficient to use in practice, as is suggested by Seydoux et al. [30] (1.1 s/km is used in this paper). Thus, if we choose gathers with the same shapes, the value of N' is a function of frequency.

Appendix A.2. The Determination of K

A statistical hypothesis test is used to determine K. At each frequency f, the eigenvalues $\{\hat{\lambda}_1, ..., \hat{\lambda}_{N'-1}\}$ of $\hat{\mathbf{R}}$ are tested sequentially at each step k using the statistic

$$\tau(k) = \frac{\hat{\lambda}_k}{\bar{\sigma}_k}, \quad (A3)$$

with

$$\bar{\sigma}_k = \frac{\sum_{i=k}^{N'} \hat{\lambda}_i}{N' - k + 1}. \quad (A4)$$

Note that $\tau(k)$ relies on the local noise characteristics of the chosen gather and may be different across gathers. The tested eigenvalue $\hat{\lambda}_k$ is rejected and suppressed to an average level at a significance level α if

$$\tau(k) > w \mathrm{P}^{-1}_{max|\hat{\mathbf{R}}_c^{N-k+1}}(1-\alpha), \quad (A5)$$

where $\mathrm{P}_{max|\hat{\mathbf{R}}_c^{N-k+1}}$ is the empirical cumulative distribution of the largest eigenvalue of an $N - k + 1$ dimensional SCM $\hat{\mathbf{R}}_c^{N-k+1}$, which can be pre-computed by 1000 Monte Carlo trials. The significance level $\alpha = 0.05$ is usually used. The weight w satisfies $0 \leq w \leq 1$ and is used to affect the selection of K. A weight selection study [22] can be referred to choose an appropriate value for the weight. When the test stops rejecting, the value k is used as K. The statistical model of the SCM $\hat{\mathbf{R}}_c$ relating to a diffuse noise field is simulated using

$$\hat{\mathbf{R}}_c = \frac{1}{M} \mathbf{R}_c \mathbf{X} \mathbf{X}^H, \quad (A6)$$

where $\mathbf{X}$ is an $N \times M$ random matrix with entries $\mathbf{X}_{ij} \sim \mathcal{CN}(0,1)$, and $\mathbf{R}_c$ is the analytical covariance matrix of isotropic noise field defined as

$$[\mathbf{R}_c]_{ij} = J_0(2\pi f \gamma \|\mathbf{r}_i - \mathbf{r}_j\|), \quad (A7)$$

in which J_0 is the zeroth order Bessel function of the first kind, and $\mathbf{r}_i$ denotes the position of the receiver i. Similarly, an estimation of the average slowness is sufficient to use in practice (1.1 s/km is used in this paper). Note that, if we choose gathers with the same shapes, the value of $\mathbf{R}_c$ is a function of frequency.

Considering these formulas from Equations (A3)–(A7), we can find that the determination of K not only relies on the frequency but also on local noise characteristics of the chosen gathers.

References

1. Campillo, M.; Paul, A. Long-range correlations in the diffuse seismic coda. *Science* **2003**, *299*, 547–549. [CrossRef] [PubMed]
2. Wapenaar, K. Retrieving the elastodynamic Green's function of an arbitrary inhomogeneous medium by cross correlation. *Physical Rev. Lett.* **2004**, *93*, 254301. [CrossRef] [PubMed]
3. Draganov, D.; Wapenaar, K.; Thorbecke, J. Seismic interferometry: Reconstructing the earth's reflection response. *Geophysics* **2006**, *71*, SI61–SI70. [CrossRef]
4. De Ridder, S.; Biondi, B. Daily reservoir-scale subsurface monitoring using ambient seismic noise. *Geophys. Res. Lett.* **2013**, *40*, 2969–2974. [CrossRef]
5. Obermann, A.; Froment, B.; Campillo, M.; Larose, E.; Planes, T.; Valette, B.; Chen, J.; Liu, Q. Seismic noise correlations to image structural and mechanical changes associated with the Mw 7.9 2008 Wenchuan earthquake. *J. Geophys. Res. Solid Earth* **2014**, *119*, 3155–3168. [CrossRef]
6. Obermann, A.; Kraft, T.; Larose, E.; Wiemer, S. Potential of ambient seismic noise techniques to monitor the St. Gallen geothermal site (Switzerland). *J. Geophys. Res. Solid Earth* **2015**, *120*, 4301–4316. [CrossRef]
7. Aki, K. Space and time spectra of stationary stochastic waves, with special reference to microtremors. *Bull. Earthq. Res. Inst.* **1957**, *35*, 415–456.

8. Claerbout, J.F. Synthesis of a layered medium from its acoustic transmission response. *Geophysics* **1968**, *33*, 264–269. [CrossRef]
9. Weaver, R.L.; Lobkis, O.I. Ultrasonics without a source: Thermal fluctuation correlations at MHz frequencies. *Phys. Rev. Lett.* **2001**, *87*, 134301. [CrossRef]
10. Weaver, R.L.; Lobkis, O.I. On the emergence of the Green's function in the correlations of a diffuse field. *J. Acoust. Soc. Am.* **2001**, *110*, 3011–3017. [CrossRef]
11. Derode, A.; Larose, E.; Tanter, M.; De Rosny, J.; Tourin, A.; Campillo, M.; Fink, M. Recovering the Green's function from field-field correlations in an open scattering medium (L). *J. Acoust. Soc. Am.* **2003**, *113*, 2973–2976. [CrossRef] [PubMed]
12. Wapenaar, K. Synthesis of an inhomogeneous medium from its acoustic transmission response. *Geophysics* **2003**, *68*, 1756–1759. [CrossRef]
13. Roux, P.; Sabra, K.G.; Kuperman, W.A.; Roux, A. Ambient noise cross correlation in free space: Theoretical approach. *J. Acoust. Soc. Am.* **2005**, *117*, 79–84. [CrossRef] [PubMed]
14. Rickett, J.; Claerbout, J. Acoustic daylight imaging via spectral factorization: Helioseismology and reservoir monitoring. *Lead. Edge* **1999**, *18*, 957–960. [CrossRef]
15. Roux, P.; Fink, M. Green's function estimation using secondary sources in a shallow water environment. *J. Acoust. Soc. Am.* **2003**, *113*, 1406–1416. [CrossRef]
16. Stehly, L.; Fry, B.; Campillo, M.; Shapiro, N.; Guilbert, J.; Boschi, L.; Giardini, D. Tomography of the Alpine region from observations of seismic ambient noise. *Geophys. J. Int.* **2009**, *178*, 338–350. [CrossRef]
17. Pilz, M.; Parolai, S.; Bindi, D. Three-dimensional passive imaging of complex seismic fault systems: evidence of surface traces of the Issyk-Ata fault (Kyrgyzstan). *Geophys. J. Int.* **2013**, *194*, 1955–1965. [CrossRef]
18. Mordret, A.; Landès, M.; Shapiro, N.; Singh, S.; Roux, P. Ambient noise surface wave tomography to determine the shallow shear velocity structure at Valhall: depth inversion with a Neighbourhood Algorithm. *Geophys. J. Int.* **2014**, *198*, 1514–1525. [CrossRef]
19. Galetti, E.; Curtis, A.; Baptie, B.; Jenkins, D.; Nicolson, H. Transdimensional Love-wave tomography of the British Isles and shear-velocity structure of the East Irish Sea Basin from ambient-noise interferometry. *Geophys. J. Int.* **2016**, *208*, 36–58. [CrossRef]
20. Wang, Y.; Lin, F.C.; Schmandt, B.; Farrell, J. Ambient noise tomography across Mount St. Helens using a dense seismic array. *J. Geophys. Res. Solid Earth* **2017**, *122*, 4492–4508. [CrossRef]
21. Weaver, R.L. Equipartition and retrieval of Green's function. *Earthq. Sci.* **2010**, *23*, 397–402. [CrossRef]
22. Wu, G.; Dong, H.; Ke, G.; Song, J. An adapted eigenvalue-based filter for ocean ambient noise processing. *Geophysics* **2019**, *85*, KS29–KS38. [CrossRef]
23. Tsai, V.C. On establishing the accuracy of noise tomography travel-time measurements in a realistic medium. *Geophys. J. Int.* **2009**, *178*, 1555–1564. [CrossRef]
24. Tsai, V.C. Understanding the amplitudes of noise correlation measurements. *J. Geophys. Res. Solid Earth* **2011**, *116*, B09311. [CrossRef]
25. Froment, B.; Campillo, M.; Roux, P.; Gouedard, P.; Verdel, A.; Weaver, R.L. Estimation of the effect of nonisotropically distributed energy on the apparent arrival time in correlations. *Geophysics* **2010**, *75*, SA85–SA93. [CrossRef]
26. Fichtner, A. Source and processing effects on noise correlations. *Geophys. J. Int.* **2014**, *197*, 1527–1531. [CrossRef]
27. Fichtner, A.; Stehly, L.; Ermert, L.; Boehm, C. Generalised interferometry-I: Theory for inter-station correlations. *Geophys. J. Int.* **2016**, *208*, 603–638. [CrossRef]
28. Olofsson, B. Marine ambient seismic noise in the frequency range 1–10 Hz. *Lead. Edge* **2010**, *29*, 418–435. [CrossRef]
29. Kim, J.; Choi, H. Response modification factors of chevron-braced frames. *Eng. Struct.* **2005**, *27*, 285–300. [CrossRef]
30. Seydoux, L.; de Rosny, J.; Shapiro, N.M. Pre-processing ambient noise cross-correlations with equalizing the covariance matrix eigenspectrum. *Geophys. J. Int.* **2017**, *210*, 1432–1449. [CrossRef]
31. McMechan, G.A.; Yedlin, M.J. Analysis of dispersive waves by wave field transformation. *Geophysics* **1981**, *46*, 869–874. [CrossRef]
32. Dosso, S.; Wilmut, M.; Lapinski, A.L. An adaptive-hybrid algorithm for geoacoustic inversion. *IEEE J. Ocean. Eng.* **2001**, *26*, 324–336. [CrossRef]

33. Saito, M. DISPER80: A subroutine package for the calculation of seismic normal mode solutions. In *Seismological Algorithms*; Cinii: Tokyo, Japan, 1980; pp. 293–319.
34. Castagna, J.P.; Batzle, M.L.; Eastwood, R.L. Relationships between compressional-wave and shear-wave velocities in clastic silicate rocks. *Geophysics* **1985**, *50*, 571–581. [CrossRef]
35. Brocher, T.M. Empirical Relations between Elastic Wavespeeds and Density in the Earth's Crust. *Bull. Seismol. Soc. Am.* **2005**, *95*, 2081–2092. [CrossRef]
36. Hicks, E.; Hoeber, H.; Houbiers, M.; Lescoffit, S.P.; Ratcliffe, A.; Vinje, V. Time-lapse full-waveform inversion as a reservoir-monitoring tool—A North Sea case study. *Lead. Edge* **2016**, *35*, 850–858. [CrossRef]
37. Zhang, X.; Hansteen, F.; Curtis, A. Fully 3D Monte Carlo Ambient Noise Tomography over Grane Field. In Proceedings of the 81st EAGE Conference and Exhibition 2019, London, UK, 3–6 June 2019; Volume 2019, pp. 1–5.

© 2020 by the authors. Licensee MDPI, Basel, Switzerland. This article is an open access article distributed under the terms and conditions of the Creative Commons Attribution (CC BY) license (http://creativecommons.org/licenses/by/4.0/).

Article

Multiparameter Elastic Full Waveform Inversion of Ocean Bottom Seismic Four-Component Data Based on A Modified Acoustic-Elastic Coupled Equation

Minao Sun [1,2] and Shuanggen Jin [2,3,*]

1. Key Laboratory of Meteorological Disaster, Ministry of Education (KLME), Joint International Research Laborotory of Climate and Environment Change (ILCEC), Colloborative Innovation Center on Forecast and Evaluation of Meteorological Disasters (CIC-FEMD), Nanjing University of Information Science and Technology, Nanjing 210044, China; minaosun@nuist.edu.cn
2. School of Remote Sensing and Geomatics Engineering, Nanjing University of Information Science and Technology, Nanjing 210044, China
3. Shanghai Astronomical Observatory, Chinese Academy of Sciences, Shanghai 200030, China
* Correspondence: sgjin@shao.ac.cn

Received: 29 July 2020; Accepted: 26 August 2020; Published: 31 August 2020

Abstract: Ocean bottom seismometer (OBS) can record both pressure and displacement data by modern marine seismic acquisitions with four-component (4C) sensors. Elastic full-waveform inversion (EFWI) has shown to recover high-accuracy parameter models from multicomponent seismic data. However, due to limitation of the standard elastic wave equation, EFWI can hardly simulate and utilize the pressure components. To remedy this problem, we propose an elastic full-waveform inversion method based on a modified acoustic-elastic coupled (AEC) equation. Our method adopts a new misfit function to account for both 1C pressure and 3C displacement data, which can easily adjust the weight of different data components and eliminate the differences in the order of magnitude. Owing to the modified AEC equation, our method can simultaneously generate pressure and displacement records and avoid explicit implementation of the boundary condition at the seabed. Besides, we also derive a new preconditioned truncated Gauss–Newton algorithm to consider the Hessian associated with ocean bottom seismic 4C data. We analyze the multiparameter sensitivity kernels of pressure and displacement components and use two numerical experiments to demonstrate that the proposed method can provide more accurate multiparameter inversions with higher resolution and convergence rate.

Keywords: elastic full waveform inversion; acoustic-elastic coupled; ocean bottom seismic; multicomponent; multiparameter

1. Introduction

Ocean bottom seismic survey is a modern platform for exploring the Earth's interior, locating seismometers at the seabed for all-weather, long-term, continuous, real-time observations. Unlike the conventional towed-streamer acquisition, OBS can record 1C pressure and 3C displacement data [1] using four-component (4C) detectors. The observed multicomponent data contain plenty of elastic properties of subsurface media, which can be used to deduct the lithology, fluid content, and pore pressure of rocks [2].

In multicomponent data processing, elastic full-waveform inversion (EFWI) plays an increasingly important role [3]. In the manner of classical FWI [4], EFWI computes parameter gradients by cross-correlating forward- and back-propagated wavefields and updates models to minimize the data misfit function. As governed by the elastic wave equation, EFWI can interpret multiple elastic wave phenomena, i.e., wave-mode conversion and AVO effects [5], and provide quantitative estimations for subsurface parameter distributions. Although it costs a large number of computing resources in wavefield simulations, its excellent performance still makes it more and more attractive [6–10].

However, the standard elastic wave equation commonly used in the conventional EFWI approaches cannot directly extract pressure components from elastic wavefields. By solving the acoustic and elastic wave equations in different computing areas, a fluid–solid coupled EFWI approach has been proposed [11–13]. It can generate pressure in the water immediately above the seabed and elastic components on the solid seabed. However, it requires to explicitly implement the correct boundary conditions, which is challenging for irregular surfaces [14]. Alternatively, Yu et al. [15] proposed an acoustic-elastic coupled (AEC) equation in elastic imaging of OBS 4C data. It introduces the physical relation between pressure and normal stress into the elastic wave equation. Thus, it can compute the pressure wavefield and avoid applying the boundary conditions. The developed 4C elastic reverse-time migration (ERTM) shows to suppress non-physical artifacts in the back-propagated wavefield and provide better-resolved subsurface images. With consideration of propagating direction and anisotropic property, this equation has been extended with an elastic vector imaging for transverse isotropy media [16,17]. However, these 4C ERTM methods aim to retrieve the subsurface structures but fail to provide quantitative parameter reconstructions.

In this study, we propose a new EFWI method based on a modified AEC equation, which can reconstruct multiple elastic parameters from OBS 4C data. Our method defines a new weighted misfit function; thus, it can adjust the weight of pressure and displacement components, and eliminate the differences in the order of magnitude. As more parameter classes and data components are involved, the blurring effects and parameter couplings [18] in the Hessian operator are prone to be more serious. To better consider the inverse Hessian operator, we reformulate the truncated Gauss–Newton-based (TGN) algorithm [19,20] in the framework of this modified AEC equation. Compared with the preconditioned conjugate gradient (PCG) algorithm, TGN can estimate a more accurate inverse Hessian and provide better parameter update directions. TGN has been widely used in multiparameter inversion for acoustic, elastic, and anisotropic media [21–23] and elastic least-squares RTM [24,25]. Besides, a pseudo-diagonal Hessian [26,27] is used as a precondition operator to remove the influences of limited observation apertures, geometry spreading, and frequency-limited wavelet.

The paper is organized as follows. In Section 2, we first review the general formulas of FWI, and then introduce the theory of the AEC-EFWI method and the implementation of the preconditioned TGN algorithm. In Section 3, we numerically analyze multiparameter sensitivity kernels of pressure and displacement components. In Section 4, we use two numerical examples to validate the effectiveness of the proposed method. Before conclusions, we discuss whether the AEC-EFWI method can invert elastic parameters using only 1C pressure data for OBS and towed-streamer acquisitions.

2. Methods

2.1. General FWI Formulation

Seismic wave equation can be expressed as

$$\mathbf{Sw} = \mathbf{f}, \tag{1}$$

where $\mathbf{w}$ denotes the subsurface wavefields, $\mathbf{f}$ indicates the source wavelet, and $\mathbf{S}$ is the parameter derivative matrix.

By taking the partial derivative of Equation (1) with respect to parameter, the sensitivity kernel $\mathbf{L}$ can be acquired,

$$\mathbf{L} = \frac{\partial \mathbf{w}}{\partial \mathbf{m}} = -\left(\mathbf{S}^{-1}\right) \frac{\partial \mathbf{S}}{\partial \mathbf{m}} \mathbf{w}. \tag{2}$$

The gradient of misfit function **g** can be obtained by applying the adjoint of sensitivity kernel to the data residuals between observed and synthetic data,

$$\mathbf{g} = \mathbf{L}^T \delta \mathbf{d}. \tag{3}$$

Based on the Newton optimization, the parameter perturbation can be estimated by solving the Newton equation

$$\mathbf{g} = \mathbf{H} \delta \mathbf{m}, \tag{4}$$

where **H** denotes the Hessian operator, which is the second-order partial derivatives of the misfit function with respect to parameter.

Thus, the $(k+1)th$ iteration model $\mathbf{m}^{(k+1)}$ can be updated by summing the $(k)th$ iteration model $\mathbf{m}^{(k)}$ and the $(k+1)th$ iteration model perturbation $\delta \mathbf{m}^{(k)}$ scaled with a suitable step-length r,

$$\mathbf{m}^{(k+1)} = \mathbf{m}^{(k)} + r \cdot \delta \mathbf{m}^{(k)}. \tag{5}$$

2.2. Acoustic-Elastic Coupled EFWI Method

Compared with the standard elastic wave equation, the original AEC equation requires one more formula to compute the pressure component from elastic wavefields. In this study, we have made some modifications to reduce the number of equations and variables, and thus provide a modified AEC equation (details referred to Appendix A), given by

$$\begin{cases} \rho \dfrac{\partial^2 u_x}{\partial t^2} - \dfrac{\partial (\tau_n^s - p)}{\partial x} - \dfrac{\partial \tau_s^s}{\partial z} = 0 \\ \rho \dfrac{\partial^2 u_z}{\partial t^2} - \dfrac{\partial \tau_s^s}{\partial x} - \dfrac{\partial (-\tau_n^s - p)}{\partial z} = 0 \\ p + (\lambda + \mu) \left(\dfrac{\partial u_x}{\partial x} + \dfrac{\partial u_z}{\partial z} \right) = f_p, \\ \tau_n^s - \mu \left(\dfrac{\partial u_x}{\partial x} - \dfrac{\partial u_z}{\partial z} \right) = 0 \\ \tau_s^s - \mu \left(\dfrac{\partial u_x}{\partial z} + \dfrac{\partial u_z}{\partial x} \right) = 0 \end{cases} \tag{6}$$

where p, u_x, and u_z denote the pressure, horizontal, and vertical particle displacement wavefields, respectively. τ_n^s and τ_s^s are the S-wave-related normal and deviatoric stress components, respectively. λ and μ are the Lamé constants, and ρ is density. f_p indicates the source function applied to the p-component. Compared with the original AEC equation, this modified one can generate OBS 4C records with same accuracy in wavefield simulation but costs less computing resources.

In this study, we define a weighted misfit function to account for both pressure and displacement components, given by

$$E(\mathbf{m}) = \frac{1}{2} \varepsilon \cdot \|\delta d_x\|^2 + \frac{1}{2} \varepsilon \cdot \|\delta d_z\|^2 + \frac{1}{2} (1 - \varepsilon) \cdot \zeta \cdot \|\delta d_p\|^2, \tag{7}$$

where δd_x, δd_z, and δd_p denote the residuals of the horizontal, vertical displacement, and pressure components, respectively. Here, ε is a weighting coefficient, satisfying $\varepsilon \in [0,1]$. A scale factor ζ is used to eliminate the differences in magnitude between pressure and displacement components.

According to the adjoint-state theory [28], the adjoint AEC equation can be given by (see Appendix B for details)

$$\begin{cases} \rho \dfrac{\partial^2 \hat{u}_x}{\partial t^2} - \dfrac{\partial [(\lambda + \mu) \hat{p}]}{\partial x} + \dfrac{\partial [\mu \hat{t}_n^s]}{\partial x} + \dfrac{\partial [\mu \hat{t}_s^s]}{\partial z} = f'_x \\ \rho \dfrac{\partial^2 \hat{u}_z}{\partial t^2} - \dfrac{\partial [(\lambda + \mu) \hat{p}]}{\partial z} - \dfrac{\partial [\mu \hat{t}_n^s]}{\partial z} + \dfrac{\partial [\mu \hat{t}_s^s]}{\partial x} = f'_z \\ \hat{p} - \left(\dfrac{\partial \hat{u}_x}{\partial x} + \dfrac{\partial \hat{u}_z}{\partial z} \right) = f'_p \\ \hat{t}_n^s + \left(\dfrac{\partial \hat{u}_x}{\partial x} - \dfrac{\partial \hat{u}_z}{\partial z} \right) = 0 \\ \hat{t}_s^s + \left(\dfrac{\partial \hat{u}_x}{\partial z} + \dfrac{\partial \hat{u}_z}{\partial x} \right) = 0 \end{cases}, \tag{8}$$

where $(\hat{u}_x, \hat{u}_z, \hat{p}, \hat{t}_n^s, \hat{t}_s^s)$ is the adjoint wavefields of $(u_x, u_z, p, \tau_n^s, \tau_s^s)$, and $\left(f'_x, f'_z, f'_p \right)$ is the multicomponent adjoint source, satisfying

$$\begin{cases} f'_x = \varepsilon \cdot \delta d_x \\ f'_z = \varepsilon \cdot \delta d_z \\ f'_p = (1 - \varepsilon) \cdot \zeta \cdot \delta d_p \end{cases}. \tag{9}$$

The gradients of the Lamé constants g_λ, g_μ and density $g_{\rho,Lame}$ can be computed by performing zero-lag cross-correlations of the adjoint wavefields (Equation (8)) and the forward wavefields (Equation (6)), given by (see Appendix B for details)

$$\begin{cases} g_\lambda = -\dfrac{p}{\lambda + \mu} \hat{p} \\ g_\mu = -\dfrac{p}{\lambda + \mu} \hat{p} - \dfrac{\tau_n^s}{\mu} \hat{t}_n^s - \dfrac{\tau_s^s}{\mu} \hat{t}_s^s \\ g_{\rho,Lame} = \dfrac{\partial^2 u_x}{\partial t^2} \hat{u}_x + \dfrac{\partial^2 u_z}{\partial t^2} \hat{u}_z \end{cases}. \tag{10}$$

Compared with the Lamé constants, the parameterization of seismic velocities is a better choice in multiparameter EFWI [29–32]. According to the chain rule, the gradients of P- (α) and S-wave velocities (β) and density can be obtained,

$$\begin{cases} g_\alpha = 2\rho\alpha \cdot g_\lambda \\ g_\beta = -4\rho\beta \cdot g_\lambda + 2\rho\beta \cdot g_\mu \\ g_{\rho,Vel} = \left(\alpha^2 - 2\beta^2 \right) \cdot g_\lambda + \beta^2 \cdot g_\mu + g_{\rho,Lame} \end{cases}. \tag{11}$$

The models of P- and S-wave velocities and density can be updated as follows,

$$\begin{bmatrix} m_\alpha^{(k+1)} \\ m_\beta^{(k+1)} \\ m_\rho^{(k+1)} \end{bmatrix} = \begin{bmatrix} m_\alpha^{(k)} \\ m_\beta^{(k)} \\ m_\rho^{(k)} \end{bmatrix} - r \cdot \mathbf{H}^{-1} \begin{bmatrix} g_\alpha^{(k)} \\ g_\beta^{(k)} \\ g_\rho^{(k)} \end{bmatrix}. \tag{12}$$

2.3. Preconditioned Truncated Gauss–Newton Algorithm

The inverse Hessian operator is estimated by a preconditioned truncated Gauss–Newton (PTGN) algorithm, as shown in Algorithm 1. In each iteration, we should perform demigration ($\mathbf{L}\mathbf{p}_k$) and migration ($\mathbf{L}^T(\mathbf{Lp}_k)$) processes to update the parameter perturbations. The migration has been illustrated in Equation (8), and the demigration of OBS 4C data can be computed through a first-order Born modeling operator, given by

$$\begin{cases} \rho\dfrac{\partial^2 \delta u_x}{\partial t^2} - \dfrac{\partial(\delta\tau_n^s - \delta p)}{\partial x} - \dfrac{\partial \delta\tau_s^s}{\partial z} = -\delta\rho\dfrac{\partial^2 u_x}{\partial t^2} \\ \rho\dfrac{\partial^2 \delta u_z}{\partial t^2} + \dfrac{\partial \delta\tau_s^s}{\partial x} - \dfrac{\partial(-\delta\tau_n^s - \delta p)}{\partial z} = -\delta\rho\dfrac{\partial^2 u_z}{\partial t^2} \\ \delta p + (\lambda + \mu)\left(\dfrac{\partial \delta u_x}{\partial x} + \dfrac{\partial \delta u_z}{\partial z}\right) = -(\delta\lambda + \delta\mu)\left(\dfrac{\partial u_x}{\partial x} + \dfrac{\partial u_z}{\partial z}\right). \\ \delta\tau_n^s - \mu\left(\dfrac{\partial \delta u_x}{\partial x} - \dfrac{\partial \delta u_z}{\partial z}\right) = \delta\mu\left(\dfrac{\partial u_x}{\partial x} - \dfrac{\partial u_z}{\partial z}\right) \\ \delta\tau_s^s - \mu\left(\dfrac{\partial \delta u_x}{\partial z} + \dfrac{\partial \delta u_z}{\partial x}\right) = \delta\mu\left(\dfrac{\partial u_x}{\partial z} + \dfrac{\partial u_z}{\partial x}\right) \end{cases} \qquad (13)$$

Algorithm 1 Preconditioned Truncated Gauss–Newton algorithm

Input: Gradient $\mathbf{g}$, Hessian precondition operator $\mathbf{H}_p$;
  Set $\mathbf{x}^{(0)} = 0$ and $\mathbf{r}^{(0)} = (\mathbf{L}^T\mathbf{L})\mathbf{x}^{(0)} - \mathbf{g}$;
  Solve $\mathbf{H}_p \cdot \mathbf{y}^{(0)} = \mathbf{r}^{(0)}$ for $\mathbf{y}^{(0)}$;
  Set $\mathbf{p}^{(0)} = -\mathbf{r}^{(0)}$ and $k = 0$;
Output: Parameter perturbation $\mathbf{x}$
1: **while** $\mathbf{r}^{(k)} > \epsilon$ **do**
2:   $s^{(k)} = \dfrac{\mathbf{r}^{(k)T}\mathbf{r}^{(k)}}{(\mathbf{Lp}^{(k)})^T(\mathbf{Lp}^{(k)})}$
3:   $\mathbf{x}^{(k+1)} = \mathbf{x}^{(k)} + s^{(k)}\mathbf{p}^{(k)}$
4:   $\mathbf{r}^{(k+1)} = \mathbf{r}^{(k)} + s^{(k)}[\mathbf{L}^T(\mathbf{Lp}^{(k)})]$
5:   Solve $\mathbf{H}_p \cdot \mathbf{y}^{(k+1)} = \mathbf{r}^{(k+1)}$ for $\mathbf{y}^{(k+1)}$
6:   $t^{(k+1)} = \dfrac{\mathbf{r}^{(k+1)T}\mathbf{y}^{(k+1)}}{\mathbf{r}^{(k)T}\mathbf{y}^{(k)}}$
7:   $\mathbf{p}^{(k+1)} = -\mathbf{y}^{(k+1)} + t^{(k+1)}\mathbf{p}^{(k)}$
8:   $k = k + 1$
9: **end while**
10: **return** $\mathbf{x}^{(k+1)}$

A diagonal pseudo-Hessian with source-side illumination is used as a precondition operator, given by

$$\mathbf{H}_p = \left(\dfrac{\partial \mathbf{S}}{\partial \mathbf{m}}\mathbf{w}\right)^T \left(\dfrac{\partial \mathbf{S}}{\partial \mathbf{m}}\mathbf{w}\right). \qquad (14)$$

According to the modified AEC equation, the diagonal blocks of the Hessian for Lamé constants and density satisfy

$$\begin{cases} H_{\lambda\lambda} = \left(\dfrac{\partial \mathbf{S}}{\partial \lambda}\mathbf{w}\right)^T \left(\dfrac{\partial \mathbf{S}}{\partial \lambda}\mathbf{w}\right) = \left(\dfrac{p}{\lambda+\mu}\right)^2 \\ H_{\mu\mu} = \left(\dfrac{\partial \mathbf{S}}{\partial \mu}\mathbf{w}\right)^T \left(\dfrac{\partial \mathbf{S}}{\partial \mu}\mathbf{w}\right) = \left(\dfrac{p}{\lambda+\mu}\right)^2 + \left(\dfrac{\tau_n^s}{\mu}\right)^2 + \left(\dfrac{\tau_s^s}{\mu}\right)^2, \\ H_{\rho\rho} = \left(\dfrac{\partial \mathbf{S}}{\partial \rho}\mathbf{w}\right)^T \left(\dfrac{\partial \mathbf{S}}{\partial \rho}\mathbf{w}\right) = \dfrac{\partial^2 u_x}{\partial t^2}\dfrac{\partial^2 u_x}{\partial t^2} + \dfrac{\partial^2 u_z}{\partial t^2}\dfrac{\partial^2 u_z}{\partial t^2} \end{cases} \quad (15)$$

and the ones for P- and S-wave velocities and density can be given by

$$\begin{bmatrix} H_{\alpha\alpha} & H_{\alpha\beta} & H_{\alpha\rho} \\ H_{\alpha\beta} & H_{\beta\beta} & H_{\beta\rho} \\ H_{\alpha\rho} & H_{\beta\rho} & H_{\rho\rho} \end{bmatrix} =$$

$$\begin{bmatrix} 2\rho & 0 & 0 \\ -4\beta\rho & 2\beta\rho & 0 \\ \alpha^2-2\beta^2 & \beta^2 & 1 \end{bmatrix} \cdot \begin{bmatrix} H_{\lambda\lambda} & H_{\lambda\mu} & H_{\lambda\rho} \\ H_{\lambda\mu} & H_{\mu\mu} & H_{\mu\rho} \\ H_{\lambda\rho} & H_{\mu\rho} & H_{\rho\rho} \end{bmatrix} \cdot \begin{bmatrix} 2\rho & 0 & 0 \\ -4\beta\rho & 2\beta\rho & 0 \\ \alpha^2-2\beta^2 & \beta^2 & 1 \end{bmatrix}^T \quad (16)$$

Thus, we have

$$\begin{cases} H_{\alpha\alpha} = 4\alpha^2\rho^2 H_{\lambda\lambda} \\ H_{\beta\beta} = 16\beta^2\rho^2 H_{\lambda\lambda} + 4\beta^2\rho^2 H_{\mu\mu} + H_{\rho\rho} \\ H_{\rho\rho} = \left(\alpha^2-2\beta^2\right)^2 H_{\lambda\lambda} + \beta^4 H_{\mu\mu} + H_{\rho\rho} \end{cases} . \quad (17)$$

3. Sensitivity Analysis

In regional and global seismic explorations, sensitivity kernels are always used to portray subsurface wavepaths [33,34]. For the elastic case, the kernels of displacement components with different parameter classes have been studied on the standard elastic equation [23,35]. In this study, we use the modified AEC equation to simulate elastic wavefields, allowing for the kernels of pressure and displacement components. The experiment is performed on the elastic Marmousi model (see Figure 1), including a shallow water layer below the sea surface. Only one shot is excited to generate a pure P-wave source at the place of (5 km, 0.03 km), and 601 4C receivers are evenly located at the seabed. The peak frequency of the source function is 8 Hz.

Observed pressure and displacement records are back-propagated from the receivers, respectively. The obtained sensitivity kernels of pressure (p) and displacement (u_x and u_z) components are presented in Figure 2. In the pressure kernels, most energy is distributed in the shallow part of the model and attenuates rapidly with the increase of depth. In contrast, the displacement kernels contain the S-wave reflection paths, thus it can enhance the illumination for the deep model. The corresponding 2D wavenumber spectrum of these kernels are shown in Figure 3. It is clear that the pressure kernels have higher resolution than the displacement ones in both vertical and horizontal directions. It is because the pressure component is computed by the spatial partial derivatives of the displacement wavefield, and these operators physically increase the frequency (or wavenumber) content of the data. Consequently, the simultaneous utilization of pressure and displacement components can provide a better characterization of subsurface structures.

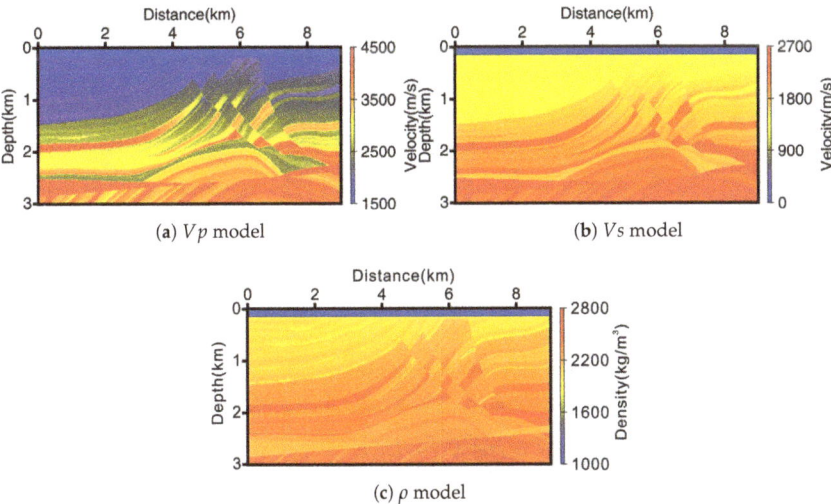

Figure 1. True parameters of the elastic Marmousi model: Vp (a), Vs (b), and ρ (c).

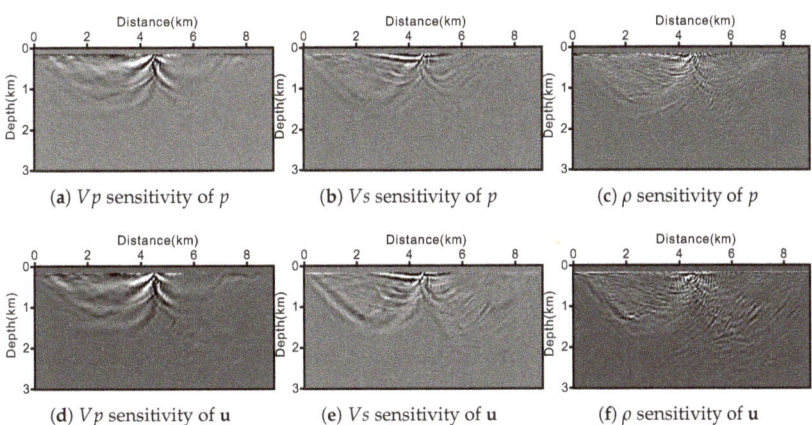

Figure 2. Sensitivity kernels of pressure (top) and displacement (bottom) components with respect to (a,d) Vp, (b,e) Vs, and (c,f) ρ.

The kernels for different parameter classes also have remarkable differences. The Vp kernels (Figure 2a,d) are mainly formed by the diving waves and reflections associated with P-wave, which show relatively isotropic distributions in the wavenumber spectrum. The Vs kernels contain more information of PS reflections (Figure 2b,e). Besides, the ρ kernels are of the highest wavenumber components and behave as migration images of subsurface interfaces. That is the reason why we can easily obtain the short-wavelength structures of density model but fail to reconstruct the background model from seismic data.

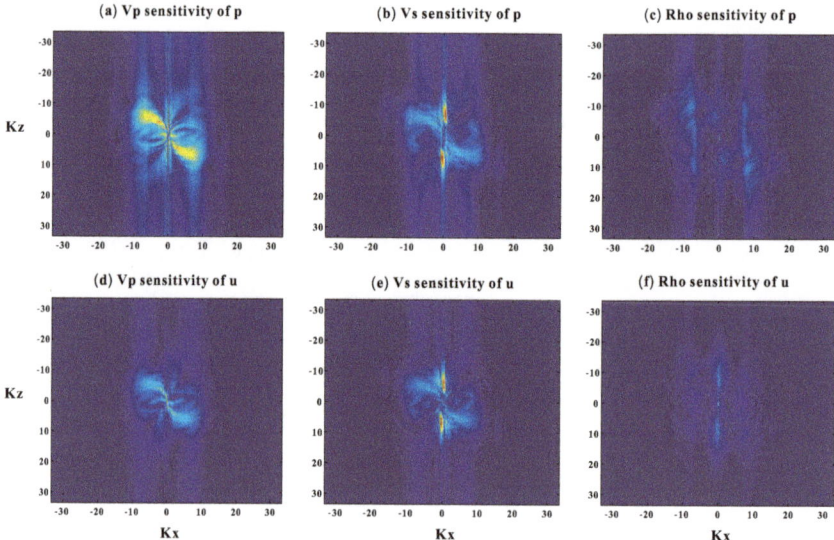

Figure 3. The 2D Wavenumber spectrum for the sensitivity kernels of pressure (top) and displacement (bottom) components with respect to (**a,d**) Vp, (**b,e**) Vs, and (**c,f**) ρ.

4. Results

We use two numerical experiments on (1) the Overthrust model and (2) the Marmousi model to validate the proposed method. An O(2, 8) time-space-domain finite-difference staggered-grid solution [36] of the modified AEC equation is used to generate both forward- and back-propagated 4C wavefields. A convolution perfectly matched layer absorbing boundary [37,38] is used around the calculation area without consideration of the sea surface. In these experiments, a Ricker wavelet is adopted to generate pure P-wave sources with a peak frequency of 8 Hz (the bandwidth in [2 Hz, 20 Hz]).

4.1. Overthrust Model Test

We first use the Overthrust model to demonstrate the effectiveness of the proposed PTGN algorithm for OBC 4C data. The true Vp and Vs models are shown in Figure 4. The density model is set to be 1000 kg/m^3 in the water layer and 2000 kg/m^3 below the seabed. The model is sampled as 801 × 166 grids with intervals of 12.5 m in both horizontal and vertical directions. The initial parameters are generated from the true ones with a smoothing window of 250 m. The acquisition geometry includes 101 shots with an interval of 100 m below the sea surface and 801 OBS receivers at the seabed for each shot. The total recording time is 4.0 s, and the temporal sampling rate is 1.0 ms. Observed seismic data are shown in Figure 5, including horizontal and vertical displacements and pressure components. As a comparison, the inversion is also performed using a preconditioned conjugate gradient (PCG). The maximum number of the loop for parameter update is 21, and that of the inner loop in the PTGN algorithm is 10.

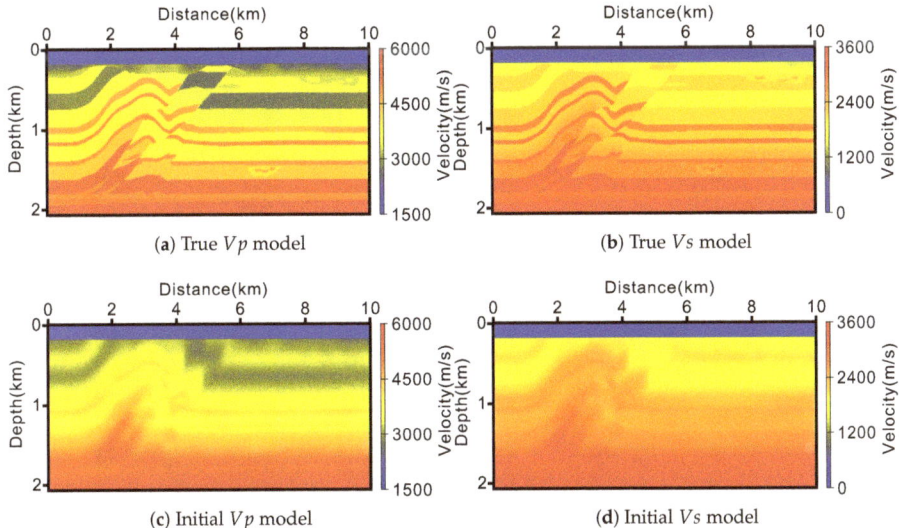

Figure 4. True and initial parameters of the Overthrust model: (a,c) Vp and (b,d) Vs.

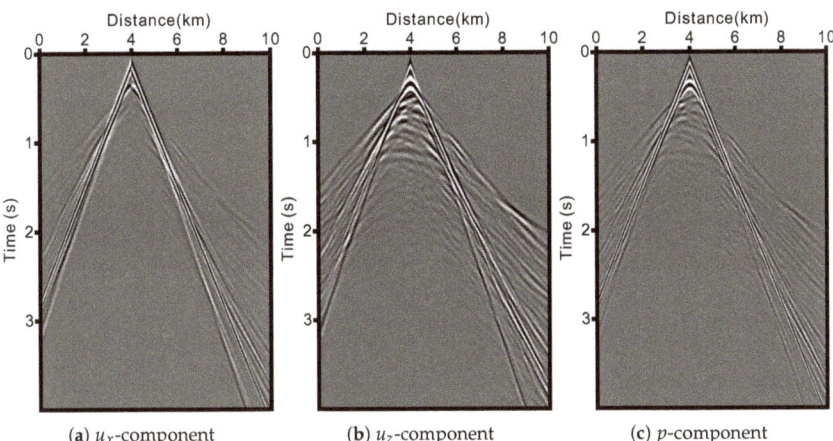

Figure 5. Observed multicomponent seismic data: (a) horizontal and (b) vertical displacement and (c) pressure components.

Figure 6 displays the multiparameter gradients of the 1*th* iteration. We observe that the PCG gradients have insufficient illuminations for the deep part of the model, while the PTGN gradients are much improved and behave as amplitude-preserving subsurface images. The final Vp and Vs models are displayed in Figure 7. The inverted Vp and Vs using PTGN give better descriptions of structural boundaries with a higher interface continuity and fewer vertical artifacts. The vertical profiles (Figure 8) and the root mean square (RMS) errors (Table 1) can further demonstrate that PTGN can provide more accurate multiparameter inversions. Multicomponent data residuals between observed and simulated are shown in Figure 9. The residuals of the PTGN method are much weaken than those of PCG, which demonstrates that the PTGN can better interpret the observed multicomponent data. The convergence curves (Figure 10) shows that PTGN has a higher decreasing rate and eventually converges to a lower misfit value.

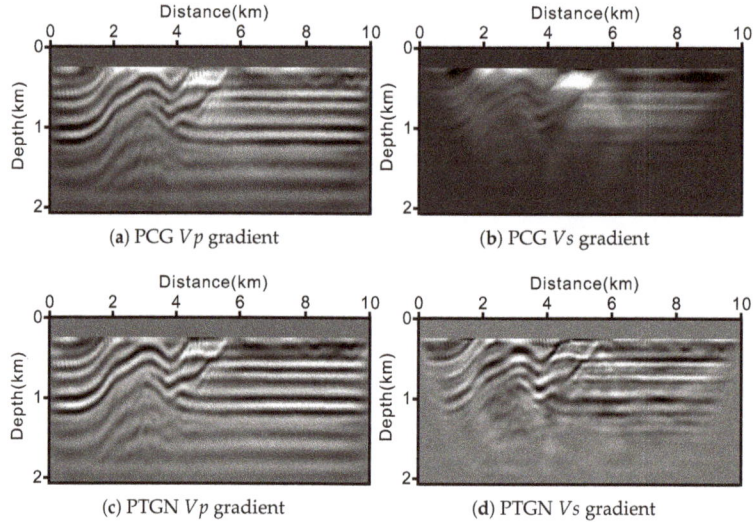

Figure 6. The multiparameter gradients of the 1*th* iteration: preconditioned conjugate gradients (PCGs) of (**a**) Vp and (**b**) Vs, and preconditioned truncated Gauss–Newton (PTGN) gradients of (**c**) Vp and (**d**) Vs.

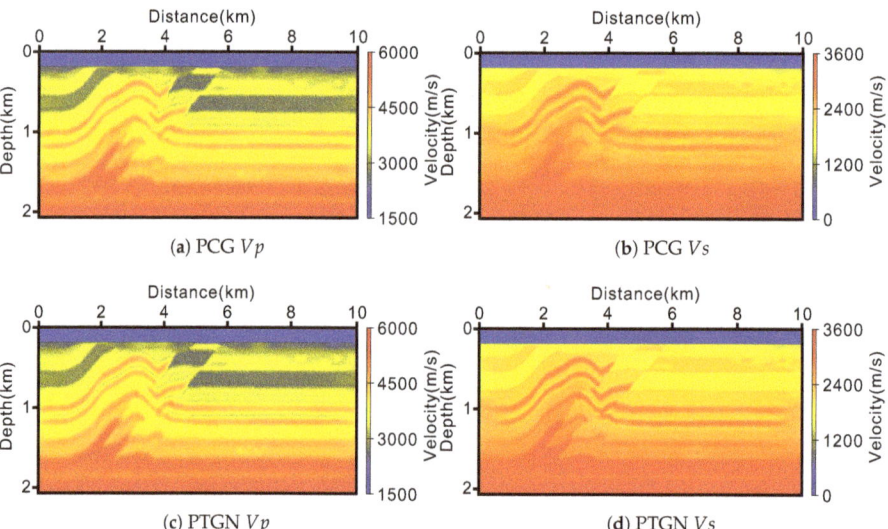

Figure 7. Multiparameter inversion results: (**a**) Vp and (**b**) Vs using PCG, and (**c**) Vp and (**d**) Vs using PTGN.

Table 1. Root mean square (RMS) errors of inversion results using PCG and PTGN.

RMS Errors (%)	Vp	Vs
PTGN	3.08	2.86
PCG	3.22	4.20

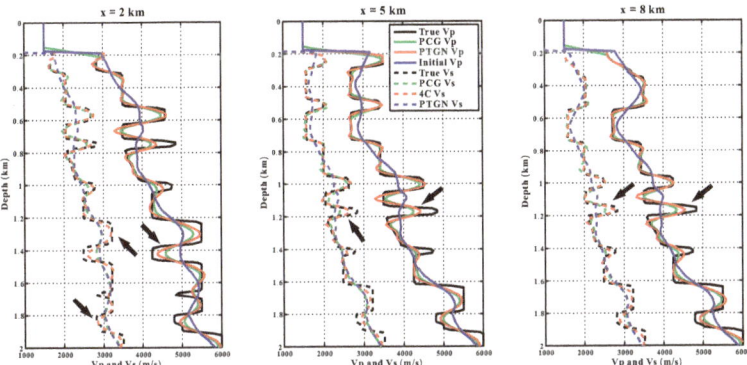

Figure 8. Vertical profiles of the inverted Vp (solid) and Vs (dashed) using the PTGN and PCG methods at horizontal distances of 2 km, 5 km, and 8 km.

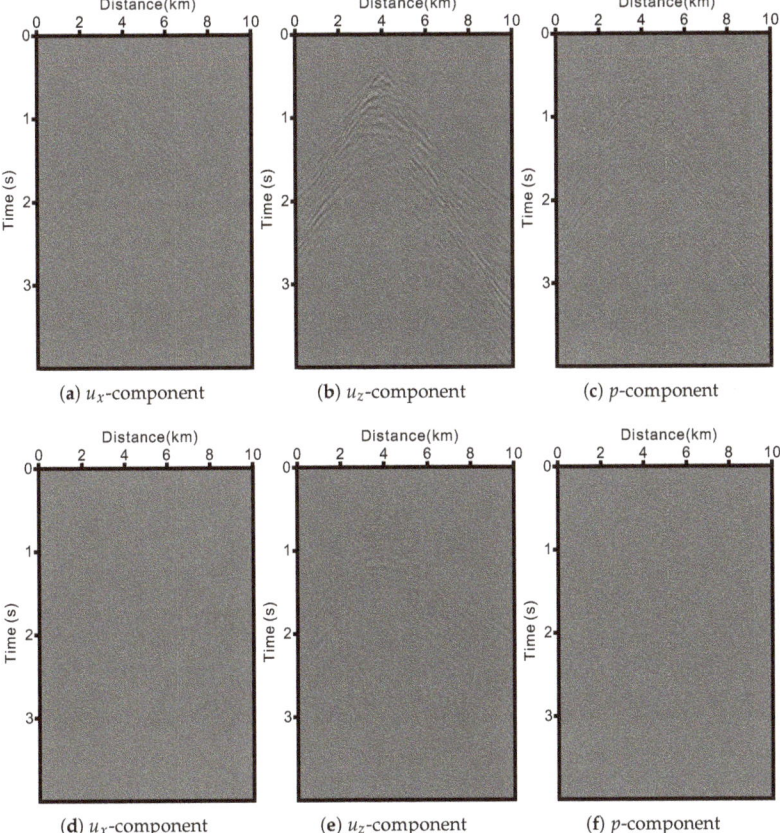

Figure 9. Multicomponent data residuals obtained by inverted models in Figure 7. The scale is consistent with the shot gathers in Figure 5 seismic data. (**a**–**c**) The residuals of PCG and (**d**–**f**) the residuals of PTGN.

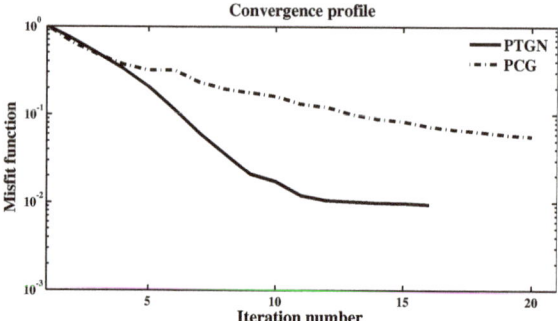

Figure 10. Convergence profiles of the misfit function using PTGN (solid) and PCG (dashed).

4.2. Marmousi Model Test

Next, we use the elastic Marmousi model to demonstrate the effectiveness of the AEC-EFWI method. Smoothed versions of true Vp, Vs, and ρ (see Figure 1) with a window of 300 m are taken as the initial models (see Figure 11). The dimension of the model is 601 × 201, and the intervals are 15 m in the horizontal and vertical directions. We have 61 shots with an interval of 150 m and 601 OBS receivers for each shot. The total recording time is 4.8 s, and the temporal sampling rate is 1.2 ms. A Ricker wavelet with a peak frequency of 8 Hz is adopted to generate a pure P-wave source. Observed multicomponent seismic data are simulated using the modified AEC equation, as shown in Figure 12. As a comparison, an EFWI method for horizontal and vertical displacement components are performed. A maximum of 15 iterations is used for the PTGN loop for both AEC-EFWI and EFWI.

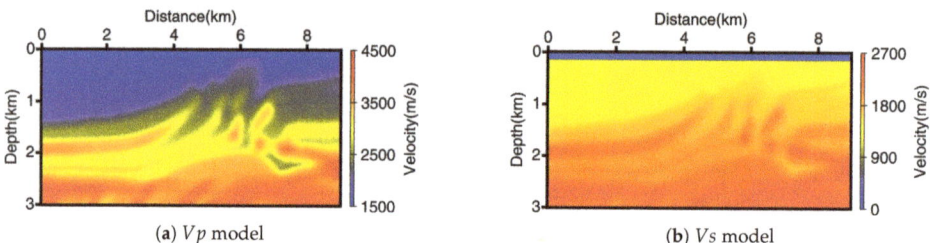

Figure 11. Initial models of Vp (a) and Vs (b).

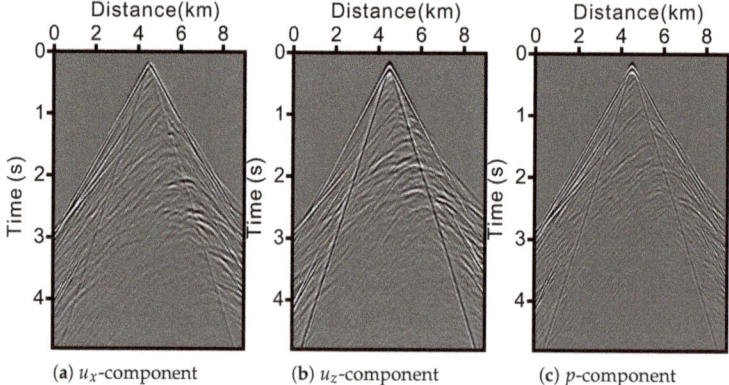

Figure 12. Observed multicomponent seismic data: (a) horizontal and (b) vertical displacement and (c) pressure components.

Figure 13 displays the inverted Vp and Vs models using the two methods. In the Vp results (Figure 13a,c), AEC-EFWI provides better-resolved structures, i.e., anticlines, faults, lithologic interfaces, and high-speed bodies. For Vs (Figure 13b,d), however, the results using the two methods are comparable. With incorporation of the RMS errors in Table 2, we can find that considering the pressure data may not take as great effects on Vs as Vp. It may be because the pressure data are more sensitive to the Vp perturbations.

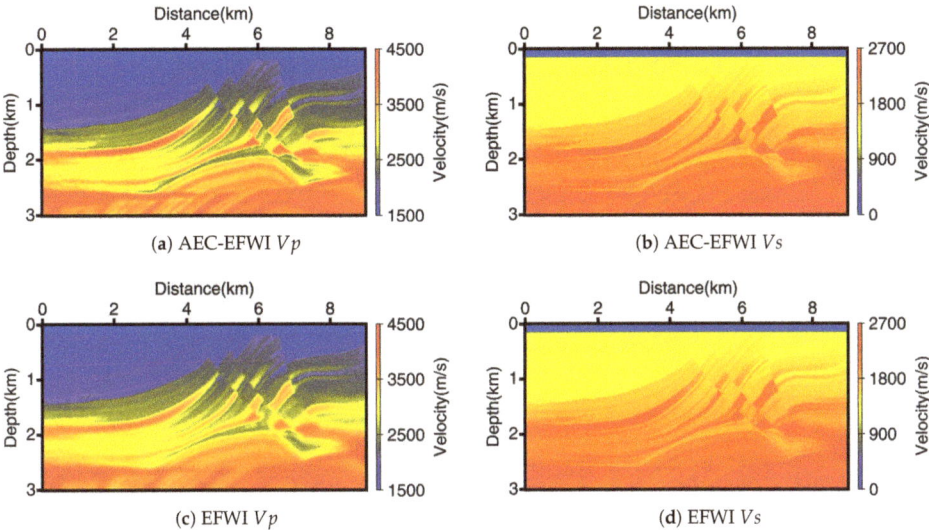

Figure 13. Multiparameter inversion results: (**a**) Vp and (**b**) Vs using modified acoustic-elastic coupled-elastic full-waveform inversion (AEC-EFWI), and (**c**) Vp and (**d**) Vs using elastic full-waveform inversion (EFWI).

The vertical profiles extracted at the horizontal distances of 3.0, 4.5, and 6.0 km are displayed in Figure 14. The AEC-EFWI results (marked by red lines) can precisely illustrate the deep reflectors with narrower sidelobes, and they are very close to the true models (marked by black lines). In contrast, as displayed in the corresponding wavenumber spectrum (Figure 15), EFWI underestimates the perturbations, especially for the interfaces with sharp parameter contrasts (see green lines in Figure 14). The data residuals simulated by the inverted results (Figure 16) and the convergence profiles of the misfit function (Figure 17) prove that the AEC-EFWI can better match the observed data and have a higher convergence rate.

Table 2. RMS errors of the inversion results using AEC-EFWI and elastic full-waveform inversion (EFWI) methods for the Marmousi model test.

RMS Errors (%)	Vp	Vs
EFWI	5.29	3.93
AEC-EFWI	4.25	3.43
RATIO	−24.5	−14.6

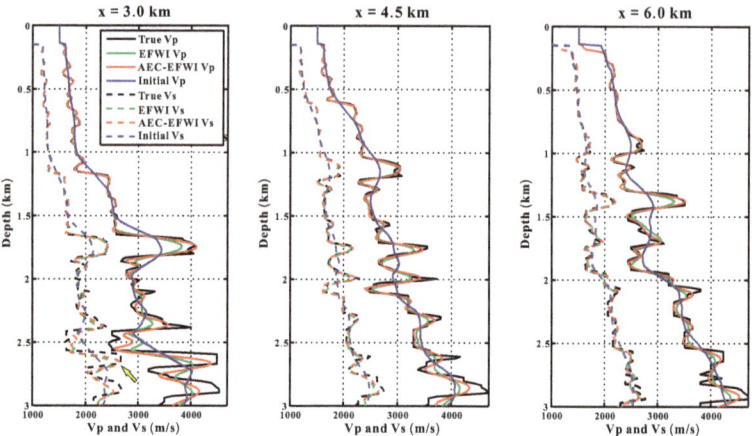

Figure 14. Vertical profiles of Vp (solid) and Vs (dashed) at the horizontal distances of 3.0, 4.5, and 6 km.

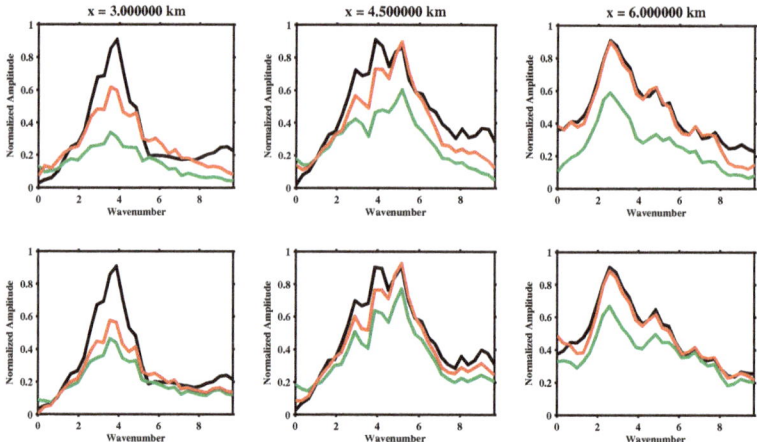

Figure 15. The 1D wavenumber spectrum of vertical profiles, corresponding to Figure 14. Black lines denote the true model, red lines indicate the AEC-EFWI results, and green lines are the EFWI results.

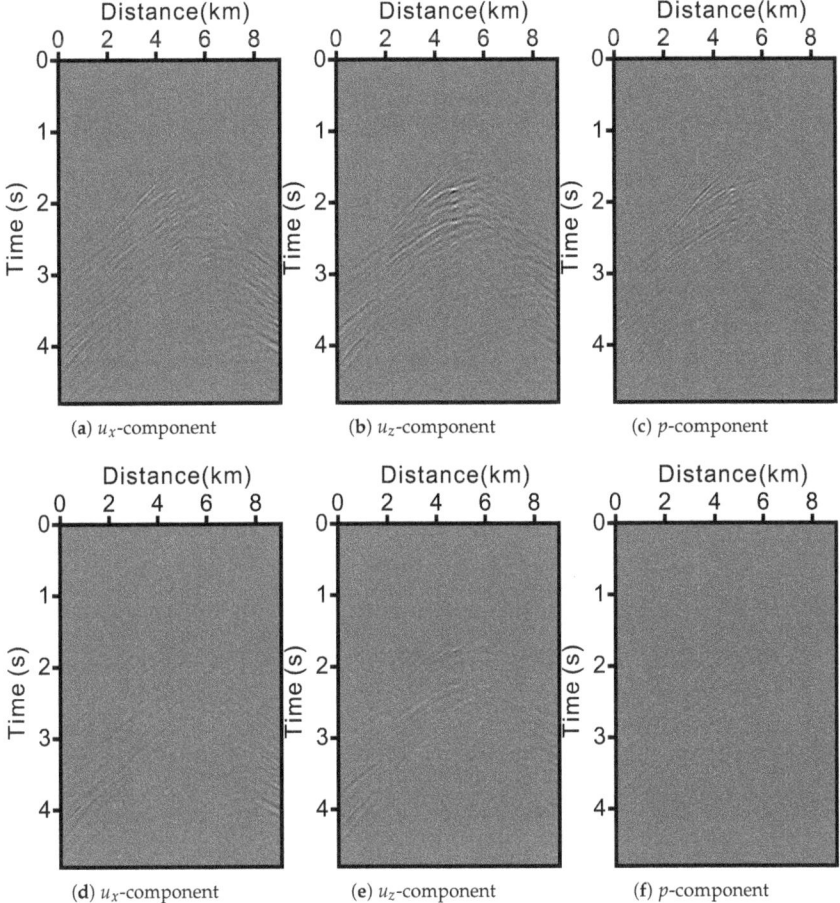

Figure 16. Multicomponent data residuals obtained by inverted models in Figure 13. The scale is consistent with the shot gathers in Figure 12. Panels (**a–c**) denote the residuals of EFWI, and panels (**d–f**) indicate those of AEC-EFWI.

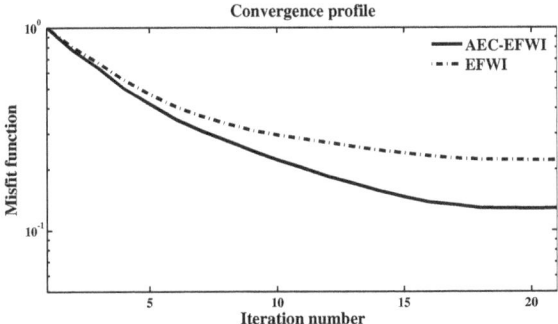

Figure 17. Convergence profiles of the AEC-EFWI (solid) and EFWI (dashed) methods.

5. Discussion

We have validated the effectiveness of the proposed AEC-EFWI method in processing OBS 4C data and improving inversion accuracy. This study mainly focuses on reconstructing the high-wavenumber components of elastic parameters from good initial models. Of course, this AEC-EFWI method also suffers from the notorious cycle-skipping and other practical issues. To alleviate this problem, this work should be further considered to combine with the reflection waveform inversion (RWI) [39–43] or the migration velocity analysis (MVA) [44–46]. In these approaches, the most critical step is to compute the PP and PS reflection paths. It can be easily accomplished by the AEC equation instead of performing a complete P/S decomposition on forward/back-propagated wavefields.

In those experiments, the weighting coefficient ε is set to be 50%. In fact, this value is determined by the difference of observed pressure and displacement components. Supposing that it reduces to zero, we wonder whether the AEC-EFWI method can still provide reasonable inversions for elastic parameters? Figure 18a shows a simple cartoon of the wave propagation process for this case. The incident P-wave excited from the source location generates both PP and PS transmissions at the seabed, and these transmissions are reflected at the interface of Layer 1. Because the water layer is assumed to be precisely known in advance, the observed PPP and PSP waves can be treated as "pseudo-first-order" reflections excited by virtual mixed sources from the seabed. Note that, this PSP wave path carries more information of subsurface Vs distribution, which makes a great contribution to Vs update.

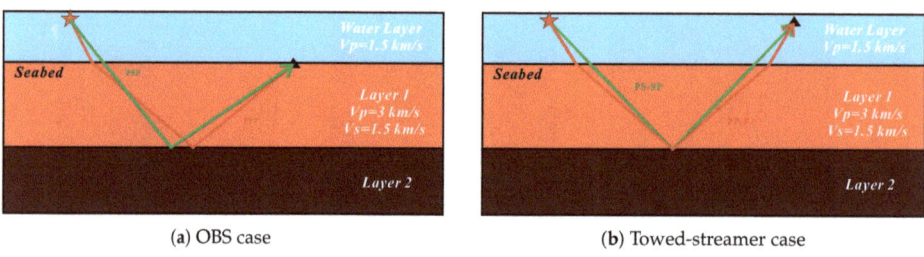

Figure 18. Cartoons of the wave propagation for 1C (**a**) ocean bottom seismic (OBS) and (**b**) towed-streamer cases.

We test the method on the Marmousi model (Figure 1), starting from the same initial models (Figure 11) with 1C pressure data. The inverted results are displayed in Figure 19. The extracted vertical profiles and the corresponding wavenumber spectrum are shown in Figures 20 and 21, respectively. Although the inversion accuracy and spatial resolution decrease to some extent, the 1C results can still provide acceptable multiparameter inversions and have a certain consistency with the true ones. It demonstrates that this AEC-EFWI method is feasible to recover elastic parameters using OBS 1C pressure data. Besides, we can find that the high-wavenumber components in the results using 1C pressure data are better reconstructed than the low-wavenumber components (see differences between the green and red lines in Figure 21), which highlights the contribution of pressure component on high-wavenumber reconstruction.

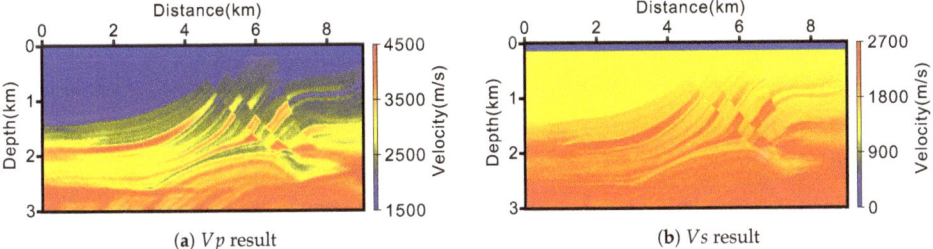

Figure 19. Inverted (**a**) Vp and (**b**) Vs results using AEC-EFWI with 1C pressure data.

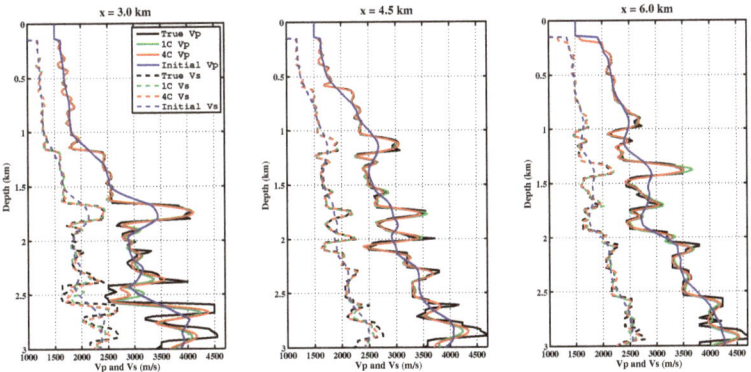

Figure 20. Vertical profiles of Vp (solid) and Vs (dashed) at the horizontal distances of 3.0, 4.5, and 6 km.

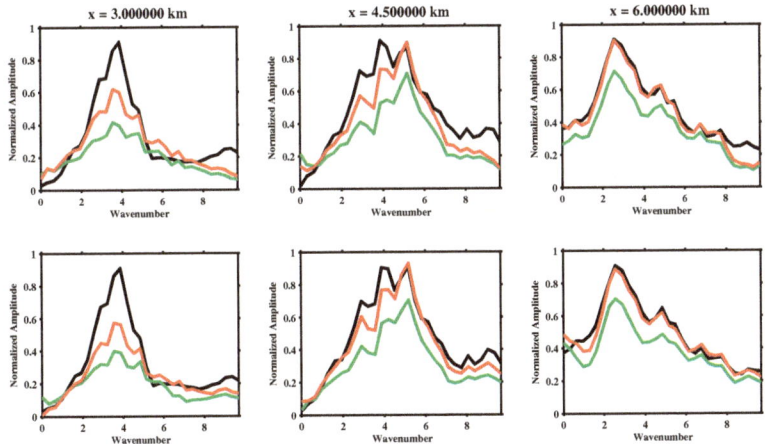

Figure 21. The 1D wavenumber spectrum of vertical profiles, corresponding to Figure 20. Black lines denote the true model, red lines indicate the results using 4C data, and green lines are the ones using 1C data.

Similarly, this method may be also applied for marine towed-streamer pressure data. As displayed in Figure 18b, the PP-PP and PS-SP wave paths help to reveal subsurface Vp and Vs distributions. Besides, other converted waves, i.e., PP-SP and PS-PP, can further enhance the illumination of Vs

model. Of course, it should be further tested with field streamer data. Although some practical problems, i.e., data preprocessing, low-frequency loss, and initial model building, have not been fully considered, the cartoons and the preliminary results can at least inspire us to use 1C pressure data for elastic parameter inversion and eventually provide a flexible method and idea for increasingly complex marine data processing.

6. Conclusions

In this study, we have proposed an elastic full-waveform inversion method based on a modified acoustic-elastic coupled equation. This method uses the modified AEC equation to simultaneously compute subsurface pressure and displacement wavefields. It adopts a weighted misfit function to quantify the contributions of pressure and displacement records. With the adjoint operator, it can simultaneously make use of OBS 4C data to reconstruct multiple elastic parameters. The preconditioned TGN algorithm with a multiparameter diagonal Hessian operator is developed to cope with unbalanced illumination and coupling effects. The sensitivity analysis and numerical experiments have validated that the AEC-EFWI method can yield a higher spatial resolution, provide more accurate elastic parameter inversions, and have a higher convergence rate. The discussion reveals the potential of this AEC-EFWI method in inverting elastic parameters using 1C pressure data for OBS and marine towed-streamer cases.

Author Contributions: Conceptualization, M.S.; Methodology, M.S.; Software, M.S.; Formal analysis, M.S.; Validation, S.J.; Writing, M.S. and S.J. All authors have read and agreed to the published version of the manuscript.

Funding: This work was supported by the National Key Research and Development Program of China Project (Grant No. 2018YFC0603502), the National Natural Science Foundation of China (Grant No. 41930105), and the National Key R&D Program of China (Grant No. 2018YFC0310100).

Acknowledgments: We thank Pengfei Yu from Hohai University for providing the code for the original acoustic-elastic coupled equation.

Conflicts of Interest: The authors declare no conflict of interest. The funders had no role in the design of the study; in the collection, analyses, or interpretation of data; in the writing of the manuscript; or in the decision to publish the results.

Appendix A. Derivation of the Modified Acoustic-Elastic Coupled Equation

We start from the standard 2D time-domain displacement–stress elastic wave equation [47], given by

$$\begin{cases} \rho \dfrac{\partial^2 u_x}{\partial t^2} = \dfrac{\partial \tau_{xx}}{\partial x} + \dfrac{\partial \tau_{xz}}{\partial z} \\ \rho \dfrac{\partial^2 u_z}{\partial t^2} = \dfrac{\partial \tau_{xz}}{\partial x} + \dfrac{\partial \tau_{zz}}{\partial z} \\ \tau_{xx} = (\lambda + 2\mu) \dfrac{\partial u_x}{\partial x} + \lambda \dfrac{\partial u_z}{\partial z} + f, \\ \tau_{zz} = (\lambda + 2\mu) \dfrac{\partial u_z}{\partial z} + \lambda \dfrac{\partial u_x}{\partial x} + f \\ \tau_{xz} = \mu \left(\dfrac{\partial u_x}{\partial z} + \dfrac{\partial u_z}{\partial x} \right) \end{cases} \quad (A1)$$

where u_x and u_z denote the horizontal- and vertical-particle displacement components, respectively; τ_{xx} and τ_{zz} are the normal stress components; and τ_{xz} is the shear stress component. λ and μ are the Lamé constants, and ρ is density. f indicates the source function that is implemented in τ_{xx} and τ_{zz} to generate a pure P-wave.

In tensor analysis, the stress tensor **T** can be decomposed into the isotropic pressure $-p\mathbf{I}$ and deviatoric τ^s parts,

$$\mathbf{T} = \tau^s - p\mathbf{I}. \quad (A2)$$

For 2D cases, the pressure wavefield satisfies

$$p = -\frac{1}{2}tr(\mathbf{T})$$
$$= -\frac{1}{2}(\tau_{xx} + \tau_{zz}) = -(\lambda + \mu)\left(\frac{\partial u_x}{\partial x} + \frac{\partial u_z}{\partial z}\right),$$
(A3)

and the deviatoric stress components become

$$\begin{cases} \tau_{xx}^s = \tau_{xx} + p = \mu\dfrac{\partial u_x}{\partial x} - \mu\dfrac{\partial u_z}{\partial z} \\ \tau_{zz}^s = \tau_{zz} + p = \mu\dfrac{\partial u_z}{\partial z} - \mu\dfrac{\partial u_x}{\partial x} \\ \tau_{xz}^s = \tau_{xz} = \mu\dfrac{\partial u_x}{\partial z} + \mu\dfrac{\partial u_z}{\partial x} \end{cases}$$
(A4)

These equations formulate the original acoustic-elastic coupled equation; we refer to the work in [15].

Note that, it is redundant to simultaneously compute the pressure and two normal stress components. Thus, we redefine

$$\tau_n^s = \tau_{xx}^s = -\tau_{zz}^s, \quad \tau_s^s = \tau_{xz}^s,$$
(A5)

where τ_n^s and τ_s^s are the redefined normal and deviatoric stress components. The simplified equation becomes

$$\begin{cases} \rho\dfrac{\partial^2 u_x}{\partial t^2} = \dfrac{\partial(\tau_n^s - p)}{\partial x} + \dfrac{\partial \tau_s^s}{\partial z} \\ \rho\dfrac{\partial^2 u_z}{\partial t^2} = \dfrac{\partial \tau_s^s}{\partial x} + \dfrac{\partial(-\tau_n^s - p)}{\partial z} = 0 \\ p = -(\lambda + \mu)\left(\dfrac{\partial u_x}{\partial x} + \dfrac{\partial u_z}{\partial z}\right). \\ \tau_n^s = \mu\left(\dfrac{\partial u_x}{\partial x} - \dfrac{\partial u_z}{\partial z}\right) \\ \tau_s^s = \mu\left(\dfrac{\partial u_x}{\partial z} + \dfrac{\partial u_z}{\partial x}\right) \end{cases}$$
(A6)

Compared with the original one, this modified equation can provide subsurface 4C elastic wavefields with same accuracy but less computing resources.

For completeness purpose, we also derive the modified AEC equation in 3D. The 3D original AEC equation is given by

$$\begin{cases} \rho\dfrac{\partial^2 u_x}{\partial t^2} = \dfrac{\partial(\tau_{xx}^s - p)}{\partial x} + \dfrac{\partial \tau_{xy}^s}{\partial y} + \dfrac{\partial \tau_{xz}^s}{\partial z} \\ \rho\dfrac{\partial^2 u_y}{\partial t^2} = \dfrac{\partial \tau_{xy}^s}{\partial x} + \dfrac{\partial(\tau_{yy}^s - p)}{\partial y} + \dfrac{\partial \tau_{yz}^s}{\partial z} \\ \rho\dfrac{\partial^2 u_z}{\partial t^2} = \dfrac{\partial \tau_{xz}^s}{\partial x} + \dfrac{\partial \tau_{yz}^s}{\partial y} + \dfrac{\partial(\tau_{zz}^s - p)}{\partial z} \end{cases}$$
(A7)

and

$$\begin{cases} p = -\left(\lambda + \dfrac{2}{3}\mu\right)\left(\dfrac{\partial u_x}{\partial x} + \dfrac{\partial u_y}{\partial y} + \dfrac{\partial u_z}{\partial z}\right) \\ \tau^s_{xx} = \dfrac{2}{3}\mu\left(2\dfrac{\partial u_x}{\partial x} - \dfrac{\partial u_y}{\partial y} - \dfrac{\partial u_z}{\partial z}\right) \\ \tau^s_{yy} = \dfrac{2}{3}\mu\left(2\dfrac{\partial u_y}{\partial y} - \dfrac{\partial u_x}{\partial x} - \dfrac{\partial u_z}{\partial z}\right) \\ \tau^s_{zz} = \dfrac{2}{3}\mu\left(2\dfrac{\partial u_z}{\partial z} - \dfrac{\partial u_x}{\partial x} - \dfrac{\partial u_y}{\partial y}\right) \\ \tau^s_{xz} = \mu\left(\dfrac{\partial u_x}{\partial z} + \dfrac{\partial u_z}{\partial x}\right) \\ \tau^s_{xy} = \mu\left(\dfrac{\partial u_x}{\partial y} + \dfrac{\partial u_y}{\partial x}\right) \\ \tau^s_{yz} = \mu\left(\dfrac{\partial u_y}{\partial z} + \dfrac{\partial u_z}{\partial y}\right) \end{cases} \tag{A8}$$

Because $\tau^s_{xx} + \tau^s_{yy} + \tau^s_{zz} = 0$, we can similarly provide a 3D modified AEC equation by replacing τ^s_{yy} as $-(\tau^s_{xx} + \tau^s_{zz})$ to reduce the number of formulas.

Appendix B. Derivation of Gradient Computation for Aec-Efwi Method

The modified AEC equation can be rewritten as

$$\mathbf{Sw} = \mathbf{f} \tag{A9}$$

where $\mathbf{w} = (u_x, u_z, p, \tau^s_n, \tau^s_s)^T$ denotes the elastic wavefields, $\mathbf{f} = (0, 0, f_p, 0, 0)^T$ denotes the source wavelet, and $\mathbf{S}$ denotes the parameter derivative matrix,

$$\mathbf{S} = \begin{pmatrix} \rho\dfrac{\partial^2}{\partial t^2} & 0 & \dfrac{\partial}{\partial x} & -\dfrac{\partial}{\partial x} & -\dfrac{\partial}{\partial z} \\ 0 & \rho\dfrac{\partial^2}{\partial t^2} & \dfrac{\partial}{\partial z} & \dfrac{\partial}{\partial z} & -\dfrac{\partial}{\partial x} \\ (\lambda+\mu)\dfrac{\partial}{\partial x} & (\lambda+\mu)\dfrac{\partial}{\partial z} & 1 & 0 & 0 \\ -\mu\dfrac{\partial}{\partial x} & \mu\dfrac{\partial}{\partial z} & 0 & 1 & 0 \\ -\mu\dfrac{\partial}{\partial z} & -\mu\dfrac{\partial}{\partial x} & 0 & 0 & 1 \end{pmatrix}. \tag{A10}$$

Substituting Equation (2) into Equation (3), the gradient satisfies

$$\dfrac{\partial E}{\partial \mathbf{m}} = \left[-\left(\mathbf{S}^{-1}\right)\dfrac{\partial \mathbf{S}}{\partial \mathbf{m}}\mathbf{w}\right]^T \mathbf{f'}. \tag{A11}$$

With consideration of the self-adjoint assumption $(\mathbf{S}^{-1})^T = (\mathbf{S}^T)^{-1}$, it can be rewritten by

$$\dfrac{\partial E}{\partial \mathbf{m}} = -\left[\dfrac{\partial \mathbf{S}}{\partial \mathbf{m}}\mathbf{w}\right]^T \left(\mathbf{S}^T\right)^{-1} \mathbf{f'}. \tag{A12}$$

Here, $\mathbf{S}^T$ is the adjoint operator of the modified AEC-equation, given by

$$\mathbf{S}^T = \begin{pmatrix} \rho\dfrac{\partial^2}{\partial t^2} & 0 & -\dfrac{\partial}{\partial x}(\lambda+\mu) & \dfrac{\partial}{\partial x}\mu & \dfrac{\partial}{\partial z}\mu \\ 0 & \rho\dfrac{\partial^2}{\partial t^2} & -\dfrac{\partial}{\partial z}(\lambda+\mu) & -\dfrac{\partial}{\partial z}\mu & \dfrac{\partial}{\partial x}\mu \\ -\dfrac{\partial}{\partial x} & -\dfrac{\partial}{\partial z} & 1 & 0 & 0 \\ \dfrac{\partial}{\partial x} & -\dfrac{\partial}{\partial z} & 0 & 1 & 0 \\ \dfrac{\partial}{\partial z} & \dfrac{\partial}{\partial x} & 0 & 0 & 1 \end{pmatrix}. \tag{A13}$$

According to the adjoint-state method, we define $\hat{\mathbf{w}} = (\mathbf{S}^{-1})^T \mathbf{f}'$ as the solution of the adjoint equation,

$$\mathbf{S}^T \hat{\mathbf{w}} = \mathbf{f}' \tag{A14}$$

where $\hat{\mathbf{w}} = (\hat{u}_x, \hat{u}_z, \hat{p}, \hat{\tau}_n^s, \hat{\tau}_s^s)^T$ is the adjoint variables as used in Equation (8).
For the parameterization of Lamé constants and density $\mathbf{m} = [\lambda, \mu, \rho]^T$, the partial derivative matrices can be given by

$$\frac{\partial \mathbf{S}}{\partial \lambda} = \begin{pmatrix} 0 & 0 & 0 & 0 & 0 \\ 0 & 0 & 0 & 0 & 0 \\ \frac{\partial}{\partial x} & \frac{\partial}{\partial z} & 0 & 0 & 0 \\ 0 & 0 & 0 & 0 & 0 \\ 0 & 0 & 0 & 0 & 0 \end{pmatrix}, \tag{A15}$$

$$\frac{\partial \mathbf{S}}{\partial \mu} = \begin{pmatrix} 0 & 0 & 0 & 0 & 0 \\ 0 & 0 & 0 & 0 & 0 \\ \frac{\partial}{\partial x} & \frac{\partial}{\partial z} & 0 & 0 & 0 \\ -\frac{\partial}{\partial x} & \frac{\partial}{\partial z} & 0 & 0 & 0 \\ -\frac{\partial}{\partial z} & -\frac{\partial}{\partial x} & 0 & 0 & 0 \end{pmatrix}, \tag{A16}$$

and

$$\frac{\partial \mathbf{S}}{\partial \rho} = \begin{pmatrix} \frac{\partial^2}{\partial t^2} & 0 & 0 & 0 & 0 \\ 0 & \frac{\partial^2}{\partial t^2} & 0 & 0 & 0 \\ 0 & 0 & 0 & 0 & 0 \\ 0 & 0 & 0 & 0 & 0 \\ 0 & 0 & 0 & 0 & 0 \end{pmatrix}. \tag{A17}$$

Substituting Equations (A14–A17) into Equation (A12), the gradients of the Lamé constants and density can be given by

$$\begin{aligned} \frac{\partial E}{\partial \lambda} &= -\frac{p}{\lambda + \mu}\hat{p} \\ \frac{\partial E}{\partial \mu} &= -\frac{p}{\lambda + \mu}\hat{p} - \frac{\tau_n^s}{\mu}\hat{\tau}_n^s - \frac{\tau_s^s}{\mu}\hat{\tau}_s^s \\ \frac{\partial E}{\partial \rho} &= \frac{\partial^2 u_x}{\partial t^2}\hat{u}_x + \frac{\partial^2 u_z}{\partial t^2}\hat{u}_z. \end{aligned} \tag{A18}$$

References

1. Gaiser, J.E.; Moldoveanu, N.; Macbeth, C.; Michelena, R.J.; Spitz, S. Multicomponent technology: The players, problems, applications, and trends: Summary of the workshop sessions. *Geophysics* **2001**, *20*, 974–977. [CrossRef]
2. Tatham, R.H.; Stoffa, P.L. Vp/Vs—A potential hydrocarbon indicator. *Geophysics* **1976**, *41*, 837–849. [CrossRef]
3. Mora, P. Nonlinear two-dimensional elastic inversion of multioffset seismic data. *Geophysics* **1987**, *52*, 1211–1228. [CrossRef]
4. Tarantola, A. Inversion of seismic reflection data in the acoustic approximation. *Geophysics* **1984**, *49*, 1259–1266. [CrossRef]
5. Sears, T.J.; Barton, P.J.; Singh, S.C. Elastic full waveform inversion of multicomponent ocean-bottom cable seismic data: Application to Alba Field, UK North Sea. *Geophysics* **2010**, *75*, R109–R119. [CrossRef]
6. Crase, E.; Pica, A.; Noble, M.; Mcdonald, J.; Tarantola, A. Robust elastic nonlinear waveform inversion; application to real data. *Geophysics* **1990**, *55*, 527–538. [CrossRef]

7. Crase, E.; Wideman, C.; Noble, M.; Tarantola, A. Nonlinear elastic waveform inversion of land seismic reflection data. *J. Geophys. Res.* **1992**, *97*, 4685–4703. [CrossRef]
8. Sears, T.J.; Singh, S.C.; Barton, P.J. Elastic full waveform inversion of multi-component OBC seismic data. *Geophys. Prospect.* **2008**, *56*, 843–862. [CrossRef]
9. Vigh, D.; Jiao, K.; Watts, D.; Sun, D. Elastic full-waveform inversion application using multicomponent measurements of seismic data collection. *Geophysics* **2014**, *79*, R63–R77. [CrossRef]
10. Oh, J.; Kalita, M.; Alkhalifah, T. 3D elastic full waveform inversion using P-wave excitation amplitude: Application to OBC field data. *Geophysics* **2018**, *83*, R129–R140. [CrossRef]
11. Choi, Y.; Min, D.; Shin, C. Two-dimensional waveform inversion of multicomponent data in acoustic-elastic coupled media. *Geophys. Prospect.* **2008**, *56*, 863–881. [CrossRef]
12. Singh, H.; Shragge, J.; Tsvankin, I. Coupled-domain acoustic-elastic solver for anisotropic media: A mimetic finite difference approach. *Seg Tech. Program Expand. Abstr.* **2019**, *2019*, 3755–3759.
13. Qu, Y.; Liu, Y.; Li, J.; Li, Z. Fluid-solid coupled full-waveform inversion in the curvilinear coordinates for ocean-bottom cable data. *Geophysics* **2020**, *85*, R113–R133. [CrossRef]
14. De Basabe, J.D.; Sen, M.K. A comparison of monolithic methods for elastic wave propagation in media with a fluid–solid interface. *Seg Tech. Program Expand. Abstr.* **2014**, *2014*, 3323–3328.
15. Yu, P.; Geng, J.; Li, X.; Wang, C. Acoustic-elastic coupled equation for ocean bottom seismic data elastic reverse time migration. *Geophysics* **2016**, *81*, S333–S345. [CrossRef]
16. Yu, P.; Geng, J.; Ma, J. Vector-wave-based elastic reverse time migration of ocean-bottom 4C seismic data. *Geophysics* **2018**, *83*, S333–S343. [CrossRef]
17. Yu, P.; Geng, J. Acoustic-elastic coupled equations in vertical transverse isotropic media for pseudo acoustic-wave reverse time migration of ocean-bottom 4C seismic data. *Geophysics* **2019**, *84*, 1–48. [CrossRef]
18. Operto, S.; Gholami, Y.; Prieux, V.; Ribodetti, A.; Brossier, R.; Metivier, L.; Virieux, J. A guided tour of multiparameter full-waveform inversion with multicomponent data: From theory to practice. *Geophysics* **2013**, *32*, 1040–1054. [CrossRef]
19. Metivier, L.; Brossier, R.; Virieux, J.; Operto, S. Full Waveform Inversion and the Truncated Newton Method. *SIAM J. Sci. Comput.* **2013**, *35*, B401–B437. [CrossRef]
20. Liu, Y.; Yang, J.; Chi, B.; Dong, L. An improved scattering-integral approach for frequency-domain full waveform inversion. *Geophys. J. Int.* **2015**, *202*, 1827–1842. [CrossRef]
21. Yang, J.; Liu, Y.; Dong, L. Simultaneous estimation of velocity and density in acoustic multiparameter full-waveform inversion using an improved scattering-integral approach. *Geophysics* **2016**, *81*, R399–R415. [CrossRef]
22. Pan, W.; Innanen, K.A.; Margrave, G.F.; Fehler, M.; Fang, X.; Li, J. Estimation of elastic constants for HTI media using Gauss-Newton and full-Newton multiparameter full-waveform inversion. *Geophysics* **2016**, *81*, R275–R291. [CrossRef]
23. Sun, M.; Yang, J.; Dong, L.; Liu, Y.; Huang, C. Density reconstruction in multiparameter elastic full-waveform inversion. *J. Geophys. Eng.* **2017**, *14*, 1445–1462. [CrossRef]
24. Ren, Z.; Liu, Y.; Sen, M.K. Least-squares reverse time migration in elastic media. *Geophys. J. Int.* **2017**, *208*, 1103–1125. [CrossRef]
25. Sun, M.; Dong, L.; Yang, J.; Huang, C.; Liu, Y. Elastic least-squares reverse time migration with density variations. *Geophysics* **2018**, *83*, S533–S547. [CrossRef]
26. Operto, S.; Virieux, J.; Dessa, J.; Pascal, G. Crustal seismic imaging from multifold ocean bottom seismometer data by frequency domain full waveform tomography: Application to the eastern Nankai trough. *J. Geophys. Res.* **2006**, *111*, B09306. [CrossRef]
27. Shin, C.; Jang, S.; Min, D.J. Improved amplitude preservation for prestack depth migration by inverse scattering theory. *Geophys. Prospect.* **2001**, *49*, 592–606. [CrossRef]
28. Plessix, R.E. A review of the adjoint-state method for computing the gradient of a functional with geophysical applications. *Geophys. J. Int.* **2006**, *167*, 495–503. [CrossRef]
29. Gholami, Y.; Brossier, R.; Operto, S.; Ribodetti, A.; Virieux, J. Which parameterization is suitable for acoustic vertical transverse isotropic full waveform inversion? Part 1: Sensitivity and trade-off analysis. *Geophysics* **2013**, *78*, R81–R105. [CrossRef]

30. Gholami, Y.; Brossier, R.; Operto, S.; Prieux, V.; Ribodetti, A.; Virieux, J. Which parameterization is suitable for acoustic vertical transverse isotropic full waveform inversion? Part 2: Synthetic and real data case studies from Valhall. *Geophysics* **2013**, *78*, R107–R124. [CrossRef]
31. Prieux, V.; Brossier, R.; Operto, S.; Virieux, J. Multiparameter full waveform inversion of multicomponent ocean-bottom-cable data from the Valhall field. Part 1: imaging compressional wave speed, density and attenuation. *Geophys. J. Int.* **2013**, *194*, 1640–1664. [CrossRef]
32. Prieux, V.; Brossier, R.; Operto, S.; Virieux, J. Multiparameter full waveform inversion of multicomponent ocean-bottom-cable data from the Valhall field. Part 2: imaging compressive-wave and shear-wave velocities. *Geophys. J. Int.* **2013**, *194*, 1665–1681. [CrossRef]
33. Woodward, M.J. Wave-equation tomography. *Geophysics* **1992**, *57*, 15–26. [CrossRef]
34. Liu, Y.; Dong, L.; Wang, Y.; Zhu, J.; Ma, Z. Sensitivity kernels for seismic Fresnel volume tomography. *Geophysics* **2009**, *74*, U35–U46. [CrossRef]
35. Gelis, C.; Virieux, J.; Grandjean, G. Two-dimensional elastic full waveform inversion using Born and Rytov formulations in the frequency domain. *Geophys. J. Int.* **2007**, *168*, 605–633. [CrossRef]
36. Li, D.; Zhang, Q.; Wang, Z.; Liu, T. Computation of lightning horizontal field over the two-dimensional rough ground by using the three-dimensional FDTD. *IEEE Trans. Electromagn. Compat.* **2013**, *56*, 143–148.
37. Komatitsch, D.; Martin, R. An unsplit convolutional Perfectly Matched Layer improved at grazing incidence for the seismic wave equation. *Geophysics* **2007**, *72*, SM155–SM167. [CrossRef]
38. Martin, R.; Komatitsch, D. An unsplit convolutional perfectly matched layer technique improved at grazing incidence for the viscoelastic wave equation. *Geophys. J. Int.* **2009**, *179*, 333–344. [CrossRef]
39. Ma, Y.; Hale, D. Wave-equation reflection traveltime inversion with dynamic warping and full-waveform inversion. *Geophysics* **2013**, *78*, R223–R233. [CrossRef]
40. Brossier, R.; Operto, S.; Virieux, J. Velocity model building from seismic reflection data by full-waveform inversion. *Geophys. Prospect.* **2015**, *63*, 354–367. [CrossRef]
41. Chi, B.; Dong, L.; Liu, Y. Correlation-based reflection full-waveform inversion. *Geophysics* **2015**, *80*, R189–R202. [CrossRef]
42. Guo, Q.; Alkhalifah, T. Elastic reflection-based waveform inversion with a nonlinear approach. *Geophysics* **2017**, *82*, R309–R321. [CrossRef]
43. Wang, T.; Cheng, J.; Guo, Q.; Wang, C. Elastic wave-equation-based reflection kernel analysis and traveltime inversion using wave mode decomposition. *Geophys. J. Int.* **2018**, *215*, 450–470. [CrossRef]
44. Biondi, B.; Symes, W.W. Angle-domain common-image gathers for migration velocity analysis by wavefield-continuation imaging. *Geophysics* **2004**, *69*, 1283–1298. [CrossRef]
45. Symes, W.W. Migration velocity analysis and waveform inversion. *Geophys. Prospect.* **2008**, *56*, 765–790. [CrossRef]
46. Wang, C.; Cheng, J.; Weibull, W.W.; Arntsen, B. Elastic wave-equation migration velocity analysis preconditioned through mode decoupling. *Geophysics* **2019**, *84*, R341–R353. [CrossRef]
47. Virieux, J. P-SV wave propagation in heterogeneous media: Velocity-stress finite-difference method. *Geophysics* **1986**, *51*, 889–901. [CrossRef]

© 2020 by the authors. Licensee MDPI, Basel, Switzerland. This article is an open access article distributed under the terms and conditions of the Creative Commons Attribution (CC BY) license (http://creativecommons.org/licenses/by/4.0/).

Technical Note

Estimation of Moisture Content in Railway Subgrade by Ground Penetrating Radar

Sixin Liu [1,2], Qi Lu [1,2,*], Hongqing Li [1,2] and Yuanxin Wang [1,2]

1. College of Geo-Exploration Science and Technology, Jilin University, No. 938 Xi MinZhu Street, Changchun 130026, China; liusixin@jlu.edu.cn (S.L.); lihq@jlu.edu.cn (H.L.); yuanxin@jlu.edu.cn (Y.W.)
2. Science and Technology on Near-Surface Detection Laboratory, Wuxi 214035, China
* Correspondence: luqi@jlu.edu.cn; Tel.: +86-431-8850-2426

Received: 26 July 2020; Accepted: 7 September 2020; Published: 8 September 2020

Abstract: China is strongly dependent on railway transportation, but the frost heaving of the subgrade in cold regions has seriously affected the safety and comfort of trains. Moisture content is an essential parameter in the subgrade frost heave. Non-destructive and efficient geophysical methods have great potential in measuring the moisture content of railway subgrade. In this paper, we use the common mid-point (CMP) measurement of ground penetrating radar (GPR) to estimate the propagation velocity of electromagnetic waves in a subgrade application. We establish a synthetic model to simulate the railway subgrade structure. The synthetic CMP gathers acquired from shallow and thin layers are seriously disturbed by multiple waves and refraction waves, which make the routine velocity analysis unable to provide accurate velocities. Through the analysis of numerical simulation results, it is found that the primary reflection waves, multiple waves, and refraction waves are dominant in different offset ranges of CMP gather. Therefore, we propose a solution of the optimal gather at a certain range of offset dominated by the primary reflection wave to calculate the velocity spectrum and extract the accurate velocities for the subgrade model. The relative dielectric constants of the corresponding layers are calculated after the stacking velocities are converted into the interval velocities. Then, the moisture content is obtained by the Topp formula, which expresses the relationship between dielectric constant and moisture content. Finally, we apply the optimal gather scheme and the above interpretation process to the GPR data acquired at the railway site, and we form a long moisture content profile of the railway subgrade. Compared with the polarizability measured by the induced polarization (IP) method, it is found that the regions with high moisture content correspond to polarizability anomalies with different strengths. The comparison shows the reliability of GPR results to some extent.

Keywords: ground penetrating radar (GPR); moisture content; velocity analysis; optimal gather

1. Introduction

China has a vast territory, a large population, and an unbalanced distribution of resources and industries. Therefore, China's railway plays an important role in economic and social development. Subgrade frost heaving is a common geological phenomenon in northern China, and its treatment is the main task of railway maintenance in winter. Frost heaving of the subgrade seriously affects the safety and comfort of train operation. Temperature, soil properties, and moisture content are the three factors of the formation of subgrade frost heaving. The main cause of soil freezing is the moisture content in soil. Therefore, it is crucial to measure the moisture content of railway subgrade quickly and accurately for solving the problem of frost heaving.

Drilling samples can obtain the most accurate information about the subgrade moisture content [1]. However, this method is time-consuming and expensive for large-scale measurement. Ground penetrating radar (GPR) is a geophysical technique for the non-destructive and efficient detection of shallow layers by electromagnetic (EM) waves, which is widely used in the field of environment and engineering. Using GPR to measure the moisture content has become an important branch of GPR applications [2]. In recent years, a lot of papers using GPR to measure soil moisture content have been published [3–12]. Some of them focus on ground water table measurement. As an example, Tsoflias et al. observe that the radar signal response is correlated to changes in the water saturation of the fracture and provide spatial information about the saturation of the fracture [5]. Saintenoy and Hopmans study the water table detection sensitivity of GPR and state a power type relationship between the reflected signal amplitude and the slope of the soil retention curve [6]. Others are concerned about the measurement of soil moisture content. Two review papers are valuable. Huisman et al. (2003) presents the basic theoretical principles of GPR and how they can be used to investigate the spatiotemporal variation of soil water content [10]. In particular, they propose four categories of methods to determine soil water content with GPR. Klotzsche et al. (2018) provide an update on the review of recent advances in vadose zone applications of GPR with a particular focus on new possibilities, e.g., for multi-offset and borehole GPR measurements [11].

A large number of GPR applications in measuring soil moisture content has benefited from the research of Topp et al., which established the relationship between the relative dielectric constant and soil moisture content [13]. It is essential to estimate the velocity of EM waves, since it is directly related to the dielectric constant of medium. The common mid-point (CMP) or wide-angle reflection and refraction (WARR) sounding mode in GPR are primarily used to obtain an estimate of the radar wave velocity versus depth in the ground by varying the antenna spacing and measuring the change in the two-way travel time [2]. The velocity analysis of CMP gather, coming from seismic exploration, is a common and accurate way to obtain wave velocity. Some scholars have carried out research studies to get a better velocity estimation in GPR. Feng et al. improved the signal to noise ratio by CMP antenna array and data processing technology and obtained a good result in velocity analysis, which they successfully applied to landmine detection [14]. Liu et al. obtain the EM wave velocity through the envelope velocity analysis method, which can monitor the underground dynamic water level [15]. Lu described the method to obtain the soil moisture content and monitored the underground water level [16]. Liu et al. used GPR profiles and moisture content to estimated hydraulic conductivity parameters and accurately distinguished a slight change of groundwater level [17]. Pue et al. proposed a modified velocity analysis method to successfully estimate depth and propagation velocity with small offset air-coupled GPR configurations by accounting for the refraction at the surface [18]. Yi et al. proposed a high-resolution velocity analysis by applying the l-1 norm regularized least-squares for pavement inspection [19]. In recent years, many other related papers have been published to improve the performance of CMP velocity analysis in the estimation of EM wave velocity [20–22]. However, there are few research studies that deal with shallow and thin multi-layered structures such as the railway subgrade. In such a case, the reliability of the routine velocity analysis process needs to be retested.

In this paper, we introduce the basic theory used here firstly. Then, two synthetic models are established to simulate the railway subgrade structure and the problems of velocity analysis in a shallow thin layer are analyzed by numerical simulation. After that, the solution of optimal gather is proposed. The proposed scheme is applied to the measured data of railway subgrade, and a long moisture content profile of railway subgrade is obtained. Finally, the polarization features from induced polarization (IP) measurement show that the proposed scheme is reliable in a certain range.

2. Methods

2.1. Velocity Analysis of CMP Data

The common mid-point (CMP) measurement in a GPR survey is a multi-offset measurement that is primarily used to obtain an estimated velocity of the EM wave in the ground. By increasing the antenna preparation symmetrically, the path of the EM wave varies while keeping the point of reflection fixed. Thus, the difference in two-way travel time enables wave velocity to be estimated [2].

The velocity analysis by CMP gather assumes that the subsurface media is horizontally layered. The two-way travel time of reflected waves can be obtained by

$$t(x_i) = \sqrt{t_0^2 + \frac{x_i^2}{v_{rms}^2}} \tag{1}$$

where x_i is the antenna offset of the i-th channel, t_0 is the two-way travel time of zero offset, and v_{rms} is the root mean square velocity.

The velocity spectrum can be obtained by transforming the data from the offset versus two-way time domain to the stacking velocity versus two-way zero-offset time domain [23]. The velocity spectrum shows the picks that correspond to the best coherency of the signal along a hyperbolic trajectory over the entire spread length of the CMP gather. The coherency can be computed in several ways [24]. However, in the velocity analysis of GPR data, the simple and effective measures are the average stacked amplitude (A) and the energy of the average stacked amplitude (E), which can suppress interference signals with large amplitude. The A and E can be defined by

$$A = \frac{1}{N} \sum_{j=1}^{M} \left| \sum_{i=1}^{N} f_{i,j+r_i} \right| \tag{2}$$

$$E = \sum_{j=1}^{M} \left(\frac{1}{N} \sum_{i=1}^{N} f_{i,j+r_i} \right)^2 \tag{3}$$

where $f_{i,j+r_i}$ is the amplitude value of the $j+r$-th point on the i-th trace in which r is a delay associated with normal moveout (NMO). N is the number of traces in the CMP gather. M is the number of sampling points in a stack time window. Each stacking velocity (v_{rms}) is scanned according to Equation (1). When the scanning velocity is equal to the v_{rms}, E reaches the maximum value, which shows a focused energy cluster in the velocity spectrum. After the stacking velocity is picked up, the interval velocity can be calculated by Dix formula [25], and the expression is as follows:

$$v_{int,n} = \sqrt{\frac{t_{0,n} v_{rms,n}^2 - t_{0,n-1} v_{rms,n-1}^2}{t_{0,n} - t_{0,n-1}}} \tag{4}$$

where $v_{int,n}$ is the interval velocity of the nth layer; $v_{rms,n}$ is the stacking velocity to the bottom of the nth layer; $t_{0,n}$ is the two-way travel time with zero offset to the bottom of the nth layer; $v_{rms,n-1}$ is the stacking velocity to the bottom of the $n-1$th layer; and $t_{0,n-1}$ is the two-way travel time with zero offset to the bottom of the $n-1$th layer.

After obtaining the interval velocities (v_{int}) of radar wave propagating in the subsurface, the relative dielectric constant ε_r in low loss medium can be obtained by

$$v_{int} = \frac{c}{\sqrt{\varepsilon_r}} \tag{5}$$

where $c = 0.3$ m/ns is the velocity of EM wave in air.

2.2. Optimal Gather in Velocity Analysis

As the offset increases, the primary reflections from the interface of each layer will be interfered by multiple reflections and refractions in CMP gathers. In the media composed of thin multi-layers, this interference is so serious that it makes the velocity spectra unable to give correct velocities of EM waves. In addition, the CMP gathers obtained in small offsets can be interfered by direct wave.

Since the layers are shallow and thin, the reflected wave is close to the direct wave, whose strong energy makes the reflected wave difficult to be identified. Therefore, it is crucial to use high-quality primary reflections from layer interfaces for velocity analysis to obtain accurate velocities.

The optimal gathers include mainly the primary reflections who suffer little interference from multi-waves, refracted waves, and direct waves. These gathers are determined and used to calculate the velocity spectrum. The reflection angle can be used to explain the operation of optimal gather.

The reflection angle θ can be calculated using the sketch shown in Figure 1. x_1 represents a small offset, and the CMP gathers acquired in x_1 are supposed to be contaminated with direct waves. x_2 represents a large offset, and the CMP gathers acquired in x_2 are supposed to be contaminated with multi-waves and refracted waves. The blue arrows in Figure 1 show the corresponding antenna locations when effective reflection from the interface can be obtained, and the reflection angle θ ranges from θ_1 to θ_2. The optimal gather for the velocity analysis should be in this angle range.

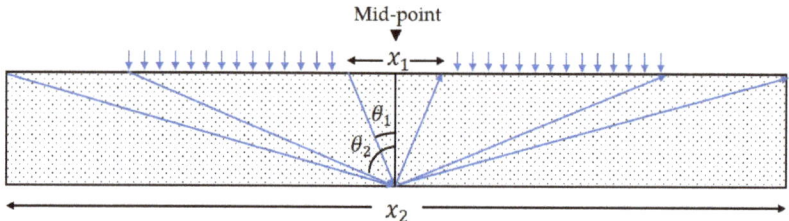

Figure 1. A description for reflection angle and optimal gather. The locations marked by arrows represent the optimal gather range.

In this paper, we can analyze the influence range of interference waves in CMP gathers through numerical simulation because the railway subgrade structure is known. It can help to determine the optimal gather in field CMP data.

2.3. Topp Formulae

A well-known empirical relationship between the relative dielectric constant and volumetric moisture content at frequencies between 1 MHz and 1 GHz was developed by Topp et al. [13]. This empirical relationship is independent of soil type, soil density, soil temperature, and soluble salt content, and it is determined by compiling data for many soils under varying moisture conditions:

$$\varepsilon_r = 3.03 + 9.3\theta + 146.0\theta^2 - 76.7\theta^3 \qquad (6)$$

where θ is the volumetric soil moisture content, which is equal to the product of the water saturation and porosity. A polynomial expression for moisture content as a function of the relative dielectric constant, ε_r, was reported as well:

$$\theta = -5.3 \times 10^{-2} + 2.92 \times 10^{-2}\varepsilon_r - 5.5 \times 10^{-4}\varepsilon_r^2 + 4.3 \times 10^{-6}\varepsilon_r^3. \qquad (7)$$

2.4. Induced Polarization

Induced polarization (IP) is a geophysical method that is mainly used in metal ore exploration and groundwater search. In recent years, with the development of sensitive instruments and methods to improve the signal-to-noise ratio, the IP method has been widely used in the field of environment and engineering [26–28]. IP may reflect the moisture content through the membrane effect, which is related to variations in the mobility of ions in fluids at grain scales in a porous medium under the influence of the electrical field. The polarizability generated by the water-bearing sand gravel layer is higher than that of non-water bearing structure.

In time domain IP, a primary current is injected in the ground for a period T using a pair of electrodes A and B. The difference of electrical potential is measured using several bipoles of electrodes M and N. The secondary voltage on a bipole MN decays over time after the shut-down of the primary current. The rate of this decay reflects the strength of the polarization effect. The simplest way to measure the IP effect with time-domain equipment is to compare the residual voltage $V(t)$ existing at a time t after the current is cut off with the steady voltage V_c during the current flow interval. It is not possible to measure potential at the instant of cutoff because of large transients caused by breaking the current circuit. On the other hand, $V(t)$ must be measured before the residual has decayed to the noise level. Therefore, the polarizability, $\eta(t)$ can be obtained by

$$\eta(t) = \frac{V(t)}{V_c} \times 100\%. \tag{8}$$

3. Velocity Analysis of CMP Models

3.1. Railway Subgrade

Figure 2 shows a typical railway subgrade structure, which can be roughly divided into 4 layers. These are constituted with ballast, graded gravel, medium to coarse-grained sand, and A, B group filling from top to bottom. There is a sheet of geotextile in the middle of the 0.2 m sand layer (marked by a red line in Figure 2).

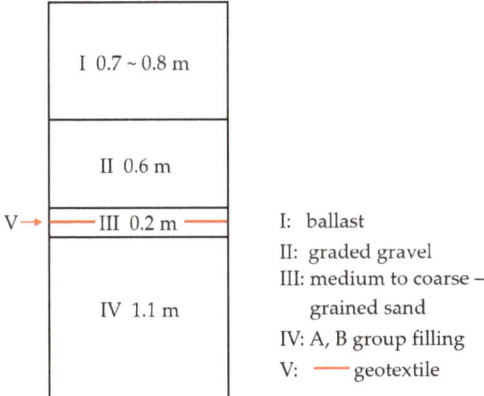

Figure 2. The profile of a typical railway subgrade.

The horizontal layered structure of the railway subgrade satisfies the assumption of velocity analysis through CMP gather. However, due to the shallow depth and the thin multi-layers, the reflected waves from the interfaces of horizontal layers will be interfered by multiple waves and refracted waves. Once the hyperbolic trajectory of the reflected wave in CMP gathers is damaged, the velocity spectrum will not obtain the picks representing the best coherency and failure to pick up the accurate velocity.

3.2. Modeling

In this section, we carry out CMP forward simulation on a simple model and a railway subgrade model to study the interference caused by multiple waves and refraction waves and propose solutions. The GPR forward modeling uses an open source software gprMax that simulates EM wave propagation by the Finite-Difference Time-Domain (FDTD) method [29].

3.2.1. A Simple Model

A simple model (Figure 3) is used firstly to analyze the sources of interference. In Figure 3, the air layer is marked by a white color with $\varepsilon_r = 1$. The thickness of the blue layer is 0.7 m with $\varepsilon_r = 4$. The velocity of EM wave 0.15 m/ns in the layer can be obtained by Equation (5). The red layer is with the $\varepsilon_r = 10$. The synthetic CMP data are generated by a 100 MHz Ricker wavelet and the grid size of 0.05 m. The mid-point of CMP gather is fixed at coordinate (15,12), as shown in Figure 3. The increment of antenna separation is 0.2 m, and 64 traces are acquired in the CMP gather.

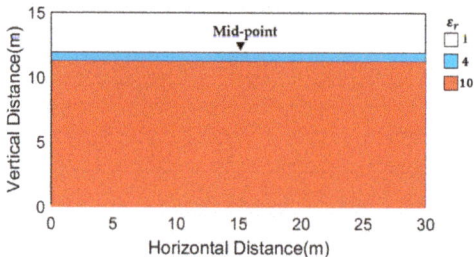

Figure 3. A simple model.

The synthetic CMP gather is shown in Figure 4. It can be seen from Figure 4a that there are several discontinuous events. In order to analyze the sources of waveforms in CMP gathers, we calculate the arrival times of reflected, multiples, and refracted waves according to the model parameters, and the calculated values are shown by curves with different color. In Figure 4b, the black dotted line marked with ① is the direct air wave that the wave directly propagates from the transmitting antenna to the receiving antenna. The black dotted lines marked with numbers ②–④ are the refracted waves propagating in the air after they are reflected twice or more by the layer interface. The red dot line in Figure 4b represents the primary reflected wave from the layer interface, which we need. However, the continuity of the event is obviously destroyed by multiple reflections. The green dotted lines in Figure 4b represent reflection multiples. Equation (1) is the signal received by the antenna after it is reflected twice by the interface, and (2)–(4) are reflected multiple times by the interface before it is received by the antenna. The above analysis shows that the existence of multiple reflections brings serious interference to velocity analysis.

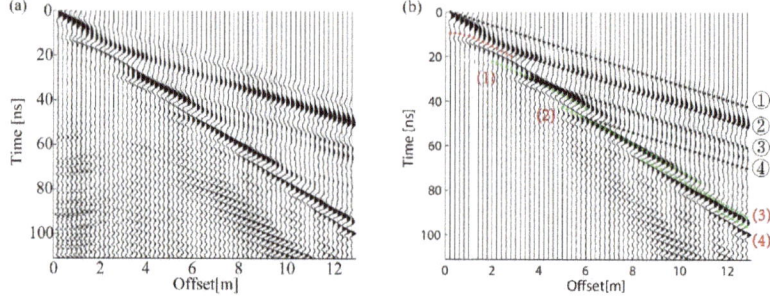

Figure 4. Synthetic common mid-point (CMP) gather for the simple model (**a**) with the direct wave, refracted waves, and reflected waves marked on (**b**).

3.2.2. A Railway Subgrade Model

The typical railway subgrade model is shown in Figure 5. The white layer is air. The thickness of the blue layer corresponding to ballast is 0.7 m with $\varepsilon_r = 4$. The next layer marked green is graded gravel that is 0.6 m thick, and the $\varepsilon_r = 5.5$. The thin yellow layer of 0.2 m thickness is medium to coarse-grained sand with $\varepsilon_r = 7$. The thickness of the orange layer is A, B group filling that is 1.1 m thick and $\varepsilon_r = 8.5$. The red part is a layer of raw soil with the ε_r of 10. The mid-point of CMP gather is fixed at coordinate (15,12), as shown in Figure 5. The increment of antenna separation is 0.2 m, and 64 traces are acquired in the CMP gather.

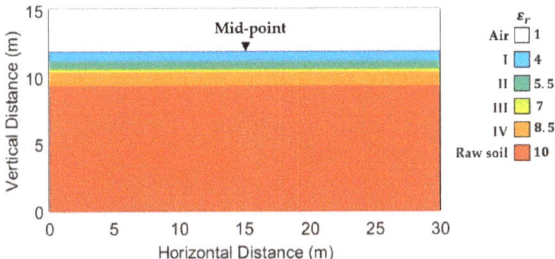

Figure 5. The typical railway subgrade model. The labels of I–IV correspond to the layers I–IV in Figure 2.

Figure 6a shows the synthetic CMP gather of the railway subgrade model. Similarly, we use the model parameters to calculate the arrival time of radar waves, and the calculated wave curves are shown with different colors in Figure 6b. The analysis for the simple model in Section 3.2.1 makes it easy to distinguish the various waves generated in this complex model.

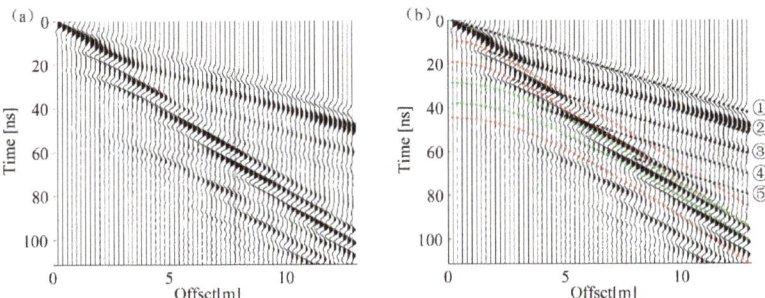

Figure 6. CMP gather for the railway subgrade model (**a**) with the direct wave, refracted waves, and reflected waves marked on (**b**).

In Figure 6b, the black dotted line marked with ① is the direct wave in air. The next four black dotted curves with the labels of ②–⑤ are refracted waves, and the amplitudes of these curves decrease in turn. There are two green dotted lines in Figure 6b that represent multiple reflections. The upper green line represents the multiple wave that crosses the first layer twice and the second layer once, and its energy is mainly obtained at the offset of 5–12.8 m. The lower green curve crosses the first layer three times and the second layer once, and its energy is mainly obtained at the offset of 9–12.8 m.

The three red dotted lines represent the primary reflections from the layer interfaces, which are the target reflections. The top red curve is reflected from the first interface, whose energy is mainly obtained at offsets less than 5 m. However, the third layer that is 0.2 m thick is too thin to be identified.

The generation of these marked reflections can be explained by the reflection angle shown in Figure 1. First, when the reflection angle is small, the reflected wave is close to the direct wave whose strong energy makes the reflected wave difficult to be identified. Second, with the increase of reflection angle, the energy of multiple reflections increase, while the primary reflections become weaker until they cannot be detected. Therefore, the reflected wave can be distinguished only when the reflection angle is within a certain range. Next, we deal with optimal gathers carefully to ensure that the velocity spectrum can provide reliable velocity.

We calculate the velocity spectra with different offsets for CMP gathers of the railway subgrade model. Since the initial offset in the field CMP data is 0.6 m (shown in Section 4), the same initial offset is used in the velocity analysis of the model. Figure 7a,c,e,g show the velocity spectra after normalization with the CMP gathers at offset 0.6–3 m, 0.6–4 m, 0.6–5 m, and 0.6–6 m, respectively. Figure 7b,d,f,h shows the curves of the average amplitude, corresponding to Figure 7a,c,e,g, which is the mean value of the amplitude of all velocities at each arrival time. The velocity can be picked up by the correspondence between the energy cluster in the velocity spectrum and the wave crest in the average amplitude curve.

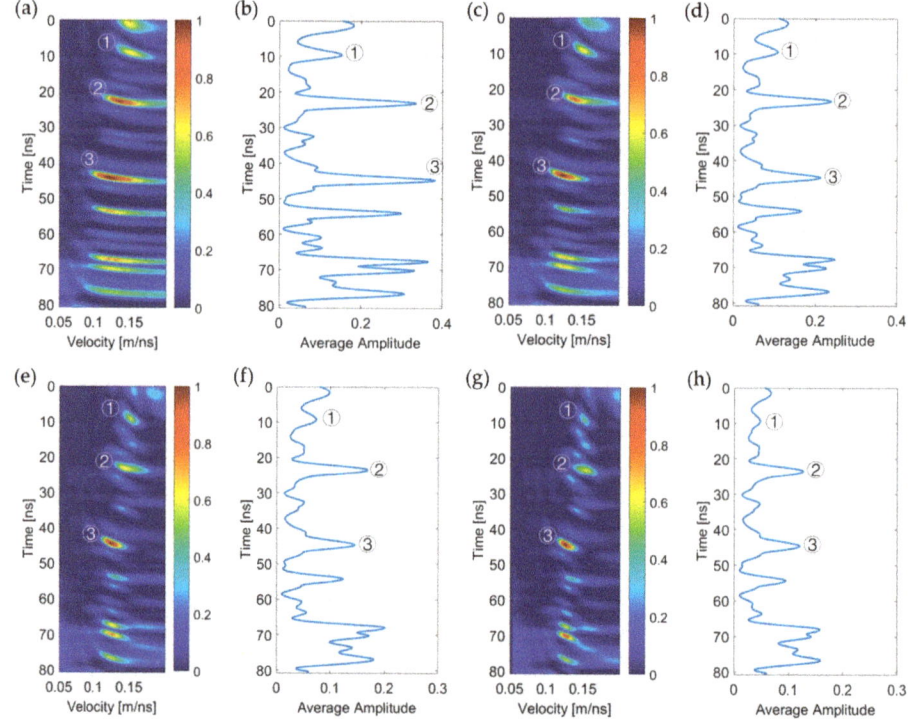

Figure 7. Velocity spectra and curves of average amplitude for the railway subgrade model are calculated using different offsets. (**a,c,e,g**) are velocity spectra with offsets of 0.6–3 m, 0.6–4 m, 0.6–5 m, and 0.6–6 m, respectively; (**b,d,f,h**) are curves of average amplitude calculated from (**a,c,e,g**), respectively.

In the eight subgraphs of Figure 7, the velocity of EM waves from each layer is represented by energy clusters in the velocity spectra and wave peaks in the average amplitude curves marked by ①, ②, and ③. Other energy clusters and peaks, such as those that appear at the arrival time greater than 50 ns, are from the contribution of multiple waves.

In Figure 7a, when the offset (0.6–3 m) is small and the number of gathers is less, the energy cluster is decentralized, which brings trouble to velocity picking. As the offset increases, the energy clusters become more and more focused in Figure 7c,e. Combined with Figure 7d,f, velocity picking becomes easier. Although Figure 7e shows that the energy clusters are more focused than that in Figure 7c, the velocity of EM waves of the first layer is clearer in Figure 7c. In Figure 7g,h, it is difficult to pick up the velocity of first layer due to the large offset, which is 0.6–6 m. Therefore, a smaller offset makes the energy decentralized, while a larger offset makes the velocity of the first layer impossible to pick up. The optimal gather in this railway subgrade model is ended at the offset between 4 and 5 m with the initial offset of 0.6 m.

Then, the velocities (v_{rms}) are picked up from the velocity spectrum in Figure 7e. The v_{rms} are converted to the v_{int} by the Dix formula of Equation (4), which is plotted by a green line in Figure 8. The black line shows the velocities that are calculated from the ε_r of the model. It shows that the estimated v_{int} in the first and second layer are basically in accordance with the model velocities. The third layer in the model is a medium to coarse-grained sand that is 0.2 m thick. GPR with the central frequency of 100 MHz failed to recognize this thin layer. Therefore, the velocity spectrum is also not able to give the velocity of this layer. It leads to the difference between the model velocity and v_{int} around 20 ns in Figure 8. The model velocity of the fourth layer is 0.103 m/ns, while the v_{int} of that is 0.106 m/ns (Table 1). The failure to pick up the v_{rms} of the third layer is part of the reason for the error. After the dielectric constant (ε_r) is obtained by Equation (5), the moisture content can be calculated by Equation (7). Table 1 shows the details of the estimated information from the velocity spectrum in Figure 7e.

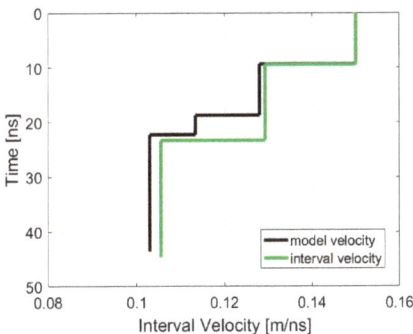

Figure 8. Comparison between the interval velocity and model velocity for the railway subgrade model.

Table 1. The detailed information estimated from the velocity spectrum in Figure 7e.

Layer Sequence	Time (ns)	v_{rms} (m/ns)	v_{int} (m/ns)	Layer Thickness (m)	Dielectric Constant	Moisture Content
1	9.4	0.15	0.15	0.705	4	5.5%
2,3	23.3	0.138	0.129	0.89	5.39	8.9%
4	44.7	0.124	0.106	1.13	8.1	15%

It can be seen from Table 1 that the estimation of the first layer is accurate. The errors occur when the second and third layers are estimated as one. It also affects the estimation of the fourth layer. Our ultimate goal is to get moisture content under the subgrade. The model moisture content of the four layers is 5.53%, 9.17%, 12.6%, and 15.8%, respectively. Compared with the estimated water content in the last column of Table 1, although the big error occurs in layers 2 and 3, the vertical variation trend of moisture content is still credible, which is essential to identify a meaningful distribution of moisture content. According to the above optimal gather method and processing flow, the moisture content estimation in the railway subgrade is conducted in the following case study.

4. Case Study

We measured the GPR data in a certain railway in Heilongjiang, which is the coldest area in China. In Figure 9, a SIR-20 GPR system that employed the shielded antenna with the central frequency of 100 MHz from GSSI (Geophysical Survey Systems, Inc.) is used in the field survey. The CMP gather is acquired with the initial offset of 0.6 m between the transmitter and receiver. The increment of antenna separation is 0.2 m and the maximum antenna transceiver distance is about 14 m. The parameter setting in acquisition is consistent with that in the model. CMP gathers from 56 mid-points are acquired, and the separation between two adjacent mid-points is 10 m. In order to avoid the influence of sleepers, CMP data are acquired along the outside of the rail (see the photo in Figure 9).

Figure 9. A photograph of the experiment site showing the CMP measurement.

The preprocessing for the original CMP gathers includes zero-time correction, gain control, and frequency domain filtering. Figure 10a,b show the original CMP gather acquired at the first CMP point and its profile after preprocessing, respectively.

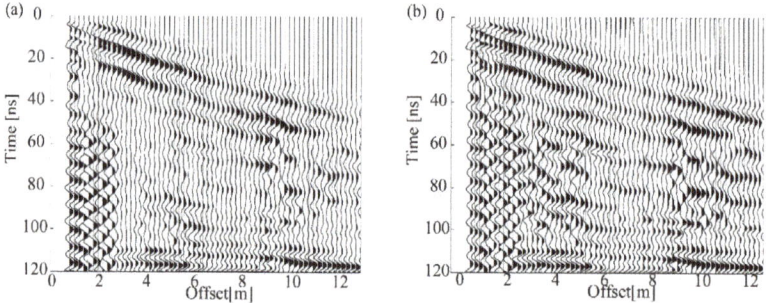

Figure 10. Original CMP gather (**a**) and the profile after preprocessing (**b**).

The CMP gathers acquired at the offset of 0.6 m to 5 m are used in velocity analysis, which is decided from the tests on the model in Section 3.2.2. Figure 11 shows the velocity analysis result for the CMP gathers in Figure 10b. In Figure 11a, the velocity spectrum shows five energy clusters which are marked by ①–③, ⓐ, and ⓑ. Only the energy clusters labeled ①–③ have corresponding peaks in the average amplitude curve, which is also marked by ①–③ in Figure 11b, while ⓐ and ⓑ cannot be identified. The hyperbolic trajectories of the signals are plotted according to the velocities picked up from the five energy clusters, as shown in Figure 12. In Figure 12a, the hyperbolic trajectories generated by velocities ①–③ correspond to relatively complete hyperbolic events, which are meaningful. On the other hand, the clusters ⓐ and ⓑ, in Figure 12b, are invalid because they produce trajectories without corresponding hyperbolic events. Therefore, the energy clusters ①–③ are the reflections from the interfaces of the railway subgrade, and we extract the v_{rms} from them to calculate the v_{int}. The v_{int} is shown in Figure 11c, and the moisture content is calculated as shown in Figure 11d.

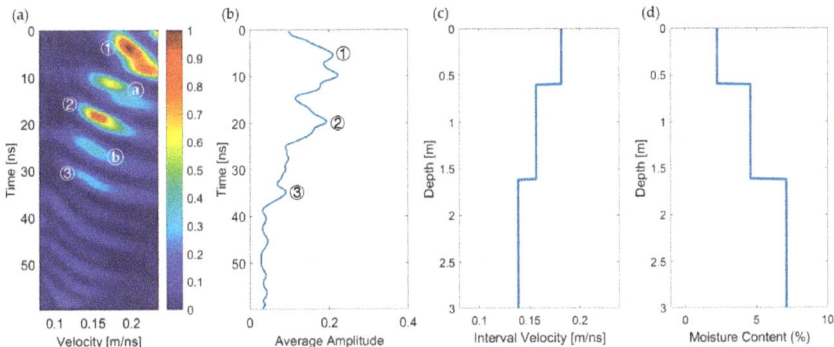

Figure 11. Velocity analysis of the CMP gather acquired at the first mid-point to obtain the moisture content of the railway subgrade. (**a**) Velocity spectrum with the offset of 0.6–5 m, (**b**) curve of average amplitude, (**c**) interval velocity, and (**d**) moisture content.

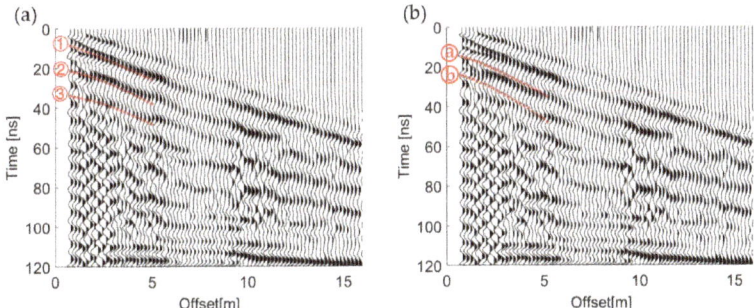

Figure 12. CMP gather with corresponding hyperbolic trajectory (red lines) which are produced by (**a**) the velocities picked up from the valid energy clusters ①, ②, and ③, and (**b**) velocities picked up from the invalid energy clusters ⓐ and ⓑ.

As shown in Figure 11d, the vertical moisture content ranges from 2.24% to 7.1% with depth, and there is no abnormal trend in moisture content change. Table 2 shows the details of related information from the estimates. The estimated total thickness is consistent with that of the actual railway subgrade structure, but some errors exist in the thickness of each layer. We think that the main reason for these errors is the velocity pick-up errors. It is also related to the resolution provided by the 100 MHz central frequency of the radar antenna, which may have a limitation in detecting thin layers such as in this case.

Table 2. The detailed information estimated from the velocity analysis in Figure 11.

Layer Sequence [1]	Time (ns)	v_{rms} (m/ns)	v_{int} (m/ns)	Layer Thickness (m)	Dielectric Constant	Moisture Content
1	6.5	0.182	0.182	0.59	2.71	2.24%
2	19.5	0.166	0.157	1.02	3.63	4.6%
3	34.5	0.155	0.139	1.05	4.63	7.1%

[1] The layers here correspond to the marks ①–③ in Figure 11a,b.

Figure 13 shows the velocity analysis for another CMP gather acquired at the 27th mid-point. Four energy clusters are considered to be valid, which are labeled by ①–④ in Figure 13a, and the corresponding wave crests can be found in the average amplitude curve (Figure 13b). The velocities, 0.2 m/ns (8 ns), 0.186 m/ns (14.5 ns), 0.150 m/ns (34 ns), and 0.135 m/ns (46 ns), are picked up.

These velocities (v_{rms}) are converted to the v_{int}, which is shown in Figure 13c, and four velocity layers are identified. The thicknesses of the first two layers, which are considered as layers of ballast and graded gravel, are accurately estimated. The third layer of medium to coarse-grained sand with the thickness of 0.2 m is not distinguished. A velocity increasing can be found in the fourth layer of the A, B group filling. In Figure 13d, the profile shows the moisture content of 11.8% in the A, B group filling, and it increases to 27.1% at the bottom of the A, B group filling at the depth of 2.47 m. We direct attention to the abnormal increase of moisture content occurring at the bottom of the A, B group filling layer.

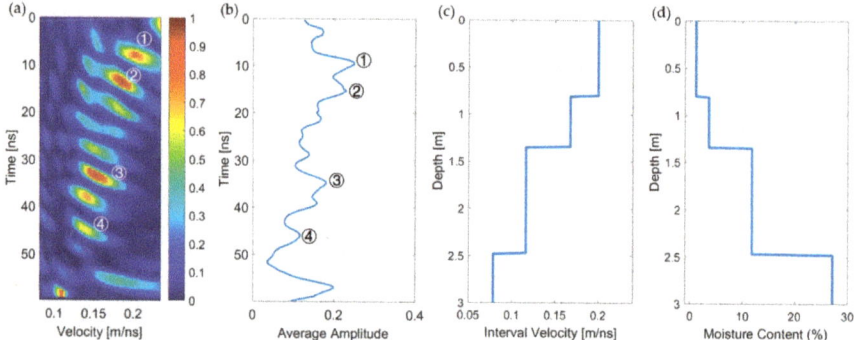

Figure 13. Velocity analysis of the CMP gather acquired at the 27th mid-point to obtain the moisture content of the railway subgrade. (**a**) Velocity spectrum with the offset of 0.6–5 m, (**b**) curve of average amplitude, (**c**) interval velocity, and (**d**) moisture content.

The total of 56 CMPs are acquired with the separation of 10 m between two adjacent mid-points. The velocity spectrum of each CMP gather is calculated by the optimal gather with an offset of 0.6 to 5 m. The valid energy clusters in the velocity spectrum are determined by judging their correspondence with the peak of the average amplitude curves and the events in CMP gathers. The stacking velocities (v_{rms}) are picked up and converted to interval velocities (v_{int}) by the Dix formula. Then, the relative dielectric constant (ε_r) can be obtained by Equation (5)—that is, $\varepsilon_r = (c/v_{int})^2$. At last, the moisture content can be calculated by Equation (7) of the Topp formula. The moisture content curves from 56 mid-points are interpolated into a profile of moisture content, as shown in Figure 14a.

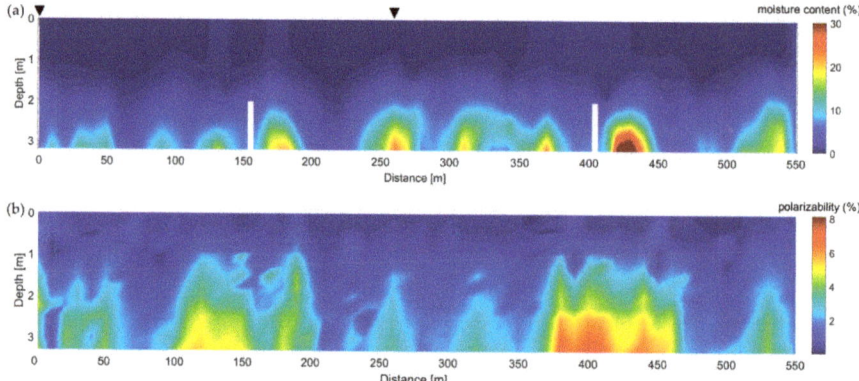

Figure 14. Comparison between moisture content obtained by velocity analysis of CMP (**a**) and polarizability obtained by induced polarization (IP) (**b**). The white boxes are culverts. The black triangles represent the locations of the first and the 27th mid-points of CMP gathers, respectively.

In Figure 14a, high moisture content areas are mainly located at around 130–140 m, 180–190 m, 260–270 m, 310 m, 370 m, 420–450 m, and 520–540 m. There are two known culverts with a size of 4 m × 4 m along this survey line. One is located between 150 and 160 m and the other is at 400–410 m with a depth of about 2 m. The positions of the culverts are marked by white boxes in Figure 14a. The culverts are filled with air, and their existence affects the moisture estimation of their surroundings. These regions feature large velocities of EM waves, small dielectric constants, and small moisture content.

IP measurements are also conducted along the same survey line. A four-electrode sounding method is used in the IP measurement. In the IP acquisition, current electrodes A and B are stainless steel electrodes, and Pb-PbCl nonpolarizable electrodes are used for the potential electrodes M and N to avoid the charge-up effect. The distance between the electrodes of M and N is fixed to 0.2 m, and the distance between electrodes A to B is gradually increased. These separations are 0.6, 0.8, 1.0, 1.2, 1.4, 1.6, 1.8, 2.0, 2.5, 3.0, 3.5, 4.0, 4.5, 5.0, 6.0, 7.0, 8.0, 9.0, and 10.0 m, respectively. The IP measuring points are consistent with the mid-points in CMP measurement, and the separation of two adjacent measuring points is 10 m. The polarizability of each measuring point is calculated by Equation (8). These 1D polarizability data are used to create a polarizability profile by interpolation, as shown in Figure 14b.

In Figure 14b, there are some anomalies of polarizability whose values are higher than that of the surroundings. Compared with Figure 14a, it is found that the positions of polarizability anomalies are consistent with those of the abnormal area of moisture content, although the level of polarizability and moisture content are not consistent. For example, there is an obvious high-water content area around 250 m in Figure 14a, but it shows a polarizability anomaly at the same position, which is weak. In Figure 14b there is an obvious polarization anomaly at 100–150 m, which is stronger than that seen at 250 m. However, the moisture content evaluated at 100–150 m is lower than that evaluated at 250 m in Figure 14a.

These observations indicate that the relationship between polarizability and moisture content is complex. There is no doubt that an IP survey can find water. However, the dominant factor controlling the strength of polarizability should be the ion concentration in water, not only the amount of water. Therefore, we do not use the strength of polarization to show the moisture content, since we have no information about the ion concentration in soil solution. We compare the positions of abnormal polarizability with that of moisture content from GPR data. We state that no matter how strong or weak, as long as anomalies of the moisture content and polarizability appear in the same position, they can be considered as representing the same target. As a result, this comparison shows that the moisture content extracted from velocity analysis is reliable in the case of railway subgrade.

5. Conclusions

In this paper, we use GPR to detect the moisture content of railway subgrade. The flow of the processing and interpretation includes: first, to obtain the stacking velocity of each layer of subgrade by the velocity analysis of CMP gathers; second, to convert the stacking velocities into interval velocities by the Dix formula and then calculate the dielectric constant; finally, to obtain the moisture content distribution of subgrade by the Topp formula. In this flow, it is the most important thing to pick up the accurate stacking velocity.

The shallow and thin multi-layers of railway subgrade make CMP gather seriously disturbed by multiple waves and refraction waves, which makes the routine velocity analysis unable to provide accurate velocity. In the railway subgrade case, we propose a solution of the optimal gather from a certain offset range dominated by the primary reflection to calculate the velocity spectrum. The extraction of the valid velocities can be completed by judging their correspondence with the peak of the average amplitude curves and the events in CMP gathers. The proposed optimal gather and interpretation flow are applied to the CMP data acquired at a railway site, and a long profile of moisture content distribution is obtained. IP measurement along the same survey line shows the abnormal regions of polarizability whose positions are consistent with large water content areas extracted from GPR data.

In the field survey, a GPR system equipped with a shielded antenna with a central frequency of 100 MHz is used. This frequency of the antenna has a limitation in the detection of the shallow and thin layer structure of the subgrade. The shielded antenna with the center frequency of 200 MHz may give a better result which is known through numerical simulation (not shown in the text). Unfortunately, we do not have such equipment. However, it is gratifying to see that the results given by this paper show that the antenna with a center frequency of 100 MHz is feasible. In conclusion, the velocity analysis of CMP gathers in GPR has great potential in quantitatively estimating the moisture content of the railway subgrade in the construction and maintenance of railway in cold regions.

Author Contributions: Conceptualization, S.L. and Q.L.; methodology, S.L., Q.L., H.L., Y.W.; formal analysis, S.L., Q.L., H.L., and Y.W.; Data processing, S.L., Q.L., H.L.; investigation, H.L., Y.W.; data curation, S.L.; writing—original draft preparation, S.L. and Q.L.; writing—review and editing, S.L., Q.L.; funding acquisition, S.L., Q.L. All authors have read and agreed to the published version of the manuscript.

Funding: This work was funded by the Natural Science Foundation of China (Grant 41874136 & 41574109), and Science and Technology on Near-Surface Detection Laboratory under Grant 61424140601.

Acknowledgments: The authors are grateful to the reviewers for their valuable suggestions.

Conflicts of Interest: The authors declare no conflict of interest.

References

1. Brown, A.C.; Dellinger, G.; Helwa, A.; El-Mohtar, C.; Zornberg, J.; Gilbert, R.B. Monitoring a Drilled Shaft Retaining Wall in Expansive Clay: Long-Term Performance in Response to Moisture Fluctuations. In Proceedings of the American Society of Civil Engineers, San Antonio, TX, USA, 17–21 March 2015.
2. Jol, H.M. (Ed.) *Ground Penetrating Radar: Theory and Applications*; Elsevier Science: Amsterdam, The Netherlands, 2009.
3. Grit, D.; Ugur, Y. Estimation of water of moisture content and porosity using combined radar and geoelectrical measurements. *Eur. J. Environ. Eng. Geophys.* **1999**, *4*, 71–85.
4. Annan, A.P.; Cosway, S.W.; Redman, J.D. Water table detection with ground penetrating radar. In Proceedings of the Extended Abstracts of 61st Annual Meeting of the Society of Exploration Geophysicists, Houston, TX, USA, 10–14 November 1991; pp. 494–496.
5. Tsoflias, G.P.; Halihan, T.; Sharp, J.M. Monitoring pumping test response in a fractured aquifer using ground penetrating radar. *Water Resour. Res.* **2001**, *37*, 1221–1229. [CrossRef]
6. Saintenoy, A.; Hopmans, J.W. Ground Penetrating Radar: Water Table Detection Sensitivity to Soil Water Retention Properties. *IEEE J. Sel. Top. Appl. Earth Obs. Remote. Sens.* **2011**, *4*, 748–753. [CrossRef]
7. Pyke, K.; Eyuboglu, S.; Daniels, J.J.; Vendl, M.; Pyke, K. A controlled experiment to determine the water table response using ground penetrating radar. *J. Environ. Eng. Geophys.* **2012**, *13*, 335–342. [CrossRef]
8. Seger, M.A. Detection of water-table by using ground penetration radar (GPR). *Eng. Tech. J.* **2011**, *29*, 554–566.
9. Yuichi, N. Estimation of groundwater level by GPR in an area with multiple ambiguous reflections. *J. Appl. Geophys.* **2001**, *47*, 241–249.
10. Huisman, J.A.; Hubbard, S.S.; Redman, J.D. Measuring soil water content with ground penetrating radar: A review. *Vadose Zone J.* **2003**, *2*, 476–491. [CrossRef]
11. Klotzsche, A.; Jonard, F.; Looms, M.C.; Kruk, J.; Huisman, J.A. Measuring soil water content with ground penetrating radar: A decade of progress. *Vadose Zone J.* **2018**, *17*, 1–9. [CrossRef]
12. Grote, K.; Hubbard, S.; Rubin, Y. GPR monitoring of volumetric water content in soils applied to highway construction and maintenance. *Lead. Edge* **2002**, *21*, 482–485. [CrossRef]
13. Topp, G.C.; Davis, J.L.; Annan, A.P. Electromagnetic determination of soil moisture content: Measurements in coaxial transmission lines. *Water Resour. Res.* **1980**, *16*, 574–582. [CrossRef]
14. Feng, X.; Sato, M.; Zhang, Y.; Liu, C. CMP antenna array GPR and signal-to clutter ratio improvement. *IEEE Geosci. Remote. Sens. Lett.* **2009**, *6*, 23–27. [CrossRef]
15. Liu, H.; Sato, M. Dynamic Groundwater Level Estimation by the Velocity Spectrum Analysis of GPR. In Proceedings of the International Conference on Ground Penetrating Radar, Shanghai, China, 4–8 June 2012; pp. 413–418.

16. Lu, Q. Quantitative Analysis for Hydrogeology and Soil Contamination by Ground Penetrating Radar. Ph.D. Thesis, Tohoku University, Sendai, Japan, 2005.
17. Liu, H.; Xie, X.; Cui, J. Groundwater Level Monitoring for Hydraulic Characterization of an Unconfined Aquifer by Common Mid-point Measurements using GPR. *J. Environ. Eng. Geophys.* **2014**, *19*, 259–268. [CrossRef]
18. Pue, J.D.; Meirvenne, M.V.; Cornelis, W.M. Accounting for surface refraction in velocity semblance analysis with air-coupled GPR. *IEEE J. Sel. Top. Appl. Earth Obs. Remote. Sens.* **2016**, *9*, 60–73. [CrossRef]
19. Yi, L.; Zou, L.; Takahashi, K.; Sato, M. High-resolution velocity analysis method using the l-1 norm regularized least-squares method for pavement inspection. *IEEE J. Sel. Top. Appl. Earth Obs. Remote. Sens.* **2018**, *11*, 1005–1015. [CrossRef]
20. Jacob, R.W.; Hermance, J.F. Assessing the precision of GPR velocity and vertical two-way travel time estimates. *J. Environ. Eng. Geophys.* **2004**, *9*, 143–153. [CrossRef]
21. Liu, H.; Sato, M. In situ measurement of pavement thickness and dielectric permittivity by GPR using an antenna array. *NDT E Int.* **2014**, *64*, 65–71. [CrossRef]
22. Liu, H.; Takahashi, K.; Sato, M. Measurement of dielectric permittivity and thickness of snow and ice on a brackish lagoon using GPR. *IEEE J. Sel. Top. Appl. Earth Obs. Remote. Sens.* **2014**, *7*, 820–827. [CrossRef]
23. Yilmaz, O. *Seismic Data Analysis: Processing, Inversion, and Interpretation of Seismic Data*; Society of Exploration Geophysicists: Tulsa, OK, USA, 2001.
24. Neidell, N.; Taner, M. Semblance and other coherency measures for multichannel data. *Geophysics* **1971**, *36*, 482–497. [CrossRef]
25. Dix, C.H. Seismic velocities from surface measurements. *Geophysics* **1955**, *20*, 68–86. [CrossRef]
26. Slater, L.; Lesmes, D. IP interpretation in environmental investigations. *Geophysics* **2002**, *67*, 77–88. [CrossRef]
27. Kemna, A.; Binley, A.; Cassiani, G.; Niederleithinger, E.; Revil, A.; Slater, L.; Williams, K.H.; Orozco, A.F.; Haegel, F.-H.; Hoerdt, A. An overview of the spectral induced polarization method for near-surface applications. *Near Surf. Geophys.* **2012**, *10*, 453–468. [CrossRef]
28. Mary, B.; Saracco, G.; Peyras, L.; Vennetier, M.; Meriaux, P.; Camerlynck, C. Mapping tree root system in dikes using induced polarization: Focus on the influence of soil moisture content. *J. Appl. Geophys.* **2016**, *135*, 387–396. [CrossRef]
29. Warren, C.; Giannopoulos, A.; Giannakis, I. gprMax: Open source software to simulate electromagnetic wave propagation for Ground Penetrating Radar. *Comput. Phys. Commun.* **2016**, *209*, 163–170. [CrossRef]

© 2020 by the authors. Licensee MDPI, Basel, Switzerland. This article is an open access article distributed under the terms and conditions of the Creative Commons Attribution (CC BY) license (http://creativecommons.org/licenses/by/4.0/).

MDPI
St. Alban-Anlage 66
4052 Basel
Switzerland
Tel. +41 61 683 77 34
Fax +41 61 302 89 18
www.mdpi.com

Remote Sensing Editorial Office
E-mail: remotesensing@mdpi.com
www.mdpi.com/journal/remotesensing

www.ingramcontent.com/pod-product-compliance
Lightning Source LLC
LaVergne TN
LVHW070206100526
838202LV00015B/2002